Egg Parasitoids in Agroecosystems with Emphasis on *Trichogramma*

Progress in Biological Control

Volume 9

Published:

Volume 1
H.M.T. Hokkanen and A.E. Hajek (eds.):
Environmental Impacts of Microbial Insecticides – Need and Methods for Risk Assessment. 2004
ISBN 978-1-4020-0813-9

Volume 2
J. Eilenberg and H.M.T. Hokkanen (eds.):
An Ecological and Societal Approach to Biological Control. 2007
ISBN 978-1-4020-4320-8

Volume 3
J. Brodeur and G. Boivin (eds.):
Trophic and Guild Interactions in Biological Control. 2006
ISBN 978-1-4020-4766-4

Volume 4
J. Gould, K. Hoelmer and J. Goolsby (eds.):
Classical Biological Control of *Bemisia tabaci* in the United States. 2008
ISBN 978-1-4020-6739-6

Volume 5
J. Romeis, A.M. Shelton and G. Kennedy (eds.):
Integration of Insect-Resistant Genetically Modified Crops within IPM Programs. 2008
HB ISBN 978-1-4020-8372-3; PB ISBN 978-1-4020-8459-1

Volume 6
A.E. Hajek, T.R. Glare and M. O'Callaghan (eds.):
Use of Microbes for Control and Eradication of Invasive Arthropods. 2008
ISBN: 978-1-4020-8559-8

Volume 7
H.M.T. Hokkanen (ed.):
Relationships of Natural Enemies and Non-Prey Foods. 2008
ISBN: 978-1-4020-9234-3

Volume 8
S.S. Gnanamanickam:
Biological Control of Rice Diseases
ISBN: 978-90-481-2464-0

For further volumes:
http://www.springer.com/series/6417

Fernando L. Cônsoli · José R.P. Parra ·
Roberto A. Zucchi
Editors

Egg Parasitoids in Agroecosystems with Emphasis on *Trichogramma*

Springer

Editors
Dr. Fernando L. Cônsoli
Universidade de São Paulo
ESALQ
Departamento de Entomologia
e Acarologia
Avenida Pádua Dias 11
13418-900, Piracicaba, São Paulo
Brazil
fconsoli@esalq.usp.br

Dr. José R.P. Parra
Universidade de São Paulo
ESALQ
Departamento de Entomologia
e Acarologia
Avenida Pádua Dias 11
13418-900, Piracicaba, São Paulo
Brazil
jrpparra@esalq.usp.br

Dr. Roberto A. Zucchi
Universidade de São Paulo
ESALQ
Departamento de Entomologia
e Acarologia
Avenida Pádua Dias 11
13418-900, Piracicaba, São Paulo
Brazil
razucchi@esalq.usp.br

ISBN 978-1-4020-9109-4 e-ISBN 978-1-4020-9110-0
DOI 10.1007/978-1-4020-9110-0
Springer Dordrecht Heidelberg London New York

Library of Congress Control Number: 2010935670

Cover pictures: Pictures on the cover are credited to BSc Heraldo Negri de Oliveira (ESALQ/USP – Departamento de Entomologia e Acarologia, Piracicaba, Brazil)

Printed on acid-free paper

Springer is part of Springer Science+Business Media (www.springer.com)

Contents

Contributors

Dirk Babendreier CABI Europe, Rue des Grillons 1, 2800 Delémont, Switzerland, d.babendreier@cabi.org

Julio S Bernal Department of Entomology, Texas A&M University, College Station, TX 77843-2475, USA, juliobernal@tamu.edu

Franz Bigler Agroscope Reckenholz-Taenikon Research Station ART, Reckenholzstr. 191, 8046 Zurich, Switzerland, franz.bigler@art.admin.ch

Ferdinando Bin Department of Scienze Agrarie e Ambientali, Faculty of Agriculture, University of Perugia, Borgo XX Giugno 74, 06121, Perugia, Italy, fbin@unipg.it

Guy Boivin Centre de Recherche et de Développement, Agriculture et Agroalimentaire Canada, 430 Boul. Gouin, St-Jean-sur-Richelieu, QC, Canada J3B 3E6, boiving@agr.gc.ca

Stefano Colazza Department of S.En.Fi.Mi.Zo., Section of Acarology, Entomology and Zoology, University of Palermo, Viale delle Scienze 13, 90128, Palermo, Italy, colazza@unipa.it

Fernando L. Cônsoli Laboratório de Interações em Insetos, Departamento de Entomologia e Acarologia, ESALQ/USP, Av. Pádua Dias 11, 13418-900, Piracicaba, São Paulo, Brazil, fconsoli@esalq.usp.br

Eric Conti Department of Scienze Agrarie e Ambientali, Faculty of Agriculture, University of Perugia, Borgo XX Giugno 74, 06121, Perugia, Italy, econti@unipg.it

R.E. Duncan Department of Entomology and Nematology, Tropical Research and Education Center, University of Florida, Homestead, FL 33031, USA, red@ifas.ufl.edu

Simon Grenier UMR INRA/INSA, Biologie Fonctionnelle Insectes et Interactions, Villeurbanne, France, simon-grenier@orange.fr

Nunzio Isidoro Faculty of Agriculture, Department of Scienze Ambientali e delle Produzioni Vegetali, Politechnic University of Marche, Via Brecce Bianche, 60131, Ancona, Italy, n.isidoro@univpm.it

Josep A. Jacas Departament de Ciències Agràries i del Medi Natural, Universitat Jaume I, E-12071 Castello de la Plana, Spain, jacas@camn.uji.es

Joop C. van Lenteren Laboratory of Entomology, Wageningen University, PO Box 8031, 6700EH Wageningen, The Netherlands, Joop.vanLenteren@wur.nl

Nick Mills Department of Environmental Science, Policy and Management, University of California, Berkeley, CA 94720-3114, USA, nmills@berkeley.edu

Renata C. Monteiro Departamento Entomologia e Acarologia, ESALQ/USP, Av. Padua Dias 11, 13418-900 Piracicaba, São Paulo, Brazil, renatathrips@yahoo.com

José Roberto Postali Parra Departamento de Entomologia e Acarologia, ESALQ/USP, Av. Pádua Dias 11, 13418-900 Piracicaba, São Paulo, Brazil, jrpparra@esalq.usp.br

Jorge E. Peña Department of Entomology and Nematology, Tropical Research and Education Center, University of Florida, Homestead, FL 33031, USA, jepe@ifas.ufl.edu

Ezio Peri Department of S.En.Fi.Mi.Zo., section of Acarology, Entomology and Zoology, University of Palermo, Viale delle Scienze 13, 90128, Palermo, Italy

John D. Pinto Department of Entomology, University of California, Riverside, CA 92521, USA, jpinto@ucr.edu

Andrew Polaszek Natural History Museum, London, UK, ap@nhm.ac.uk

Ranyse B. Querino Embrapa Meio-Norte, Av. Duque de Caxias 5650, 64006-220 Teresina, Piauí, Brazil, ranyse@cpamn.embrapa.br

Roberto Romani Department of Scienze Agrarie e Ambientali, Faculty of Agriculture, University of Perugia, Borgo XX Giugno 74, 06121, Perugia, Italy, rromani@unipg.it

James E. Russell Department of Biology, University of California, Riverside, CA 92521, USA, jeruss002@gmail.com

Gianandrea Salerno Department of Scienze Agrarie e Ambientali, Faculty of Agriculture, University of Perugia, Borgo XX Giugno 74, 06121, Perugia, Italy, Salerno@unipg.it

Richard Stouthamer Department of Biology, University of California, Riverside, CA 92521, USA, richard.stouthamer@ucr.edu

Serguei Triapitsyn Department of Entomology, University of California, Riverside, CA 92521, USA, serguei.triapitsyn@ucr.edu

Bryan J. Ulmer Department of Entomology and Nematology, Tropical Research and Education Center, University of Florida, Homestead, FL 33031, USA

S. Bradleigh Vinson Texas A&M University, College Station, TX 77843-2475, USA, bvinson@tamu.edu

Eric Wajnberg INRA – 400, Route des Chappes, BP 167 – 06903, Sophia Antipolis, Cedex, France, wajnberg@sophia.inra.fr

Roberto A. Zucchi Departamento Entomologia e Acarologia, ESALQ/USP, Av. Pádua Dias 11, 13418-900 Piracicaba, São Paulo, Brazil, razucchi@esalq.usp.br

Chapter 1
Reproduction and Immature Development of Egg Parasitoids

Guy Boivin

Contents

1.1 Introduction

Although several egg parasitoid species are among the most commonly used biological control agents (van Lenteren 2003), the data available on the reproduction and immature development of these organisms are scarce. The reasons for this will be obvious to anyone working with egg parasitoids. Their small size, the difficulty to obtain significant field data and the unresolved taxonomic status of even well known species will come to mind. However, the past decades have seen a number of interesting developments and we now have a better understanding of the constraints faced by these species and which adaptations have evolved.

Given the difficulty facing anyone looking at the reproduction or the immatures of egg parasitoids, why would anyone choose egg parasitoids as model organisms? The choice of a model organism responds to a number of reasons, one of them being the curiosity and amazement an animal inspires (Parker 2001).

G. Boivin (✉)
Centre de Recherche et de Développement, Agriculture et Agroalimentaire Canada, 430 Boul. Gouin, St-Jean-sur-Richelieu, QC, Canada J3B 3E6
e-mail: boiving@agr.gc.ca

F.L. Consoli et al. (eds.), *Egg Parasitoids in Agroecosystems with Emphasis on Trichogramma*, Progress in Biological Control 9, DOI 10.1007/978-1-4020-9110-0_1,

Egg parasitoids are among the smallest animals and the smallest insect known is a Mymaridae, *Dicopomorpha echmepterygis* Mockford, that parasitizes eggs of Psocoptera (Mockford 1997, Gahlhoff 1998). The males of this species measure about 140 μm which is smaller than some species of paramecia, a unicellular organism (Hewett 1980). The fact that these small egg parasitoids possess a morphology and a range of behaviors similar to much larger species is by itself something than can arouse interest. Such small size also implies that throughout their life, egg parasitoids face important constraints. Their lifetime is in the range of days, even hours for some species, their energetic reserve is at best limited and their capacity to engage in any meaningful movement within the habitat is also problematic. To face such challenges, egg parasitoids are likely to have evolved a series of adaptations that are exacerbated by the magnitude of the constraints they face. Their small size also implies that although manipulating them in the laboratory is difficult, they can be reared in large numbers using little resources and it is possible to gather lifetime data on aspects such as reproduction, fitness gain or resource use. Finally, egg parasitoids are defined based on their lifestyle but they come from different taxonomic groups. Because these groups share common constraints, they may have evolved similar adaptations in response to common situations and can thus be used to demonstrate cases of convergent evolution.

In this chapter I will review the recent advances on egg parasitoids reproduction and immature development. Some developments bring new lights on the morphology and physiology of egg parasitoids but the most challenging facts come from the evolutionary ecology and behavioral ecology fields (see also Chapter 5). I will therefore concentrate on these aspects, giving the basic data necessary to understand the importance of the new information. Although egg parasitoids are found in several Hymenoptera families (including Trichogrammatidae, Mymaridae, Eulophidae, Scelionidae, Aphelinidae, Encyrtidae), most of the information comes from the few families used in biological control programs. This chapter will therefore be biased on these families although I will present data on other families when they are available.

1.2 Reproduction

The difficulties mentioned above do apply to the study of reproduction in egg parasitoids. Their small size makes physiological studies and field observations difficult but several studies have managed to overcome these difficulties and have come out with interesting insight into the mating behavior and strategies displayed by certain groups of egg parasitoids. I will cover here aspects of mate finding, mating behavior and mating strategies. Most of the data available deal with females, a bias that is not unique to egg parasitoids. Males seem to be the forgotten sex possibly because they are seen as having an unlimited supply of cheap gametes, the spermatozoids, and that maximizing their number of mating is thought to be the only answer to maximize their lifetime fitness. This may not be always true and among the few studies that looked at reproduction from the male's point of view some have used egg parasitoids as models. This information will also be included in this chapter.

Looking at the reproduction of egg parasitoids is important for several reasons including the fact that some species of Trichogrammatidae, Mymaridae and Scelionidae are widely used in biological control programs as inundative agents and as such are reared in large numbers (van Lenteren 2003, see also Chapters 10, 13, and 14). Rearing a species without having a clear understanding of its reproduction is at best risky and biocontrol producers could certainly gain by having access to data enabling them to optimize their rearing facilities. Factors such as apparent or cryptic mate choice, sperm competition and sperm quality can all affect mass rearing.

Some unique characteristics of egg parasitoid reproduction also make them ideal models in behavioral ecology. Most egg parasitoids reproduce by arrhenotokous parthenogenesis, the dominant mode of sex determination in the Hymenoptera, where an unfertilized egg develops into a haploid male and a fertilized egg into a diploid female (Heimpel and de Boer 2008). Virgin females can thus reproduce although they are constrained to produce only male progeny (Quicke 1997). Females of arrhenotokous species can therefore allocate the sex of their progeny. The study of sex allocation in parasitoids has been very successful in producing models predicting the optimal behavior that females should express to maximize their lifetime fitness. Some egg parasitoid species also reproduce through thelytokous parthenogenesis where only females are produced and both genetic and endosymbiont-induced forms of thelytoky are known in Hymenoptera (Heimpel and de Boer 2008).

Several factors that are known to influence the sex allocation of females are amplified in egg parasitoid females. Once again the small size of egg parasitoids limits their energy reserve and individuals are thus strongly selected to optimize their behaviors accordingly. Gamete production, that is known to be a limit for female parasitoids, can also be a limit for males of egg parasitoids. Males that are facing such constraints can be studied by optimality models and the behavioral options available to these constrained males open several interesting questions.

Insect eggs are often found in discrete patches and most egg parasitoids mate on the emergence patch. Not surprisingly, local mate competition, inbreeding and patch time allocation of egg parasitoids have received much attention (Boivin et al. 2004, Martel and Boivin 2004a, Wajnberg 2006). Some outpatch mating occurs in most species of egg parasitoids (Loch and Walter 2002, Martel and Boivin 2004b) suggesting that both sexes can apply optimal strategies in their exploitation of the emergence patch.

1.2.1 Males

The insect male reproductive system comprises a pair of testes from which the sperm leaves through vas deferens (Fig. 1.1). The sperm is then stored in the seminal vesicles that are swollen regions of the vas deferens. Paired accessory glands,

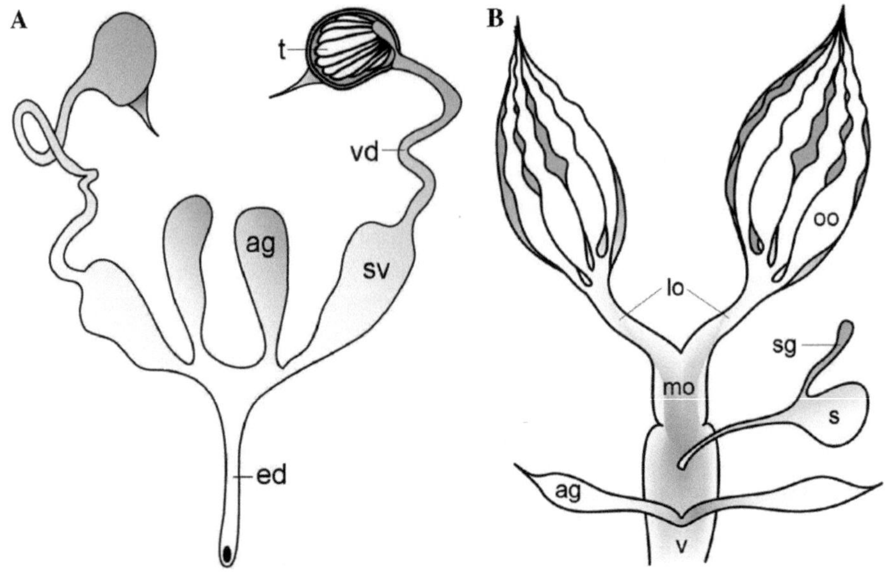

Fig. 1.1 (**a**) Reproductive structures of male insects. t: testis follicles, vd: vas deferens, sv: seminal vesicles, ag: accessory glands, ed: ejaculatory duct (Drawing by Franz Vanoosthuyse); (**b**) Reproductive structures of female insects. o: oocyte in ovariole, lo: lateral oviduct, mo: median oviduct, s: spermatheca, sg: spermathecal gland, ag: accessory glands, v: vagina (Drawing by Franz Vanoosthuyse)

that produce substances that are transferred with the ejaculate, are also present. At copulation the sperm exit the seminal vesicles, along with various substances, by the ejaculatory duct to the aedageus (Quicke 1997).

The male reproductive system of most egg parasitoids is typical of Hymenoptera (Fig. 1.2) but the testes of male egg parasitoids are often of reduced size or empty (Quicke 1997). The capacity of adult males to produce sperm has an important

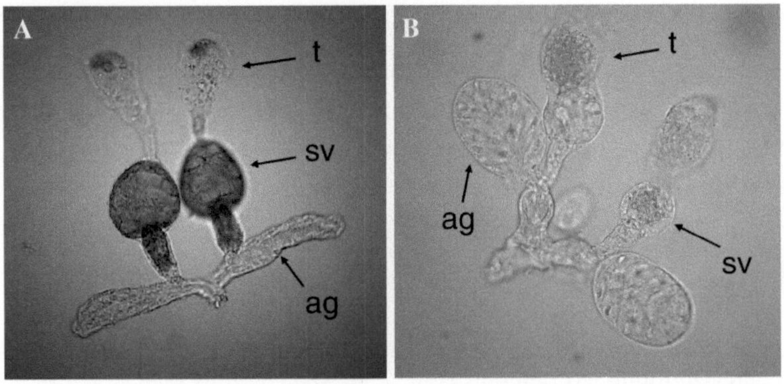

Fig. 1.2 Reproductive structure of male *Trichogramma turkestanica* (**a**) and *Anaphes victus* (**b**). t: testis follicles, sv: seminal vesicles, ag: accessory glands

impact on the male reproductive strategy and the mating structure of a species. In *Trichogramma turkestanica* Meyer (= *evanescens*) the seminal vesicles are divided into a large anterior part and a smaller vesicular pocket (Fig. 1.2a) that play a role in the transfer of sperm to the female at mating (see below) (Damiens and Boivin 2005).

The sperm of parasitic Hymenoptera is composed of an elongated nucleus and a motile tail with a pair of mitochondrial derivatives (Quicke 1997). In the few species of egg parasitoids where sperm length has been measured, it varies between 115 μm (*Anaphes listronoti* Huber) and 275 μm (*T. turkestanica*) (Boivin and Martel, unpublished data).

In Hymenoptera, although it was assumed that adult males were unable to produce sperm (Gerling and Legner 1968, Chihrane and Laugé 1994, Ayasse et al. 2001), sperm production has been shown in males of several parasitoid species (Gerling and Rotary 1974, Hogge and King 1975, Nadel and Luck 1985, Ramadan et al. 1991). Sperm production in adult male parasitoids varies depending on the status of the testes at emergence. When the testes are depleted by the time of emergence, no sperm production occurs during the adult life while in other species the testes are still maturing at emergence and sperm production starts after adult emergence. These differences in sperm production in male parasitoids are described by the spermatogeny index (Boivin et al. 2005) that is a measure of the degree to which sperm production is concentrated into early adult life. Species where males emerge with their full sperm complement and produce none during their lifetime have a spermatogeny index of one (1), while species where males emerge with no sperm but produce some later in life have a spermatogeny index of zero (0). Most species have a spermatogeny index of intermediate value. The level of spermatogeny is linked to several other life-history traits such as longevity, size, distribution of mating opportunities and dispersion before and after mating. Not surprisingly, species that are small and short-lived, such as egg parasitoids, are expected to have a high spermatogeny index because most mating opportunities occur early in life, advantaging males that have their full sperm complement at emergence.

Spermatogeny index have been measured in some egg parasitoids. Indices of 1 have been measured in *Trichogramma brassicae* Bezdenko (Chihrane and Laugé 1994) and *T. turkestanica* (Damiens and Boivin 2005). In *A. listronoti*, the number of sperm present in the seminal vesicles increases after emergence for about 24 h resulting in a spermatogeny index of 0.70 (G Boivin unpublished data). This suggests that even within egg parasitoids the spermatogeny index may vary according to the ecology of the species. Although both *Trichogramma* and *Anaphes* are short lived, *Anaphes* species typically parasitize hosts found in smaller patches than hosts parasitized by *Trichogramma* species. Smaller host patches may imply that outpatch mating is more important in *Anaphes* species and thus that a lower spermatogeny index would be optimal.

Egg parasitoids show a large phenotypic variability according to the host in which they have developed and to the number of individuals that have developed per host. The capacity of males to produce and store sperm is likely affected by their size. As expected, small males do produce and store fewer sperm. In *A. listronoti*, the size of males (expressed as tibia length) decreases from 275.9 to 203.6 μm for males that developed singly or as triplets in eggs of *Listronotus oregonensis* (Le Conte) (G Boivin unpublished data). The number of sperm present in the seminal vesicles of these males 24 h after emergence also varied from 918.9 to 556.9 sperm. Thus, developing alone or as triplets decreased the size of males by 26% and reduced their sperm supply by 39%. A similar pattern was found in males *T. turkestanica* that developed in host eggs of different sizes. Males that developed in eggs of *Trichoplusia ni* (Hübner) and *Plutella xylostella* (L.) (egg volume of 0.62 and 0.13 mm^3 respectively) (Godin and Boivin 2000) show 38% decrease in size (tibia size of 186.7 and 115.3 μm for *T. ni* and *P. xylostella*, respectively), but the number of sperm stored in the seminal vesicles of males decreased by 76% (3330.9 and 786.9 sperm for *T. ni* and *P. xylostella* hosts respectively) (Martel and Boivin unpublished data).

Males of egg parasitoids may find mates either on the emergence patch or later after their dispersion from the emergence site. Males generally emerge shortly before females (Pompanon et al. 1995, Doyon and Boivin 2006) and wait on the emergence patch for females to emerge. Most females that leave the patch have been mated (Forsse et al. 1992, Pompanon et al. 1995, Martel and Boivin 2004b), but a small percentage of virgin females do leave the patch suggesting some level of genetic exchange between subpopulations. Males respond to both long-range (> 2 cm) and short-range sex pheromones emitted by the female (Keeling et al. 2004). Sex volatile pheromones have been mentioned in *T. brassicae* and *A. listronoti* (Pintureau and Toonders 1983, Cormier et al. 1998). In *T. brassicae*, a substrate-borne sex pheromone responsible for mate finding at short distance elicits arrestment and male courtship behavior (Pompanon et al. 1997, Delpuech et al. 1998). The presence of such substrate-borne pheromone with low volatility induces mating and influences the retention of males on the patch, modifying their patch residence time. Not surprisingly, insecticides that affect neuronal pathways do change sex pheromone detection by males, some products decreasing pheromone detection while others increasing it (Delpuech et al. 1998, 1999, 2001).

Most male insects are expected to maximize their lifetime fitness gain by maximizing the number of females they mate (Thornhill and Alcock 2001). Maximization of the number of mating is optimal if mating has no cost for males, which is unlikely to be true in egg parasitoids where males are short-lived and could be sperm-limited. When mating does incur costs, models predict that some form of mate selection could be optimal. Mate choice is predicted to evolve when mating or mate searching involve costs or when there is some variability in mate quality (Bonduriansky 2001). Because mate choice also implies costs such as searching time, loss of mating opportunities and increased risks associated with mate searching (Martel et al. 2008a), the benefits obtained by being choosy must be important.

Male mate choice has been studied in few parasitoid species (King et al. 2005), only one of them being an egg parasitoid. In general, males should favor females that bring them a larger fitness gain either because they live longer, are more fecund or can find or subdue better hosts. In *T. turkestanica* (=*evanescens*), males prefer more fecund, large females (Boivin and Lagacé 1999), and they prefer to copulate with virgin over mated females (Martel et al. 2008a). This preference for virgin females reflects the fact that in *T. turkestanica* there is first male sperm precedence and therefore males that mate with an already mated female may expect to father only 10–20% of the progeny of that female (Damiens and Boivin 2006). These data suggest that males are optimizing their mating strategy rather than simply maximizing the number of mates they acquire. Optimality models could eventually be used to describe the strategies used by males to choose their mate.

Once a female has been chosen, the male will initiate his courtship behavior. In the case of some gregarious egg parasitoids, mating can occur within the host egg even prior to emergence, as in *Trichogramma dendrolimi* Matsumara, or males can force females to emerge as in *Trichogramma papilionis* Nagarkatti (Suzuki and Hiehata 1985). In these cases no or minimal courtship behavior is expected. When mating occurs after emergence, a certain number of steps are common to most parasitoid species including Trichogrammatidae (Al-Wahaibi et al. 2005), Eulophidae (Ruberson et al. 1988, Hamerski and Hall 1989) and Mymaridae (Collins and Grafius 1986). A male generally faces the female and fans his wings for a moment. When the female is responsive, the male goes behind the female and mounts her, holding her mesothorax with his front legs and touching her antennae with his. Wing fanning can be used to disperse a short-distance volatile pheromone produced by the male or to produce a sound that is part of the courtship (Hardy et al. 2007).

However there are many variations in this pattern. The males of both the Trichogrammatidae *Lathromeroidea* that parasitizes eggs of the aquatic Hemiptera Gerridae, and the Scelionidae *Telenomus solitus* Johnson do not exhibit wing fanning and they proceed directly to mounting, touching the female's antennae with their own during copulation (Navasero and Oatman 1989, Henriquez and Spence 1993). Contact between the antennae of males and females may be used for sexual recognition. In the Scelionidae *Trissolcus basalis* (Wollaston), the male mounts the female and curls his antennae around the female's antennae. The glands present on the A5 segment of the male's antenna come in contact with the sensillae on segments A6 to A11 of the female's antenna (Isidoro et al. 1996, see also Chapter 3).

In gregarious and quasi-gregarious species, competition among males for females can result in fights as in Scelionidae (Waage 1982). In *T. basalis*, the males emerge first and compete for access to the egg mass. One male becomes dominant, guards the egg mass and mates with as many females as possible as they emerge from the egg mass (Loch and Walter 2002). Males perform a brief precopulatory antennation where both male and female antennae touch frequently, followed by copulation. During copulation males continue antennation and generally chase the female off the egg mass following copulation (Loch and Walter 2002). The males of the Trichogrammatidae *Ufens* fight using their head and antennae to push competing males out of the emerging patch (Al-Wahaibi et al. 2005).

Polygyny is common in Hymenoptera parasitoids (Hardy et al. 2007) and males that mate with several females in succession may temporarily (synspermatogenic species) or definitely (prospermatogenic species) deplete their sperm supply

(Ode et al. 1997, Damiens and Boivin 2005). The amount of sperm transferred to females gradually decreases until no sperm are transferred at mating (Damiens and Boivin 2005), although males that are sperm-depleted continue to mate (Damiens and Boivin 2006). The amount of sperm transferred is determined in *T. turkestanica* by the relative sizes of the vesicular pocket and the anterior part of the seminal vesicle (Damiens and Boivin 2005). The valve between the two chambers of the seminal vesicle stays open between copulations, allowing continuous movement of sperm between the two chambers. During copulation, that valve closes and the two vesicular pockets are emptied. The number of sperm transferred is initially constant, being limited by the size of the vesicular pockets but when the total number of sperm per seminal vesicle reaches a level where the vesicular pocket is not full, the number of sperm transferred begins to decrease in a curvilinear manner (Damiens and Boivin 2005).

Sperm-depleted males that continue to mate are not transferring any sperm to females, but they still obtain some fitness by doing so. In *T. turkestanica*, a virgin female that has mated with a sperm-depleted male needs to mate with three virgin males in order to obtain the number of sperm that a single mating with a virgin male would have provided in the first place (Damiens and Boivin 2006). This is probably due to the seminal liquid transferred by the sperm-depleted male that fills the female's spermatheca. Females that mated with a sperm-depleted male are likely to deposit a constrained sex ratio composed of only or mostly male progeny. By doing so, sperm-depleted males decrease the reproductive opportunities of competing males, increasing their representation in the next generation. They also increase the proportion of males in the next generation, thus facilitating access to males to their own daughters.

Sperm competition refers to the competition between the sperm from two or more males for the fertilization of a given set of ova (Parker 1998). Sperm competition often results in sperm precedence which is the non random utilization of sperm from one of several males to fertilize eggs (Simmons and Siva-Jothy 1998). First male precedence was shown in *T. turkestanica* (Damiens and Boivin 2006) where the second male fathers only 15–20% of the progeny. Males do respond to the risk of sperm competition. *Trichogramma turkestanica* males that live in group transfer fewer sperm than when alone in response to their perception of an increased probability of sperm competition (Martel et al. 2008b).

1.2.2 Females

The typical reproductive system of Hymenoptera females comprises a pair of ovaries (in which eggs are produced) and lateral oviducts, followed by a common oviduct and the bursa copulatrix, or vagina, that receives the male copulatory organs (Fig. 1.1b). The spermatheca connects to the bursa or the common oviduct. Glands such as the venom gland, the Dufour's gland, and sometimes a calyx gland are present (Quicke 1997). The female reproductive system found in egg parasitoids is typical of Hymenoptera, but in groups such as the Scelionidae, the venom glands are lost, and their role taken by a glandular oviduct (Rosi et al. 1995, Quicke 1997).

The spermatheca holds the sperm after copulation and consists of a small reservoir that opens in a narrow duct leading to the anterior end of the common oviduct. Not all sperm transferred to females at mating are kept in the spermatheca by the female. About 50% of the sperm is found in the spermatheca of females of *T. turkestanica* 2 h after mating (Damiens and Boivin 2005). This level of retention is rather high compared to the few other parasitoids species for which data is available (Bressac and Chevrier 1998). Egg parasitoid females use sperm parsimoniously, releasing one or a few sperm each time a female progeny is about to be laid. In *T. turkestanica*, the female stores about 50 sperm in her spermatheca, close to her average lifetime production of about 48 daughters (Martel et al. 2008b).

The temporal distribution of oocyte production is described by an ovigeny index (Jervis et al. 2001, 2008). Females that have all their eggs ready to oviposit at emergence and do not produce more during their life are termed proovigenic (ovigeny index of 1), while those that have no egg ready to oviposit at emergence and mature them later are termed synovigenic (ovigeny index of 0). Most parasitoid species however are between these two extremes and are considered as moderately synovigenic. Egg parasitoids, which have been often considered as proovigenic, are in fact moderately synovigenic. Jervis et al. (2001) reported that eight species of Trichogrammatidae are moderately synovigenic, while six species of Mymaridae are strictly proovigenic. The ovigeny index of six species of *Trichogramma* (*T. cacoeciae* Marchal, *T. sibericum* Sorkina, *T. chilonis* Ischii, *T. minutum* Riley, *T. pintoi* Voegelé and *T. leptoparameron* Dyurich) varied between 0.09 and 0.43 (unpublished data). However, the Mymaridae *Anaphes victus* Huber is also moderately synovigenic with an ovigeny index of 0.62 (unpublished data). The relatively high ovigeny index in egg parasitoids reflects the fact that these species are short-lived and therefore that it is important for the female to have a large proportion of eggs ready to be oviposited early in life (Jervis et al. 2008).

As in most Hymenoptera species, it is mostly the female that attracts the male from a distance. Most female Hymenoptera produce sex pheromones (Keeling et al. 2004) and females of egg parasitoids produce long and short distance volatile pheromones and short-distance non-volatile pheromones (Pintureau and Toonders 1983, Pompanon et al. 1997, Cormier et al. 1998, van Beek et al. 2005). *T. turkestanica* females produce a substrate-borne pheromone made of two components produced at a rate of 1–3 pg/h per virgin female (van Beek et al. 2005). Following mating, females of *T. brassicae* stop their production of pheromone (Pompanon et al. 1997) and their receptivity to males also decrease.

In insects, it is generally assumed that a single mating supplies enough sperm to fertilize all eggs so that females should be monoandrous (Halliday and Arnold 1987). Although sperm quantity is in most cases sufficient, multiple mating by females is not uncommon in parasitoids (Quicke 1997). In *T. turkestanica*, the 50 sperm stored in the spermatheca is enough for the total number of daughters produced over the female lifetime (Damiens and Boivin 2005, Martel et al. 2008b), but some *Trichogramma* species that live longer may face sperm depletion (King 1987, Leatemia et al. 1995) and polyandry is likely to increase the reproductive success of these females (van den Assem 1986). There are at least 30 species among 9 genera that are polyandrous (Ridley 1993, Quicke 1997, Chevrier and Bressac 2002), the majority of these being gregarious (Ridley 1993). Females of gregarious species mate mostly with polygynous males that are more prone to become sperm-depleted (Godfray 1994). Ridley (1988, 1993) also noted that polyandrous-gregarious parasitoids had higher longevity and fecundity compared to monoandrous-solitary species, supporting the idea that multiple mating might be needed to replenish sperm supply in longer lived and more fecund species. Polyandry is expected to be relatively rare and unlikely in species with important sib-mating at emergence (Ridley 1993, Godfray 1994, Quicke 1997). Because of that, females of egg parasitoids are often considered monoandrous. However, polyandry has been shown in a number of species: *T. minutum* (Leatemia et al. 1995), *T. turkestanica* (Pintureau et al. 1997, Jacob and Boivin 2005), *T. dendrolimi, T. papilionis* (Suzuki and Hiehata 1985) and *T. pretiosum* (Kazmer and Luck 1991).

Mate choice by females should also be optimal when mating or mate searching involves costs and when there is variability in mate quality (Bonduriansky 2001). Few data on mate choice by females are available for egg parasitoids. When *T. turkestanica* females were placed in presence of small and large males, 88% of all mating were done by large males (Boivin and Lagacé 1999), but it is not clear whether this difference is due to a preference by the female or a better competitive ability of large males. *T. turkestanica* females do not select mate based on their age (Doyon and Boivin 2006).

1.3 Immature Development

Egg parasitoids are defined by the fact that all their immature stages develop within host eggs. It is thus the immature stages that have to deal with the constraints imposed by the host (see Chapter 2). Most egg parasitoid immatures, because of their minute size and their special habitat, are considered as morphologically specialized when compared to other Hymenoptera immatures (Hagen 1964, Jarjees et al. 1998).

All egg parasitoids are idiobiont, killing the host egg at the beginning of its development, and the developing immature has thus a finite reserve of food. The size of the host egg has a major effect on the available resource and ultimately on the size and fitness of the emerging parasitoid. There are both upper and lower size limits for the hosts of egg parasitoids (Strand et al. 1988), but even when host size enables development, it has a major effect on several fitness parameters (Roitberg et al. 2001). In *T. turkestanica*, the fitness of both males and females decreases when they are reared on eggs of *P. xylostella* a smaller host than *Anagasta kuehniella* (Zeller) (Boivin and Lagacé 1999).

The host egg is not a simple container, and the egg chorion plays an important role in gaseous exchanges between the immature parasitoid and its surrounding environment. The incapacity of parasitoid immatures to develop in sterile host eggs has been reported in *Telenomus heliothidis* Ashmead, *T. pretiosum* (Strand 1986, Strand et al. 1986) and *A. victus* (Picard et al. 1991). The absence of the endochorion layers that are normally produced by the embryo explains this incapacity. The serosal endocuticle plays an active role in water regulation in fertile hosts and its absence probably causes desiccation of the parasitoid immature. While *A. victus* females readily parasitize *L. oregonensis* eggs less than 24 h old, the survival of the parasitoid immature is lower than in older eggs. This period corresponds to the formation of the serosal membrane that consists of a series of layers (Nénon et al. 1995). The serosal membrane comprises 90 layers 72 h after oviposition, but less than 6 at 24 h. *Anaphes victus* immatures are unable to complete development in a host egg with an incompletely formed serosal membrane.

A special adaptation to obtain oxygen within the host egg has been found in several groups of egg parasitoids. In *Edovum puttleri* Grissell (Eulophidae), *Avetianella longoi* Siscaro (Encyrtidae) and some *Platystasius* sp. and *Fidiobia* sp. (Platygasteridae), the late instar larva pierces the host egg chorion with its mandibles to gain access to atmospheric air (Bin et al. 2000). The mandibles of some of these species are exodont (they face outward) to enable the larva to pierce the chorion.

1.3.1 Eggs

Several egg types are found in Hymenoptera parasitoids and we will follow in this section the types described by Hagen (1964). Most eggs of Trichogrammatidae are hymenopteriform in shape (Voegelé et al. 1974, Pak and Oatman 1982, Dahlan and Gordh 1996, Jarjees and Merritt 2002), although stalked eggs are found in the genus *Poropoea* Förster (Clausen 1940). At oviposition, the eggs of the genus *Trichogramma* have a length of 100–140 μm and a width of 30–50 μm (Tanaka 1985a, Manweiler 1986, Saakian-Baranova 1990, Dahlan and Gordh 1996, Jarjees and Merritt 2002), but smaller eggs are laid by *T. chilonis* females (75 μm × 23 μm) (Tanaka 1985a). The eggs of Trichogrammatidae are hydropic (Jarjees and Merritt 2002) and they increase in size soon after oviposition as they absorb nutrient directly from the host through the egg membranes. The eggs of *T. chilonis* are laid in or near the peripheral yolk of the host egg (Tanaka 1985a), while those of *T. dendrolimi* are laid in the host egg yolk (Takada et al. 2000).

Most scelionids have stalked eggs (Navasero and Oatman 1989, Sousa 1999). Egg size at oviposition varies from 49 μm by 14 μm in *T. basalis* (Volkoff and Colazza 1992) to 700 μm by 100 μm in *Scelio calopteni* Riley (Pickford 1964). Most species however have eggs in the range 100–150 μm by 40–60 μm (Rothschild 1970, Gerling 1972). No increase in size has been reported in Scelionidae eggs. The eggs of *T. heliothidis* are generally oviposited next to the host embryo (Strand et al. 1986), while those of *Telenomus remus* Nixon are found both within and outside the host embryo (Gerling 1972).

Stalked eggs are common in the Mymaridae (Hagen 1964, Yeo et al. 1990, Bosco and Arzone 1991, Boivin et al. 1993). The egg of *Caraphractus cinctus* Walker is hymenopteriform (Jackson 1961). Egg size at oviposition varies between 100 and 200 μm in length and 30 and 60 μm in width (Meyerdirk and Moratorio 1987, Bosco and Arzone 1991, Boivin et al. 1993). Hydropy has been observed in the eggs of Mymaridae.

The eggs of the Encyrtidae are encyrtiform with a dumbbell shape (Maple 1947, Hagen 1964, Brown 1984). These encyrtiform eggs are deposited in the host egg with their anterior egg stalk protruding from the host chorion (Maple 1947). The stalk and a portion of the egg body bear an aeroscopic plate that is used by the embryo and the larva for respiration (Masutti et al. 1993, Brown 1984).

1.3.1.1 Hatching and Teratocytes

In two groups of egg parasitoids, the Trichogrammatidae and Scelionidae, hatching is a two-step process. The egg chorion is shed early in the embryogenesis and the embryo retains a membrane, either an embryonic exuvia in the case of the Trichogrammatidae or the serosal membrane in the Scelionidae, and continues to develop. Eventually, the first instar larva frees itself of this membrane to start feeding. The early shedding of the chorion may be linked to life within a host egg. The immature is within its food source and faces no immediate danger, therefore the protective role of the chorion may have little evolutionary value. Casting the egg chorion may ease transfer of oxygen or nutrient. More details on hatching in other groups of egg parasitoids could bring interesting data to support these hypotheses.

Eighteen hours after oviposition the embryo of Trichogrammatidae is surrounded by two envelopes: the chorion and an inner embryonic exuvia. Between these two envelopes, free cells with large nuclei can be observed (Volkoff et al. 1995). Soon after, the late-stage embryo, still surrounded by the embryonic exuvia, shows vigorous movements of its anterior section (Jarjees et al. 1998) that eventually ruptures the chorion. The embryo is then released within the host egg together with the cells that were observed between the chorion and the embryonic exuvia. At this stage no feeding takes place and this stage has been termed a "free living embryo" (Volkoff et al. 1995, Cônsoli et al. 1999). A similar stage enclosed in its serosal membrane has also been described from Scelionidae. During that stage the embryo is still growing and the first peristaltic movements start. The mandibles gradually appear and are pushed against the embryonic exuvia. At the end of that stage the embryonic exuvia is broken by the mandibles at its anterior end and the larva emerges (Volkoff et al. 1995). The embryonic membrane remains attached to the posterior end of the larva. Such an attachment could be adaptive by anchoring the larva against the movement of its anterior portion and by preventing the free floating membrane from clogging the mouthparts during feeding (Jarjees et al. 1998).

No teratocytes are produced in *Trichogramma.* The embryonic exuvia takes the form of 1–3 spheres that seem to have a digestive function in *Trichogramma brasiliensis* Ashmead and *T. chilonis* (Voegelé et al. 1974, Tanaka 1985b). These spheres are present throughout the larval development and measure 20–25 μm in *Trichogramma maidis* Pintureau and Voegelé (Hawlitzky and Boulay 1982). Strand

(1986) observed two polar bodies liberated at the emergence of *T. pretiosum* and these polar bodies may have been confounded with teratocytes in other species.

As mentioned earlier, the immatures of Scelionidae rupture their chorion during their development and continue to develop within the serosal membrane. The serosa originates from the polar nuclei and associated cytoplasm of the original division of the oocyte (Strand et al. 1985). The rupture of the serosal membrane releases both an actively feeding larva and teratocytes. Large portions of the serosa autolyse, while some cytoplasmic areas remain undigested, which becomes teratocytes (Strand et al. 1986). The teratocytes vary in number and size according to species. In both *T. heliothidis* and *T. remus* there are 20–30 teratocytes per larva. These teratocytes measure 10 μm in diameter at eclosion and increase to 35 μm at the beginning of the third instar (Gerling and Orion 1973, Strand et al. 1985, Strand et al. 1988). In *T. basalis*, about 80 teratocytes, that are polyhedric at release, but become spherical within 1 h, are released (Volkoff and Colazza 1992, Cônsoli et al. 2001). They average 15 μm in diameter upon release and reach their maximum size of 50–80 μm in diameter later during the first instar.

Little information is available on hatching in Mymaridae. In *Anagrus optabilis* (Perkins), mandibles are not involved in rupturing the egg chorion and it is the pressure of the embryo that causes a rupture of the chorion below the pedicel (Sahad 1984). In *Caraphractus cinctus* Walker, the pharate first instar "force a hole" also just below the pedicel (Jackson 1961). No teratocytes are present in Mymaridae.

1.3.2 Larvae

Larvae of most species of egg parasitoids show some level of simplification, either in the number of instars or in their external morphology. Most egg parasitoid larvae undergo hypermetamorphosis (first instar differs in form from later instars) (Hagen 1964), except in groups where the larval immature is very much simplified, such as most *Trichogramma* species. In this section we will use the types described by Hagen (1964). Some features are common to several groups of egg parasitoids. The presence of a caudal appendage, long setae, cephalic process and lack of marked segmentation has been reported in species of Mymaridae, Scelionidae and Trichogrammatidae. These convergent traits are a result of the exploitation of similar habitats (insect eggs) and represent homoplasy (Gibson 1986).

1.3.2.1 Number of Instars

The plesiomorphic (primitive) condition for the number of larval instars in the Hymenoptera is five instars as the Symphyta have five instars (Hagen 1964, Quicke 1997). A reduction in the number of larval instars is often the result of specialization, such as life in the enclosed and well protected habitat of the host egg.

The number of instars in some groups is still controversial. The reasons are probably the small size of egg parasitoid larvae and the fact that the egg chorion is sometimes shed before the end of embryogenesis. The free living embryo, still enclosed in a vitelline membrane or an embryonic membrane, is present for some time in the host egg and it has probably been confused with the first larval instar in several studies. Until more detailed studies are made on more species, the number of instars reported in the literature must be treated with caution especially so in the Trichogrammatidae.

The number of instars reported for Trichogrammatidae varies from one to five instars. Bakkendorf (1934) already reported differences in the number of instars recognized by authors, and considered that the genera *Ophioneurus* and *Poropoea* had 2 instars while the genera *Chaetostricha*, *Trichogramma* and *Prestwichia* had only one instar. However, later authors reported a variety of number of instars within the genus *Trichogramma*. The instars varied from two in *T. dendrolimi* (Takada et al. 2000), three in *T. buesi* Voegelé (Abbas 1989), *T. evanescens* (Saakian-Baranova 1990), *T. embryophagum* Htg. (Saakian-Baranova 1990), *T. pintoi* (Saakian-Baranova 1990), *T. semblidis* (Aurivillius) (Saakian-Baranova 1990), *T. minutum* (Manweiler 1986), *T. platneri* Nagarkatti (Saakian-Baranova 1990), *T. brassicae* Bezdenko (Wu et al. 2000), *T. maidis* (Hawlitzky and Boulay 1982) and four in *T. brevicapillum* Pinto and Platner (Pak and Oatman 1982). *Trichogramma australicum* Girault, *T. brassicae* and *T. cacoeciae* were reported to have three larval instars (Brenière 1965, Saakian-Baranova 1990), but later studies reported only one instar (Volkoff et al. 1995, Dahlan and Gordh 1996, 1997, Jarjees et al. 1998, Wu et al. 2000, Jarjees and Merritt 2002). Recent studies suggest that only one larval instar is present in *Trichogramma* based on the fact that the shape and size of the first (and only) larval instar change considerably during larval development because of important ingestion of food early in its development (see later). When mandibles are measured, including sections buried in tissue, no increase in size is recorded (Volkoff et al. 1995). Although the data presented for *T. cacoeciae* and *T. australicum* are convincing, it is still too early to generalize their finding to the whole *Trichogramma* genus.

In the genus *Trichogrammatoidea,* while *Trichogrammatoidea bactrae* Nagaraja is reported to have two instars (Hutchison et al. 1990), *T. armigera* Nagaraja and *T. simmondsi* Nagaraja have three instars (Manjunath 1972, Zongo et al. 1993). Whether or not the same problems of measurement apply to this genus is unknown.

The number of instars in Scelionidae varies from two (Gerling 1972, Volkoff and Colazza 1992) to five instars (Safavi 1968). Most commonly however, there are two or three instars. The fact that a free living embryo occurs probably explains some of the differences found in the number of instars reported. In the genus *Telenomus*, two instars are described in the species *T. remus* (Gerling and Orion 1973) and *Telenomus solitus* Johnson (Navasero and Oatman 1989), while three instars are reported in *T. heliothidis* (Strand et al. 1986, Strand et al. 1988), *T. basalis* (Volkoff and Colazza 1992), *Telenomus rowani* (Gahan) and *Telenomus dignus* (Gahan) (Rothschild 1970).

There are mostly two or three instars described in the Mymaridae. In the genus *Anagrus* the number of instars is reported as two for *Anagrus giraulti* Crawford (Meyerdirk and Moratorio 1987), *Anagrus optabilis* (Perkins) (Sahad 1984) and *Anagrus incarnatus* Haliday (Yeo et al. 1990) and as three for *Anagrus epos*

Girault (McKenzie and Beirne 1972). Three instars are reported in *Anaphes victus* (Boivin et al. 1993, Nénon et al. 1995, van Baaren et al. 1997) and in *A. listronoti* (van Baaren et al. 1997). Bakkendorf (1934) recognizes only two instars in *Alaptus minimus* Haliday, *Lymaenon effusi* Bakkendorf, *Ooctonus heterotomus* Förster, *Polynema pusillus* Halliday and *Polynema ovulorum* (L.). A total of four instars are reported only in *Caraphractus cinctus* Walker (Jackson 1961).

In the Encyrtidae, the number of larval instars in the genus *Ooencyrtus* has been reported to vary from three to five, but was later determined to be five based on measurements of the larva mandibles (Takasu and Hirose 1989). The Pteromalidae *Teleas* sp. has four instars (Ayers 1883). The Aphelinidae *Centrodora scolypopae* (Valentine), that parasitize eggs of orthopterous and homopterous insects, has three larval instars (Gerard 1989). The Proctotrypidae *Phanurus angustatus* Thomson, a parasitoid of heteropterous eggs, has two larval instars as the Entedonidae *Anellaria conomeli* Bakkendorf that parasitizes eggs of Cicadellidae (Bakkendorf 1934).

1.3.2.2 Morphology and Feeding

In egg parasitoids, most of the variability in larvae is found in the first instar. The later instars are quite uniform in shape and external features and are generally referred to as hymenopteriform (Quicke 1997). The variety of forms in larval instars has been described by Clausen (1940) and Hagen (1964), and more than a dozen names have been proposed to describe these various forms. In this section we will follow the terminology used by Hagen (1964).

Quite early, the difficulty in separating larval instars in the Trichogrammatidae has been recognized and two groups of genera were proposed (Bakkendorf 1934, Clausen 1940). In one group, comprising genera such as *Ophioneurus* and *Poropoea*, two larval instars were found, the first one being of the mymariform type. The first instar larvae of these genera have a cephalic region that is rostriform, with the mouth at the tip and a distinct segmentation. There is a caudal appendage with a ventral tooth, segmentally arranged hairs and a long dorsal spine on the last segment. These descriptions are very similar to those of first instar larvae of Mymaridae. The description of the first instar of *Poropoea stollwerckii* Förster provided by Silvestri in 1916 is reported by Clausen (1940, Fig. 49). The first instar measures 280 μm and its abdomen consists of seven segments, the last one being curved ventrally. All segments except the last one bear setae. The second group includes genera such as *Trichogramma*, *Chaetostrichia*, *Oligosita* and *Prestwichia*. Only one larval instar is described, these larvae being sacciform (bag-shaped) with little external features and minute mandibles.

The more recent descriptions of Trichogrammatidae immatures concern mostly the genus *Trichogramma*. The larva is sacciform, being ovoid without segmentation, lacks setae, is apneustic and mandibles are visible. The mandibles are fully formed at the beginning of the larval development and are simply covered by tissue until later in the larval development (Volkoff et al. 1995, Dahlan and Gordh 1996). The embryonic membrane is generally attached to the posterior end of the larva. Some authors report that the first instar larva is vermiform at hatching but takes a sacciform shape as it starts to feed (Hawlitzky and Boulay 1982, Volkoff et al. 1995, Cônsoli et al. 1999). Whether authors reported one or several instars, the size of the first instar larvae varied. In general at hatching the first instar larva measures between 150–300 μm in length and 80–120 μm in width and grows to up to 500 μm in length and 300 μm in width.

Feeding by *Trichogramma australicum* and *T. brassicae* has been described in details (Brenière 1965, Jarjees et al. 1998, Wu et al. 2000, Jarjees and Merritt 2002). The feeding apparatus occupies an important part of the newly hatched larva that has a vermiform shape. The oral opening is 10–12 μm in diameter and is positioned ventrally at the anterior extremity of the larva, flanked by a pair of mandibles. At rest, the mouth opening is occluded by the epipharynx, a lobe-shaped extension of the dorsal wall of the pharynx. When feeding, the pharynx (stomodaeum in Wu et al. 2000) dilates and constricts with a frequency of 40–60 times per minute. These dilatations and constrictions are caused by contractions of four dorsal and one ventral dilator muscles. With each contraction, food enters the pharynx and is pumped back and forth until the food particles are small enough to pass through the cardiac valve into the midgut. The larva feeds continuously until it has absorbed the totality of the host egg content, which occurs less than 10 h after hatching. The larva is able to ingest not only disorganized host content but also solid particles (Wu et al. 2000). Salivary glands are connected near the oral opening by salivary ducts. The midgut rapidly increases in volume, its cells, flattened and highly vacuolated, become thin and elongated as the midgut wall extends. The posterior section of the midgut is occluded by a group of rosette cells, but it forms a continuous tube from the midgut though the hindgut. The prevention of defecation by the developing larva is thus due to the presence of these rosette cells rather than by the fact that the midgut is not connected to the hindgut.

The Scelionidae have an exceptional uniformity in the morphological characters of their larvae (Clausen 1940). The first instar larva of Scelionidae is of the teleaform type (Hagen 1964). These larvae have a complete lack of segmentation, but the body is divided by a sharp constriction. The mandibles are external, large and sickle-shaped. The cephalothorax bears no external structure except for the mandibles. The abdomen terminates in a caudoventral tail with a sharp point. There are also a variable number of setae near the anterior margin of the abdomen that sometimes make a complete transverse ring around the abdomen. Larvae of *Telenomus* measure 100–150 μm in length and 80–90 μm in width at emergence, are teleaform with strong mandibles and a bifurcated caudal tail (Strand et al. 1988, Volkoff and Colazza 1992). Setae are generally present on the abdomen (Rothschild 1970, Gerling 1972).

Three types of first instar larvae of Mymaridae are recognized (Jackson 1961): mymariform, sacciform and elongated. The mymariform larvae are characterized by a cephalic structure with a frontal process above the mouth aperture and mandibles. The caudal appendage is slender and bears several setae that are used by the larva to move within the host egg. In this group are found genera such as *Anaphes, Polynema, Ooctonus* (Bakkendorf 1934, Boivin et al. 1993, van Baaren et al. 1997). Other genera, such as *Anagrus*, have sacciform larvae that lack most of the structure of the mymariform larvae (Meyerdirk and Moratorio 1987). Finally elongated larvae with no visible segmentation and no mandible are found in the genus *Caraphractus* (Jackson 1961).

The mymariform larvae use their mandible to fight with other larvae in either conspecific or interspecific superparasitism (or multiparasitism) (Nénon et al. 1995, van Baaren et al. 1997, Boivin and Brodeur 2006). These mymariform larvae are present in solitary species, such as *A. victus*, but also in facultative gregarious species such as *A. listronoti* (van Baaren et al. 1997). These two species are closely related and are both parasitoids of Curculionidae eggs (Landry et al. 1993). When a larva of both species are present in a host, each has a survival probability of about 50% indicating that the two species have similar fighting abilities (Boivin and van Baaren 2000). When the solitary *A. victus* is present with varying numbers of *A. listronoti* immatures, its probability of survival matches the predictions of the immobility hypothesis (Boivin and van Baaren 2000). Direct observations showed that the solitary *A. victus* larvae indeed move ten times more than the gregarious *A. listronoti* larvae.

The larvae of male *Encarsia porteri* (Mercet) (Aphelinidae) develop in eggs of Lepidoptera and possess a sculptured cuticle, bear long spines along the venter and have horn-like projections on the head capsule (Hunter et al. 1996, Hunter and Woolley 2001). The Proctotrypidae *Phanurus angustatus* Thomson has first-instar larvae that are round and constricted and show two large mandibles (Bakkendorf 1934).

The Encyrtidae *Ooencyrtus* are egg parasitoids and their larvae have a unique respiratory system within the Hymenoptera. At eclosion the eggshell ruptures and wrinkles back on the larvae. The two spiracles of the larvae stay attached to the egg band and the larva breathes through the egg stalk that protrudes from the egg chorion (Maple 1947). These Encyrtidae larvae thus breathe atmospheric air.

1.4 Concluding Remarks

The egg parasitoids are model organisms that will continue to be used extensively in behavioral ecology and functional morphology. As new methodologies become available to study the morphology, physiology and behavior of these organisms, new hypotheses will be tested on these animals.

Two issues remain to be solved however. The first one is how representative egg parasitoids are to other Hymenoptera parasitoids and ultimately to insects in general. While their special characteristics can be used to test specific hypotheses, generalization of the results could be problematic if the adaptations of egg parasitoids to their small size hosts make them specialized organisms with little in common with other groups. The second point that will need data in the near future is how laboratory results relate to field data. The few data available at this point just cannot be used to assess the value of laboratory data on reproductive strategies, mating structure or dispersion.

References

Abbas MST (1989) Studies on *Trichogramma buesi* as a biocontrol agent against *Pieris rapae* in Egypt. Entomophaga 34:447–451

Al-Wahaibi AK, Owen AK, Morse JG (2005) Description and behavioural biology of *Ufens* species (Hymenoptera: Trichogrammatidae), egg parasitoids of *Homalodisca* species (Hemiptera: Cicadellidae) in southern California. Bull Entomol Res 95:275–288

Ayasse M, Paxton RJ, Tengo J (2001) Mating behavior and chemical communication in the order Hymenoptera. Annu Rev Entomol 46:31–78

Ayers H (1883) On the development of *Oecanthus niveus* and its parasite, *Teleas*. Mem Boston Soc Nat Hist 3:225–281

Bakkendorf O (1934) Biological investigations on some Danish Hymenopterous egg-parasites, especially in Homopterous and Heteropterous eggs, with taxonomic remarks and descriptions of new species. Ent Medd 19:1–135

Bin F, Roversi PF, Meierrose C, Isidoro N, Romani R (2000) Ventilation holes in host egg shell: a parasitoid adaptation to a host constraint. 7th European Workshop on Insect Parasitoids, Haarlem, The Netherlands

Boivin G, van Baaren J (2000) The role of larval aggression and mobility in the transition between solitary and gregarious development in parasitoid wasps. Ecol Lett 3:469–474

Boivin G, Brodeur J (2006) Intra- and interspecific interactions among parasitoids: mechanisms, outcomes and biological control. In: Brodeur J, Boivin G (eds), Trophic and guild interactions in biological control. Springer, Dordrecht, The Netherlands, pp 123–144

Boivin G, Lagacé M (1999) Effet de la taille sur la fitness de *Trichogramma evanescens* (Hymenoptera: Trichogrammatidae). Ann Soc Entomol France 35(suppl.):371–378

Boivin G, Picard C, Auclair JL (1993) Pre-imaginal development of *Anaphes* nsp. (Hymenoptera: Mymaridae), an egg parasitoid of the carrot weevil (Coleoptera: Curculionidae). Biol Control 3:176–181

Boivin G, Fauvergue X, Wajnberg E (2004) Optimal patch residence time in egg parasitoids: innate versus learned estimate of patch quality. Oecologia 138:640–647

Boivin G, Jacob S, Damiens D (2005) Spermatogeny as a life-history index in parasitoid wasps. Oecologia 143:198–202

Bonduriansky R (2001) The evolution of male mate choice in insects: a synthesis of ideas and evidence. Biol Rev 76:305–339

Bosco D, Arzone A (1991) Indagini sui parassitoidi oofagi di *Lindbergina aurovittata* (Douglas) e *L. spoliata* (Horvath) (Homoptera Auchenorrhyncha). Redia 74:147–162

Brenière J (1965) Les trichogrammes parasites de *Proceras sacchariphagus* Boj. borer de la canne à sucre à Madagascar. 2- Etude biologique de *Trichogramma australicum* Gir. Entomophaga 10:99–117

Bressac C, Chevrier C (1998) Offspring and sex ratio are independent of sperm management in *Eupelmus orientalis* females. J Insect Physiol 44:351–359

Brown MW (1984) Literature review of *Ooencyrtus kuvanae* (Hym.: Encyrtidae), an egg parasite of *Lymantria dispar* (Lep.: Lymantriidae). Entomophaga 29:249–265

Chevrier C, Bressac C (2002) Sperm storage and use after multiple mating in *Dinarmus basalis* (Hymenoptera: Pteromalidae). J Insect Behav 15:385–398

Chihrane J, Laugé G (1994) Incidences de chocs de températures élevées sur la lignée germinale mâle de *Trichogramma brassicae* (Hym: Trichogrammatidae). Entomophaga 39:11–20

Clausen CP (1940) Entomophagous insects. McGraw-Hill, New York

Collins RD, Grafius E (1986) Courtship and mating behavior of *Anaphes sordidatus* (Hymenoptera: Mymaridae), a parasitoid of the carrot weevil (Coleoptera: Curculionidae). Ann Entomol Soc Am 79:31–33

Cônsoli FL, Rossi MM, Parra JRP (1999) Developmental time and characteristics of the immature stages of *Trichogramma galloi* and *T. pretiosum* (Hymenoptera: Trichogrammatidae). Rev Bras Entomol 43:271–275

Cônsoli FL, Conti E, Dangott LJ, Vinson SB (2001) *In vitro* culture of the teratocytes of *Trissolcus basalis* (Hymenoptera, Scelionidae) and their requirements for host-derived components. Biol Control 22:176–184

Cormier D, Royer L, Vigneault C, Panneton B, Boivin G (1998) Effect of female age on daily cycle of sexual pheromone emission in gregarious egg parasitoid *Anaphes listronoti*. J Chem Ecol 24:1595–1610

Dahlan AN, Gordh G (1996) Development of *Trichogramma australicum* Girault (Hymenoptera: Trichogrammatidae) on *Helicoverpa armigera* (Hübner) eggs (Lepidoptera: Noctuidae). Austral J Entomol 35:337–344

Dahlan AN, Gordh G (1997) Development of *Trichogramma australicum* (Hym.: Trichogrammatidae) at low and high population density in artificial diet. Entomophaga 42: 525–536

Damiens D, Boivin G (2005) Male reproductive strategy in *Trichogramma evanescens*: sperm production and allocation to females. Physiol Entomol 30:241–247

Damiens D, Boivin G (2006) Why do sperm-depleted parasitoid males continue to mate? Behav Ecol 17:138–143

Delpuech JM, Froment B, Fouillet P, Pompanon F, Janillon S, Bouletreau M (1998) Inhibition of sex pheromone communications of *Trichogramma brassicae* (Hymenoptera) by the insecticide chlorpyrifos. Environ Toxicol Chem 17:1107–1113

Delpuech JM, Legallet B, Terrier O, Fouillet P (1999) Modifications of the sex pheromonal communication of *Trichogramma brassicae* by a sublethal dose of deltamethrin. Chemosphere 38:729–739

Delpuech JM, Legallet B, Fouillet P (2001) Partial compensation of the sublethal effect of deltamethrin on the sex pheromonal communication of *Trichogramma brassicae*. Chemosphere 42:985–991

Doyon J, Boivin G (2006) Impact of the timing of male emergence on mating capacity of males in *Trichogramma evanescens* Westwood. BioControl 51:703–713

Forsse E, Smith SM, Bourchier RS (1992) Flight initiation in the egg parasitoid *Trichogramma minutum:* effects of ambient temperature, mates, food, and host eggs. Entomol Exp Appl 62:147–154

Gahlhoff JE (1998) ook of insect records – University of Florida Press. http://ufbir.ifas.ufl.edu/chap38.htm. Accessed 9 Jan 2008

Gerard PJ (1989) Biology and morphology of immature stages of *Centrodora scolypopae* (Hymenoptera: Aphelinidae). NZ Entomol 12:24–29

Gerling D (1972) The developmental biology of *Telenomus remus* Nixon (Hym., Scelionidae). Bull Entomol Res 61:385–388

Gerling D, Legner EF (1968) Developmental history and reproduction of *Spalangia cameroni*, parasite of synantrophic flies. Ann Entomol Soc Am 61:1436–1443

Gerling D, Orion T (1973) The giant cells produced by *Telenomus remus* (Hymenoptera: Scelionidae). J Invertebr Pathol 21:164–171

Gerling D, Rotary N (1974) Structure and function of the seminal vesicles and the spermatheca in *Bracon hebetor*. Int J Insect Morphol Embryol 3:159–162

Gibson GAP (1986) Evidence of monophyly and relationships of Chalcidoidea, Mymaridae, and Mymarommatidae (Hymenoptera: Terebrantes). Can Entomol 118:205–240

Godfray HCJ (1994) Parasitoids: behavioral and evolutionary ecology. Princeton University Press, Princeton

Godin C, Boivin G (2000) Effects of host age on parasitism and progeny allocation in Trichogrammatidae. Entomol Exp Appl 97:149–160

Hagen KS (1964) Developmental stages of parasites. In: DeBach P (ed), Biological control of insect pests and weeds. Chapman & Hall, London, pp 168–246

Halliday T, Arnold SJ (1987) Multiple mating by females: a perspective from quantitative genetics. Anim Behav 35:939–941

Hamerski MR, Hall RW (1989) Adult emergence, courtship, mating, and ovipositional behavior of *Tetrastichus gallerucae* (Hymenoptera: Eulophidae), a parasitoid of elm leaf beetle (Coleoptera: Chrysomelidae). Environ Entomol 18:791–794

Hardy ICW, Ode PJ, Siva-Jothy MT (2007) Mating behavior. In: Jervis MA (ed), Insects as natural enemies: a practical perspective. Springer, Dordrecht, The Netherlands, pp 219–260

Hawlitzky N, Boulay C (1982) Régimes alimentaires et développement chez *Trichogramma maidis* Pintureau et Voegele (Hym. *Trichogrammatidae*) dans l′oeuf *d′Anagasta kuehniella Zeller* (Lep. *Pyralidae*). Colloques l′INRA 9:101–106

Heimpel GE, de Boer JG (2008) Sex determination in the Hymenoptera. Annu Rev Entomol 53:209–230

Henriquez NP, Spence JR (1993) Studies of *Lathromeroidea* sp. nov. (Hymenoptera: Trichogrammatidae), a parasitoid of gerrid eggs. Can Entomol 125:693–702

Hewett SW (1980) The effect of prey size on the functional and numerical responses of a protozoan predator to its prey. Ecology 61:1075–1081

Hogge M, King PE (1975) The ultrastructure of spermatogenesis in *Nasonia vitripennis* (Walker) (Hymenoptera: Pteromalidae). J Submicrosc Cytol 7:81–96

Hunter MS, Woolley JB (2001) Evolution and behavioral ecology of Heteronomous Aphelinid parasitoids. Annu Rev Entomol 46:251–290

Hunter MS, Rose M, Polaszek A (1996) Divergent host relationships of males and females in the parasitoid *Encarsia porteri* (Hymenoptera: Aphelinidae). Ann Entomol Soc Am 89:667–675

Hutchison WD, Moratorio M, Martin JM (1990) Morphology and biology of *Trichogrammatoidea bactrae* (Hymenoptera: Trichogrammatidae), imported from Australia as a parasitoid of pink bollworm (Lepidoptera: Gelechiidae) eggs. Ann Entomol Soc Am 83:46–54

Isidoro N, Bin F, Colazza S, Vinson SB (1996) Morphology and antennal gustatory sensilla and glands in some parasitoid Hymenoptera with hypothesis on their role in sex and host recognition. J Hymenopt Res 5:206–239

Jackson DJ (1961) Observations on the biology of *Caraphractus cinctus* Walker (Hymenoptera: Mymaridae), a parasitoid of the eggs of Dytiscidae (Coleoptera). Parasitology 51:269–294

Jacob S, Boivin G (2005) Costs and benefits of polyandry in the egg parasitoid *Trichogramma evanescens* Westwood (Hymenoptera: Trichogrammatidae). Biol Control 32:311–318

Jarjees EA, Merritt DJ (2002) Development of *Trichogramma australicum* Girault (Hymenoptera: Trichogrammatidae) in *Helicoverpa* (Lepidoptera: Noctuidae) host eggs. Austral J Entomol 41:310–315

Jarjees E, Merritt DJ, Gordh G (1998) Anatomy of the mouthparts and digestive tract during feeding in larvae of the parasitoid wasp *Trichogramma australicum* Girault (Hymenoptera: Trichogrammatidae). Int J Insect Morphol Embryol 27:103–110

Jervis MA, Heimpel GE, Ferns PN, Harvey JA, Kidd NAC (2001) Life-history strategies in parasitoid wasps: a comparative analysis of "ovigeny". J Anim Ecol 70:442–458

Jervis MA, Ellers J, Harvey JA (2008) Resource acquisition, allocation, and utilization in parasitoid reproductive strategies. Annu Rev Entomol 53:361–385

Kazmer DJ, Luck RF (1991) The genetic- mating structure of natural and agricultural populations of *Trichogramma*. Colloques l'INRA 56:107–110

Keeling CI, Plettner E, Slessor KN (2004) Hymenopteran semiochemicals. Top Curr Chem 239:133–177

King BH (1987) Offspring sex ratios in parasitoid wasps. Q Rev Biol 62:367–396

King BH, Saporito KB, Ellison JJ, Bratzke RM (2005) Unattractiveness of mated females to males in the parasitoid wasp *Spalangia endius*. Behav Ecol Sociobiol 57:350–356

Landry BS, Dextrase L, Boivin G (1993) Random amplified polymorphic DNA markers for DNA fingerprinting and genetic variability assessment of minute parasitic wasp species (Hymenoptera: Mymaridae and Trichogrammatidae) used in biological control programs of phytophagous insects. Genome 36:580–587

Leatemia JA, Laing JE, Corrigan JE (1995) Production of exclusively male progeny by mated, honey-fed *Trichogramma minutum* Riley (Hym., Trichogrammatidae). J Appl Entomol 119:561–566 van Lenteren JC (2003) Commercial availability of biological control agents. In: van Lenteren J (ed), Quality control and production of biological control agents: theory and testing procedures. CABI Publishing, Wallingford, pp 167–179

Loch AD, Walter GH (2002) Mating behavior of *Trissolcus basalis* (Wollaston) (Hymenoptera: Scelionidae): potential for outbreeding in a predominantly inbreeding species. J Insect Behav 15:13–23

Manjunath TM (1972) Biological studies on *Trichogrammatoidea armigera* Nagaraja, a new dimorphic egg parasite of *Heliothis armigera* (Hübner) in India. Entomophaga 17:131–147

Manweiler SA (1986) Developmental and ecological comparisons of *Trichogramma minutum* and *Trichogramma platneri* (Hymenoptera: Trichogrammatidae). Pan-Pac Entomol 62:128–139

Maple JD (1947) The eggs and first instar larvae of Encyrtidae and their morphological adaptations for respiration. Publ Entomol 8:1–122

Martel V, Boivin G (2004a) Impact of competition on sex allocation by *Trichogramma*. Entomol Exp Appl 111:29–35

Martel V, Boivin G (2004b) Premating dispersion in the egg parasitoid *Trichogramma* (Hymenoptera: Trichogrammatidae). Environ Entomol 33:855–859

Martel V, Damiens D, Boivin G (2008a) Male mate choice in *Trichogramma turkestanica*. J Insect Behav 21:63–71

Martel V, Damiens D, Boivin G (2008b) Strategic ejaculation in the egg parasitoid *Trichogramma turkestanica* (Hymenoptera: Trichogrammatidae). Ecol Entomol 33:357–361

Masutti L, Battisti A, Milani N, Zanata M, Zanazzo G (1993) In vitro rearing of *Ooencyrtus pityocampae* (Hym., Encyrtidae), an egg parasitoid of *Thaumetopoea pityocampa* (Lep., Thaumetopoeidae). Entomophaga 38:327–333

McKenzie LM, Beirne BP (1972) A grape leafhopper, *Erythroneura ziczac* (Homoptera: Cicadellidae), and its mymarid (Hymenoptera) egg parasite in the Okanagan Valley, British Columbia. Can Entomol 104:1229–1233

Meyerdirk DE, Moratorio MS (1987) Biology of *Anagrus giraulti* (Hymenoptera: Mymaridae), an egg parasitoid of the beet leafhopper, *Circulifer tenellus* (Homoptera: Cicadellidae). Ann Entomol Soc Am 80:272–277

Mockford EL (1997) A new species of *Dicopomorpha* (Hymenoptera: Mymaridae) with diminutive, apterous males. Ann Entomol Soc Am 90:115–120

Nadel H, Luck RF (1985) Span of female emergence and male sperm depletion in the female-biased, quasi-gregarious parasitoid, *Pachycrepoideus vindemiae* (Hymenoptera: Pteromalidae). Ann Entomol Soc Am 78:410–414

Navasero RC, Oatman ER (1989) Life history, immature morphology and adult behavior of *Telenomus solitus* (Hymenoptera: Scelionidae). Entomophaga 34:165–177

Nénon JP, Boivin G, Allo MR (1995) Fine structure of the egg envelopes in *Listronotus oregonensis* (LeConte) (Coleoptera: Curculionidae) and morphological adaptations to oviposition sites. Int J Insect Morphol Embryol 24:333–342

Ode PJ, Antolin MF, Strand MR (1997) Constrained oviposition and female-biased sex allocation in a parasitic wasp. Oecologia 109:447–455

Pak GA, Oatman ER (1982) Biology of *Trichogramma brevicapillum*. Entomol Exp Appl 32:61–67

Parker GA (1998) Sperm competition and the evolution of ejaculates: towards a theory base. In: Birkhead TR, Moller AP (eds), Sperm competition and sexual selection. Academic, San Diego, pp 3–53

Parker GA (2001) Golden flies, sunlit meadows: a tribute to the Yellow Dungfly. In: Dugatkin (ed), Model systems in behavioral ecology. Princeton University Press, Princeton, pp 3–26

Picard C, Auclair JL, Boivin G (1991) Response to host age of the egg parasitoid *Anaphes* n.sp. (Hymenoptera: Mymaridae). Biocontrol Sci Technol 1:169–176

Pickford R (1964) Life history and behaviour of *Scelio calopteni* Riley (Hymenoptera: Scelionidae), a parasite of grasshopper eggs. Can Entomol 96:1167–1172

Pintureau B, Toonders FB (1983) Quelques résultats concernant l'étude de l'attraction des mâles par les femelles vierges chez *Trichogramma maidis* (Hym., Trichogrammatidae). Bull Mens Soc Linn Lyon 52:81–87

Pintureau B, Calvin MPI, Grenier S (1997) Effectiveness of the second mating in a bisexual *Trichogramma* species and the first mating in a thelytokous *Trichogramma* species (Hymenoptera: Trichogrammatidae). Can Entomol 129:35–41

Pompanon F, Fouillet P, Bouletreau M (1995) Emergence rhythms and protandry in relation to daily patterns of locomotor activity in *Trichogramma* species. Evol Ecol 9:467–477

Pompanon F, de Schepper B, Mourer Y, Fouillet P, Bouletreau M (1997) Evidence for a substrate-borne sex pheromone in the parasitoid wasp *Trichogramma brassicae*. J Chem Ecol 23:1349–1360

Quicke DLJ (1997) Parasitic wasps. Chapman and Hall, London

Ramadan MM, Wong TTY, Wong MA (1991) Influence of parasitoid size and age on male mating success of opiines (Hymenoptera: Braconidae), larval parasitoids of fruit flies. Biol Control 1:248–255

Ridley M (1988) Mating frequencies and fecundity in insects. Biol Rev Cambridge Philos Soc 63:509–549

Ridley M (1993) Clutch size and mating frequency in parasitic Hymenoptera. Am Nat 142:893–910

Roitberg BD, Boivin G, Vet LEM (2001) Fitness, parasitoids, and biological control: an opinion. Can Entomol 133:429–438

Rosi MC, Isidoro N, Colazza S, Bin F (1995) Functional anatomy of the female reproductive system of *Trissolcus basalis* (Woll.). Colloques l'INRA 73:101–103

Rothschild GHL (1970) Parasites of rice stemborers in Sarawak (Malaysian Borneo). Entomophaga 15:21–51

Ruberson JR, Tauber MJ, Tauber CA (1988) Reproductive biology of the two biotypes of *Edovum puttleri*, a parasitoid of Colorado potato beetle eggs. Entomol Exp Appl 46:211–219

Saakian-Baranova AA (1990) Morphological study of preimaginal stages of six species of *Trichogramma* Westwood (Hymenoptera, Trichogrammatidae). Entomol Obozrenie 2:257–263

Safavi M (1968) Etude biologique et écologique des Hyménoptères parasites des oeufs des punaises des céréales. Entomophaga 13:381–495

Sahad KA (1984) Biology of *Anagrus optabilis* (Perkins) (Hymenoptera, Mymaridae), an egg parasitoid of Delphacid planthoppers. Esakia 22:129–144

Simmons LW, Siva-Jothy MT (1998) Sperm competition in insects: mechanisms and the potential for selection. In: Birkhead TR, Moller AP (eds), Sperm competition and sexual selection. Academic Press, San Diego, pp 341–342

Sousa JM (1999) Development of *Tiphodytes gerriphagus* (Hymenoptera: Scelionidae) in *Limnoporus dissortis* eggs (Hemiptera: Gerridae). Can Entomol 131:219–228

Strand MR (1986) The physiological interactions of parasitoids with their hosts and their influence on reproductive strategies. In: Waage J, Greathead D (eds), Insect parasitoids. Academic Press, London, pp 97–136

Strand MR, Quarles JM, Meola SM, Vinson SB (1985) Cultivation of teratocytes of the egg parasitoid *Telenomus heliothidis* (Hymenoptera: Scelionidae). In Vitro Cell Dev Biol 21:361–367

Strand MR, Meola SM, Vinson SB (1986) Correlating pathological symptoms in *Heliothis virescens* eggs with development of the parasitoid *Telenomus heliothidis*. J Insect Physiol 32:389–402

Strand MR, Vinson SB, Nettles WC, Xie ZN (1988) In vitro culture of the egg parasitoid *Telenomus heliothidis* the role of teratocytes and medium consumption in development. Entomol Exp Appl 46:71–78

Suzuki Y, Hiehata K (1985) Mating systems and sex ratios in the egg parasitoids, *Trichogramma dendrolimi* and *T. papilionis* (Hymenoptera: Trichogrammatidae). Anim Behav 33:1223–1227

Takada Y, Kawamura S, Tanaka T (2000) Biological characterisics: growth and development of the egg parasitoid *Trichogramma dendrolimi* (Hymenoptera: Trichogrammatidae) on the cabbage armyworm *Mamestra brassicae* (Lepidoptera: Noctuidae). Appl Entomol Zool 35:369–379

Takasu K, Hirose Y (1989) The number of larval instars in *Ooencyrtus* species (Hymenoptera, Encyrtidae). Jpn J Entomol 57:398–401

Tanaka M (1985a) Early embryonic development of the parasitic wasp, *Trichogramma chilonis* (Hymenoptera, Trichogrammatidae). In: Ando H, Miya K (eds), Recent advances in insect embryology in Japan. ISEBU, Tsukuba, Japan, pp 171–179

Tanaka M (1985b) Embryonic and early post-embryonic development of the parasitic wasp, *Trichogramma chilonis* (Hymenoptera, Trichogrammatidae). In: Ando H, Miya K (eds), Recent advances in insect embryology in Japan. ISEBU, Tsukuba, Japan, pp 181–189

Thornhill R, Alcock J (2001) The evolution of insect mating systems. 1Universe.com Inc, Lincoln, NE, USA

van Baaren J, Boivin G, Le Lannic J, Nénon JP (1997) The male and female first instar larvae of *Anaphes victus* and *A. listronoti* (Hymenoptera, Mymaridae). Zoomorphology 117:189–197

van Beek TA, Silva IMMS, Posthumus MA, Melo R (2005) Partial elucidation of *Trichogramma* putative sex pheromone at trace levels by solid-phase microextraction and gas chromatography-mass spectrometry studies. J Chromatogr A 1067:311–321

van den Assem J (1986) Mating behavior in parasitic wasps. In: Waage J, Greathead D (eds), Insect parasitoids. Academic, London, pp 137–167

Voegelé J, Brun P, Daumal J (1974) Les Trichogrammes 1 – Modalités de la prise de possession et de l′élimination de l′hôte chez le parasite embryonnaire *Trichogramma brasiliensis* (Hym. Chalcidoidea). Ann Soc Entomol France 10:757–761

Volkoff AN, Colazza S (1992) Growth patterns of teratocytes in the immature stages of *Trissolcus basalis* (Woll.) (Hymenoptera: Scelionidae), an egg parasitoid of *Nezara viridula* (L.) (Heteroptera: Pentatomidae). Int J Insect Morphol Embryol 21:323–336

Volkoff AN, Daumal J, Barry P, François MC, Hawlitzky N, Rossi MM (1995) Development of *Trichogramma cacoeciae* Marchal (Hymenoptera: Trichogrammatidae): time table and evidence for a single larval instar. Int J Insect Morphol Embryol 24:459–466

Waage JK (1982) Sib- mating and sex ratio strategies in scelionid wasps. Ecol Entomol 7:103–112

Wajnberg E (2006) Time allocation strategies in insect parasitoids: from ultimate predictions to proximate behavioral mechanisms. Behav Ecol Sociobiol 60:589–611

Wu ZX, Cohen AC, Nordlund DA (2000) The feeding behavior of *Trichogramma brassicae*: new evidence for selective ingestion of solid food. Entomol Exp Appl 96:1–8

Yeo YS, Chang YD, Goh HG (1990) A morphological observation of an egg parasitoid, *Anagrus incarnatus* Haliday (Hymenoptera: Mymaridae), of the rice planthoppers. Korean J Appl Entomol 29:1–5

Zongo JO, Vincent C, Stewart RK (1993) Biology of *Trichogrammatoidea simmondsi* (Hym.: Trichogrammatidae) on sorghum shoot fly, *Atherigona soccata* (Dipt.: Muscidae) eggs. Entomophaga 38:207–212

Chapter 2
Nutritional Ecology of Insect Egg Parasitoids

S. Bradleigh Vinson

Contents

2.1 Introduction

As noted by Slansky and Rodriguez (1987), the consumption, utilization, and allocation of resources (foods) are essential components of all animal lives and differences in these components result in the differences among animal life-styles. In fact, it has been noted that selection pressures on these three components have in most cases led to the evolution of the different life-styles (Brues 1946, Beck 1972, Calow 1977, Southwood 1977, Montgomery 1978, Lawton and McNeil 1979, Crawley 1983, Kim 1984, McNab 1984, Slansky and Scriber 1985). Thus, when considering the nutritional ecology of an organism one must examine the ecology, behavior, physiology and evolution within a nutritional context of food location, consumption,

S.B. Vinson (✉)
Texas A&M University, College Station, TX 77843-2475, USA
e-mail: bvinson@tamu.edu

F.L. Consoli et al. (eds.), *Egg Parasitoids in Agroecosystems with Emphasis on Trichogramma*, Progress in Biological Control 9, DOI 10.1007/978-1-4020-9110-0_2,

utilization and allocation of the resources to various aspects of the organism's life stile. Such a background can provide some insight into an organism's evolutionary biology and the various life strategies.

As noted above, the nutritional ecology of an organism involves many factors and deciding where to begin can be a challenge. For this reason it may be important to contrast the basic nutritional ecology of the different arthropod feeding strategies. For some arthropods the decision to locate and feed on an item is the decision of the arthropod that will be feeding. For example, for most hemimetabolous species the adult and immature feed on similar items, but it is their decision based on their genetics and experience (Sehnal 1985). For example, both adult and immature grasshoppers chose what they will feed on (Chambers et al. 1996). In other cases a female lays eggs and the immature that emerges may be able to make some choices, but these choices may be limited. This is particularly true of some holometabolous species such as the Lepidoptera, where the adults generally feed on nectar and the larvae feed on a plant where the female has laid her eggs, but the larvae of these can often make some choice such as *Heliothis* sp., with the larvae choosing to feed on leaves, flowers and/or fruit (Farrar and Bradley 1985). However, there is one strategy used by a few insects in which the decision is not made by the one feeding and they have no choice. These are the parasitoids and internal seed predators – example, a few species of Coleoptera, such as *Callosobruchus maculatus* (F) – that lay an egg in a seed which is then consumed (Guedes et al. 2007), and thus these can be considered comparable to the parasitoids, i.e., seed parasitoids. In the case of seed parasitoids or insect parasitoids where the young are unable to choose their food or avoid competition with conspecifics if multiple eggs are deposited (Colegrave 1994), the ovipositing female makes the choice alone. As a result the egg parasitoids and the internal single seed predators (seed parasitoids) noted above are the most nutritionally restricted.

One of the major key differences in the nutritional ecology of egg parasitoids over many other species is that the focus of their nutritional ecology is the eggs of other insects. A major issue for an egg parasitoid is locating, recognizing and accepting the host (oviposition), followed by the progeny hatching and utilizing the host (feeding), followed by pupation and emerging as an adult. But as noted by Hawkins (1994), locating a host egg, regardless if it is exposed or concealed, is a major challenge.

The general nutritional ecology of egg parasitoids is summarized in Fig. 2.1. Although there are a number of places to begin, one major issue is the different strategies that have evolved by egg parasitoids to locate hosts. The next step is to understand the type of cues that a parasitoid might use and the sequence of cues that could lead to a host (long range, short range and contact cues). If accepted both female-derived fluids and an egg (or eggs) are injected. The female then marks the egg which may impact its acceptance by other females as well as the female that has just laid an egg (Nufio and Papaj 2001, Rosi et al. 2001). Now the egg of the egg parasitoid is on its own and must deal with the host egg with some help supplied by the ovipositing female, but this is largely up to the parasitoid egg and hatching larva. This is followed by larval development, preparation for and pupation, and then

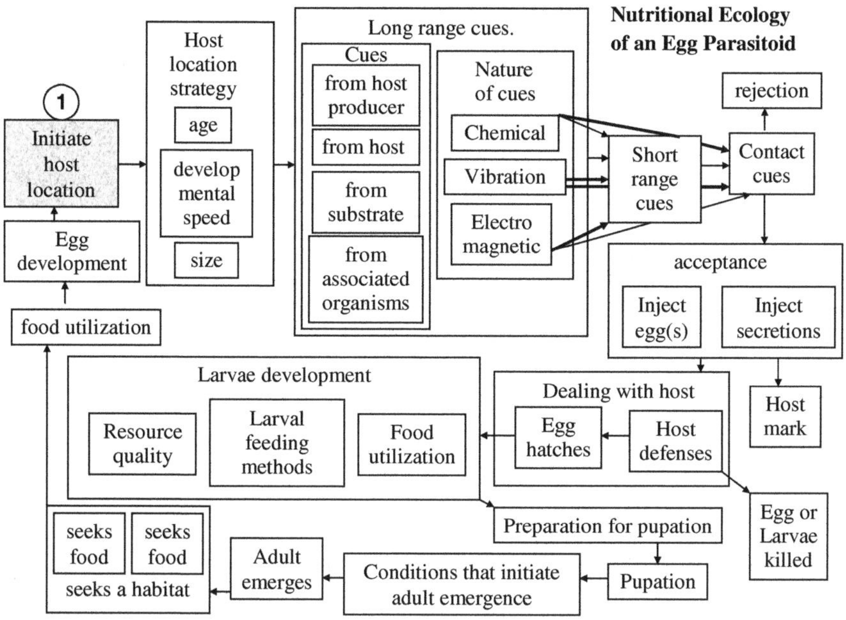

Fig. 2.1 General schematic of the nutritional ecology of an egg parasitoid beginning at "1" and ending at "1"

followed by waiting for the right conditions to initiate adult emergence. The adult may need to locate a habitat, find a mate and food. If food is required, then there are issues with egg maturation and female longevity that may play a role as to when it initiates host location that completes the cycle (Fig. 2.1). All of these issues are important to the nutritional ecology of a female egg parasitoid, but there are others as well, which I have not included. For example, the issue of sex allocation, needs of the adult male, impact of sex altering entities (such as *Wolbachia*), or competition, as a few examples.

It is not possible to cover all of the aspects of the nutritional ecology of egg parasitoids due to a time and space constraint, but several of the topics are covered elsewhere in this book. It is also not the intention here to review the extensive literature concerning the various aspects of the nutritional ecology of egg parasitoids, but rather to focus on and point out various issues in regard to *i*) the host egg and its nutritional constraints that influence the strategies used by egg parasitoids to locate a host, *ii*) some of the challenges that egg parasitoids face in dealing with eggs that the host producer is trying to protect and *iii*) the potential source and characteristics of factors involved in host location. Some other issues such as the feeding behavior and nutrition are covered elsewhere in this book (see Chapter 1). Another important aspect of the nutritional ecology of egg parasitoids is adult feeding and nutrition that can include nectar, honeydew (Jervis et al. 1993), and host feeding (Jervis and Kidd 1986), although the host feeding by *Trichogramma turkestanica* Meyer decreased longevity possibly due to increased carbohydrate cost of additional eggs (Ferracini et al. 2006). Also issues in regards to locating such food and decisions in regard to oviposition versus host feeding occur, but are probably rarer for egg parasitoids. While these are important nutritional ecology issues, the nutritional ecology of the adult stage will not be covered here.

The purpose here is to examine the host *egg* or *oocyte* in regards to it as a resource as well as other issues that the egg content can have on a parasitoid. The second issue is to take a look at egg age that influences the strategy used by egg parasitoids to find hosts. The third issue is to look at just at the host oocyte and consider factors such as size, the importance of being dispersed or clumped (and number) and issues regarding location (access) and protection. The last issue that needs to be considered is how parasitoids locate hosts followed by the sources and nature of cues that egg parasitoids may use to locate hosts.

2.2 The Host Egg

The host egg (Fig. 2.2) consists of two basic components, the oocyte and the egg shell or chorion that surrounds the oocyte and provides protection (the chorion will be discussed later). The oocyte is provisioned with nutrients that function to support the developing insect embryo until the embryo matures and a larvae and nymph emerges (Trougakos and Margaritis 2002). Initially, the nutrients consist of primarily yolk proteins produced by the mothers' fat body (Bownes 1994, Raikhel and Snigirevskaya 1998, Lee et al. 2000, Vinson et al. 2008) or in part by the ovarian follicle cells (Bilinski 1979, Tsuchida et al. 1992, Melo et al. 2000), and are accumulated in the egg (Raikhel and Dhadialla 1992). The proteins along with some lipids,

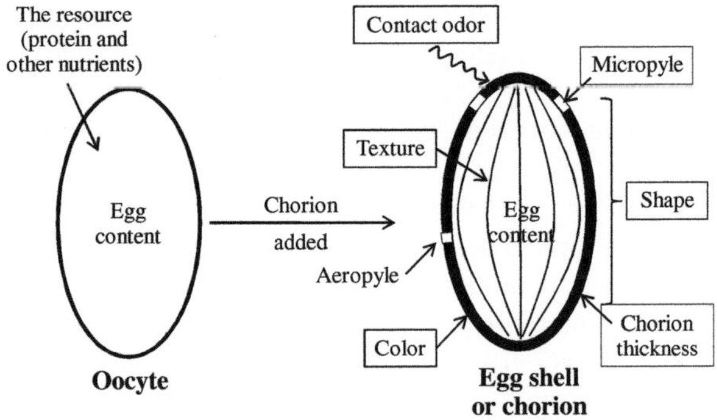

Fig. 2.2 An egg consists of 2 components, the oocyte and the egg shell or chorion. This figure illustrates that the oocyte is the parasitoids developmental resource. The egg shell provides protection for the oocyte and the parasitoid, if parasitized, but it also provides recognition cues (in rectangle) to a parasitoid that is trying to determine if it is a potential host. The micropile provides sperm entrance and the aeropile allows oxygen penetration

carbohydrates and organelles from the female make up the oocyte. This is the nutritional starting material for any egg parasitoid and its composition and amount play a major role regarding the egg's suitability as a host. It is the oocyte that is the key to the nutritional ecology of egg parasitoids. But in considering the nutritional ecology of an egg parasitoid a number of things need to be taken into account.

One is the different parasitoid groups. *Trichogramma* are generally considered to be generalists (Castaneda-Samayoa and Holst 1990, Smith 1996, Saour 2004, Makee 2005), although some have clear host preferences (Guang and Oloo 1990, Monje et al. 1999, Mansfield and Mills 2003). The Mymaridae and Scelionidae appear to be more specialized (Huber 1995, Austin and Field 1997), as are the few egg parasitoids in other families.

For the more generalists the parasitism rates and number of wasps produced/egg can vary greatly (Puterka et al. 1985, Corrigan and Laing 1994, Sá and Parra 1994, Hoffmann et al. 2001, Honda and Luck 2001). This can be due to some eggs being better for development than others (Kivan and Kilic 2002, 2004, Pacheco and Corrêa-Ferreira 1998, Roriz et al. 2005) or due to differences in the parasitoid's ability to respond to the diversity of host location cues that lead to an acceptable host (Mansfield and Mills 2003).

Different hosts can affect the developing parasitoids body size and fecundity (Kazmer and Luck 1991, Parra 1997), longevity (Kuhlmann and Mills 1999) and sex ratio (Corrigan and Laing 1994). Although rearing on different hosts results in a number of effects on the adult, as noted by Brotodjojo and Walter (2006), many of these studies have involved factitious or artificial hosts (Salt 1940, Consoli and Parra 1996, 1999, Grenier et al. 2001, Kölliker-Ott et al. 2003, Makee 2005). Studies have shown that egg volume (Bai et al. 1992), age (Calvin and Losey 1991, Reznik et al. 1997, Monje et al. 1999, Garcia 2000), and nutrition (Barrett and Schmidt 1991) can all have an effect on the adult parasitoid.

Another consideration is that eggs may contain compounds that are not of nutritional value may be found in eggs of some insects (Eisner et al. 2002). For example, defensive compounds appear to be sequestered in eggs of some insect species (Ferguson and Metcalf 1985, Ferguson et al. 1985, Rothschild 1992, Pasteels et al. 1986, Nishida and Fukami 1989, 1990, Blum and Hilker 2002 and references therein), which may determine if the egg is suitable as a host for a certain species. Thus, parasitoids attacking such eggs would be under pressure to evolve to adapt to these toxic compounds or another possibility is to avoid these toxic hosts. For example, eggs of some species of Coccinellidae, which are toxic have compounds on their surface that advertise their toxic nature (Hemptinne et al. 2000). Such eggs often posses aposematic coloration that may signal egg toxicity (Stamp 1980). Also, as Stamp (1980) noted, eggs laid in clusters often have longer incubation periods than singly laid eggs and thus, may more likely be protected in some way. However, protection may not be the only reason for laying eggs in clusters (Clark and Feath 1998).

Another factor is the ability of an egg to mount a defense against the egg parasitoids' egg or larvae. Although insect eggs are able to mount a defense against bacteria and fungi through antimicrobial proteins (Gorman et al. 2004), an immune response of insect eggs towards the egg or larvae of egg parasitoids had not been reported (Strand 1986, Strand and Pech 1995) until the embryo reached gastrulation, which corresponds to the moment hemocytes are formed (Abrams et al. 1993, Strand and Pech 1995). However, an encapsulation response by an egg against a parasitoid has been recently reported (Reed et al. 2007).

While the egg also has a shell or chorion (discussed later), it is the oocyte that is of major importance to the nutritional ecology of the parasitoid. The growth and development of the parasitoid, with the exceptions noted above, is dependent on the amount and quality of the materials within the oocyte. Certainly host size is an issue, but there is also another very important issue that can influence quality which is host age.

2.3 Host Age and Quality

Although an oocyte consists of a lot of proteins (vitellogenins), these proteins are converted to more complex molecules over time (Anderson 1972a, b, Jura 1972) resulting in various embryonic tissues. Although this change and its importance to egg parasitoids was noted by Agrell and Lundquist (1973) and Sander et al. (1985). Pak et al. (1986) working with *Trichogramma* suggested that the nutritional resource and energy that an egg contains remains constant throughout embryonic development. However, this may not be the case for all egg parasitoids. As discussed elsewhere (see Chapter 1), some *Trichogramma* species appear to hatch quickly and the newly hatched larva, which is basically a sack, engulfs the egg contents within a short time, and then completes the process of digestion as the parasitoid larval tissues develop.

In contrast, a number of studies have provided evidence that the success of parasitism of eggs by some egg parasitoids declines as the eggs age (Lewis and Redlinger 1969, Marston and Ertle 1969, Leibee et al. 1979, Houseweart et al. 1982, Powell and Shepard 1982, Strand and Vinson 1983a,b, Reznik and Umarova 1990, Lopes and Parra 1991, Beserra et al. 2002). Ruberson et al. (1987) challenged the view of Pak et al. (1986) and showed that *Edovum puttleri* Grissel (Eulophidae) was less likely to oviposit in older eggs, but if they did the larvae were less likely to survive. Although host age is important, it is not just host age, but the speed of the developmental process. The speed of embryonic development and hatching is one of the more variable factors that a potential host goes through and may range from several days for some eggs of dipterans to a year with some Phasmatidae (Hinton 1981). This has important implications for parasitoids as suggested and illustrated by Vinson (1998), which is modified here as Figs. 2.3 and 2.4 and illustrates a generalized normal egg development period from the time of oviposition to the time of hatching on the *X* axis. The *Y* axis shows the quality of the host oocyte that begins to decrease at point "*W*" to the time through development when the larva "hatches". Due, in part, to these changes, various parasitoids have evolved different strategies to deal with the lower and more complex host resource that occurs over time.

But there are a number of interacting factors involved. One is whether a parasitoid can use a particular host and whether a host can change (in evolutionary

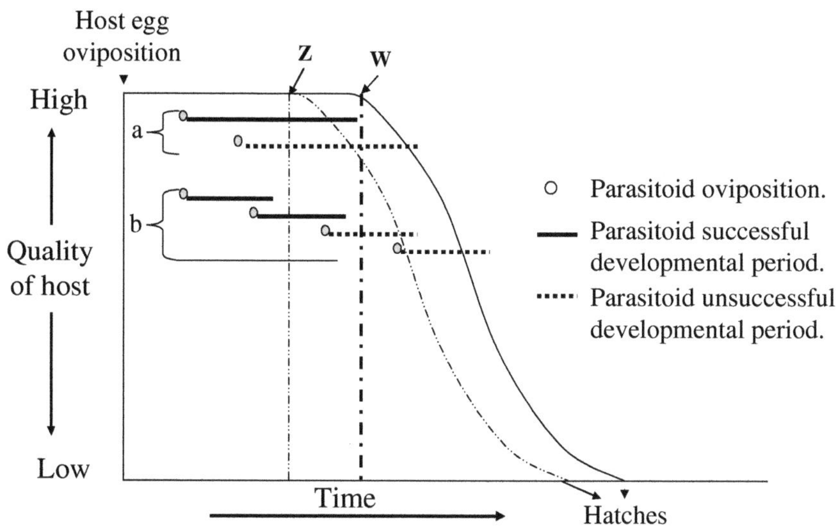

Fig. 2.3 The effects of host age or developmental speed on the quality of the egg as a resource for parasitoids (W = normal, Z = accelerated host tissue development). This assumes that at some point the egg can not meet the developing parasitoids needs. (**a**) The top line shows a species with a long development time that needs to be in a young host, if it oviposits later it does not have the time to complete its development (Lower line). (**b**) By shortening their development time, the top 2 lines, a parasitoid could increase the age that they could successfully develop, but if the host shortens its development time in this case the parasitoid still extends it period of successful attack. The point is that there are evolutionary issues that can be very important

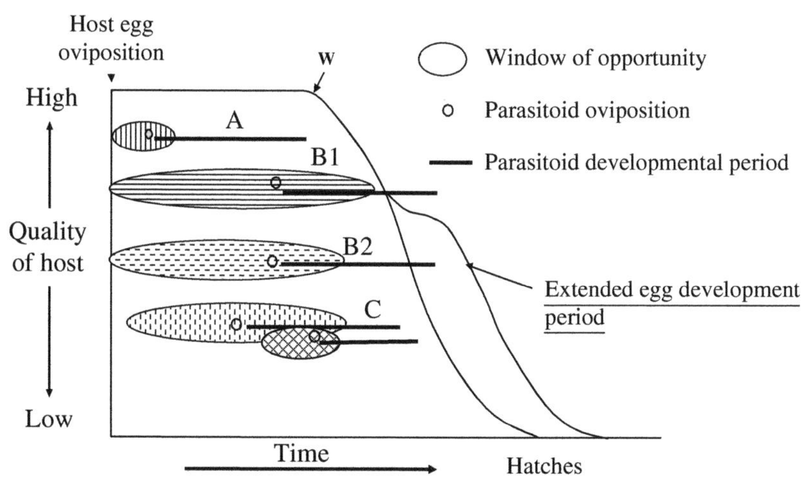

Fig. 2.4 Theoretical strategies used by egg parasitoids to exploit an egg as the oocyte develops into an immature insect and hatches. (**A**) An egg parasitoid that only oviposits in young hosts. (**B1**) An egg parasitoid that cannot or (**B2**) that can manipulate the host's development (The parasitoid extends the egg developmental period). (**C**) An egg parasitoid that kills the embryo and develops more as an internal predator and as a result has flexibility in its development time

terms) in ways that reduces a parasitoid's survival. In particular; can the interactions between decreasing host size and speeding up host development result in the more rapid reduction in the nutritional quality of the host? And can this challenge be met by parasitoids by decreasing their size, increasing their developmental speed and broadening their nutritional restrictions?

As shown in Fig. 2.3, the nutritional quality of a host begins to decline at some point (W in Fig. 2.3) and a parasitoid (a – upper developmental line) that oviposits in a young host egg has the time to complete its development, but if it attacks an older host egg (a.- lower developmental line) it does not have the time. But this also depends on the developmental speed of the egg in question. As noted above, the eggs of some species hatch within a few days while others require months (Hinton 1981). But if a parasitoid "a" attacks a faster developing host species whose quality declines faster (Fig. 2.3 – decreasing host quality begins at "z"), it also does not have the time to develop. An option for some parasitoids is to evolve to successfully attack smaller eggs, but this means less resource (b in Fig. 2.3) which results in smaller parasitoids, but there are limits. In fact, the smallest egg parasitoid may be a trichogrammattid, *Megaphragma caribea* Delvare (170 μm in length), which parasitize thrips eggs (Delvare 1993) or a mymarid, *Dicopomorpha echmepterygis* (139 μm in length) which parasitize psocid eggs (Mockford 1997), however size in this last species only refers to the males. The real question is what are the smallest eggs that can produce a female parasitoid? Such data is not often reported.

While small size and rapid development may provide some parasitoids opportunities to attack small and rapid developing host eggs this also has some negative effects. Parasitoids must search for hosts that are often dispersed, but being very small limits the ability to fly in search of hosts. Further, some egg parasitoids defend egg patches (Calbert and Keller 1998, Field and Calbert 1998, Field et al. 1998), where being larger can be a benefit. Being small is not ways the best option.

So what are the strategies?

2.4 Parasitoid Strategies to Deal with Different Host Quality and Protection Issues

2.4.1 Strategy A

One approach is for the female parasitoid to only oviposit in young hosts, but there are several approaches and possible mechanisms that would insure that only young hosts were attacked (Vinson 1988).

2.4.1.1 Strategy A1

This approach is to just locate host eggs of any age, but only attack young hosts or reject older hosts. This assumes that a female can determine if a potential host is too old to allow her egg to successfully develop. There are a number of reports that older eggs are rejected (Marston and Ertle 1973, Pak 1988, Strand and Vinson 1983c) and, as noted by Strand et al. (1986), differences in age can be important to parasitoid development. But what are the factors that result in rejection? As suggested by Vinson (1998), there may be at least seven factors namely:

Changes in Physical Recognition Cues, Such as Shape, Color, or Size

Of the physical recognition cues mentioned, shape has been implicated. An example is provided by Strand and Vinson (1983b), who reported that the eggs of *Heliothis virescens* (F) are nearly spherical when oviposited, but become more conical with time and these conical shaped eggs are rejected by *Telenomus heliothidis* Ashmead. Pak (1988) also noted the importance of host shape in rejection.

Changes in Physical Condition of the Egg Shell or Membranes

The increased handling time of *Uscana lariophaga* Steffan attacking older host eggs of a bruchid was attributed to increased resistance to ovipositor penetration in older eggs (van Huis et al. 1994). Leibee et al. (1979) suggested that the inability of *Anaphes diana* Girault (=*Patasson lameerei*) to attack older hosts was due to a hardened chorion.

Changes in Chemical Recognition Cues

Changes, such as a decrease in kairomones or a change in relative concentrations over time can be important (Shu and Jones 1985, 1988).

Changes in Chemical Acceptance Cues

Host rejection following ovipositor insertion has been suggested (Salt 1937, Vinson and Iwantsch 1980), which could be due to a build up of more complex compounds as the embryo develops or due to the absence of ovipositional releasers or stimulants as reported for several egg parasitoids (Wu and Qin 1982, Xie et al. 1991).

Presence of Competitors

Rejection due to the presence of eggs or larvae of competitors has been suggested (Ables et al. 1981, Kartsev 1985).

Rejection Due to the Detection of Toxins

Toxins occur in some insect eggs (noted above), and these appear to result in rejection (Bezzerides et al. 2004), but their role in host rejection over time is unclear and is likely to decrease. If so, the toxins might play a more important role in protecting young "host" eggs.

2.4.1.2 Strategy A2

The second strategy to insure attacking only young hosts is to only search for young hosts. This can be considered to be an ambush approach (Vinson 1985), which is to "be there before the host egg is oviposited" (Fig. 2.4A). But there are three approaches to accomplish this. One is the phenomena of phoresy (A2a), another is to respond to the host producing males or females preparing to mate and then wait for oviposition, a spy approach (A2b), and the third is a stealth approach (A2c), which is to search for host food or feeding adults and wait for eggs to appear.

Phoresy

The term "phoresie" was proposed by Lesne in 1896 (Bulletin Entomol. Soc. France, March 25). Phoresy (as used by Howard 1927) refers to the transport of certain insects on the bodies of other insects for purposes other than direct parasitism, and was first reported for egg parasitoids by Warner (1903). Phoresy is a common practice among a number of Trichogrammatidae, Scelionidae and Mymaridae (Clausen 1976). Basically, it is a parasitoid that locates a host producer (*hp*), climbs onto the *hp* and remains there until the *hp* starts to oviposit. The parasitoid then leaves the *hp* and parasitizes the newly oviposited eggs (Tabata and Tamanuki 1940, Bin and Johnson 1982). But phoretic egg parasitoids must locate females instead of eggs. There is evidence that some phoretic egg parasitoids are attracted to sex pheromones of their host producer (Aldrich et al. 1984, Arakaki et al. 1996). Once the phoretic host is located the parasitoid mounts and settles in on the host (Malo 1961). In the case of egg parasitoids of moths, the parasitoid attaches to the wings while in the case of grasshoppers the parasitoid attaches to the edges of the abdominal segments (van Vŭůren 1936, Kolomiyets 1957, Bin and Johnson 1982, Orr et al. 1986). In some cases, such as the scelionid *Mantibaria manticida* (Kieff). that attacks the eggs of the European mantis *Mantis religiosa* (L), the parasitoid removes her wings once on the female mantis and may feed on exudates from small wounds the parasitoid causes (Bin 1985). When the mantis lays her eggs within a frothy coating, the wingless parasitoid enters the froth, parasitizes some eggs and remounts the mantis, allowing the parasitoid to parasitize successive egg masses (Chopard 1923, Couturier 1941). Clausen (1976) pointed out that phoresy provides the egg parasitoid with immediate access to an egg prior to much embryonic development, and provides access to eggs oviposited in widely dispersed locations or eggs that are enclosed in various coverings.

Spy Approach

The spy approach to an ambush strategy is to respond to sex or aggregation pheromones that indicate that mating and/or oviposition may occur in the near future. A good example is provided by the scelionid *Telenomus remus* Nixon which attacks the eggs of the moth *Spodoptera frugiperda* (JE Smith), where the parasitoid was found to be attracted to abdominal extracts (Lewis et al. 1982, Nordlund et al.

1983) and to the sex pheromones (Nordlund et al. 1983) that were identified from the abdominal extracts (Sekul and Cox 1967, Sekul and Sparks 1967, Sparks 1980).

Stealth Approach

The stealth approach to an ambush strategy is for the egg parasitoid, particularly of monophagous herbivores, to respond to the food (usually plants) of the host producer or to aggregations of the host producer. With many insects, both the adult and the immature insects feed on the same plant, thus to respond to such plants even before eggs are laid is a reasonable strategy and is used by some larval parasitoids which respond to odors of the food of their host (Steidle et al. 2001, Mbata et al. 2004). An example of responding to an aggregation of host producers was provided by Leal et al. (1995) who isolated and identified an aggregation pheromone of the male bean bugs *Riptortus clavatus* (Thunberg), that also attracted *Ooencyrtus nezarae* Ishii, an egg parasitoid of the bean bug.

2.4.2 Strategy B

Parasitoids evolving towards strategy B (Fig. 2.4B) are less likely to find young hosts yet they may need a similar period of time to develop. But such species do occur and there are two approaches to deal with such hosts. One is to kill the egg and consume the inside as a predator, an idiobiont approach (Strategy B1) and the other is to regulate the egg host's development, a koinobiont approach (Strategy B2). In either case this increases the window of opportunity for the parasitoid, and such parasitoids tend to search for cues associated with egg oviposition (discussed later).

2.4.2.1 Strategy B1

For this strategy the parasitoid kills the embryo and feeds on dead tissues and stored material in the egg. It appears that some *Trichogramma* may fit this situation as they rapidly consume the host and then they can take time to develop (see Chapter 1). Such parasitoids tend to be smaller, which means that their development time may be shorter. This may be important as killing the host without consuming the dead tissue could lead to problems as the host tissue may degrade and become less valuable as a resource with time. Many of these parasitoids are attracted to many cues associated with oviposition (see Chapter 4).

2.4.2.2 Strategy B2

In contrast, this strategy is to slow the development of the host (extend the development period), giving parasitoids the time needed to complete their development. It appears that some species do not kill the host, but can manipulate the egg or embryo in way's that slows the development of the tissues, thus preserving the resource.

While there is no real difference externally to these two "B" strategies, or in the strategies these parasitoids use to locate a host, it does have some implications for their use of the host resources and whether the term idiobiont is appropriate for all egg parasitoids. This is an important issue that needs to be covered.

As noted by Askew and Shaw (1986), there are a number of terms used to categorize parasitoids. These include "generalist and specialist" or "endoparasitoids and ectoparasitoids" which have been used for years (Sweetman 1963), "r- or K-selected" (MacArthur and Wilkson 1967), "koinophytes or idiophytes" (Haeselbarth 1979), and "host regulators" (Vinson and Iwantsch 1980) or "conformers" (Lawrence 1986, 1990). Askew and Shaw (1986) in an effort to develop an ecological concept modified the terms of Haeselbarth (op cite) to *koinobionts* that benefit from the continued life of the host and *idiobionts* that do not. They went on to note that idiobionts include the ectoparasitoids that permanently paralyze or kill their host following oviposition plus the egg and pupal parasitoids. They state that the host is consumed in the location and state it is in when attacked. From an ecological point of view they are correct. But over the last decade, the terms "host regulators and host conformers" have become synonymous with koinobiont and idiobiont. This creates problems. For example, egg parasitoids are considered idiobionts, and while some kill or stop the development of their host (Jarjees and Merrit 2004), others appear to regulate their host in the early stages of parasitism (Strand and Vinson 1980, Strand 1986). The same is true of pupal parasitoids, some regulate their host (Rivers and Denlinger 1994). The point is that all of these terms have limitations and the terms idiobiont and koinobiont are fine from an ecological perspective, but not from a physiological perspective.

2.4.3 Strategy C

The fourth strategy (Fig. 2.4C) is to speed up development if deposited in an older host. With gregarious larval parasitoids this can occur. As shown by Capinera and Lilly (1975) and Beckage and Riddford (1978), the greater the parasitoid load, the more rapidly the adults emerge. However, the adults from high load hosts tend to be smaller and less fecund (Wylie 1965, Azab et al. 1967).

There may be some other strategies such as laying male eggs in less suitable hosts, such as older or very small eggs.

2.5 Host Size

In addition to host age and quality being an issue, host size is also an issue as the smaller the egg the less the resource. Most egg parasitoids are influenced by size (Brotodjojo and Walter 2006) in several ways. Some prefer a host size that is about the same size as their natal host (Nurindah et al. 1999). Others, particularly the host specific species, are restricted to certain sizes or shapes (Buleza and Mikheev 1978).

As a result some eggs may be too small and may not have the needed resources for the developing egg parasitoids. For example the dipteran *Acnemia amoena* Winnertz lay eggs that are about 140 microns in diameter (Mazzini and Santini 1983) and there may not be many egg parasitoids that could develop on such a limited resource. While any response will depend on the egg having the right cues, such possibilities as being too small or too big do exist. For example, in making artificial eggs for *Trichogramma* some very small, but otherwise identical, spheres were sometimes produced that elicited interest by *Trichogramma evanescens* Westwood, although they did not try to oviposit (Nettles and Vinson unpublished) (see also Chapter 11). Some parasitoids may be less discriminate and an egg can be too big. An example is provided by Boldt et al. (1973) who reported that the eggs of some host species can be too big. For example, a solitary egg parasitoid ovipositing one egg in a large host egg may result in the parasitoid larvae being unable to use the extra resource which can interfere with emergence of the parasitoid or impact the regulation of the host egg's humidity or oxygen levels. Further, left over resourse can result in other problems even if the parasitoid reaches the pupal stage. In contrast, large eggs are important to gregarious species, but they can also have a problem with too much resource unless they measure and adjust the egg number to use the resource efficiently.

For many egg parasitoids the shape and size are important (see discussion by Schmidt 1991) and these parameters can be measured by the parasitoid in several ways (see Schmidt and Smith 1985, 1986, 1987a, 1989). The ability to measure attests to the importance of adjusting the parasitoid load to the resource.

2.6 Host Access

Many egg parasitoids disperse and may move (Antolin and Strong 1987, Keller et al. 1985, Newton 1988), thus it is important to consider if eggs are single, clumped, dispersed, exposed or hidden. These issues can influence a parasitoids access to these resources. It is important to first just consider the nutritional resource, that is the oocyte, without considering the egg shell or the various coating that also influence access and location (to be covered briefly in the next section).

1. Eggs (oocyte) can be exposed, partially embedded, completely embedded or submerged in soil or water. All of these situations influence the evolution of parasitism with exposed eggs being exposed to a greater number of parasitoids than eggs that are submerged, which often requires some specialization on part of the parasitoid in order to reach the egg.
2. Eggs can be laid singly or in groups that can influence a parasitoids access to the egg. If an egg is considered a cube then there are six sides that can be approached and if the egg is suspended or supported by a thread, e.g., some *Chrysopa* (Parker and Rudall 1957), the six sides are available. But if an egg is placed on a leaf, one side is removed from direct access (Fig. 2.5), and as more eggs are laid, some are

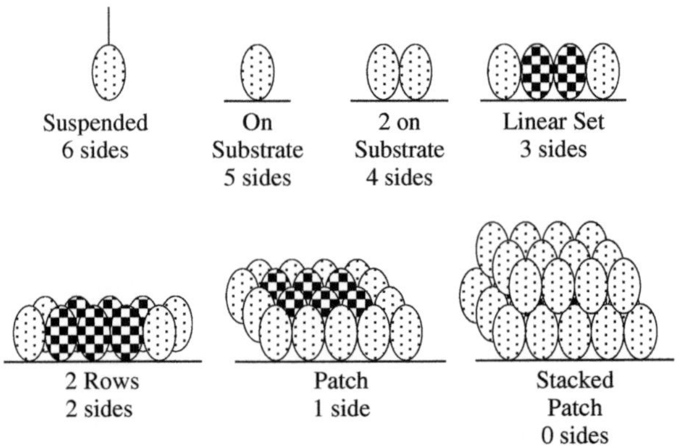

Fig. 2.5 Another issue regarding resource accessibility is the number and arrangement of eggs that can also influence their vulnerability. Here the number of sides or surface area exposed goes from 6 sides (full) in the case of a suspended egg, such as some neuropteran, to a patch in which only the top of the inside eggs (checkered) are exposed and if the patches are stacked the inside eggs are not accessible unless the parasitoid has evolved a long ovipositor

surrounded by other eggs that reduce the vulnerability of the center eggs to the top side only. If eggs are stacked the inner eggs are no longer available, unless there are species with a longer ovipositor.

3. Groups of eggs can be laid in chambers or covered, which may restrict parasitoid access to all or some of these eggs.
4. Eggs can also be spread out. Spacing of hosts can play a role (Schmidt and Smith 1987b, Casas 1991, Mills and Lacan 2004, Mills and Kuhlmann 2004). For example, if two or several eggs are placed together and found, there is a good chance they all will be attacked. If separated, but on the same leaf, there is a slight reduction that the second will be found by the same parasitoid, but if eggs are dispersed (on a different leaf of the plant or a different plant) there is a decrease in the possibility that the separated eggs will be found by the same parasitoid.

2.7 Host Location

As noted above an egg or oocyte provides very few cues for parasitoids to use in host location, but there are many other sources of cues that are associated with the egg. But in considering possible cues, it is important to keep seven issues in mind. These include (1) the type of cue, (2) the range or distance the cue operates over, (3) the association of the cue with the host egg, (4) the sequence within the host selection chain that the cue operates, (5) the sensory systems of the parasitoid involved, (6) the characteristics of the female parasitoid and (7) the characteristics of the host insect's biology.

The first issue concerns the types of cues that include (a) electromagnetic (light, color, heat, patterns, movement), vibration (sound, vibrations, movement), chemical (compounds of different functional groups, different volatility or solubility) and physical (shape, texture, spines, pits).

The second issue is the range that the cue operates in and the environment. For example, chemical cues can be very volatile (soluble) acting over a long distance (in the order of hundreds of meters) depending on the compounds volatility (solubility), temperature and wind (or water current). They can be weakly volatile compounds operating in the range of a meter or two or if not volatile, they act as contact chemicals (Bennet-Clark 1998, Stumpner and Meyer 2001). In the case of vibrations, they can be modulated and transmitted in air as sound that can be transmitted over meters or within various substrates as vibration patterns. For example, vibrations produced by an insect stridulating on a leaf can be transmitted to the stem and up and down the plant, but not to the next plant (Cocroft and Rodriguez 2005).

The third issue is the association of the cue with the host egg. Here the host location strategy needs to be considered. This issue is not as clear because the association does not always mean proximity of the cue with the egg in time or space for some parasitoid host location strategies. In many cases, particularly for strategy A, the parasitoid must get to the egg either before or soon after the egg is laid.

The fourth issue is the sequence within the host selection chain that the cue operates (see Vinson 1998). When a female parasitoid is ready to initiate the process of host location she will likely need to locate a potential host habitat or location and respond to cues that operate over distance in meters, such as volatile chemicals, sounds, light or heat (early factors in the host selection process). Once a host is contacted, other factors that operate at close range or on contact become important and the issues of the cues working over a distance are no longer important. The female now needs to confirm it is a potential host, sting and oviposit. But oviposition may depend on cues within the host which may be water soluble compounds.

The fifth issue is to understand the sensory systems of the parasitoid that are involved (Bin et al. 1989). This also depends on the parasitoid and the level of the host selection sequence being considered. For long range cues the eyes (detect electromagnetic factors such as color, movement, light-dark), antennae (detects volatile chemicals) or legs or thorax (detect sound or vibrations) are important. Once the vicinity of a host egg has been reached the antennae and tarsi become increasingly important in picking up host producer traces (such as contact or slightly volatile chemicals and sometimes shape such as silk threads (Weseloh 1980), and the antennae and tarsi remain important once the host is contacted (now in addition to contact chemicals, texture or shape are important). If everything is acceptable, the host is stung and the ovipositor (detecting some physical cues and water soluble chemicals) become more important.

The sixth issue is the parasitoid. Some egg parasitoids are reasonably large and robust, thus they can fly while others are very small and subject to slight air movement and float in the air or resort to walking. But another issue regarding the parasitoid is that when they emerge they can be considered to be naive, but preimaginal induction and imprinting on factors associated with the egg as they emerge may influence parasitoid behavior (Caubet et al. 1992). They also gain experience once a host is located and parasitized, and are able to find a second host faster (Cardé and Lee 1989, McAuslane et al. 1991, Kester and Barbosa 1992) and will respond to new cues associated with the host (Carew and Sahley 1986, Ding et al. 1989, Lewis et al. 1991). These issues can be important at all levels of the "host selection chain" as shown by Kaiser et al. (1995) (see Chapter 4).

The seventh issue is the host biology. Some insect species lay eggs on a plant and the resulting larvae or nymphs and adults all feed on the same plant for several generations. A parasitoid could cue on the plant or on almost any insect stage. This is very different from other species that lay their eggs on various plants on which the larvae or nymphs can feed while the adult either feeds elsewhere or does not feed. In this case a parasitoid should cue on factors involved with egg laying. Some insects place eggs under ground or under water, which would require the parasitoid to have the ability to craw under ground or move under water.

2.8 Source of Cues and a "Host-Egg-Unit"

As discussed by Randlkofer et al. (2007), a female insect must balance protecting her eggs from predators and parasitoids but must also insure that her hatching progeny have access to food. While the eggs are reasonably defenseless (see Blum and Hilker 2002 for exceptions), they are a poor source of cues that an egg parasitoid can use. The cues that parasitoids can use come from many sources, but these must be associated with the egg to be of use. The challenge is to understand where these cues could come from and to have some idea of the association of these cues with the host. One way to do this is to look at more than just the egg but all that is associated with the egg, which has been referred to as a "host-egg-unit" (Conti et al. 2000, 2003). The best place to begin to develop a host-egg-unit (Fig. 2.6) is to look at the substrate and the potential parent insects (male and female); but these two factors often interact.

a) *Substrate*: As shown in Fig. 2.6, the substrate (A) can be inert (cues rarer) or alive (cues possible, *see* Host food) and the substrate may just support an egg or a group of eggs or the egg or eggs may be embedded. What role the substrate will play in the host location sequence depends on the strategy that the parasitoid in question uses to locate a host.
b) *Potential parents*: Parents must be considered next due to traces of their presence as a host producer on the substrate (Fig. 2.6B), but in other cases they must be considered as a first step, particularly for parasitoids employing strategy A (see previous discussion).

Most insects must find a mate that usually involves attracting the opposite sex. This can involve sex pheromones, sounds or visual displays as examples. Although not an egg parasitoid one of the early examples regarding sound concerned a parasitoid attacking adult crickets that are attracted to male cricket mating calls

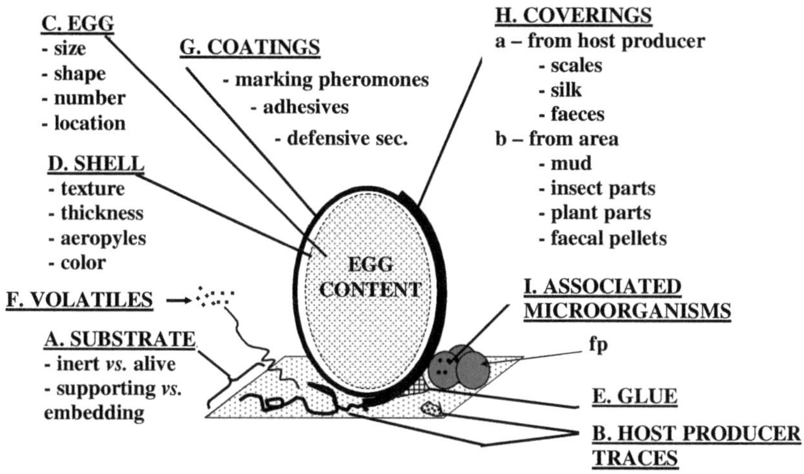

Fig. 2.6 (**A**) "Host Egg Unit" begins with considering the substrate (A) followed by cues that are from or associated with the host producer (**B**). Although egg contents are important nutritionally their importance for host finding is low (**C**). The egg shell (**D**) provides important contact cues. Eggs are either glued to the substrate or inserted into it with lubricants (**E**). If the substrate is living the act of ovipositing and factors associated with it may induce volatiles (**F**). Many eggs are coated with oviposition associated factors (**G**) and also often covered with factors separate from oviposition but associated with the female (Ha) or parts of other organism associated with the female (Hb). Yet other materials from the environment may be placed around or near-by the eggs such as fecal pellets (fp) that may harbor microorganisms (**I**)

(Cade 1975). In the case of parasitoids employing strategy "A2" cuing on the parents is an essential first step in host location often involving sex or aggregation pheromones (Colazza et al. 1997), but sound can be involved as well (Borges et al. 1999). But the parents can be important to parasitoids using strategy B. The importance of the parents in relationship to the different strategies is illustrated in Fig. 2.7. In cases where host eggs are placed on an inert substrate or scattered, then strategy A2a (phoresy) that involves responding to sex pheromones of the parents, then identify and mount (accept) the female is important. Parasitoids using strategy A2b (spy approach) involves the parasitoid responding to aggregation or sex pheromones (2.7A2b), and if the substrate is a plant and plant damage is caused by relatives of the parent, where volatile induction could occur before the parent lays eggs, then the parasitoid is attracted to the location and waits for host eggs (2.7A1c).

But the parents are also important for parasitoids using strategy B. In this case the parasitoid may respond to cues that precede or occur soon after oviposition. In this situation the parasitoid may respond to induced plant volatiles due to oviposition or feeding (chewing or piercing) and then is stimulated to search the area due to a variety of cues associated with the host producer that may have been left (Fig. 2.7B). These cues can include salivary enzymes and associated plant damage, footprints, body traces, scales, silk, feces, and defensive secretions.

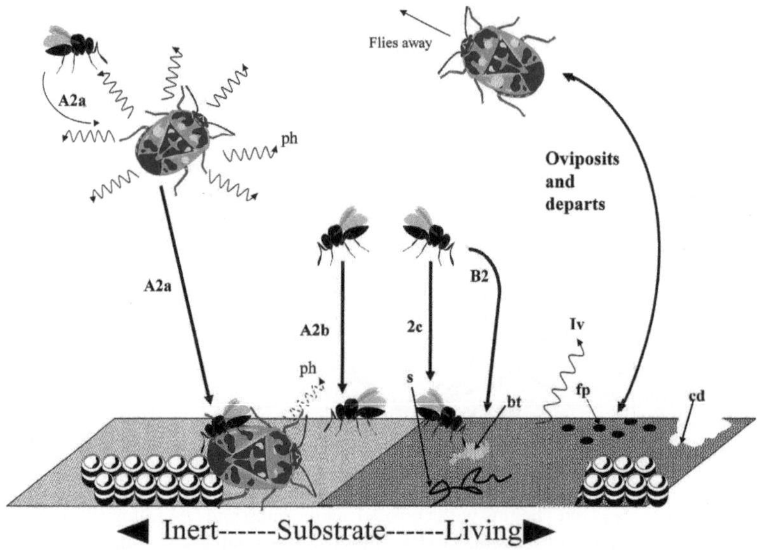

Fig. 2.7 Strategies used to locate hosts. Parasitoids using strategy A2a or A2b are attracted from a distance to pheromones, sound or other long range cues. A2a then must recognize the target female using close range and contact cues as she mounts the host producer to wait for hosts. A2b responds to host producer long range cues (ph) and then waits in the area for hosts to be produced. A2c parasitoids responding to host producer associates respond to plant and host long range cues and wait for hosts to be produced. Parasitoids using strategy B are attracted from a distance by sound, pheromones (ph) and volatile induction (Iv) due to oviposition, chewing (cd) or piercing. They are stimulated to stay and search the area by foot prints (fp), body traces (bt) or other host producer traces such as silk or scales (s)

In fact, the host producer (ovipositing female) has been frequently reported as a source of cues. Laing (1937) reported that *T. evanescens* increased their residence time when in areas contaminated with host traces. There have been a number of reports of egg parasitoid responses to contaminated areas (Lewis et al. 1971, 1972, 1975a, b, Beevers et al. 1981, Smits 1982, Gardner and van Lenteren 1986) and some of the responsible compounds have been identified such as hydrocarbons that have some volatility (Jones et al. 1973, Shu and Jones 1985) and organic acids that tend to stick to surfaces (Gueldner et al. 1984). Even the footprints of the parents can be detected by parasitoids (Borges et al. 2003). Responses to host volatiles have also been reported (Noldus and van Lenteren 1985, Mattiacci et al. 1991, 1993) (see Chapter 4). While these are important cues, they primarily play a role in host location by egg parasitoids employing strategy "B".

Host food: As noted earlier many insect are specialist, feeding as both immature as a well as adults on a particular plant. Many larval parasitoids are attracted to non-infested plants (Vinson et al. 1994, Steidle et al. 2001, Mbata et al. 2004), but they respond more to infested plants.

In recent years it has become clear that plants release volatiles in response to herbivores or plant damage (Dicke and van Loon 2000, Turlings et al. 2000, Hilker and Meiners 2002, 2006, Moraes et al. 2005) and it is clear that the volatiles released play a role in attracting the third trophic level (Vinson and Williams 1991, Vinson 1993, 1998, Turlings et al. 1995, Rose et al. 1998, Vet 1999). It has also been suggested that this communication between the first and third trophic levels plays a key role balancing the tritrophic system (Vinson 1999, 2005). Thus, it is not surprising that insect species that insert their eggs into plant tissue stimulate the release of volatiles that attract egg parasitoids. One example is the elm leaf beetle *Xanthogaleruca luteola* (Muller), which scratches the leaf with its mouth and then glues an egg in the scratch with an oviduct secretion. The secretion when applied to an artificial scratch induces the emission of synomones attractive to egg parasitoid *Oomyzus gallerucae* (Fonscolombe) (Meiners and Hilker 1997, 2000, Meiners et al. 2000). In another case, the pine sawfly *Diprion pini* (L.) inserts its eggs into a pine needle with its ovipositor along with the release of an oviduct secretion that itself can induce plant synomones, which in turn attract the egg parasitoid *Chrysonotomyia ruforum* (Krausse) (Hilker et al. 2002a). It is clear that the oviposition of insect eggs can result in the plants release of an early herbivore alert (Hilker and Meiners 2006). The question is whether these inductions favor parasitoids that must locate fresh eggs as the two examples above require 24–72 h for the volatiles to be detected (Hilker et al. 2002b).

The oocyte: The oocyte is not likely to release compounds other than some CO_2 and it is not moving, feeding, defecating or releasing complex chemicals and thus may not be very important to location of a host but may play a role in recognition due to its size, shape, and possibily number (Fig. 2.6C). But the oocyte is important in that it can be accepted or rejected depending on chemical cues picked up by the ovipositor. These cues are likely water-soluble compounds detected by the ovipositor, although defensive compounds present in the egg may cause rejection. Yet, previous parasitized or diseased eggs may also be rejected. These issues can all influence oocyte acceptance.

The egg shell: The oocyte is enclosed by a vitelline membrane produced by the oocyte and follicle cells (Chauvin and Barbier 1979, Mouzaki and Margaritis 1994), which is followed by the secretion of the chorion (Fig. 2.6D) that can consist of several different layers (Trougakos and Margaritis 2002). The egg shell is primarily composed of proteins (Regier and Kafatas 1985), although wax layers may exist (Papassideri and Margaritis 1986). Its primary functions, as noted by Margaritis (1985) and Margaritis and Mazzini (1998), are to protect the embryo from environmental hazards, such as humidity, temperature changes, as well as biological hazards such as attack by bacteria, fungi, predators and parasites. When an egg is first oviposited the shell often has some elasticity but usually hardens for protection. The egg shell does have some holes to allow for sperm (micropyle) and air penetration (aeropyle) (Fig. 2.2). The egg chorion can provide a number of cues for parasitoids (Fig. 2.6D) that may include shape, texture, and special structures such as spines or ridges that can be important in contact recognition. However, the chorion or egg shell can vary in thickness between species and it can also harden with time which can reduce the ability of a parasitoid to penetrate (Leibee et al. 1979, Pak et al. 1990, van Huis et al. 1994). While the main function of the chorion is to protect the oocyte, it may also provide some contact chemical cues, although many of the cues isolated from eggs tend to be coatings.

Glue or inserted eggs and induced volatiles: Obviously if an egg is inserted into a plant, the plant is damaged and may release volatiles, as discussed previously, but the glue (Fig. 2.6E) may also be absorbed and induce a plant to release attractive volatiles (Fig. 2.6F). However, the glue itself can also release both volatiles and contact chemicals (Bin et al. 1993, Conti et al. 2004, Santis et al. 2007) and may play a role for eggs placed on inert substrates (Fig. 2.6).

Coatings: Coatings (Fig. 2.6G) can be many things, such as marking pheromones, repellents and other defensive secretions, adhesives that may serve to hold the next group, the coverings; but a key is that the compounds come from glands associated with the reproductive system of the female.

Coverings: These are materials (Fig. 2.6H) that may be associated with the female and can be placed into two groups. The first group (a) is made up of items from the female that are not associated with its reproductive system and would include hairs or scales, silk, or secretions from other body regions. It would also include fecal material. The second group is made up of materials not associated with the female (Fig. 2.6Hb), with an exception. The exception is that the foreign material is handled by the female and often is contaminated with salivary secretions and thus can serve as a reliable cue. These items include mud, fecal material of another insect species, insect or plant parts.

Associated organisms: In a few cases the female host producer may place particular items near her eggs to protect them and these may contain associated organisms, such as bacteria.

Egg guarding: Some insects that lay egg clutches that are guarded by the male or female depending on the species. For example, male assassin bug of the genus *Zelus* guard an egg clutch (Ralston 1977). Since such eggs could be older, the egg parasitoid's strategy should be B, but the initial cues may be associated with the adult guards.

2.9 Concluding Remarks

The nutritional ecology of egg parasitoids is complex due in large part to the fact that the producer of the resource has evolved to maximize the survival of its progeny by insuring access to resources for its progeny following hatching, while at the same time reducing the period of time the egg is available, reducing the amount and quality of resource the egg represents, and trying to hide or protect the egg in a number of different ways.

In spite of these efforts the host producer provides a trail of cues that the egg parasitoid has evolved to take advantage of. In addition, there is a third party in many of these interactions, the plant, which is involved in manipulating the herbivore and the parasitoid (Vinson 2005). As a result, the nutritional ecology of an egg parasitoid is complex, primarily in the location of a quality resource by the host producer. Following egg acceptance, the developing parasitoid egg and larva has no choice. The larval feeding, nutrition, growth and development are topics covered in Chapter 1.

References

Ables JR, Vinson SB, Ellis JS (1981) Host discrimination by *Chelonus insularis* (Hymenoptera: Braconidae), *Telenomus heliothidis* (Hymenoptera: Scelionidae), and *Trichogramma pretiosum* (Hymenoptera: Trichogrammatidae). Entomophaga 26:149–156

Abrams JM, White K, Fessler LL, Steller H (1993) Programmed cell death during *Drosophila* embryogenesis. Development 117:29–43

Agrell I, Lundquist AM (1973) Physiological and biochemical changes during insect development. In: Rockstein M (ed) The physiology of the Insecta, vol 1. Academic, New York, pp 159–249

Aldrich JR, Kochanasky JP, Abrams CB (1984) Attractant for a beneficial insect and its parasitoids: pheromone of the predatory spined solider bug. *Podisus maculiventris* (Hemiptera: Pentatomidae). Environ Entomol 13:1031–1036

Anderson DT (1972a) The development of hemimetabolous insects. In: Counce J, Waddington CH (eds) Developmental system: Insects, vol 1. Academic, New York, pp 95–163

Anderson DT (1972b) The development of holomelabolous insects. In: Counce J, Waddington CH (eds) Developmental system: Insects, vol 1. Academic, New York, pp 165–242

Antolin MF, Strong DR (1987) Long-distance dispersal by a parasitoid (*Anagrus delicatus*, Mymaridae) and its host. Oecologia 73:288–292

Arakaki N, Wakamura S, Yasuda T (1996) Phoretic egg parasitoid, *Telenomus euproctidis* (Hymentoptera: Scelionidae), uses sex pheromone of Tussock moth *Euproctis taiwana* (Lepidoptera: Lymantriidae) as a kairomone. J Chem Ecol 22:1079–1085

Askew RR, Shaw MR (1986) Parasitoid communities: their size, structure and development. In: Waage J, Greathead D (eds) Insect parasitoids. Academic, London, pp 225–287

Austin AD, Field SA (1997) The ovipositor system of scelionid and platygastrid wasps (Hymenoptera: Platygastoidea): comparative morphology and phylogenetic implications. Invertebr Taxon 11:1–87

Azab AK, Tawfik MFS, Awadallah AK (1967) Biology of *Nasonia vitripennis* Walker. Bull Soc Entomol Egypte 51:469–82

Bai B, Luck RF, Forster L, Stephens B, Janssen JAM (1992) The effect of host size on quality attributes of the egg parasitoid *Trichogramma pretiosum*. Entomol Exp Appl 64:37–48

Barrett M, Schmidt JM (1991) A comparison between the amino acid composition of an egg parasitoid wasp and some of its hosts. Entomol Exp Appl 59:29–41

Beck SD (1972) Nutrition, adaptation and environment. In: Rodriguez JG (ed) Insect and mite nutrition. North Holland, Amsterdam, pp 1–6

Beckage NE, Riddford LM (1978) Developing interactions between *Manduca* and its parasitoid *Apanteles congregatus*. Entomol Exp Appl 23:139–151

Beevers M, Lewis WJ, Gross Jr. HR, Nordlund DA (1981) Kairomones and their use for management of entomophagous insects: X. Laboratory studies on manipulation of host-finding behavior of *Trichogramma pretiosum* Riley with a kairomone extracted from *Heliothis zea* (Boddie) moth scales. J Chem Ecol 7:635–648

Bennet-Clark HC (1998) Size and scale effects as constraints in insect sound communication. Philos Trans R Soc London 353B:407–419

Beserra EB, Dias CTDS, Parra JRP (2002) Distribution and natural parasitism of *Spodoptera frugiperda* (Lepidoptera: Noctuidae) eggs at different phenological stages of corn. Florida Entomol 85:588–593

Bezzerides A, Yong TH, Bezzerides J, Husseini J, Ladau J, Eisner M, Eisner T (2004) Plant-derived pyrrolizidine alkaloid protects eggs of a moth (*Utetesia ornatrix*) against a parasitoid wasp (*Trichogramma ostriniae*). Proc Natl Acad Sci USA 101:9029–9032

Bilinski S (1979) Ultrastructural study of yolk formation in *Porcellio scaber* Latr. (Isopoda). Cytobios 26:123–130

Bin F (1985) Foresia in un parassitoide oofago: *Mantibaria seefelderiana* (de Stef.Perez) (Hymenoptera: Scelionidae). *Atti XIV Congreso Nazionale Italiani di Entomologie*, Palermo, Erice, Bagheria, p 38

Bin F, Johnson NF (1982) Some new species of *Telenomus* (Hymenoptera: Scelionidae) egg parasitoids of tropical pyralid pests (Lepidoptera: Pyralidae). *Redia* 65:229–252

Bin F, Colazza S, Isidoro N, Solinas M, Vinson SB (1989) Antennal chemosensilla and glands, and their possible meaning in the reproductive behaviour of *Trissolcus basalis* (Woll.) (Hymenoptera: Scelionidae). Entomologia 24:33–97

Bin F, Vinson SB, Strand MR, Colazza S, Jones WA (1993) Source of an egg, kairomone for *Trissolcus basalis*, a parasitoid of *Nezara viridula*. Physiol Entomol 18:7–15

Blum MS, Hilker M (2002) Chemical protection of insect eggs. In: Hilker M, Meiners T (eds) Chemical ecology of insect eggs and egg deposition. Blackwell, Berlin, pp 61–90

Boldt PE, Marston N, Dickerson WA (1973) Differential parasitism of several species of lepidopteran eggs by two species of *Trichogramma*. Environ Entomol 2:1121–1122

Borges M, Costa MLM, Sujii ER, Cavalcanti MG, Redigolo GF, Resck IS, Vilela EF (1999) Semiochemical and physical stimuli involved in host recognition by *Telenomus podisis* (Hymenoptera: Scelionidae) toward *Euschistus heros* (Heteroptera: Pentatomidae). Physiol Entomol 24:227–233

Borges M, Colazza S, Ramirez-Lucas P, Chauhan KR, Moraes MCB, Aldrich JR (2003) Kairomonal effect of walking traces from *Euschistus heros* (Heteroptera: Pentatomidae) on two strains of *Telenomus podisi* (Hemenoptera: Scelionidae). Physiol Entomol 28:349–356

Bownes M (1994) The regulation of the yolk protein genes, a family of sex differentiation genes in *Drosophila melanogaster*. BioEssays 16:745–752

Brotodjojo RRR, Walter GH (2006) Oviposition and reproductive performance of a generalist parasitoid (*Trichogramma pretiosum*) exposted to host species that differ in their physical characteristics. Biol Control 39:300–312

Brues CT (1946) Insect dietary. Harvard University Press, Cambridge

Buleza VV, Mikheev AV (1978) Factors determining the choice and infestation of the host in *Trissolocus grandis* and *T. viktorovi*. Zool Zh 57:1162–1168

Cade W (1975) Acoustically orienting parasitoids: fly phonotaxis to cricket song. Science 190:1312–1313

Calbert G, Keller MA (1998) Patch defense in the parasitoid wasp *Trissolcus basalis* (Insecta: Scelionidae): the time structure of pairwise contests, and the "waiting game". Ethology 104:821–840

Calow P (1977) Ecology, evolution and energetics: a study in metabolic adaptation. Adv Ecol Res 10:1–6

Calvin DD, Losey JE (1991) Preference of *Trichogramma pretiosum* (Hym., Trichogrammatidae) for three age classes of southwestern corn borer eggs. Colloques l'NRA 56:59–62

Capinera JL, Lilly JH (1975) *Tetrastichus asparagi* (Hymenoptera: Eulophidae) parasitoid of asparagus beetle (Coleoptera: Chrysomelidae) some aspects of host– parasitoid interaction. Ann Entomol Soc Am 68:595–596

Cardé RT, Lee H (1989) Effect of experience on the responses of the parasitoid *Brachymeria intermedia* (Hymenoptera: Chalcididae) to its host, Lymnatria *dispar* (Lepidoptera: Lymantriidae), and to kairomone. Ann Entomol Soc Am 82:653–657

Carew TJ, Sahley CL (1986) Invertebrate learning and memory. Annu Rev Neurosci 9:435–487

Casas J (1991) Density dependent parasitism and plant architecture. Redia 74:217–222

Castaneda-Samayoa O, Holst H (1990) Biological control of grape berry moth with egg parasites of the genus *Trichogramma*. Bull OILB/SROP 13:62–65

Caubet Y, Jaison P, Lenor A (1992) Preimaginal induction of adult behavior in insects. Q J Exp Psychol, Sect B – Comp Physiol Psychol 44B:165–178

Chambers P, Sword G, Angel J, Behmer S, Bernays EA (1996) Foraging by generalists grasshoppers: dietary mixing and the role of crypsis. Anim Behav 52:155–165

Chauvin G, Barbier R (1979) Morphogenese de l'enveloppe vitelline, ultrastructure duchorion et de la cuticule serosale chez *Korscheltellus lupulinus* L. (Lepidoptera: Hepialidae). Int J Insect Morphol Embryol 8:375–386

Chopard L (1923) Les parsites de la *Mante religreuse. Ann Soc Entomol France* 91:249–272

Clark B, Feath S (1998) The evolution of egg clustering in butterflies: a test of egg desiccation hypothesis. Evol Ecol 12:543–552

Clausen CP (1976) Phoresy among entomophagous insects. Annu Rev Entomol 21:343–368

Cocroft RB, Rodriguez RL (2005) Behavioral ecology of insect vibrational communication. Bioscience 55:323–334

Colazza S, Rosi MC, Clemente A (1997) Response of egg parasitoid *Telenomus busseolae* to sex pheromone of *Sesamia nonagrioides*. J Chem Ecol 23:2437–2444

Colegrave N (1994) Game-theory models of competition in closed systems – asymmetries in fighting and competitive ability. Oikos 71:499–505

Consoli FL, Parra JRP (1996) Biology of *Trichogramma galloi* and *T. pretiosum* (Hymenoptera: Trichogrammatidae) reared in vitro and in vivo. Ann Entomol Soc Am 89:828–834

Consoli FL, Parra JRP (1999) Development of an artificial host egg for *in vitro* egg laying of *Trichogramma galloi* and *T. pretiosum* using plastic membranes. Entomol Exp Appl 91:327–336

Conti E, Bin F, Vinson SB (2000) Host range in egg parasitoids: a tentative approach through the analysis of the host unit. In: Antonie van Leeuwenhoek Symposium, 7th European Workshop on Insect Parasitoids, Haarlem, The Netherlands, Oct. 1–6, p 32

Conti E, Salerno G, Bin F, Williams HJ, Vinson SB (2003) Chemical cues from *Murgantia histrionica* eliciting host location and recognition in the egg parasitoid *Trissolcus brochymenae*. J Chem Ecol 29:115–130

Conti E, Salerno G, Bin F, Williams HJ, Vinson SB (2004) The role of semiochemicals in parasitoid specificity: a case study with *Trissolcus brochymenae* and *Trissolcus simony* on pentatomid bugs. Biol Control 29:435–444

Corrigan JE, Laing JE (1994) Effects of the rearing host species on the host species attacked on performance by *Trichogramma minutum* Riley (Hymenoptera: Trichogrammatidae). Environ Entomol 23:755–760

Couturier A (1941) Noubelles observations sur *Rielica manticida* Kief., Hymenoptereparasite de la *Mante religieuse*. II Comportemente de L'insecta parfait. Rev Zool Agri Appl 40:49–62

Crawley MJ (1983). Herbivory. Studies in biology, vol 10, University of California Press, Berkeley

Delvare G (1993) Sur Les Megaphragma de Guadeloupe avec la description d'uneespèce nouvelle (Hymenoptera, Trichogrammatidae). Rev French Entomol 15:149–152

Dicke M, van Loon JJA (2000) Multitrophic effects of herbivore-induced plant volatiles in an evolutionary context. Entomol Exp Appl 97:237–249

Ding DC, Swedenborg PD, Jones RL (1989) Plant odor preferences and learning in *Marcocentrus grandii* (Hymenoptera: Braconidae), a larval parasitoid of the European Corn Borer, *Ostrinia numbilalis* (Lepidoptera: Pyralidae). J Kansas Entomol Soc 62:164–176

Eisner T, Rossini C, Gonzalez A, Iyengar VK, Siegler MVS, Smedley SC (2002) Paternal investment in egg defense. In: Hilker M, Meiners T (eds) Chemical ecology of insect eggs and egg deposition. Blackwell, Berlin, pp 91–116

Farrar RR, Bradley JR (1985) Within-plant distribution of *Heliothis* spp. (Lepidoptera: Noctuidae) eggs and larvae on cotton in North Carolina. Environ Entomol 14:205–209

Ferguson JE, Metcalf RL (1985) Cucurbitacins: plant-derived defence compounds for the Diabroticites (Coleoptera: Chrysomelidae). J Chem Ecol 11:311–319

Ferguson JE, Metcalf RL, Fischer DC (1985) Disposition and fate of cucurbitacin B in five species of Diabroticites. J Chem Ecol 11:1307–1321

Ferracini C, Boivin G, Alma A (2006) Costs and benefits of host feeding in the parasitoid wasp *Trichogramma turkestanica*. Entomol Exp Appl 121:229–234

Field SA, Calbert G (1998) Patch defence in the parasitoid wasp *Trissolcus basalis*: when to begin fighting. Behaviour 135:629–642

Field SA, Calbert G, Keller MA (1998) Patch defence in the parasitoid wasp *Trissolcus basalis* (Insecta: Scelionidae): the time structure of the pairwise contests, and the "waiting game". Ethology 104:821–840
Garcia P (2000) Biologia de *Trichogramma cordubensis* Vargas & Cabello (Hym., Trichogrammatidae) numa perspective de controlo biológico. PhD Dissertation, Departmento de Biologia, Universidade dos Acores, 238p
Gardner SM, van Lenteren JC (1986) Characterization of the arrestment responses of *Trichogramma evanescens*. Oecologia 68:265–270
Gorman MJ, Kankanala P, Kanost MR (2004) Bacterial challenge stimulates innate immune responses in extra-embryonic tissues of tobacco worm eggs. Insect Mol Biol 13:19–24
Grenier S, Grille G, Basso C, Pintureau B (2001) Effects of the host species and the number of the parasitoids per host on the size of some *Trichogramma* species (Hymenoptera: Trichogrammatidae). Biocontrol Sci Technol 11:21–26
Guang LQ, Oloo GW (1990) Host preference studies on *Trichogramma* sp. nr. *mwanzai* Schulten and Feijen (Hymenoptera: Trichogrammatidae) in Kenya. Insect Sci Appl 11:757–763
Guedes RCN, Guedes NNP, Smith RH (2007) Larval competition within seeds: from the behavioral process to the ecological outcome in the seed beetle *Callosobruchus maculatus*. Austral Ecol 32:697–707
Gueldner RC, Nordlund DA, Lewis WJ, Thean JE, Wilson DM (1984) Kairomones and their use for management of entomophagous insects. XV. Identification of several acids in scales of *Heliothis zea* moths and comments on their possible role as kairomones for *Trichogramma pretiosum*. J Chem Ecol 10:245–252
Haeselbarth E (1979) Zur Parasitierung der Puppen von Foreule (*Panolis flammea* [Schiff.]), Kiefernspanner (*Bupalus piniaius* [L.J]) und Heidelberrspanner (*Boarmia bistortana* [Goezel]) in bayerischen Kiefernwäldren. Z Angew Entomol 87:186–202
Hawkins BA (1994) Patterns and processes in host-parasitoid interactions. Press Syndicate University Cambridge, Cambridge
Hemptinne JL, Lognay G, Gauthier C, Dixon AFG (2000) Role of surface chemical signals in egg cannibalism and interguild predation in ladybirds (Coleoptera: Coccinellidae). Chemoecology 10:123–128
Hilker M, Meiners T (2002) Induction of plant responses towards oviposition and feeding of herbivorous arthropods: a comparison. Entomol Exp Appl 104:181–192
Hilker M, Meiners T (2006) Early herbivore alert: insect eggs induce plant defense. J Chem Ecol 32:1379–1397
Hilker M, Kobs C, Varama M, Schrank K (2002a) Insect egg deposition induces *Pinus* to attract egg parasitoids. J Exp Biol 205:455–461
Hilker M, Rohfritsch O, Meiners T (2002b) The plants response towards insect egg deposition. In: Hilker M, Meiners T (eds) Chemical ecology of insect eggs and egg deposition. Blackwell, Berlin, pp 295–233
Hinton HE (1981) Biology of insect eggs. Pergamon, Oxford
Hoffmann MP, Ode PR, Walker DL, Gardner J, van Nouhuys S, Shelton AM (2001) Performance of *Trichogramma ostriniae* (Hymenoptera: Trichogrammatidae) rearing on factitious hosts, including the target host, *Ostrinia nubilalis* (Lepidoptera: Cranbidae). Biol Control 21:1–10
Honda Y, Luck RF (2001) Interactions between host attributes and wasp size: a laboratory evolution of *Trichogramma platneri* as an augmentative biological control agent for two avocado pests. Entomol Exp Appl 100:1–13
Houseweart MW, Southard SG, Jennings DT (1982) Availability and acceptability of spruce budworm eggs to parasitism by the egg parasitoid, *Trichogramma minutum* (Hymenoptera: Trichogrammatidae). Can Entomol 114:657–666
Howard LO (1927) Concerning phoresy in insects. Entomol News 38:145–147
Huber JT (1995) Mymaridae. In: Hanson PE, Gauld LD (eds) The Hymenoptera of Costa Rica. Oxford University Press, Oxford, pp 344–349
Huis van A, Schutte C, Cools MH, Fanget PH, van der Hoek H, Piquet SP (1994) The role of semiochemicals in host location by *Uscana lariophaga*, egg parasitoid of *Callosobruchus*

maculatus. In: Highley E, Wright EJ, Banks HJ, Champ BR (eds) Stored product protection, vol 2. CAB International, Wallingford, pp 1158–1164

Jarjees EA, Merrit DJ (2004) The effect of parasitization by *Trichogramma australicum* on *Heliocoverpa armigera* host eggs and embryos. J Invertebr Pathol 85:1–8

Jervis MA, Kidd NAC 1986. Host-feeding strategies in hymenopteran parasitoids. Biol Rev 61:395–434

Jervis MA, Kidd NAC, Fitton MG, Huddleston T, Dawah HA (1993) Flower-visiting by hymenopteran parasitoids. J Nat Hist 27:67–105

Jones RL, Lewis WJ, Beroza M, Bierl BA, Sparks AN (1973) Host seeking stimulants (Kairomones) for the egg parasite *Trichogramma evanescens*. Environ Entomol 2:593–596

Jura C (1972) Development of opterygote insects. In: Counce J, Waddington DN (eds) Developmental systems: Insects, vol 1. Academic, New York, pp 47–94

Kaiser L, Bartheye C, Kerguelen V, Phamdeleque MH (1995) Odour conditioning of ovipositor probing in a parasitic wasp. Ethol, Ecol Evol 7:245–255

Kartsev VM (1985) The mechanism of recognition of the host infested egg parasites of the genus *Trissolcus* (Hymenoptera: Scelionidae). Zool Zh 65:1318–1326

Kazmer DJ, Luck RF (1991) Female body size, fitness biological control quality: field experiments with *Trichogramma pretiosum*. Colloques l'INRA 56:37–40

Keller MA, Lewis WJ, Stinner RE (1985) Biological and practical significance of movement by *Trichogramma* species: a review. Southwest Entomol 106(Suppl. 8):138–155

Kester KM, Barbosa P (1992) Effects of postemergence experience on searching and landing responses of the insect parasitoid, *Cotesia congregata* (Say) (Hymenoptera, Branconidae), to plants. J Insect Behav 5:301–320

Kim KC (1984) Coevolution of parasitic arthropods and mammals. Wiley, New York

Kivan M, Kilic N (2002) Host preference: parasitism, emergence and development of *Trissolcus semistriatus* (Hym., Scelionidae) in various host eggs. J Appl Entomol 126:395–399

Kivan M, Kilic N (2004) Parastiism and development of *Trissolcus simoni* in eggs of different host species. Phytoparasitica 32:57–60

Kölliker-Ott UM, Bigler F, Hoffmann AA (2003) Does mass rearing of field collected *Trichogramma brassicae* parasitoids influence acceptance of European corn borer eggs? Entomol Exp Appl 109:197–203

Kolomiyets NG (1957) New data on the parasites of the Siberian silkworm moth. Lesn Kh-vo 7:57–58 (Biological Abstracts 35, No. 30566)

Kuhlmann U, Mills NJ (1999) Comparative analysis of the reproductive attributes of three commercial-produced *Trichogramma* species (Hymenoptera: Trichogrammatidae). Biocontrol Sci Tech 9:335–346

Laing J (1937) Host-finding by insect parasites. I. Observations on the finding of hosts by *Alysia manducator, Mormoniella vitripennis* and *Trichogramma evanescens*. J Anim Ecol 6:298–317

Lawrence PO (1986) Host-parasite hormonal interactions: an overview. J Insect Physiol 32:295–298

Lawrence PO (1990) The biochemical and physiological effects of insect hosts on the development and ecology of their parasites: an overview. Arch Insect Biochem Physiol 3:217–228

Lawton JH, McNeil S (1979) Between the devil and the deep blue sea: on the problem of being a herbivore. In: Anderson RM, Turner BD, Taylor LR (eds) Population dynamics. Blackwell, Melbourne, pp 223–244

Leal WS, Higuchi H, Mizutani N, Nakamori H, Kadosawa T, Ono M (1995) Multifunctional communication in *Riptortus clavatus* (Heteroptera: Alydidae): Conspecific nymphs and egg parasitoid *Ooencyrtus nezarae* use the same adult attractant pheromone as chemical cue. J Chem Ecol 21:973–985

Lee JM, Nishimori Y, Hatakeyama M, Bae TW, Oishi K (2000) Vitellogenin of the cicada *Graptopsaltria nigrofuscata* (Homoptera): analysis of its primary structure. Insect Biochem Mol Biol 30:1–7

Leibee GL, Pass BC, Yeargan KV (1979) Developmental rates of *Patsson lameerei* (Hymenoptera: Myamridae) and the effect of host egg age on parasitism. Entomophaga 24:345–348

Lewis WJ, Redlinger LM (1969) Suitability of eggs of the almond moth, *Cadra cautella*, of various ages to parasitism by *Trichogramma evanescens*. Ann Entomol Soc Am 62:1482–1485

Lewis WJ, Sparks AN, Redlinger LM (1971) Moth odor: a method of host finding by *Trichogramma evanescens*. J Econ Entomol 64:557–558

Lewis WJ, Jones RL, Sparks AN (1972) Host seeking stimulant for the egg-parasite *Trichogramma evanescens*: its source and a demonstration of its laboratory and field study. Ann Entomol Soc Am 65:1087–1089

Lewis WJ, Jones RL, Nordlund DA, Sparks AN (1975a) Kairomones and their use for management of entomophagous insects: I. Evaluation for increasing rates of parasitization by *Trichogramma* spp. J Chem Ecol 1:343–347

Lewis WJ, Jones RL, Nordlund DA, Gross HRJ (1975b) Kairomones and their use for management of entomophagous insects: II. Mechanisms causing increase in rate of parasitization by *Trichogramma* spp. J Chem Ecol 1:343–347

Lewis WJ, Nordlund DA, Gueldner RC, Teal PEA, Tumlinson JN (1982) Kairomones and their use for management of entomophagous insects XIII. Kairomonal activity for *Trichogramma* spp. of abdominal tips, excretion, and a synthetic sex pheromone blend of *Heliothis zea* (Boddie) moths. J Chem Ecol 8:1323–1331

Lewis WJ, Tumlinson JH, Krasnoff S (1991) Chemically mediated associative learning – an important function in the foraging behavior of *Microplitis croceipes* (Cresson). J Chem Ecol 17:1309–1325

Lopes JRS, Parra JRP (1991) Effect of egg age from natural and factitious host on the development and parasitism of two *Trichogramma* species. Rev Agric 66:221–244

MacArthur R, Wilkson EO (1967) The theory of island biogeography. Princeton University Press, Princeton

Makee H (2005) Factors influencing the parasitism of codling moth eggs by *Trichogramma cacoeciae* March and *T. principium* Sug. et Sor. (Hymen. Trichogrammatidae). J Pest Sci 78:31–30

Malo F (1961) Phoresy of *Xenufens*, a parasite of *Caligo eurilochus*. J Econ Entomol 54: 465–466

Mansfield S, Mills NJ (2003) A comparison of methodologies for the assessment of host preference of the gregarious egg parasitoid *Trichogramma platneri*. Biol Control 29:332–340

Margaritis LH (1985) Structure and physiology of the eggshell. In: Gilbert LI, Kerkut GA (eds) Comprehensive insect physiology, biochemistry and pharmacology, vol 1. Pergamon, Oxford, pp 151–230

Margaritis LH, Mazzini M (1998) Structure of the egg. In: Harrison FW, Locke M (eds) Microscopic anatomy of invertebrates, vol 11C – Insecta. Wiley Liss, New York, pp 995–1037

Marston N, Ertle LR (1969) Host age and parasitism by *Trichogramma evanescens* (Hymenoptera: Trichogrammatidae). Ann Entomol Soc Am 57:570–574

Marston N, Ertle LR (1973) Host influence on the bionomics of *Trichogramma minutum*. Ann Entomol Soc Am 66:1155–1162

Mattiacci L, Vinson SB, Howard WH, Aldrich JR, Colazza S, Bin F (1991) Kairomones for the egg parasitoid *Trissolcus basalis* (Woll.): isolation and identification of a compound from the metathoracic exocrine glands of *Nezara viridula* (L.). Redia 124:167–168

Mattiacci L, Vinson SB, Williams HJ, Aldrich JR, Bin F (1993) A long range attractant kairomone for the egg parasitoid *Trissolcus basalis*, isolated from defensive secretion of its host, *Nezara viridula*. J Chem Ecol 19:1067–1181

Mazzini M, Santini L (1983) Sulla fine struttura del micropilo negli insetti. XVII. L'uovo di *Acnemia amoena* Winnertz (Diptera Mycetophilidae, Sciophilinae). Frustula Entomol 6: 1–12

Mbata GN, Shu S, Phillips TW, Ramaswamy SB (2004) Semiochemical cues used by *Pteromalus cerealellae* (Hymenopters: Pteromalidae) to locate its host, *Callosobruchus maculates* (Coleoptera: Bruchidae). Ann Entomol Soc Am 97:353–360

McAuslane JH, Vinson SB, Williams HJ (1991) Influence of adult experience on host microhabitat location by the generalist parasitoid, *Campoletis conorensis* (Hymenoptera: Ichneumonidae). J Insect Behav 4:101–113
McNab BK (1984) Energetics – The behavioral and ecological consequences of body size. Florida Entomol 67:68–73
Meiners T, Hilker M (1997) Host location in *Oomyzus gallerucae* (Hymenoptera: Eulophidae), an egg parasitoid of the elm leaf beetle *Xanthogaleruca luteola* (Coleoptera: Chrysomelidae). Oecologia 112:87–93
Meiners T, Hilker M (2000) Induction of plant synomones by oviposition of a phytophagous insect. J Chem Ecol 26:221–232
Meiners T, Westerhaus C, Hilker M (2000) Specificity of chemical cues used by a specialist egg parasitoid during host location. Entomol Exp Appl 95:151–159
Melo AC, Valle D, Machado EA, Salerno AP, Paiva-Silva NL, de Souza W, Masuda H (2000) Synthesis of vitellogenin by the follicle cells of *Rhodnius prolixus*. Insect Biochem Mol Biol 30:549–557
Mills NJ, Lacan I (2004) Ratio dependence in the functional response of insect parasitoids: evidence from *Trichogramma minutum* foraging for eggs in small host patches. Ecol Entomol 29:208–216
Mills NL, Kuhlmann U (2004) Oviposition behavior of *Trichogramma platneri* Nagarkatti and *Trichogramma pretiosum* Riley (Hymenoptera: Trichogramatidae) in patches of single and clustered eggs. Biol Control 30:42–51
Mockford, EL (1997) A new species of dicomorpha (Hymenoptera: Mymaridae) with dimunitive, apterous males. Ann Entomol Soc Am 90:115–120
Monje JC, Zebitz CPW, Ohnesorge B (1999) Host and host age preference of *Trichogramma galloi* and *T. pretiosum* (Hym: Trichogrammatidae) reared on different hosts. J Econ Entomol 92:97–103
Montgomery GG (1978) Ecology of arboreal folivores. Smithsonian Institution, Washington
Moraes MCB, Laumann RA, Sujii ER, Pires CSS, Borges M (2005) Induced volatiles in soybean and pigeon pea plants artificially infested with the Neotropical brown stink bug, *Euschistus heros*, and their effect on the egg parasitoid, *Telenomus podisi*. Entomol Exp Appl 115:227–237
Mouzaki DG, Margaritis LH (1994) The eggshell of the almond wasp *Eurytoma amygdale* (Hymenoptera, Eurytomidae)-1. Morphogenesis and fine structure of the egg shell layers. Tissue Cell 26:559–568
Newton PJ (1988) Movement and impact of *Trichogrammatidae cryptophlebiae* Nagaraja (Hymenoptera: Trichogrammatidae) in citrus orchards after inundative releases against the false codling moth, *Cryptophlebia leucotreta* (Meyrick) (Lepidoptera: Tortricidae). Bull Entomol Res 78:85–99
Nishida R, Fukami H (1989) Oviposition stimulant of an Aristolochiaceae-feeding swallowtail butterfly, *Atrophaneura alcinous*. J Chem Ecol 15:2565–2575
Nishida R, Fukami H (1990) Sequestration of distasteful compounds by some pharmacophagous insects. J Chem Ecol 16:151–164
Noldus LPJJ, van Lenteren JC (1985) Kairomones for the egg parasite *Trichogramma evanescens* Westwood I. Effect of volatile substances released by two of its hosts, *Pieris brassicae* L. and *Pieris rapae* L. J Chem Ecol 11:793–791
Nordlund DA, Lewis WJ, Gueldner RC (1983) Kairomones and their use for management of entomophagous insects XIV. Response of *Telenomus remus* to abdominal tips of *Spodoptera frugiperda*, (Z)-9-Tetradecene-1-ol acetate and (Z)-9-Dodecene-1-ol acetate. J Chem Ecol 9:695–701
Nufio CR, Papaj DR (2001) Host marking behavior in phytophagous insects and parasitoids. Entomol Exp Appl 99:273–293
Nurindah, Cribb BW, Gordh G (1999) Influence of rearing hosts on host size acceptance by *Trichogramma australicum*. Biocontrol 44:129–141

Orr DB, Russin JS, Boethel DJ (1986) Reproductive biology and behavior of *Telenomus calvus* (Hymenoptera: Scelionidae), a phoretic egg parasitoid of *Podisus maculiventris* (Hemiptera: Pentatomidae). Can Entomol 118:1063–1072

Pacheco DP, Corrêa-Ferreira BSS (1998) Reproductive potential and longevity of the parasitoid *Telenomus podisi* Ashmead in eggs of different stink bugs species. Ann Soc Entomol Brasil 27:585–591

Pak GA (1988) Selection of *Trichogramma* for inundative biological control. Ph.D. Dissertation, Wageningen Agricultural University, Wageningen, The Netherlands

Pak GA, Buis CEM, Heck ICC, Hermans MLG (1986) Behavioural variations among strains of *Trichogramma* spp. Host age selection. Entomol Exp Appl 40:247–258

Pak GA, van Dalen A, Kaashoek N, Dijkman H (1990) Host egg chorion structure influencing host suitability for the egg parasitoid *Trichogramma* Westwood. J Insect Physiol 36:869–875

Papassideri IS, Margaritis LH (1986) Specific secretion of wax by the follicular cells of *Drosophila melanogaster*. Cell Biol Int Rep 10:963–968

Parker KD, Rudall KM (1957) The silk of the egg-stalk of the green lace-wing fly. Nature 179:905–906

Parra JRP (1997) Técnicas de criaçã de *Anagasta kuehniella*, hospedeiro alternativo papa produção de *Trichogramma*. In: Parra JRP, Zucchi RA (eds) Trichogrammae o Controle Biológico Aplicado. FEALQ, Piracicaba, SP, Brazil, pp 121–150

Pasteels JM, Daloze D, Rowell-Rahier M (1986) Chemical defence in chrysomelid eggs and neonate larvae. Physiol Entomol 11:29–37

Powell JE, Shepard M (1982) Biology of Australian and United States strains of *Trissolcus basalis*, a parasitoid of the green vegetable bud, *Nezara viridula.* Austral J Ecol 7:81–186

Puterka GJ, Slosser JE, Price JR (1985) Parasites of *Heliothis* spp. (Lepidoptera: Noctuidae): parasitism and seasonal occurrence for host crops in the Texas Rolling Plains. Environ Entomol 14:441–446

Raikhel AS, Dhadialla TS (1992) Accumulation of yolk proteins in insect oocytes. Annu Rev Entomol 37:217–251

Raikhel AS, Snigirevskaya ES (1998) Vitellogenesis. In: Harrison FW, Locke M (eds), Microscopic anatomy of invertebrates, vol 11C – Insecta. Wiley Liss, New York, pp 933–955

Ralston JS (1977) Egg guarding by male assassin bugs of the genus *Zelus* (Hemiptera: Reduviidae). Psyche 84:103–107

Randlkofer B, Obermaier E, Meineers T (2007) Mother choice of the oviposition site: balancing risk of egg parasitism and food supply for the progeny with an infochemical shelter? Chemoecology 17:177–186

Reed DA, Luhring KA, Stafford CA, Hansen AK, Millar JC, Hanks LM, Paine TD (2007) Host defensive response against an egg parasitoid involves cellular encapsulation and melanization. Biol Control 41:214–222

Regier JC, Kafatos FC (1985) Molecular aspects of chorion formation. In: Gilbert LI, Kerkut GA (eds) Comprehensive insect physiology, biochemistry, and pharmocology, vol 1. Pergamon, Oxford, pp 113–151

Reznik SY, Umarova TY (1990) The influence of host age on the selectivity of parasitism and fecundity of *Trichogramma.* Entomophaga 35:31–37

Reznik SY, Umarova TY, Voinovich ND (1997) The influence of previous host age on current host acceptance in *Trichogramma.* Entomol Exp Appl 82:153–157

Rivers DB, Denlinger DL (1994) Developmental fate of the flesh fly, *Sarcophaga bullata*, envenomated by the pupal parasitoid, *Nasonia vitripennis*. J Insect Physiol 40:121–127

Roriz V, Oliveira L, Garcia P (2005) Host suitability and preference studies of *Trichogramma cordubensis* (Hymenoptera: Trihogrammatidae). Biol Control 36:331–336

Rose USR, Lewis WJ, Tumlinson JH (1998) Specificity of systemically released cotton volatiles as attractants for specialist and generalist parasitic wasps. J Chem Ecol 24:303–319

Rosi MC, Isidoro N, Colazza S, Bin F (2001) Source of the host marking pheromone in the egg parasitoid *Trissolcus basalis* (Hymenoptera: Scelionidae). J Insect Physiol 47:989–995

Rothschild M (1992) Egg protection by the Atala hairstreak butterfly (*Eumaeus atala florida*). Phytochemistry 31:1959–1960

Ruberson JR, Tauber MJ, Tauber CA (1987) Biotypes of *Edovum puttleri* (Hymenoptera: Eulophidae): responses to developing eggs of the Colorado potato beetle (Coleoptera: Chrysomelidae). Ann Entomol Soc Am 80:451–455

Sá LAN, Parra JRP (1994) Natural parasitism of *Spodoptera frugiperda* and *Helicoverpa zea* (Lepidoptera: Noctuidae) eggs in corn by *Trichogramma pretiosum* (Hymenoptera: Trichogrammatidae) in Brazil. Florida Entomol 77:185–188

Salt G (1937) Experimental studies in insect parasitism. V. The sense used by *Trichogramma* to distinguish between parasitized and unparasitized hosts. Proc R Soc Lond 122B:57–75

Salt G (1940) Experimental studies in insect parasitism. VII. The effect of different host on parasite *Trichogramma evanescens* West. (Hym.: Chalcidoidea). Proc R Soc Lond 15A:81–95

Sander K, Gutzeit JO, Jäckle H (1985) Insect embryogenesis: morphology, physiology, genetic and molecular aspects. In: Kerkut GA, Gilbert LI (eds) Comprehensive insect physiology, biochemistry and pharmacology, vol 1. Pergamon, Oxford, pp 319–385

Santis FD, Conti E, Romani R, Salerno G, Parillo F, Bin F (2007) Colletrial glnds of *Sesamia nonagrioides* as a source of the host-recognition kairomone for the egg parasitoid *Telenomus busseolae*. Physiol Entomol 33:7–16

Saour G (2004) Efficacy assessment of some *Trichogramma* species (Hymenoptera: Trichogrammatidae) in controlling the potato tuber moth *Phthorimaea operculella* Zell. (Lepidoptera: Gelechiidae). J Pest Sci 77:229–234

Schmidt JM (1991) The role of physical factors in tritrophic interactions. Redia 74:43–93

Schmidt JM, Smith JJ (1985) Host volume measurement by the parasitoid wasp *Trichogramma minutum*: the roles of curvature and surface area. Entomol Exp Appl 39:213–221

Schmidt JM, Smith JJ (1986) Correlations between body angles and substrate curvature in the parasitoid wasp *Trichogramma minutum*: a possible mechanism of host radius measurement. J Exp Biol 129:151–164

Schmidt JM, Smith JJ (1987a) Measurement of host curvature by the parasitoid wasp *Trichogramma minutum*: and its effect on host examination and progeny allocation. J Exp Biol 129:151–164

Schmidt JM, Smith JB (1987b) The effect of host spacing on the clutch size and parasitization rate of *Trichogramma minutum*. Entomol Exp Appl 43:125–131

Schmidt JM, Smith JJ (1989) Host examination walk and oviposition site selection of *Trichogramma minutum*: studies on spherical hosts. J Insect Behav 2:143–171

Sehnal F (1985) Growth and life cycles. In: Kerkut GA, Gilbert LI (eds) Comprehensive insect physiology, biochemistry and pharmacology, vol 1. Pergamon, Oxford, pp 1–86

Sekul AA, Cox HC (1967) Sex pheromone in the fall army worm *Spodoptera frugiperda* (JE Smith). Bioscience 15:670–671

Sekul AA, Sparks AN (1967) Sex pheromone in the army worm moth – isolation, identification and synthesis. J Econ Entomol 60:1270–1273

Shu SQ, Jones RL (1985) Laboratory studies of the host-seeking behavior of a parasitoid, *Trichogramma nubilale* and a kairomone from its host, *Ostrinia nubilale*. Les Colleques l'INRA 43:249–265

Shu SQ, Jones RL (1988) Kinetic effects of a kairomone in moth scales of the European corn borer on *Trichogramma nublale* Ertle and Davis (Hymenoptera: trichogrammatidae). J Insect Behav 2:123–131

Slansky F, Rodriguez JG (1987) Nutritional ecology of insects, mites, spiders, and related invertebrates. Wiley, New York

Slansky F, Scriber JM (1985) Food consumption and utilization. In: Kerkut GA, Gilbert LI (eds) Comprehensive insect physiology, biochemistry and pharmacology, vol 4. Pergamon, Oxford, pp 87–163

Smith SM (1996) Biological control with *Trichogramma*: advances, successes, and potential of their use. Annu Rev Entomol 41:375–406

Smits PH (1982) The influence of kairomones of *Mamestra brassicae* on the searching behavior of *Trichogramma evanescens*. Colloques I'INRA 9:139–150
Southwood TRE (1977) Habitat, the template for ecological strategies. J Anim Ecol 46:337–365
Sparks AN (1980) A review of the biology of the fall armyworm. Florida Entomol 62:82–87
Stamp NE (1980) Egg deposition patterns in butterflies: why do some species cluster their eggs rather than deposit them singly? Am Nat 115:367–380
Steidle JLM, Steppuhn A, Reinhard J (2001) Volatile cues from different host complexes for host location by the generalist parasitoid *Lariophagus distinguendus* (Hymenoptera: Pteromalidae). Basic Appl Ecol 29:131–143
Strand MR (1986) The physiological interactions of parasitoids with their influence on reproductive strategies. In: Waage JK, Greathead DJ (eds) Insect parasitoids. Academic, London, pp 97–136
Strand MR, Vinson SB (1980) Maternally induced host regulation by the egg parasitoid *Telenomus heliothidis*. Physiol Entomol 8:469–475
Strand MR, Vinson SB (1983a) Factors affecting host recognition and acceptance in the egg parasitoid, *Telenomus heliothidis*. Environ Entomol 12:1114–1119
Strand MR, Vinson SB (1983b) Host acceptance behavior of *Telenomus heliothidis* towards *Heliothis virescens*. Ann Entomol Soc Am 76:781–785
Strand MR, Vinson SB (1983c) Factors affecting host recognition and acceptance in the egg parasitoid *Telenomus heliothidis*. Environ Entomol 12:1114–1119
Strand MR, Pech LL (1995) Immunological basis for compatibility in parasitoid-host relationships. Annu Rev Entomol 40:31–56
Strand MR, Meola SM, Vinson SB (1986) Correlating pathological symptoms in *Heliothis virescens* eggs with development of the parasitoid *Telenomus heliothidis*. J Insect Physiol 32:389–402
Stumpner A, Meyer S (2001) Songs and the function of song elements in four duetting bushcricket species (Ensifera, Phaneropteridae, Barbitistes). J Insect Behav 14:511–534
Sweetman HL (1963) The principles of biological control. Wm. C. Brown, Dubugue
Tabata S, Tamanuki K (1940) On the hymenopterous parasites of the pine caterpillar, *Dendrolimus sibiricus albolineatus* Mats. In southern Sakhalin Cent Exp Sta Sakhalin Rep 33 Forest. (In Japanese) (Abstract In: Rev Appl Entomol A 29:95)
Trougakos IP, Margaritis LH (2002) Chemoecology of insect eggs. In: Hilker M, Meiners T (eds) Chemical ecology of insect eggs and egg deposition. Blackwell, Berlin, pp 3–36
Tsuchida K, Kawooya JK, Law JH, Wells MA (1992) Isolation and characterization of two follicle-specific proteins from eggs of *Manduca sexta*. Insect Biochem Mol Biol 22:89–98
Turlings TCJ, Loughrin JH, McCall PJ, Röse USR, Lewis WJ, Tumlinson JH (1995) How caterpillar-damaged plants protect themselves by attracting parasitic wasps. Proc Natl Acad Sci USA 92:4169–4174
Turlings TCJ, Alborn HY, Loughrin JH, Tumlinson JH (2000) Volicitin, an elicitor of maize volatiles in oral secretion of *Spodoptera exigua*: isolation and bioactivity. J Chem Ecol 26:189–202
Vet LEM (1999) From chemical to population ecology: infochemical use in an evolutionary context. J Chem Ecol 25:31–49
Vinson SB (1985) The behavior of parasitoids. In: Kerkut GA, Gilbert LI (eds) Comprehensive insect physiology, biochemistry and pharmacology, vol 9. Pergamon, Oxford, pp 417–469
Vinson SB (1988) Comparison of host characteristics that elicit host recognition behavior of parasitoid Hymenoptera. In: Gupta V (ed) Advances in parasitic hymenoptera research. E. J. Brill, Amsterdam, pp 285–291
Vinson SB (1993) Parasitoid attraction to plants. In: Anais do 14° Congresso Brasileiro de Entomologia, SEB, Piracicaba, SP, Brasil. Fundação de Estudos Agrários Luiz de Queiroz (FEALQ). Piracicaba, SP, pp 29–40
Vinson SB (1998) The general host selection behavior of parasitoid hymenoptera and a comparison of initial strategies utilized by larvaphagous and oophagous species. Biol Control 11:79–96

Vinson SB (1999) Parasitoid manipulation as a plant defense strategy. Ann Entomol Soc Am 92:812–828

Vinson SB (2005) A new look at the coevolution of the insect–plant relationship and its' relevance to agriculture. In: Heinz K, Frisbie RE, Brogán C (eds) Entomology at the Land Grant University: perspectives from the Texas A&M University Department Centenary. Texas A&M, College Station, pp 165–192

Vinson SB, Iwantsch GF (1980) Host suitability for insect parasitoids. Annu Rev Entomol 25:397–419

Vinson SB, Williams HJ (1991) Host selection behavior of *Campoletis sonorensis*: a model system. Biol Control 1:101–117

Vinson SB, Williams HJ, Lu J (1994) Identification of different compounds from different plants responsible for the orientation of *Campoletis sonorensis* host sites. Norw J Agric Suppl 16:207–210

Vinson SB, Pietrantonio PV, Lu HL, Coates CJ (2008) The physiology of reproduction in the imported fire ant, *Solenopsis invicta* Buren. In: Liu HN (ed) Recent advances in insect physiology, toxicology and molecular biology. Research Signposts, Kerala, India, pp 153–171

Vŭüren van L (1936) Waarnemingen omtrent *Pharnurus beneficiens* (Ze, Oxox, UK.hmt.) (Hymenoptera: Scelionidae) op *Schoenobius bipunctifer* Walk. Entol Meded Nederl, Indie Buitenzorg 1:29–33

Warner WV (1903) Notes on the habits of *Secilo*. Proc Entomol Soc Washington 5:308–309

Weseloh RM (1980) Behavioral changes in *Apanteles melanoscelus* females exposed to gypsy moth silk. Environ Entomol 9:345–349

Wu Z, Qin J (1982) Ovipositional response of *Trichogramma dendrolimi* to the chemical content of artificial eggs. Acta Entomol Sin 25:363–372

Wylie HG (1965) Effects of superparasitism on *Nasonia vitripennis* (Walker). Can Entomol 97:326–331

Xie Z, Nettles Jr WC, Vinson SB (1991) Identification of ovipositional stimulants for *Anastatus japonicus*. Colloques I'INRA 56:101–114

Chapter 3
Antennal Structures Used in Communication by Egg Parasitoids

Roberto Romani, Nunzio Isidoro, and Ferdinando Bin

Contents

3.1 Introduction

Egg parasitoids are considered the most effective biocontrol agents of crop pests among parasitic wasps, since they are able to remove the herbivore from the agroecosystem before larval eclosion. The use of arthropod eggs as hosts by parasitoids has evolved in the Hymenoptera at least in 14 families. Of these, only Trichogrammatidae, Scelionidae and Mymaridae are composed entirely by species with this life style. These egg parasitoids are associated with hosts belonging to 15 insect Orders. The huge number of host-parasitoid associations generated by the

R. Romani (✉)
Department of Scienze Agrarie e Ambientali, Faculty of Agriculture, University of Perugia, Borgo XX Giugno 74, 06121, Perugia, Italy
e-mail: rromani@unipg.it

F.L. Consoli et al. (eds.), *Egg Parasitoids in Agroecosystems with Emphasis on Trichogramma*, Progress in Biological Control 9, DOI 10.1007/978-1-4020-9110-0_3,

common target, the host egg, and the varieties of taxonomic entities, thousand host species, produce a great diversity which can be evaluated and measured in different ways. The antennae are posed here due to their importance in taxonomy and parasitoid behavior, such as habitat location, sex recognition and reproductive isolation, host recognition, inter- and intra-specific marking discrimination.

A number of studies has been carried out on different aspects of the biology, physiology and ecology of egg parasitoids in order to implement their use in integrated pest management. Nevertheless, some aspects dealing with the ultrastructure of sensory structures and their sensory physiology have received less attention: this is most likely due to the reduced body size of many species, which imposes severe restrictions on the implementation of well-established research tools (i.e. transmission electron microscopy, TEM; electroantenna-detection, EAD; single cell recording, SCR).

The extant data however revealed extraordinary adaptations for several species of egg parasitoids, finely tuned to allow the location and recognition of the mate (in the context of mating behavior) and of the host (in the context of host location and recognition). These adaptations reached their highest level in those organs directly involved in the interaction with the environment, i.e. the antennae and the ovipositor.

While antennal sensory structures are respectively used by males and females to locate the mate and the host, and to recognize them through behavioral sequences involving several cues (for host location and recognition, see Chapter 4), the ovipositor is used by females to finally accept a potential suitable host. The ovipositor is a highly specialized structure evolved by hymenopteran parasitoids through which they are able to oviposit within various host stages. The three valves that compose the ovipositor are equipped with specialized sensory structures that provide females with information on the host internal status (host suitability – i.e. whether it is too old or already parasitized). Despite the importance of this structure, in egg parasitoids there are scant data on the ultrastructural organization of the sensilla associated with the ovipositor (Cônsoli et al. 1999). Therefore, this chapter will focus on the structures carried by the main groups of egg parasitoids on their antennae.

3.2 General Features of Antennae in Egg Parasitoids

Among the appendages characterizing the different body parts of insects, antennae are considered those involved primarily in the perception of stimuli from the environment.

Antennae are "segmented appendages that function primarily as chemosensory and mechanosensory structures" (Loudon 2003). This is in agreement with the fact that the antennal surface is covered by several small organs (in some cases more than 17000, see Steinbrecht 1970) termed "sensilla".

Antennae are homologous appendages inserted at various, mainly frontal positions, on the head capsule of mobile stages of insects (both larvae and adults). However, the role of antennae becomes more important in the adults, due to the different relationship with the environment mediated by a plethora of cues that need specialized receptor organs.

Insect antennae occur in a large variety of shapes and sizes, but a general scheme can be outlined considering these appendages divided into three parts: (i) scape (or scapus), the basal segment, is articulated with the head capsule through the torulus, where it connects via an elastic joint membrane. The base of the scape is inserted into a socket where extrinsic muscles are attached, (ii) pedicel, the second antennal segment, is articulated proximally with the scape and distally with the rest of the antenna. In most insects, the pedicel houses the Johnston's organ, an auditory organ, and (iii) flagellum, all of the remaining antennal segments, is the main part of the antenna. The flagellum is composed by a different number of segments, termed "flagellomeres", bearing most of the sensilla. In Pterygote (higher insects), since no muscles have been found within the pedicel and flagellum, antennal movements are made possible by articulations that connect the scape-pedicel and pedicel-1st flagellomere joints through extrinsic muscles.

In hymenopteran egg parasitoids, antennae are of the geniculate type, i.e. the scape is usually longer and forms an angle of about 90° with the pedicel. The flagellum varies in length, but in most cases is reduced to about 10–12 flagellomeres. In females, the last flagellomeres (the number is variable, but usually does not exceed 5) can be noticeably enlarged and swollen, forming a club, which can result from the fusion of adjacent flagellomeres. Furthermore, antennae are sexually dimorphic, either due to differences in antennomere number (higher in males) or antenna shape (in males the antennae are usually filiform and seldom form an apical club). General features of different antennal types in the main groups of egg parasitoids are illustrated in here (Fig. 3.1).

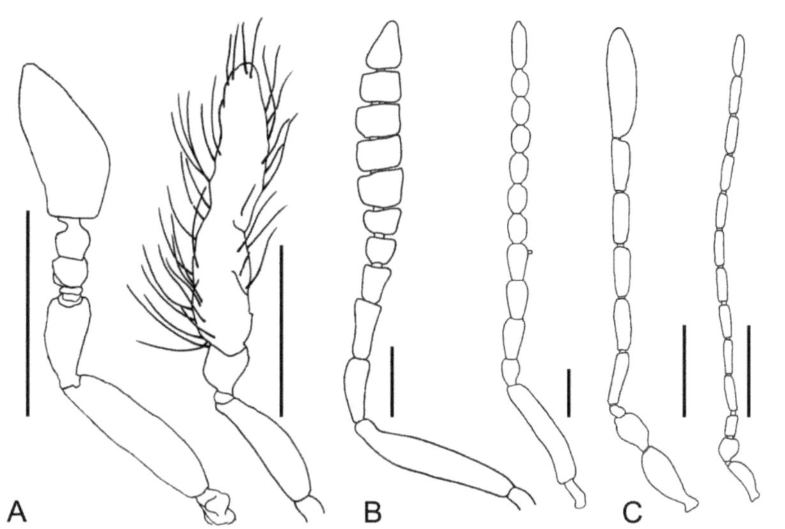

Fig. 3.1 Some examples of egg parasitoids antennae. (**a**) *Trichogramma* (Trichogrammatidae); (**b**) *Trissolcus* (Scelionidae); (**c**) *Anaphes* (Mymaridae). For each group, female antenna on the left and male antenna on the right. Bar scale in all: 100 μm

3.3 Antennal Structures in Egg Parasitoids: Sensilla and Glands

Historically, insect antennae have been considered the sensory center of the smell and taste, as well as the perception of airborne or contact vibration (Zacharuk 1985, Keil 1999). These sensory functions are made possible by the presence of small organs, known as sensilla. A sensillum is defined as "... a well-defined complex of bipolar ... receptor neurons, auxiliary cells, and cuticular elements" (Keil 1999). In the last 25 years, a series of studies dealing with morphological and behavioral aspects of insect parasitoids revealed that in addition to sensilla, antennae are the site of different structures as well (Isidoro et al. 1996, Bin et al. 1999a). Particularly, the presence of antennal glands, in most cases found in male antennae, were discovered in several families of parasitic wasps, as well as other hymenopterans, with different functional significance (Bin et al. 1999b, Isidoro et al. 2000, Kaltenpoth et al. 2005, Romani et al. 2003, 2005, 2006, 2008, Goettler et al. 2007). In this chapter, morphological and functional aspects of antennal structures that are involved in the communication between egg parasitoids will be covered. This review will be restricted to the families in which ultrastructural studies (using SEM and/or TEM techniques) have been carried out.

3.4 Antennal Sensilla in Egg Parasitoids

Insect sensilla are small functional units, each comprising several different parts built around the true "sensors", i.e. the sensory neurons. These specialised cells fall into the insect's peripheral nervous system, together with motor neurons. Sensory neurons are afferent cells that bring the signal from the extreme periphery of the insect's body to the main processing centers, located within the head capsule, mainly the protocerebrum and deutocerebrum. A sensory neuron can be divided into a cellular part (perikaryon), an efferent projection (axon) and an afferent dendritic process. This latter is further divided into an inner and outer dendritic segment. The outer dendritic segment begins to appear as a cilium at the level of the ciliary constrictions, where the dendrite assumes the structure of a cilium with the typical arrangement of nine doublets of microtubules, lacking however the central doublet (Thurm 1970). Microtubules originate at the level of two distinct basal bodies (Gaffal and Bassemir 1974).

One or more sensory neurons are associated with differently shaped cuticular structures whereby stimulation (either mechanical, physical or chemical) of the dendrite projections is allowed. Sensory neurons are ensheathed by three or more auxiliary cells that take part during the morphogenesis of the different sensillar parts (socket, dendrite sheath) as well as assure functionality of the sensillum.

3.4.1 General Features of Antennal Sensilla

Insect sensilla are based on a stereotyped scheme, and according to this basis, they have been defined as "kleinorgan" (Henke 1953) or "organule" (Lawrence

1966). Each sensillum is made up of an external cuticular part and several cellular components (Fig. 3.2). The cuticular part, that in most cases houses the sensory neurons projections, can have different shapes, although the differentiation in hair or hair-like structures seems to be a common feature. Each sensillum's cuticular wall exhibits either a single (or few) pore located in distinct parts (i.e. the tip) or numerous pores located on defined parts or evenly distributed over the sensillum surface, or they may have no pores at all. In cases where the sensillum itself is completely embedded below the antennal wall, the sensillum's cuticular part may be absent (i.e. scolopidia). Although the degree of innervation (i.e. the number of sensory neurons associated per sensillum) varies, some rules can be applied.

The first attempts of classification for insect sensilla were based on light microscopy, and the shape of the external cuticular part was exclusively considered (Schneider 1964). The development of electron microscopy techniques after WW2 made it possible to discover new, unexpected features, leading to new classifications based on ultrastructural details, particularly the structure of the sensilla wall (Altner

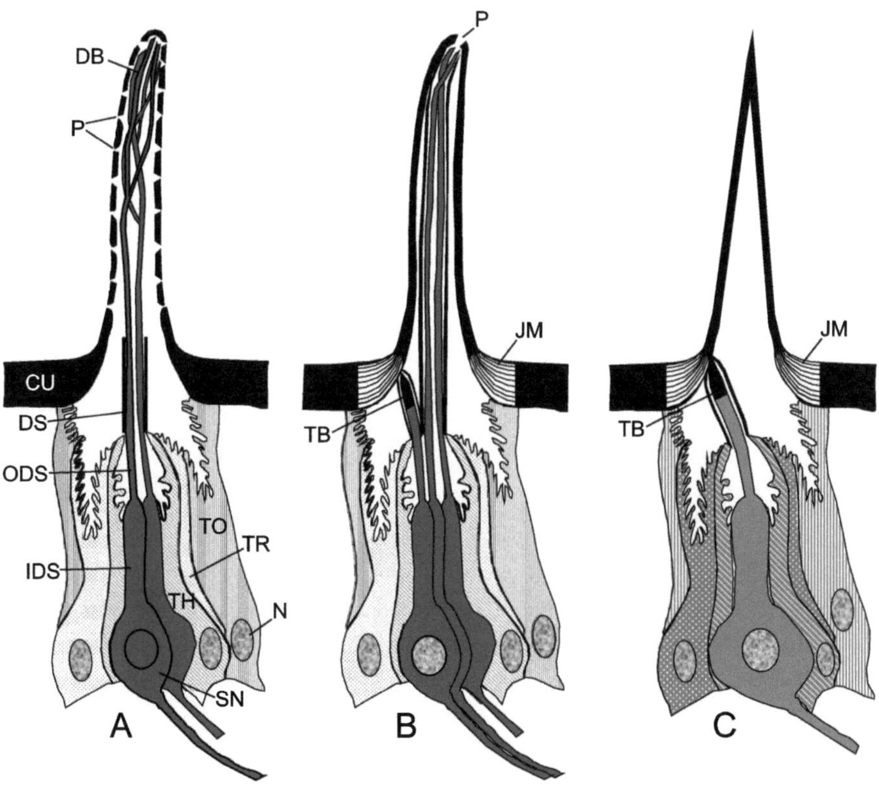

Fig. 3.2 Schematic drawing of different type of sensilla. (**a**) multiporous olfactory sensillum; (**b**) uniporous gustatory sensillum; (**c**) mechanosensory hair. CU – cuticle, DB – dendritic branches, DS – dendrite sheath, IDS – inner dendritic segments, JM – joint membrane, N – nucleus, ODS – outer dendritic segments, P – pores, SN – sensory neurons, TB – tubular body, TH – thecogen cell, TO – tormogen cell, TR – trichogen cell

1977, Altner and Prillinger 1980). As a general consideration, we have to admit that none of them are completely conclusive, although we will adopt the one proposed by Altner because of its functional significance. Electrophysiological recordings from receptor neurons are conclusive in order to assign a defined function to different sensilla, although some sensilla can respond according to a bi-modal or multimodal pattern (Altner et al. 1981, Pophof 1997, Hunger and Steinbrecht 1998).

Based on the wall pore structure, as it can be observed through transmission electron microscope, Altner classified insect sensilla as:

- No-pore (aporous) sensilla: these sensilla have no permeable pores on the sensilla wall. Different morphological types have been reported, and their functional significance is related with mechanosensitivity;
- Tip-pore (uniporous) sensilla: mostly of these sensilla appear as hairs or pegs, and possess a single pore in apical position. When they occur in form of hair or pegs, they are inserted on the antenna wall through a flexible socket, thus acting as a bi-modal mechano-chemosensory structure;
- Wall-pore (multiporous) sensilla: the sensilla wall is perforated by numerous pores that are all localized on only some parts of the sensillum or are evenly distributed over the sensory cuticle. Although the main function of these sensilla is the olfaction, in some cases there have been reports they acted as bi-modal receptors, combining olfaction and thermo-perception (Altner et al. 1983, Hansson et al. 1996).

In egg parasitoids, a few exceptions are known and they will be later described (see Section 3.4.4).

In this chapter we will focus on the different sensilla on the basis of the stimulus-conducting systems, in terms of pores: the presence or absence of, their relative abundance and location (see Table 3.1).

3.4.2 Aporous Sensilla

They are characterized by the absence of evident cuticular pores, although a pore canal system that connects the lumen of the sensillum with the environment may occur (Zacharuk 1985). Aporous (or no-pore) sensilla can be represented according to different morphological types, i.e. mechanosensory hairs, styloconic sensilla, campaniform sensilla, coeloconic sensilla and scolopidia. Here we will report only the different classes of aporous sensilla that are know to egg parasitoids.

Mechanosensory hair. Mechanosensory hairs possess a cuticular shaft of varied length that allows the perception of the flexible basal socket's relative distortion, through which they are inserted into the antennal wall (Keil 1997). The cuticular shaft can either be made of a solid cuticle or appears empty, but the presence of sensory neurons has never been observed. These sensilla are composed of a single sensory neuron and a number of two to three accessory cells (McIver 1985)

Table 3.1 Number of different antennal sensilla type and associated sensory neurons on male and female antennae of egg parasitoids

Type of sensilla	Species	N. sensilla per individual		N. sensory neurons per sensillum		References
		Female	Male	Female	Male	
APOROUS						
Mechanosensory hair	*Tetrastichus hagenowii*	n.i.	n.i.	1	1	Barlin et al. (1981)
	Anaphes victus *A. listronoti*	510	Numerous	n.i.	n.i.	van Baaren et al. (1999)
	Trichogramma evanescens *T. maltbyi* Nagaraja and Nagarkatti *T. brasiliensis*	60	n.i.	n.i.	n.i.	Voegelé et al. (1975)
	T. australicum	294	26	n.i.	n.i.	Amornsak et al. (1998)
	T. nubilale	36–48	n.i.	n.i.	n.i.	Olson and Andow (1993)
	T. minutum	38	n.i.	n.i.	n.i.	Schmidt and Smith (1987)
Campaniform	*Trissolcus basalis*	10	10	1	1	Bin et al. (1989)
	Trichogramma australicum	8	8	n.i.	n.i.	Amornsak et al. (1998)
Coeloconic	*Trissolcus basalis*	4	4	4	4	Isidoro (1992)
	Trichogramma australicum	4	4	n.i.	n.i.	Amornsak et al. (1998)
	T. galloi *T. pretiosum*	4	n.i.	n.i.	n.i.	Cônsoli et al. (1999)

Table 3.1 (continued)

Type of sensilla	Species	N. sensilla per individual		N. sensory neurons per sensillum		References
UNIPOROUS						
Uniporos Gustatory	*Trissolcus basalis*	54	54	5	5	Bin et al. (1989)
	Telenomus reynoldsi	Several	Several	n.i.	n.i.	Cave and Gaylor (1987)
	Trichogramma evanescens *T. maltbyi* *T. brasiliensis* (Ashmead)	2	–	n.i.	–	Voegelé et al. (1975)
	T. australicum	2	–	n.i.	–	Amornsak et al. (1998)
	T. nubilale	2	–	n.i.	–	Olson and Andow (1993)
	T. galloi *T. pretiosum*	2	–	n.i.	–	Cônsoli et al. (1999)
	Anaphes victus *A. listronoti*	20–40	8	n.i.	n.i.	van Baaren et al. (1999)
MULTIPOROUS						
Single walled						
Trichoideum	*Trichogramma evanescens* *T. maltbyi* *T. brasiliensis*	16	n.i.	n.i.	n.i.	Voegelé et al. (1975)
	T. nubilale	16	Numerous	n.i.	n.i.	Olson and Andow (1993)

Table 3.1 (continued)

Type of sensilla	Species	N. sensilla per individual		N. sensory neurons per sensillum		References
	T. australicum	16	120	n.i.	n.i.	Amornsak et al. (1998)
	T. galloi	Several	Numerous	n.i.	n.i.	Cônsoli et al. (1999)
	T. pretiosum					
	Tetrastichus hagenowii	Several	Several	15–16	4	Barlin et al. (1981)
Sickle-shaped	*Trissolcus basalis*	72	340	28	18	Bin et al. (1989)
	Telenomus reynoldsi	40	300	n.i.	n.i.	Cave and Gaylor (1987)
Multiporous plate	*Trichogramma evanescens*	Several	Several	n.i.	n.i.	Barlin et al. (1981)
	Tetrastichus hagenowii	Numerous	Numerous	>50	>50	Barlin and Vinson (1981)
	Torymus warreni	”	”	>50	>50	
	Trichogramma evanescens	10	n.i.	n.i.	n.i.	Voegelé et al. (1975)
	T. maltbyi					
	T. brasiliensis					
	T. nubilale	10	n.i.	n.i.	n.i.	Olson and Andow (1993)
	T. australicum	10	6	n.i.	n.i.	Amornsak et al. (1998)
	T. galloi	10	n.i.	n.i.	n.i.	Cônsoli et al. (1999)
	T. pretiosum					
	Anaphes victus	36	100	n.i.	n.i.	van Baaren et al. (1999)
	A. listronoti					

Table 3.1 (continued)

Type of sensilla	Species	N. sensilla per individual		N. sensory neurons per sensillum		References
Multiporous Gustatory						
	Trichogramma evanescens	56	—	n.i.	—	Voegelé et al. (1975)
	T. maltbyi	40		"		
	T. brasiliensis	44		"		
	T. nubilale	Several	–	n.i.	–	Olson and Andow (1993)
	T. australicum	70	–	n.i.	–	Amornsak et al. (1998)
	T. brassicae	40	–	10	–	Isidoro et al. (1996)
	T. galloi	62	—	n.i.	—	Cônsoli et al. (1999)
	T. pretiosum	54		"		
	Telenomus reynoldsi	12	–	n.i.	–	Cave and Gaylor (1987)
	T. busseolae	12	–	200	–	Isidoro et al. (2001)
	Trissolcus basalis	18	–	400	–	Bin et al. (1989)
	Anaphes victus	4	-	n.i.	-	van Baaren et al. (1999)
	A. listronoti					
Double walled						
Multiporous Grooved Peg	*Trissolcus basalis*	12	12	4	4	Bin et al. (1989)
	Telenomus reynoldsi	Several	Several	n.i.	n.i.	Cave and Gaylor (1987)

Table 3.1 (continued)

Type of sensilla	Species	N. sensilla per individual		N. sensory neurons per sensillum		References
	Trichogramma evanescens *T. maltbyi* *T. brasiliensis*	14	n.i.	n.i.	n.i.	Voegelé et al. (1975)
	T. nubilale	14	n.i.	n.i.	n.i.	Olson and Andow (1993)
	T. australicum	14	14	n.i.	n.i.	Amornsak et al. (1998)
	T. galloi *T. pretiosum*	several	n.i.	n.i.	n.i.	Cônsoli et al. (1999)
	Anaphes victus *A. listronoti*	24	24	n.i.	n.i.	van Baaren et al. (1999)

n.i. not investigated; – not present

(Fig. 3.2c). The dendritic projection of the sensory neuron is typically modified into a tubular body (Thurm 1964), i.e. a variable number of microtubules closely packed together and arranged in parallel. Typically, the tubular body is located on the distal end of the outer dendritic segment and it appears circular in a cross section, and very electrondense because of the material lying between the microtubules. The tubular body is accepted as the site of sensory transduction (Gnatzy and Tautz 1980, Thurm 1983). Mechanosensory bristles (or hairs) possess a modified base (socket) that allows mechanical distortion and stimulation of the sensory neuron. Depending on the body cuticle's relative hardness, this socket in invariably elastic and is characterized by specialized cuticular elements that connects to the external hair, a joint membrane, a socket septum and suspension fibers (Gaffal et al. 1975). The tubular body is mostly found inserted at the very base of the hair shaft. In egg parasitoids, mechanosensory hairs were reported on the antennae of several species belonging to different families (Amornsak et al. 1998, Cônsoli et al. 1999, van Baaren et al. 1999). They usually represent the most abundant class of antennal sensilla, and their function is related to the touching of different surfaces scanned by insect's antennae.

Trichoid sensilla can be found in distinct functional groups known as "hair plates". These proprioreceptive position detectors can be found at the joint region of different body parts, i.e. legs, abdomen, wings. Antennal hair plates were described in *Trichogramma minutum* Riley as proprioreceptors involved in the calculation of the host egg surface curvature (Schmidt and Smith 1987a, 1987b). The antennal hair plates have been found at the level of the head-scape and scape-pedicel joint, and are represented by two to four relatively short pegs. These pegs work as proprioreceptors, keeping the parasitoid informed of the relative position of the scape in respect to the head, and the pedicel in respect to the scape, allowing no-stimulation for angles between 70–80° and 130–150° (Schmidt and Smith 1986). A lower angle is related to the reduced body size of the potential host egg, while a higher angle is consistent with a large host, so the parasitoid can make a decision on the optimal clutch size to be laid. A similar situation has been reported for the scelionid *Trissolcus basalis* (Woll.) (Isidoro 1991), with a similar arrangement and external morphology of the hair plates (Fig. 3.3). Ultrastructural details revealed the mechanosensory features for these sensilla.

Campaniform sensilla. Campaniform sensilla, that owe their name to their bell- or dome- shape, occur on different body regions. In particular, these sensilla are closely located to the joint structures, where a mechanical deformation of the cuticle takes place (Moran et al. 1971, Gnatzy et al. 1987, Grünert and Gnatzy 1987).

A typical campaniform sensillum is made up of an external cuticular part in the form of a flattened, or slightly raised smooth circular cuticular disk. This is elastically inserted into the cuticular wall through a modified joint membrane system. As a rule, these sensilla are innervated by a single sensory neuron, which terminate at the distal end of its outer dendritic segment, with the tubular body. Campaniform sensilla respond to strains of the cuticle (Pringle 1961, Zill and Moran 1981). In regards to egg parasitoids, there have only been a few descriptions of the occurrence of campaniform sensilla on the antenna. In the scelionid *Trissolcus basalis*, campaniform sensilla are associated with male and female dorsal antennal glands (Bin et al.

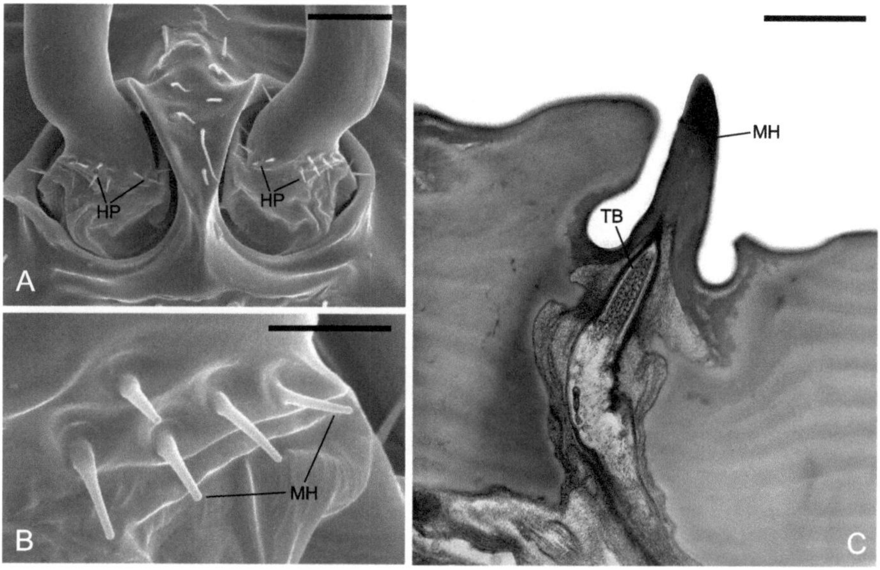

Fig. 3.3 Hair plates at the head-scape joint in *Trissolcus basalis*. (**a, b**) SEM showing the hair plates (HP) at the base of the scape and the external morphology of the mechanosensory hairs (MH). Note the different orientation of each sensillum; (**c**) TEM longitudinal section of one of the mechanosensory hairs with evidence of the tubular body (TB). Bars: a=20 μm, b=5 μm, c= 1 μm

1989). The fine structure revealed the presence of a hemispherical, non-porous cap embedded within the antennal cuticle, innervated by a single sensory neuron and embedded by a single accessory cell (Fig. 3.4a–d). Functionally speaking, campaniform sensilla could be involved in the release of the secretion of the antennal glands. Campaniform sensilla were also reported in Trichogrammatidae: in *Trichogramma australicum* Girault, 4 convex, disk-like structures occur on the distal part of the pedicel, where articulation with the first flagellomere occurs (Amornsak et al. 1998) (Fig. 3.4b).

Coeloconic sensilla. These poreless sensilla are reported as short pegs set in pits. The cuticular peg can be either completely embedded within the antennal wall, therefore communicating with the external environment through a narrow opening, or it can stand on the antennal surface, resemblig basiconic or styloconic-like sensilla (Altner and Loftus 1985). Coeloconic sensilla are usually in relatively low numbers when compared with the other antennal sensory structures, with a defined distribution on the antennal segments. Their main striking ultrastructural features are (i) the peg with a thick, continuous wall, (ii) three associated sensory neurons as a rule, among which only two enter the peg, completely filling the lumen and leaving no space for sensillum lymph space, (iii) the third neuron typically ends at the peg base, where it develops into numerous, characteristically lamellated, dendritic branches (Altner and Prillinger 1980).

Electrophysiological investigations revealed a role in the perception of changes in temperature and humidity for coeloconic sensilla (Tichy 1979, Altner et al. 1981,

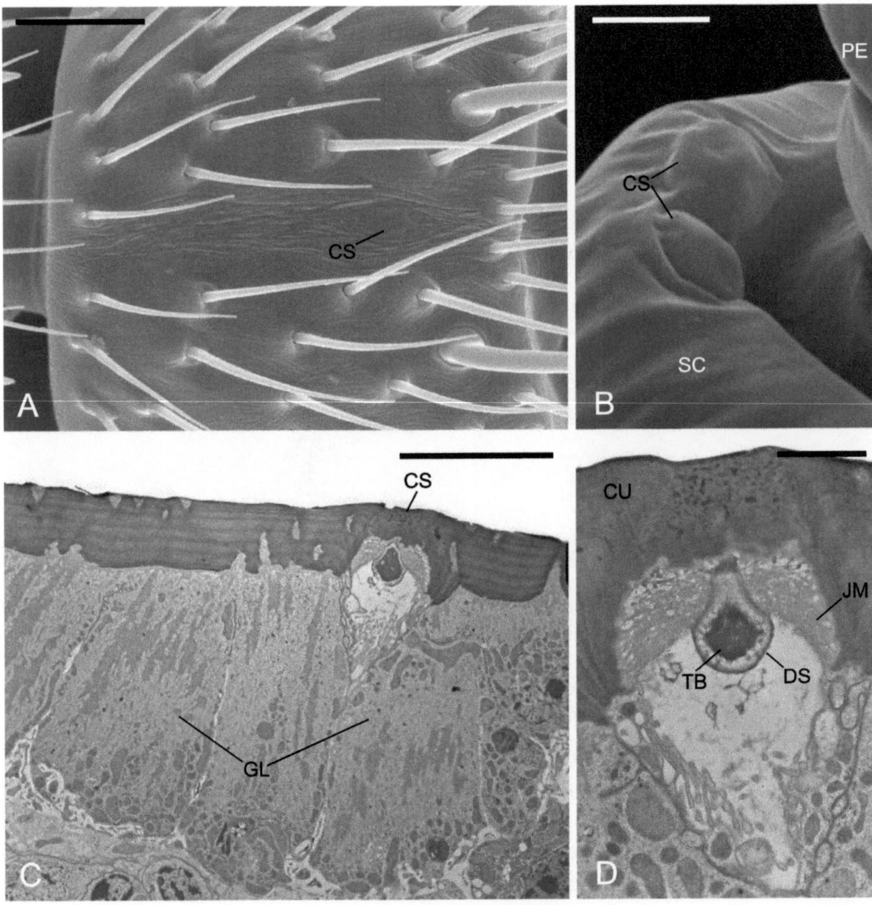

Fig. 3.4 (**a, b**) SEM pictures showing campaniform sensilla (CS) in *Trissolcus basalis* and *Trichogramma brassicae* female antennae. In *T. basalis* (**a**) the CS is located dorsally on the antennomere, where a dorsal depression is found. In *T. brassicae* (**b**) two CS are positioned on the distal edge of the scape (SC), close to the pedicel (PE). (**c, d**) TEM pictures of *T. basalis* CS surrounded by the glandular epithelium (GL) of the female dorsal gland. The apical tubular body (TB) is inserted below the cuticle (CU). DS – dendrite sheath, JM – joint membrane. Bars: A=10 μm, B=2 μm, C= 5 μm, D=1 μm

Yokohari 1983). Coeloconic sensilla are poorly studied in egg parasitoids, either because of their seldom occurrence or their small size, which make them difficult to detect in such small insects. They were described in Trichogrammatidae, but only using the SEM, therefore no ultrastructural data are available (Fig. 3.5c). In *Trichogramma evanescens* Westwood females, the *sensilla campaniformia* located on the second funicular segment and proximally on the club (Voegelé et al. 1975) may actually be the coeloconic sensilla. Olson and Andow (1993) described a similar situation in *Trichogramma nubilale* Ertle and Davis, where three

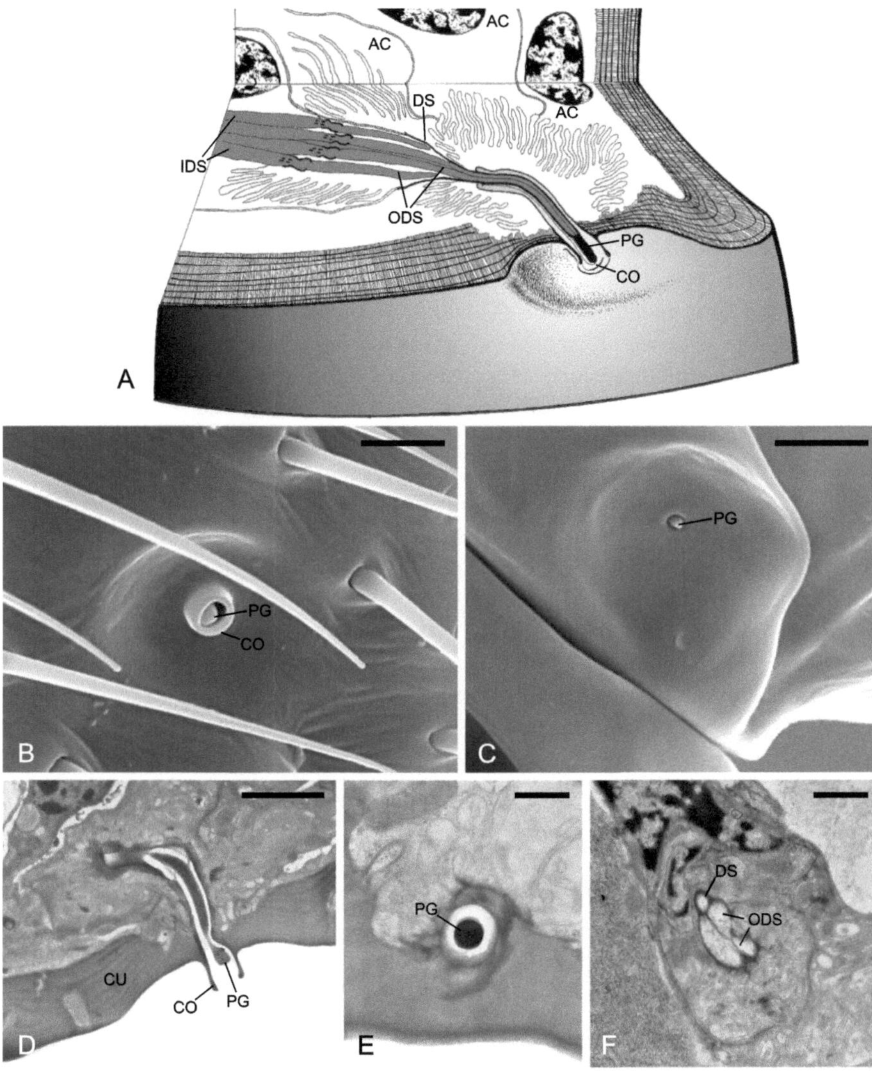

Fig. 3.5 (**a**) Schematic drawing of the coeloconic sensillum on the antennae of *Trissolcus basalis*. (**b, c**) SEM close-up views of the coeloconic sensillum in females of *T. basalis* (**b**) and *Trichogramma brassicae* (**c**). In both cases the cuticular peg (PG) is aporous and almost completely embedded within the antennal wall, from which only the peg tip is exposed; (**d–f**) TEM pictures showing ultrastructural details of the coeloconic sensillum in *T. basalis*; (**d**) londitudinal section of the aporous, non-flexible peg surrounded by a cuticular collar (CO) that extends inside the antennal lumen; (**e**) cross section of the peg; (**f**) cross section at the outer dendritic segments (ODS) level showing the four dendrites innervating the sensillum enclosed by the dendrite sheath (DS). AC – accessory cells, CU – cuticle, IDS – inner dendritic segment. Bars: b–d =2 μm, e=0.5 μm, f=1 μm

campaniform-like structures were observed at the level of the second and fourth flagellomeres and on the ventro-proximal side of the club. These structures are the same described more correctly as coeloconic sensilla in *T. australicum* (Amornsak et al. 1998), *Trichogramma galloi* Zucchi and *Trichogramma pretiosum* Riley (Cônsoli et al. 1999).

Coeloconic sensilla are also known in Scelionidae as belonging to *sensilla styloconica* in *Telenomus reynoldsi* Gordh and Cork (Cave and Gaylor 1987). These sensilla were found on antennomere 4 and 10 in females, and 7 and 9 in males. They appear as tiny pegs inserted in a smooth depression around a collar emerging from the surrounding surface. Although no ultrastructural data were provided, a possible role as thermo-hygroreceptors was hypothesized. To date, the most comprehensive ultrastructural study of a coeloconic sensillum in egg parasitoids is the one by Isidoro (1992) on *T. basalis*. In this species, coeloconic sensilla occur in very low number (two per antenna), consisting of a small, inflexible cuticular peg without pores, set in a pit (Fig. 3.5a–e). They are innervated by four sensory neurons; two of them enter the peg lumen while the other two terminate basally (Fig. 3.4f). Based on the ultrastructural homology with sensilla coeloconica from other species proven to be thermo-hygroreceptors (Altner et al. 1981, Yokohari et al. 1982 Hansson et al. 1996), it is likely that these sensilla have the same function in egg parasitoids.

In Mymaridae, coeloconic sensilla-like structures were reported in *Anagrus atomus* (L.) (Chiappini et al. 2001), occurring only two per antenna (one on the apical club and the other on the second flagellomere), and were defined as *sunken peg sensilla*. Ultrastructural data revealed similar cuticular and cellular features to that of the coeloconic sensillum of *T. basalis*, consistent with a possible thermo-hygrosensitivity.

3.4.3 Uniporous Sensilla

Uniporous sensilla are characterized by the presence of a single, in most cases apical, cuticular pore, which opens onto a cuticular shaft usually appearing as a hair or a peg of various lengths (Fig. 3.2b). The external pore may be simple or be the result of elaborated cuticular projections surrounding or giving rise to the external opening. Furthermore, the pore may either be exposed or plugged by electron-dense material. These sensilla are usually innervated by four to six sensory neurons that run unbranched inside the peg lumen, reaching the apical pore (Zacharuk 1985). The lumen is sometimes divided into two chambers: the first chamber (the dendritic chamber) is occupied by the dendrites enveloped by the dendrite sheath, while the other (the sensillar chamber) sends pore tubules or filaments to the external pore. Usually, uniporous sensilla are bi-modal sensilla, since they combine mechanosensory and chemosensory functions in a single sensory unit. In fact, in most cases one of the sensory neurons does not enter the peg lumen, thus ending at the base of the sensillum where a typical mechanosensory tubular body is formed. Mechanical stimulation is achieved through a flexible socket, similar to what has been reported for mechanosensory hairs. In regards to the chemosensory function, uniporous sensilla are consider "gustatory" units, i.e. chemical signals are perceived through contact with the substrate.

In Scelionidae, putative uniporous sensilla occurring on both male and female antennae were termed "sensilla chetica" in *Te. reynoldsi* (Cave and Gaylor 1987) and *Tr. basalis* (Bin et al. 1989). While in the former species only external features were investigated, in the latter, ultrastructural analysis revealed details of the general scheme of uniporous sensilla, i.e. the presence of an outstanding hair shaft with a single apical pore inserted in a flexible socket and a set of 4+1 sensory neurons of chemosensory and mechanosensory type, respectively (Fig. 3.6a–e). These sensilla were later named "Uniporous Gustatory Sensilla" (UGS) (Isidoro et al. 1996), and different hypotheses on their biological significance were given: (i) courtship behavior and sex recognition in males and (ii) host searching behavior and recognition in females.

In *T. galloi* and *T. pretiosum* a single apical sensillum on the female club called "uniporous sensillum trichoideum" by Cônsoli et al. (1999) [formerly known as *sensillum type e* (Voegelé et al. 1975), *uniporous pitted sensillum* (UPP, Olson and Andow 1993) or *sensillum styloconicum* (Amornsak et al. 1998)], can be referred to UGS, since it is reported as having a single apical pore and it bents towards to the base, therefore allowing stimulation through contact (Fig. 3.6f, g). Sen et al. (2005) discussed a similar hypothesis for this sensillum in *Trichogramma chilonis* (Ishii) females.

Sensilla chaetica type 1 (SC1), 3 (SC3) and 4 (SC4) were described in the mymarids *Anaphes listronoti* Huber and *Anaphes victus* Huber which occur on female antennae on the ventral side of the last two antennomeres forming the club (van Baaren et al. 1999). SC1 is a single sensillum on the antennal tip, while SC3 and SC4 are present in number of 3 and 8, respectively. SC4 possess a peculiar structure at the level of the apical pore, which is positioned sub-apically and can be exposed or hidden by means of a finger-like cuticular projection above it. SC1 and SC3 present the typical apical pore. Despite some differences in number, form of the cuticular shaft and in the structure of the apical pore, these sensilla can be considered as three different types of UGS.

3.4.4 Multiporous Sensilla

Multiporous sensilla are characterized by the presence of numerous, well evident cuticular pores opening on the sensory cuticle (Steinbrecht 1997).The presence of these pores is sometimes hard to evidence when an external examination is made using SEM, while through TEM they appear as distinct pores. Based on the external appearance (i.e. the morphology of the external cuticular apparatus), multiporous sensilla usually occur in the form of hairs (of various lengths), pegs or plates (Fig. 3.2a). The pores can be found on the whole sensory cuticle of the sensillum, or they can open on specific regions. Altner (1977) classified these sensilla as wall-pore sensilla, further divided into single-walled and double-walled, assessing a predominant olfactory function.

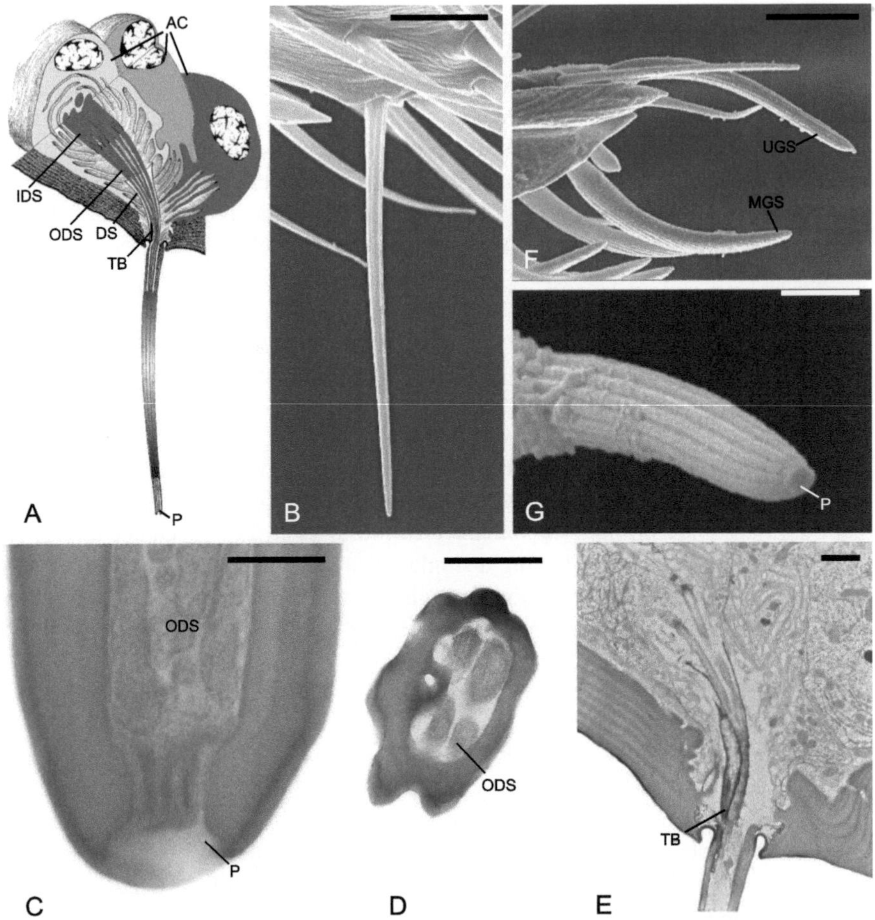

Fig. 3.6 (**a**) Schematic drawing of the uniporous gustatory sensillum (UGS) on *Trissolcus basalis* female antennae; (**b**) SEM micrograph of the UGS. (**c–e**) TEM micrographs showing in (**c**) the apical pore (P), in (**d**) a cross section of the hair shaft with four unbranched outer dendritic segments (ODS), in (**e**) longitudinal section at the socket level with one dendrite ending in a tubular body (TB); (**f–g**) SEM pictures of the apical UGS in female *Trichogramma brassicae*. AC – accessory cells, DS – dendrite sheath, IDS – inner dendritic segment, MGS – multiporous gustatory sensillum. Bars: b–f =5 μm, c=0.25 μm, d=0.5 μm, e–g=1 μm

Moreover, the sensory transduction system in olfactory sensilla can be of two different modalities, depending on the specialization of cuticular structures allowing the transport of the volatile molecules from the external aerial environment to the receptor cells. In single-walled sensilla, the cuticle is perforated by numerous tiny pores. Ultrastructural details of the pores revealed that they consist of a small chamber opening below the external pore, from which several small cuticular channels (pore channels) develop throughout the antennal wall. These pore channels house specialized stimulus-transport structures, known as pore-tubules that reach the sensillar lumen and sometimes establish contact with the olfactory neurons (Locke 1965, Steinbrecht and Müller 1971, Keil 1982). Hydrophobic molecules from the external environment are therefore trapped by "waxy filaments" within the pore tubules and carried inside the sensillar lumen, where olfactory neurons are bathed by the sensillum lymph. Here, the molecules are selectively trapped by odorant binding proteins (OBPs), a class of water-soluble proteins which were discovered in several insect orders (Vogt and Riddiford 1981, Klein 1987, McKenna et al. 1994, Dickens et al. 1995, Tuccini et al. 1996, Paesen and Happ 1995). OBPs, which have been further divided into pheromone binding proteins (PBPs), general odorant binding proteins (GOBPs) and antennal binding proteins (ABPx), are believed to be responsible for the transport of stimulus molecules to the dendritic membrane, where specific receptors are then activated, therefore triggering the response to the olfactory stimulus (Vogt et al. 1990, Ziegelberger 1995).

In double-walled sensilla, the sensillum lymph cavity is divided in an innermost and an outermost space, which arise from the external cuticular finger of the sensilla, and are partly fused together (Hunger and Steinbrecht 1998). The sensory neurons are located exclusively in the innermost space, and volatile molecules reach the outer dendritic segment passing through "spoke channels", i.e. small canals crossing both walls and typically arranged like spokes of a wheel (Altner and Prillinger 1980). In this case, pore-tubules were never observed.

Although no electrophysiological single sensillum recording data are available for egg parasitoids to date, the ultrastructural features of multiporous sensilla deal with the proposed olfactory function. However, some exceptional structures with a possible gustatory function have been found.

3.4.4.1 Single-Walled Sensilla

The sensilla belonging to this group posses a single cuticular wall that can be of the "thin" or "thick" type, depending on the relative thickness of the wall itself (Altner 1977). They are mainly characterized by an inflexible socket, i.e. the socketed base of the sensillum is quite rigid allowig only slight deformation of the base, therefore they never combine chemosensory and mechanosensory functions. These sensilla were found in several egg parasitoids, and different morphological classes can be found in egg-parasitoid antennae.

Sensilla Trichoidea

These are usually long sensilla, ending in a fine, sharp tip. They have an inflexible socket, and are innervated by a relatively low number of sensory neurons. These sensilla usually possess a thick wall pierced by evenly distributed minute pores. Sensilla trichoidea were reported in several groups of egg parasitoids. In Chalcidoidea, the ultrastructure of this sensillum was studied in Trichogrammatidae and Eulophidae. *Trichogramma* has been extensively studied in regards to the morphology of the antennal sensory structures because of its great importance as a biocontrol agent. However, all available data derived from SEM external observations, which led to misinterpretations or ambiguous attempts to force the diverse sensillar type into newly created *ad hoc* morphological classes, giving as a result more confusion (see Amornsak et al. 1998). Here, at least in regards to the function of some antennal sensory structures, we will try to establish some order based on ultrastructural data.

Sensilla trichoidea have been described in *Trichogramma* males and females by several authors (Voegelé et al. 1975, Olson and Andow 1993, Amornsak et al. 1998), and have been classified by Amornsak et al. (1998) in five different classes (i.e. TS1, TS2 and so on), yet only on the basis of their external morphology. Among this proliferation of different sensilla classes, it is advisable to be more careful, until further ultrastructural and electrophysiological data becomes available. Multiporous sensilla trichoidea, dorsally located, were reported in females of *T. galloi* and *T. pretiosum* (Cônsoli et al. 1999), for which an olfactory function was proposed. Preliminary TEM investigation on *Trichogramma brassicae* Bezdenko females revealed the presence of pores opening on a "thick" sensory wall, and the sensillum lumen housing several dendritic projections (Isidoro et al. unpublished data) (Fig. 3.7a–c). Therefore, these sensilla could be involved in the perception of volatiles released by the hosts or by the plants exploited by the host.

Multiporous non-socketed hairs of the short and long type were described on female and male antennae, respectively, of the eulophid *Tetrastichus hagenowii* (Ratzeburg) (Barlin et al. 1981). The presence of outer dendritic segments within the cuticular shaft (15–16 in the female, 4 in the male), an inflexible socket and

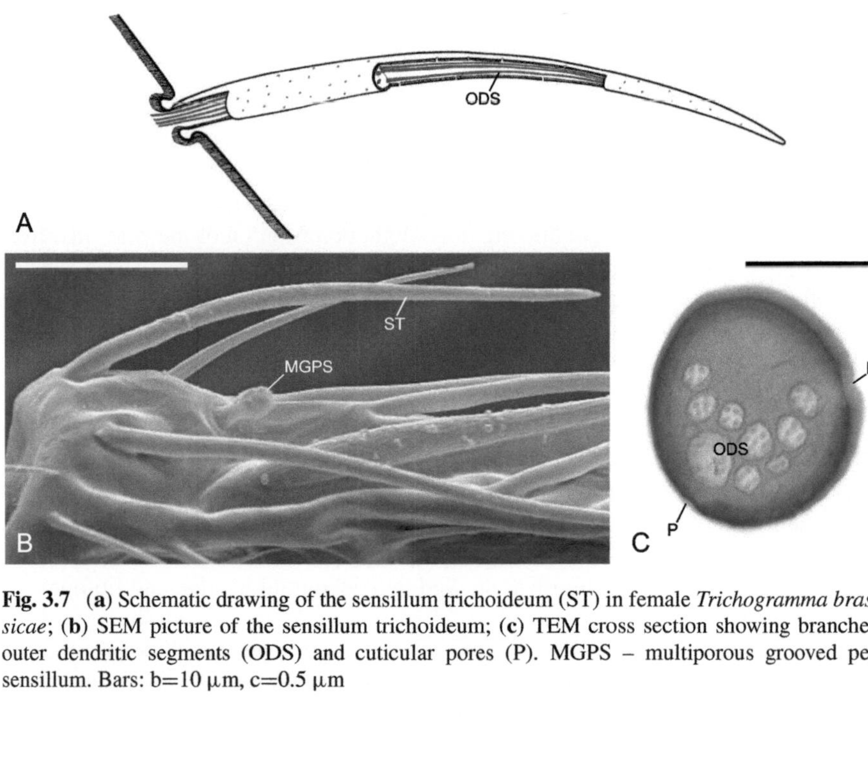

Fig. 3.7 (**a**) Schematic drawing of the sensillum trichoideum (ST) in female *Trichogramma brassicae*; (**b**) SEM picture of the sensillum trichoideum; (**c**) TEM cross section showing branched outer dendritic segments (ODS) and cuticular pores (P). MGPS – multiporous grooved peg sensillum. Bars: b=10 μm, c=0.5 μm

numerous pores opening on the thick wall strongly suggest an active involvement in volatile perception.

Multiporous Plate Sensilla (MPS)

These were mainly described in Hymenoptera (Navasero and Elzen 1991, Basibuyuk and Quicke 1998) and Coleoptera (Allsopp 1990). They are characterized by a thin cuticular structure, often in the form of a circular or oval disk, but also as long plaques, flattened at the level of the antennal wall or slightly arising from it. The cuticle is perforated by numerous tiny pores concentrated in a part of the plate or evenly distributed onto the sensillum cuticular surface. The number of associated sensory neurons is variable, although in some cases many neurons have been found projecting dendritic branches below the sensory cuticle. In some cases, up to six accessory cells can be associated per sensillum. For these sensilla, an olfactory function has been proved (Lacher 1964).

MPS were comparatively described in Chalcidoidea by Barlin and Vinson (1981), particularly in egg parasitoids of Trichogrammatidae, Eulophidae and Torymidae families. In these groups, MPS occurred on both male and female antennae according to two different morphological types: MPS1 possessed a relatively thin sensillum wall where numerous pores opened in both male and female, MPS2 have a thicker wall with fewer pores and occurred only in females. Both types are elongated, non-socketed plates, longitudinally oriented on the antenna, with the distal part free from the antennal wall. These sensilla are innervated by up to 50 sensory neurons [*T. hagenowii* and *Torymus warren* (Coquerel)] (Barlin and Vinson 1981, Barlin et al. 1981). It is remarkable that both MPS1 and MPS2 occurred in females, while males only exhibited MPS1. This difference was related to a different tuning of MPS1 and 2 for specific substances (MPS2 is more generalist than MPS1), although this hypothesis would need to be verified by electrophysiological data.

In Trichogrammatidae, MPS have been reported in females of *T. evanescens* (*sensilla type h*) (Voegelé et al. 1975) occurring in number of five, with a possible olfactory role. This was later confirmed by Olson and Andow (1993) in *T. nubilale*, where the same number of MPS were counted on the female antennae, without differences in their morphological features, either in terms of size or pore distribution (about 28 per μm^2). In a more recent study by Amornsak et al. (1998), MPS sensilla were renamed as "placoid sensilla" (PS) in *T. australicum* and on the basis of external SEM features, they reported two different types of PS (PS1 and PS2), using as discriminatory feature differences in the pore density. Cônsoli et al. (1999) also described MPS on male and female antennae of *T. galloi* and *T. pretiosum*. Since the classification proposed by Amornsak et al. (1998) is based only on external observations, therefore making difficult to use pore-related features to be used without the support of TEM investigations, we propose that these sensilla be named "Multiporous Olfactory Plate Sensilla" (MOPS), combining both morphological and functional features more comprehensively (Fig. 3.8a–c).

In the mymarids *A. listronoti* and *A. victus*, *placoid sensilla* were described in all antennomeres of the flagellum in both males and females (van Baaren et al. 1999),

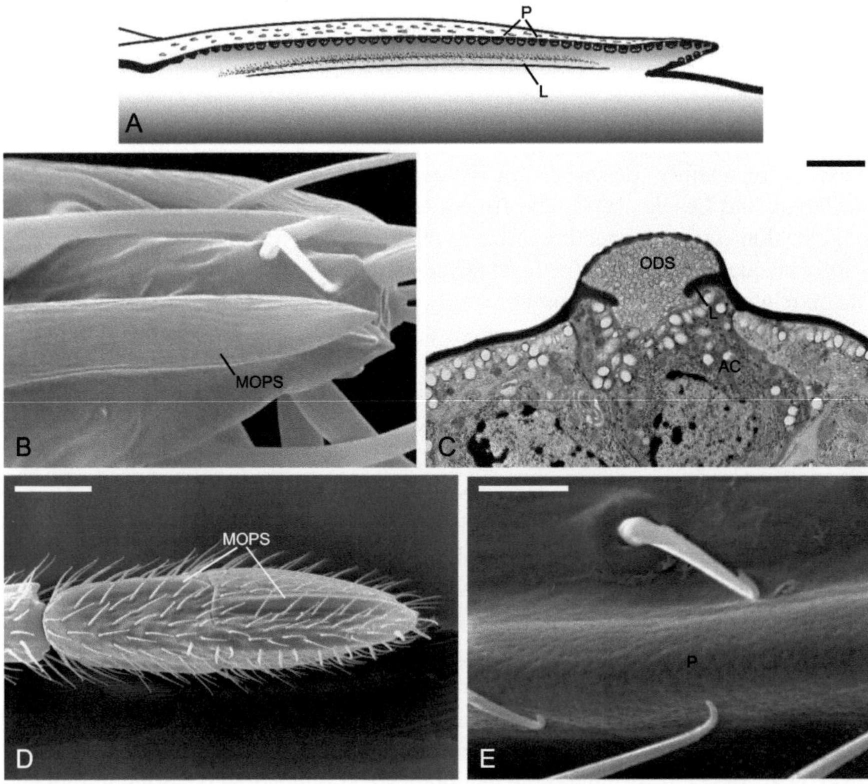

Fig. 3.8 (**a**) Schematic reconstruction of the multiporous olfactory plate sensillum (MOPS) cuticular parts on female *Trichogramma brassicae*, based on TEM longitudinal section; (**b**) SEM picture of the MOPS distal part; (**c**) TEM cross section of the MOPS showing numerous branches of outer dendritic segments (ODS); (**d–e**) SEM micrographs of apical antennomere in female *Anaphes victus*, latero-ventral view. AC – accessory cells, L – ledge, P – pores. Bars: b=2 μm, c=1 μm, d=20 μm, e=2 μm

with different distribution between males (where there are five sensilla per antennomere) and female (2 sensilla per antennomere, except for the club counting 4) (Fig. 3.8d, e).

MPS have never been observed in Platygastroidea. However, different types of multiporous olfactory sensilla were described which are closely related to Chalcidoidea MPS either in their ultrastructural features as in functional significance.

In the scelionid *Tr. basalis*, *sickle-shaped* sensilla occur on both male and female antennae (Bin et al. 1989). They show a remarkable sexual dimorphism as total number, being as much as 170 on each male antenna. On the other hand, females exhibit only about 40 per antenna. These sensilla possess a rigid shaft (inflexible socket), about 15 μm in length, running almost parallel with the antennal wall and slightly diverging at the tip. In addition, the shaft becomes sharp at the tip (Fig. 3.9a, b). The

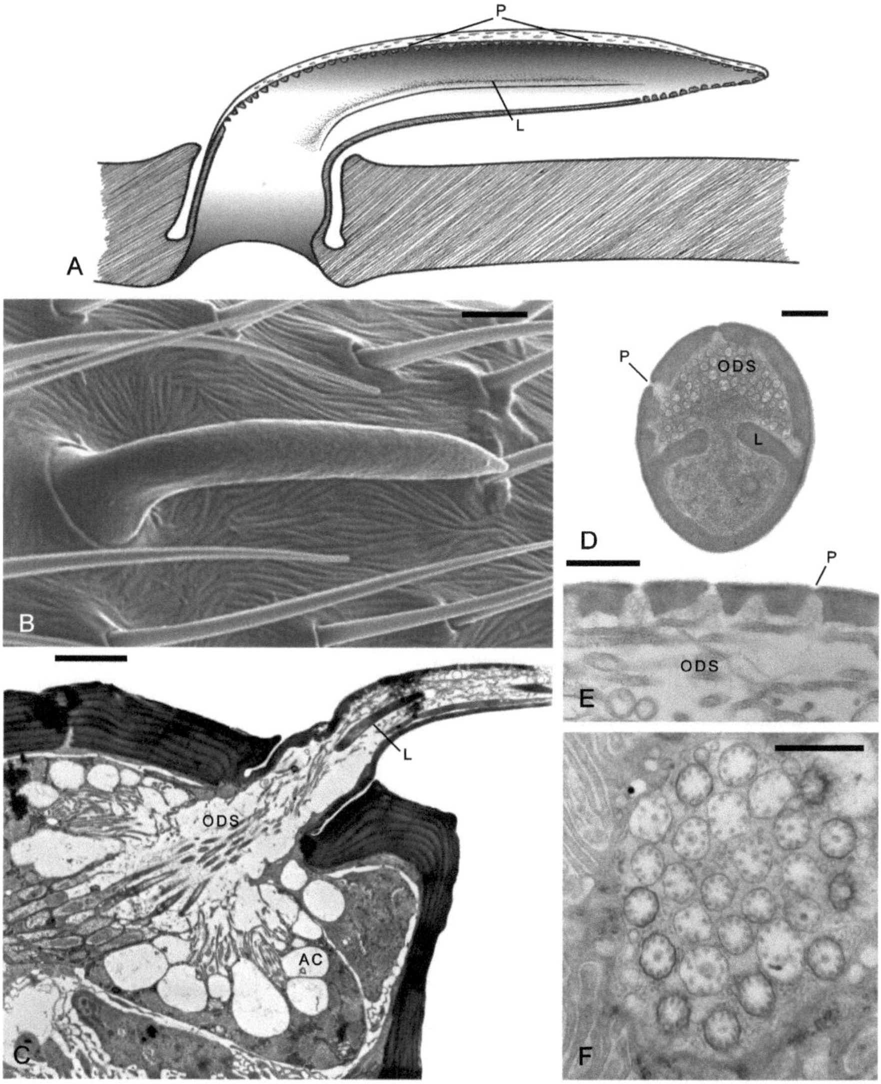

Fig. 3.9 (**a**) Schematic reconstruction of the multiporous olfactory sickle-shaped sensillum (MOSS) cuticular shaft on female *Trissolcus basalis*, based on TEM longitudinal section; (**b**) SEM picture of the MOSS; (**c**) TEM longitudinal section of the MOSS at the socket level; (**d, e**) TEM cross and longitudinal sections of the MOSS hair shaft; (**f**) Cross section of the sensory neurons at the ciliary constrictions level. AC – accessory cells, L – ledge, P – pores, ODS – outer dendritic segments. Bars: b=2 μm, c=5 μm, d–e=0.5 μm, f=1 μm

pores are typically absent on the side facing the antennal wall, but they are present elsewhere. In cross sections, the sensillar lumen presents two cuticular reinforcements, which divide the lumen into two chambers, with one of them occupied by the outer dendritic segments (Fig. 3.9c–e). About 28 sensory neurons were observed per each sensillum (Fig. 3.9f). Similar sensilla where described in ants (Fresneau 1979, Hashimoto 1990) and in another scelionid, *Te. reynoldsi* (Cave and Gaylor 1987) with the term "sensilla trichodea curvata", with a proposed olfactory function. We propose for these sensilla the name "Multiporous Olfactory Sickle-shaped Sensilla" (MOSS).

Multiporous Gustatory Sensilla

Although the presence of multiporous areas on the sensillum wall strongly supports the olfactory function, there are some exceptions to this generally accepted scheme in egg parasitoids. Multiporos gustatory sensilla (MGS) were first described by Bin (1981) in Scelionidae as *plate sensilla*, and were used to define a claval formula, a feature of taxonomic value. Such sensilla only occur on the ventral side of female antenna in both Scelionidae and Platygastridae (Bin et al. 1989, Isidoro et al. 1996, 2001) (Fig. 3.10a–d). For these structures, the hypothesis of a gustatory function is strongly supported on the following basis: (i) MGS are facing the ventral side of the antenna, i.e. the one contacting the substrate during antennation, differently from olfactory sensilla that are always facing the antennal dorsal and/or lateral side; (ii) the Scelionidae MGS possess a cuticular shaft in the form of a truncate cone. The apical part is flattened and crossed by transversal ridges in different numbers (seven to eight in *Tr. basalis*, three in *Telenomus busseolae* Gahan) (Fig. 3.10c). The apical part is pierced by numerous tiny pores, which can be observed only through TEM, since the pores are hidden below cuticular "liftable lobes" (Bin et al. 1989, Isidoro et al. 1996, 2001) (Fig. 3.10e). This sensory cuticle is reached by the unbranched outer dendritic segments of an impressive number of sensory neurons, ranging from about 200 in *Te. busseolae* to more than 400 in *Tr. basalis* (Bin et al. 1989, Isidoro et al. 2001) (Fig. 3.10f). In addition, Scelionidae MGS are equipped with a pair of unusual accessory glands (see Section 3.5.3 for structural details), which release their secretion at the sensillum base. The secretion may play a role during host recognition as a dissolving medium for the contact kairomone (the glue used by the host to attach the eggs to the substrate) (Bin et al. 1989, 1993, Isidoro et al. 1996, 2001). There are two types of cuticular structures involved in the protection of the multiporous area: liftable lobes and spherical projections. In Scelionidae, the first case seems to predominate although an intra-sensillar difference has been found (the apical antennomere has spherical projections while the others have liftable lobes). In *Mantibaria anomala* Kirby, all the MGS show only spherical projections (Bin et al. unpublished). Whether this could be due to a difference in the host range or to the perception of specific semiochemicals still needs an investigation.

In *Trichogramma* females, the ventral area of the antennae is characterized by the presence of peculiar sensilla, which are not found elsewhere. These sensilla, first described by Voegelé et al. (1975) as *sensilla type i*, make up a functional area called "sole chercheuse" (searching sole) because of its importance during host recognition (Cônsoli et al. 1999). They appear as "sickle-shaped" or recurved sensilla trichoidea arranged in almost parallel rows on the distal half of the apical antennomere (Fig. 3.11a–c). Among *Trichogramma*, the number differs, ranging from

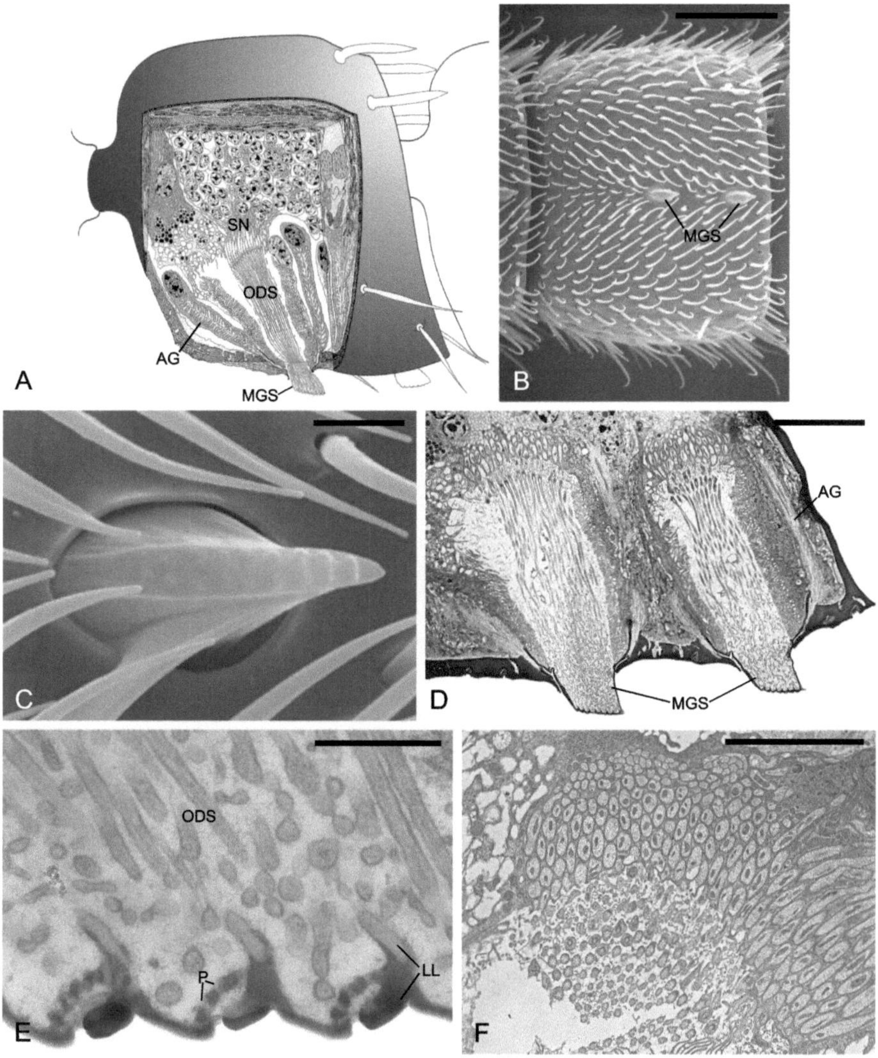

Fig. 3.10 (**a**) Tridimensional drawing of the multiporous gustatory sensillum (MGS) on *Trissolcus basalis* female antennae; (**b**) SEM picture ventral view of the 9th female antennomere showing two MGS; (**c**) Detail of the previous micrograph showing the ventral multiporous area; (**d, e**) TEM longitudinal section of MGS showing in (**d**) the whole sensilla, in (**e**) the apical part with pores (P) covered by cuticular liftable lobes (LL) and unbranched outer dendritic segments (ODS); (**f**) Oblique TEM section just below the ciliary constrictions level showing the impressive number of sensory neurons (SN). AG –accessory glands, CU – cuticle, IDS – inner dendritic segment. Bars: b=20 μm, c=2 μm, d, f=5 μm, e=0.5 μm

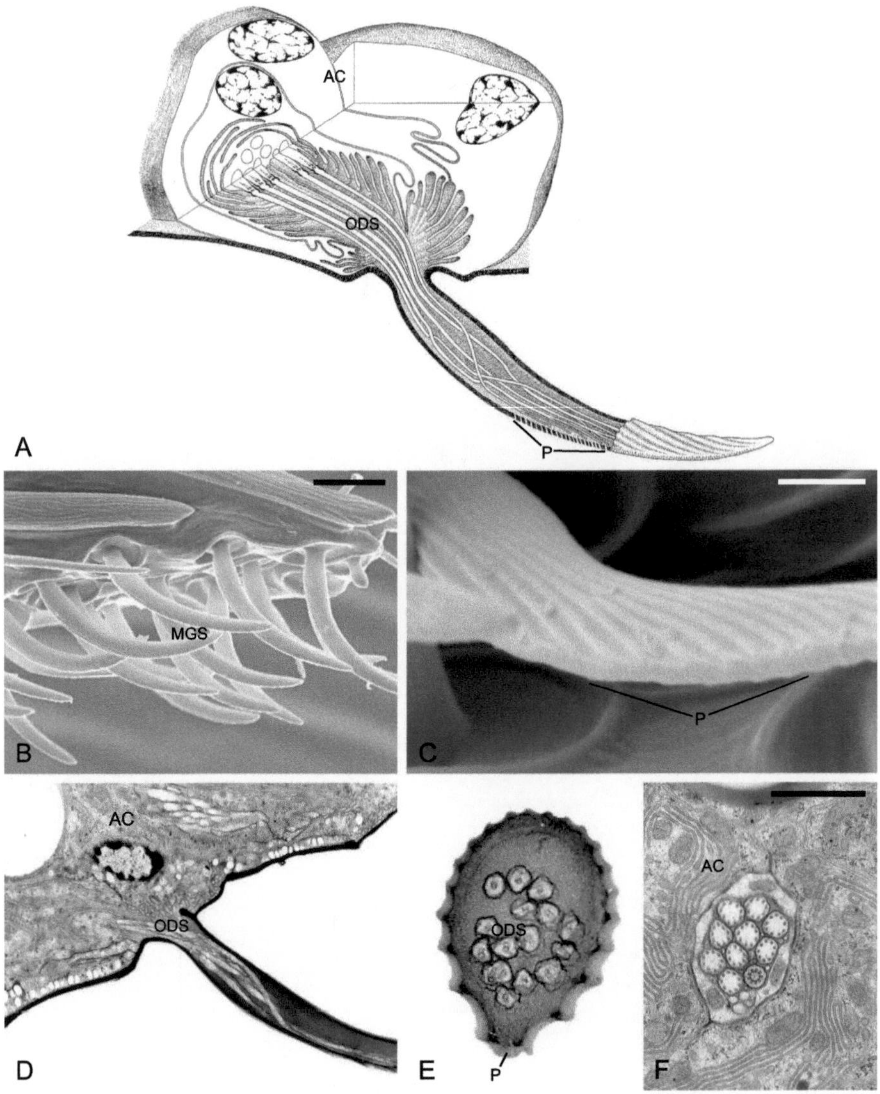

Fig. 3.11 (**a**) Tridimensional drawing of the multiporous gustatory sensillum (MGS) on *Trichogramma brassicae* female antennae; (**b**) SEM picture, ventro-lateral view of the apical antennomere showing several MGS; (**c**) Detail of an MGS showing the ventral margin with pores (P); (**d**) TEM longitudinal section of MGS at the socket level; (**e**) Cross section of the cuticular shaft showing ventral pores and branched outer dendritic segments (ODS); (**f**) TEM cross section at the ciliary constrictions level. AC – accessory cells. Bars: b=2 μm, c=1 μm, d=5 μm, e=0.5 μm, f=1 μm

20 in *T. brassicae* (Isidoro et al. 1996) to nearly 35 in *T. australicum* (Amornsak et al. 1998). These sensilla are laterally flattened, ending with a sharp tip and they show numerous grooves running parallel on both sides up to the ventral margin, where they converge. For these sensilla, different functional hypothesis have been proposed; from bi-modal mechano-chemosensory units (Voegelé et al. 1975) to olfactory organs (Sen et al. 2005). TEM investigations on *T. brassicae* (Isidoro et al. 1996) revealed the following ultrastructural features: (i) a multiporous area located along the ventral margin (Fig. 3.11e); (ii) 10 sensory neurons innervating each sensillum (Fig. 3.11f); and the absence of a tubular body (Fig. 3.11d).

These data are consistent with a possible gustatory function of these sensilla, therefore making them a different case of multiporous gustatory sensilla (MGS).

In Mymaridae, sensilla chaetica 2 (SC2), occurring in number of two apically on the ventral margin of the female club, could be ascribed to MGS because of the apical multiporous area (van Baaren et al. 1999). TEM data are needed to verify this hypothesis, though.

3.4.4.2 Double-Walled Sensilla

Double-walled sensilla occur as short pegs inserted on the antennal wall with an inflexible socket. The pegs have been found to outstand from the surrounding cuticle, being inserted into a shallow depression, deeply sunken, or completely embedded within the antennal wall. Therefore, they could communicate with the environment through narrow openings. They are characterized by a double, concentrically-arranged cuticular wall. In addition, they are defined by two sensillar spaces: an innermost cavity, containing the dendritic processes, and an outermost cavity that can be filled with vesicles or other secretory products from the accessory cells. Externally, the peg is grooved in most of its length (sometimes only on the distal half). However, numerous pores open at the bottom of the grooves, where pore tubules were never observed. Given these features, we will refer to these sensilla as *multiporous grooved peg sensilla* (MGPS). Functionally, MGPS were identified as olfactory sensilla (Altner et al. 1977, Altner and Prillinger 1980, Pophof 1997, Diehl et al. 2003, Pophof et al. 2005), bi-modal chemo-thermoreceptors (Altner et al. 1981, Altner and Loftus 1985) and thermo-hygroreceptors (Waldow 1970).

The first study showing ultrastructural details of MGPS was that by Barlin et al. (1981) on *Te. hagenowii*. In this species, it was reported the presence of a *multiporous peg* with bulbous distal end and scalloped wall, occurring singly on the apical antennomere and in number of 15 on the other flagellomeres.

In Scelionidae, on both male and female *Tr. basalis* there are six *grooved pegs* per antenna (Bin et al. 1989). Externally, they are typical MGPS with a smooth proximal peg wall and a distal grooved surface (Fig. 3.12a, b). Internally, the double-walled organization is evident. Furthermore, there are four sensory neurons associated and large sheath cells actively producing electronlucid secretory vesicles (Fig. 3.12c–e). Based exclusively on its ultrastructural features, both olfactory and gustatory functions were proposed for this sensillum. MGPS was previously reported for male

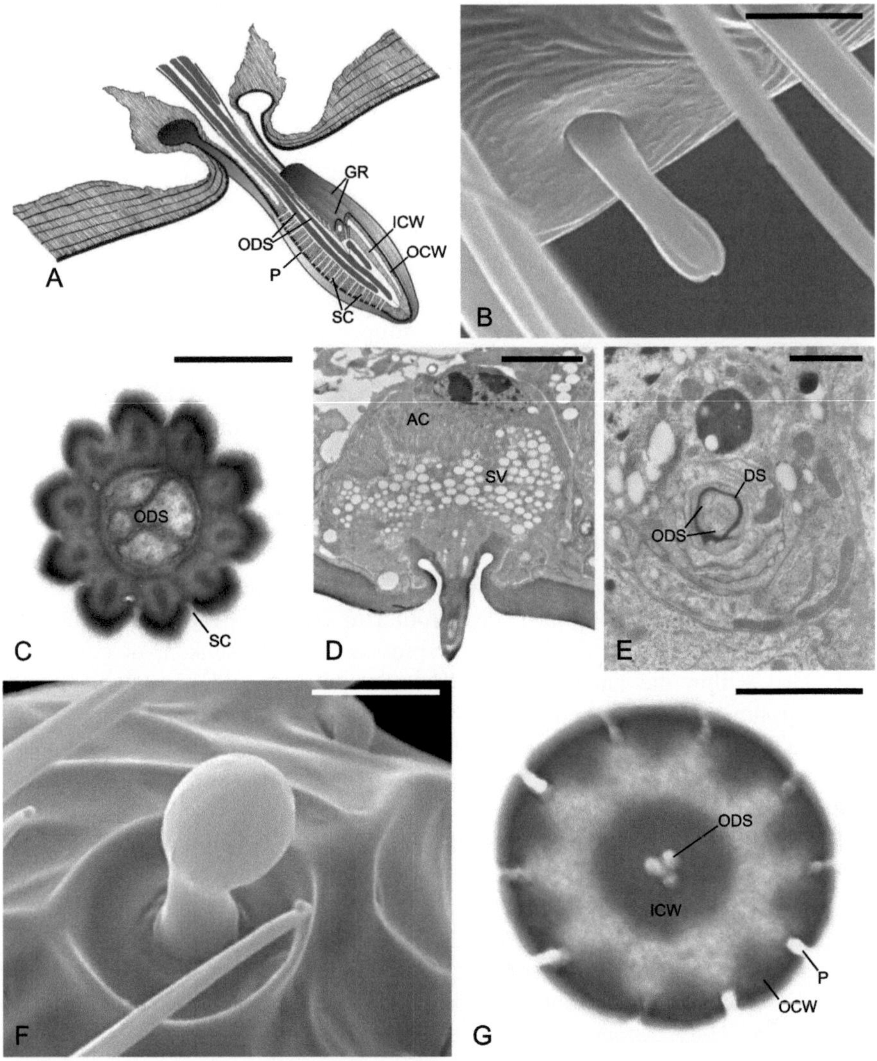

Fig. 3.12 (**a**) Schematic tridimensional drawing of the multiporous grooved peg sensillum (MGPS) on *Trissolcus basalis* female antennae; (**b**) SEM close-up view of the MGPS; (**c–e**) TEM micrographs showing in (**c**) cross section of the MGPS double-walled peg with outer dendritic segments (ODS) and spoke-channels (SC), in (**d**) the peg base with the accessory cells (AC) producing secretory vesicles (SV), and (**e**) the four outer dendritic segments enclosed by the dendrite sheath (DS); (**f**) SEM high magnification picture of the MGPS in female *Trichogramma brassicae*, and (**g**) TEM cross section of the peg showing the inner cuticular wall (ICW), the outer cuticular wall (OCW), pores (P) and three outer dendritic segments. Bars: b=2 μm, c=0.5 μm, d=2 μm, e=1 μm, f=2 μm, g=0.5 μm

and female *Te. reynoldsi* as "multiporous grooved peg" (Cave and Gaylor 1987), yet only providing SEM details and no functional hypothesis.

In *Trichogramma*, MGPS have been described as *sensilla ampullacea* (Voegelé et al. 1975), *multiporous grooved sensilla basiconica C* (Olson and Andow 1993), *basiconic capitate peg sensilla 1 and 2* (Amornsak et al. 1998) and *multiporous grooved basiconica sensilla* (Cônsoli et al. 1999). These sensilla, which occur in number of seven per antenna in females and males, are short, stalked, bulbous pegs with longitudinal grooves. They can be referred to as MGPS, as revealed by Isidoro et al.'s TEM preliminary work (Fig. 3.12f, g). These sensilla could be involved in volatile perception, even if a possible bi-modal function as chemo- and thermoreceptors could not be ruled out.

In Mymaridae, MGPS with a possible olfactory function `were found on both male and female antennae of *A. victus* and *A. listronoti* (van Baaren et al. 1999): the authors refer to them as *basiconic sensilla*.

3.5 Antennal Glands in Egg Parasitoids

Insect communication is often mediated by stimuli of chemical nature, generally defined as semiochemicals. These stimuli are made up of different classes of biologically active compounds, either released by the insects themselves or by other organisms (plants, hosts), acting at intraspecific or interspecific level. Insects release chemicals through specialized cuticular areas due to the presence of cuticular pores or by a very thin cuticle, where produced substances are released through. These secretory products are produced by specialised secretory cells, defined as glands, either isolated or organized in glandular complexes. These glands have been the subject of extensive studies, because of the extraordinary biological relevance of their products. Insect exocrine glands were reported to occur on almost every part of the body, which is particularly true in social hymenopterans (Billen 1991).

Nevertheless, the presence of glands on some body parts, such as the antennae, was neglected for a long time, probably because the dogma that the antennae are mainly sensory appendages (therefore sensilla are the most abundant structures), and antennal release sites are sometimes not so evident or of a small size.

Despite early descriptions of antennal glands in myrmecophilous beetles (Wasmann 1903, Mou 1938, Cammaerts 1974), only late in the 1980s morphological studies on the antennae of parasitic wasps highlighted the presence of specialised antennal glands. This discovery was brought on mostly by studies on the male antennae, functionally related with mating behavior (Dahms 1984, Bin and Vinson 1986). Later, antennal glands were also described from several families of Hymenoptera, including Terebrantia (Bin et al. 1999b, Isidoro and Bin 1995, Isidoro et al. 1996, 1999, Battaglia et al. 2002) and Aculeata (Isidoro et al. 2000, Romani et al. 2002, 2003, 2005, 2006), as well as from other insect orders (Bartlet et al. 1994, Weis et al. 1999, Giglio et al. 2005), with different biological significance. In some groups of egg parasitoids, antennal glands can be associated with both males and females specialized antennomeres, as well as closely associated with peculiar sensory structures. In the following section will be reviewed different cases.

3.5.1 General Features and Classification of Glands

Insect epidermal glands, i.e. secretory cells derived from modified epidermal cells, were reported in many different body parts, leading to classifications based on the topography of the glands (cephalic, sternal, abdominal, etc.). However, these classification methods were not exhaustive because of differences that occurred for the same glands in different insect orders. Other classifications, dealing with the biological significance of the released secretions were not comprehensive, since in different castes of social insects the same gland can release different compounds.

At present, the most generally accepted classification for insect epidermal glands takes into account the relationship between secretory cells and the cuticle that the secretion must cross in order to be released. Noirot and Quennedey (1974) and Quennedey (1998) classified insect epidermal glands into three different classes as follows (Fig. 3.13):

(a) Class 1 – the gland cells are directly in contact with the cuticle. The produced secretion is released after having passed the cuticular layer, where minute pore canals are present;

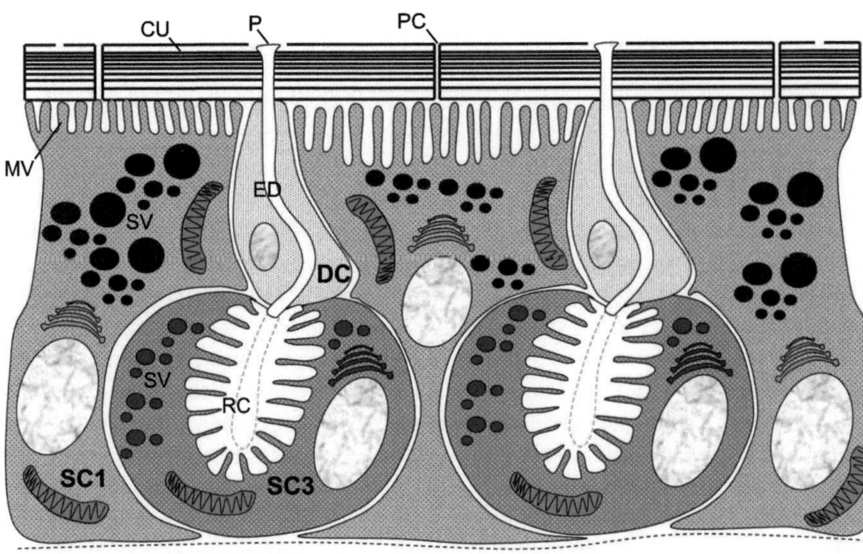

Fig. 3.13 Schematic drawing of secretory cells Class 1 (SC1) and Class 3 (SC3), according to Noirot & Quennedey (1974). CU – cuticle, DC – duct cell, ED – evacuating duct, MV – microvilli, P – pores, PC – pore canals, RC – receiving canal, SV – secretory vesicles. Modified from Quennedey (1998)

(b) Class 2 – secretory cells are not directly in contact with the cuticle, discharging the secretion to the surrounding Class 1 cells, that are in charge of releasing the secretion;
(c) Class 3 – characterised by more than one cell-forming secretory unit connected to external cuticular pores by means of specialized conducting canals.

Within the Hymenoptera, only Class 1 and Class 3 secretory cells occur on different body regions (Billen and Morgan 1998). In egg parasitoids, both classes of secretory cells are present and differently represented between the two sexes. Here we will discuss male and female antennal glands separately, referring to the studies where a detailed description of the presence of glands was provided.

3.5.2 Male Antennal Glands

Bin and Vinson (1986) were the first to describe male antennal glands in egg parasitoids. They reported the occurrence of a secretory epithelium associated with the sexually dimorphic fifth antennomere of *Tr. basalis* males. The ventral side of the antennomere is smooth and presents an outstanding, basiconic process with 8–10 apical openings (Fig. 3.14c–e). Ultrastructural investigations revealed the presence of a glandular epithelium made up of bicellular secretory units, therefore belonging to the Class 3 (Fig. 3.14f). The secretion could be made of a proeinaceous matrix (which is removed from the pores after neutrase enzymatic treatment) embedding other unknown, biologically active compounds. Furthermore, this secretion was hypothesized to be involved in the courtship behavior (Bin and Vinson 1986, Isidoro et al. 1996).

A second type of antennal glands was described in this same species (Bin et al. 1989, Isidoro et al. 1996). This gland occurs on the dorsal side of the antennae, from the 6th to the 11th antennomere, lying underneath a cuticular depression (Fig. 3.14a, b). Each glandular complex is made by about 20 secretory cells belonging to the Class 1, and is less developed in males than in females. Cytological features (abundance of mitochondria and ribosomes and the highly microvillated apical region of the cells) are consistent with a strong secretory activity, possibly involving a biological significance during the elaborated mating behavior. A sensillum campaniform has always been found associated with these glands (see Section 3.4.2). Besides Scelionidae, other descriptions of male antennal glands in egg parasitoids are lacking. However, the following cases in non egg-parasitoid species belonging to families where there are also egg parasitoids are consistent with a more widespread presence of these structures: in the eulophid *Melittobia australica* Girault, Dahms (1984) reported a highly modified scape most likely bearing a glandular complex; in the platygastrid *Amitus spiniferus* (Brethes), a glandular complex, that is made up of Class 3 secretory cells, is found in the modified 5th antennomere (Isidoro and Bin 1995); in the encyrtids *Leptomastix dactylopii* (Howard), *Asitus phragmitis* (Ferrière) and *Rhopus meridionalis* (Ferrière), the modified and sexually dimorphic apical (in the first two species) and sub-apical (the last species)

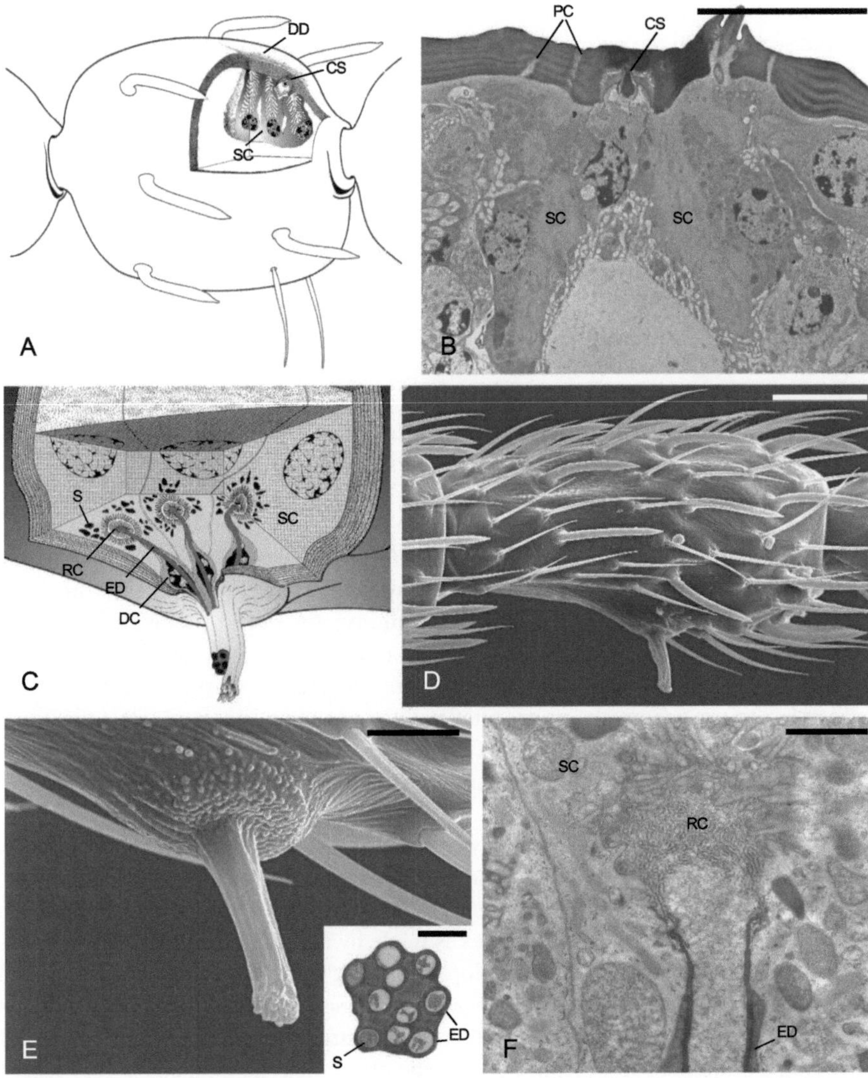

Fig. 3.14 (**a**) Tridimensional schematic drawing of *Trissolcus basalis* male antennomere 9 showing, in the removed part, the male dorsal gland underneath the dorsal depression (DD), pore canals (PC) and a campaniform sensillum (CS); (**b**) TEM cross section of the male dorsal gland with campaniform sensillum; (**c**) Tridimensional schematic drawing of *T. basalis* male ventral gland; (**d, e**) SEM pictures of the 5th antennomere with the cuticular peg; (**f**) Longitudinal section of a secretory cell (SC) of male ventral gland. The secretion (S) is released through evacuating ducts (ED), which form a bundle within the peg (inset in **e**). DC – duct cell, RC – receiving canal. Bars: b=5 μm, d=20 μm, e=5 μm, f=1 μm, inset=2 μm

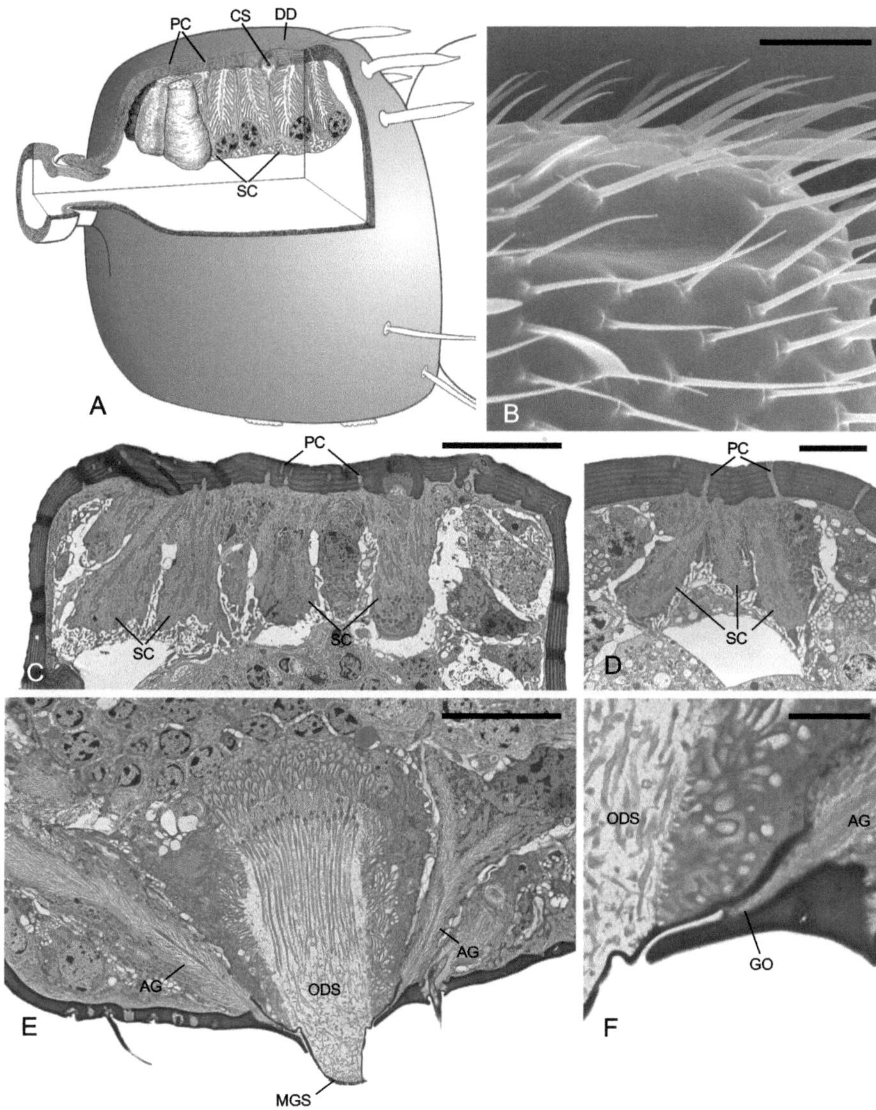

Fig. 3.15 (**a**) Tridimensional schematic drawing of *Trissolcus basalis* female antennomere 9 showing, in the removed part, the female dorsal gland underneath the dorsal depression (DD), pore canals (PC) and a campaniform sensillum (CS); (**b**) SEM detail of the dorsal depression; (**c, d**) TEM micrographs showing, in (**c**) a longitudinal section of the female dorsal gland, in (**d**) a cross section of the same; (**e**) TEM longitudinal section of female *T. basalis* apical antennomere through the multiporous gustatory sensillum (MGS) and the two accessory glands (AG); (**f**) Close-up view of the accessory gland outlet (GO) at the MGS socket level. ODS – outer dendritic segments, SC – secretory cells. Bars: b, c=10 μm, d, e=5 μm, f=0.5 μm

antennomere bears Class 1 glands associated with variously shaped (scale-like or peg-like) release structures (Guerrieri et al. 2001); and in the aphelinid *Aphytis melinus* DeBach the apical antennomere bears a small area which is the release site of Class 1 glands (Romani et al. 1999).

3.5.3 Female Antennal Glands

Although female antennal glands have been reported in several families of Hymenoptera so far (Isidoro et al. 2000, Kaltenpoth et al. 2005, Romani et al. 2006), only a few cases have been reported in egg parasitoids, mostly in Scelionidae. In *Tr. basalis*, females are equipped with a set of dorsal glands very similar (although much more developed) to those previously described in males (Fig. 3.15a–d). These glands occur at the level of cuticular depressions in various antennomeres (from A7 to A11), and are associated with a campaniform sensillum (Bin et al. 1989). A similar situation was reported in *Trissolcus simoni* (Mayr), where the dorsal depressions extend from the 4th to the 11th antennomere and appear like rounded cavities (Isidoro et al. 1996).

The second type of antennal glands found on female egg parasitoids are those associated with the multiporous gustatory sensilla (MGS, see Section 3.4.4) in *Tr. basalis* (Bin et al. 1989). In this case, two to three secretory cells Class 1 are strongly associated with each sensillum, releasing their secretion right at the socket level so that the released products partially cover the sensillum tip (Fig. 3.15e, f). Ultrastructural investigations revealed electrondense secretory vesicles near a large microvillate space, through which the secretion reaches the gland outlet. The secretion released by these glands was hypothesized to be involved in host recognition by dissolving/carrying the contact kairomone present on the eggs, therefore taking part in the stimulus-transporting mechanism (Bin et al. 1989, Isidoro et al. 1996). Within the Scelionidae, glands associated with MGS were described in *Te. busseolae*, while they were not found in the platygastrid *Amitus spiniferus* (Isidoro et al. 2001).

3.6 Concluding Remarks

Antennal structures in egg parasitoids are of extreme importance for these organisms, due to their involvement in different aspects of the reproductive behavior. Due to their peculiar adaptations to the parasitic lifestyle, egg parasitoids have evolved unique sensory structures to exploit cues from the host. This is particularly true in female egg parasitoids, which have to face the problem of host location in the environment. Arthropod eggs that may effectively be utilized by female egg parasitoids are small (in terms of biomass). In addition, chemical cues released by the eggs themselves are available in small amounts and for a relatively short time. Given that, egg parasitoid females have to be extremely efficient in the use of these signals from the host. The evolution of highly specialized sensory structures (i.e. multiporous gustatory sensilla described in Scelionidae and Trichogrammatidae) seems to deal with this need. Scelionidae MGS are equipped with hundreds of sensory neurons, a peculiar feature in the context of insect sensilla. Moreover, these sensilla are unique, due to the presence of associated accessory glands.

Besides the few studies listed in this chapter, egg parasitoids antennal structures have received little attention so far, unless we consider the numerous SEM investigations carried out for taxonomic purposes. Furthermore, if we exclude the studies where only SEM were used, data on sensilla ultrastructure based on TEM could be counted on the fingers of one hand. This fact led to misinterpretation of the sensilla functional aspects, as well as the use of new names to define the same sensilla. In *Trichogramma*, one of the most important and studied families of egg parasitoids, a comprehensive ultrastructural study on the antennal sensilla is still lacking. The implementation of functional investigations based on sensory physiology technique is far from its application in the functional properties study of egg parasitoid antennal sensilla. TEM investigations could also be necessary to reveal the functional features of yet unknown antennal structures. The discovery of antennal glands is a good example in this context. So far, considering only egg parasitoids, antennal glands were described in species belonging to the Platygastroidea superfamily, although we believe that this biological feature could be much more widespread.

Future developments of the research could be focused on the followings: (i) ultrastructural characterization of antennal sensory structures, especially in those groups where no or few data are available; (ii) further research on the the presence of antennal glands (either on male or female) in egg parasitoid families other than Scelionidae; and (iii) development of sensory physiology set-up specifically designed to record electric activity from very small sensilla.

It is certain that the answers to these points will be the bases for better understanding many basic biological aspects of host and sex selection processes in egg parasitoids. These data could be of great help to enhance the use of egg parasitoids in integrated pest management strategies in agroecosystems.

Acknowledgements We would like to thank the CUME (University Centre for Electron Microscopy), Perugia University, where all pictures presented on this chapter were made. Our lab researches have been financially supported by the Italian MiUR through PRIN projects, and the CRUI (Vigoni projects).

References

Allsopp PG (1990) Sexual dimorphism in the adult antennae of *Antitrogus parvulus* Britton and *Lepidiota negatoria* Blackburn (Coleoptera: Scarabaeidae: Melolonthinae). J Austral Entomol Soc 29:261–266

Altner H (1977) Insect sensillum specificity and structure: an approach to a new typology. Olfaction Taste 6:295–303

Altner H, Loftus R (1985) Ultrastructure and function of insect thermo- and hygroreceptors. Annu Rev Entomol 30:273–295

Altner H, Prillinger L (1980) Ultrastructure of invertebrate chemo-, thermo- and hygroreceptors and its functional significance. Int Rev Cytol 6:69–139

Altner H, Sass H, Altner I (1977) Relationship between structure and function of antennal chemoreceptive, hygroreceptive, and thermoreceptive sensilla in *Periplaneta americana*. Cell Tissue Res 176:389–405

Altner H, Routil C, Loftus R (1981) The structure of bimodal chemoreceptive, thermoreceptive, and hygroreceptive sensilla on the antenna of *Locusta migratoria*. Cell Tissue Res 215:289–308

Altner H, Schallerselzer L, Stetter H, Wohlrab I (1983) Poreless sensilla with inflexible sockets – a comparative study of a fundamental type of insect sensilla probably comprising thermoreceptors and hygroreceptors. Cell Tissue Res 234:279–307

Amornsak W, Cribb B, Gordh G (1998) External morphology of antennal sensilla of *Trichogramma australicum* Girault (Hymenoptera: Trichogrammatidae). Int J Insect Morphol Embryol 27:67–82

Baaren J van, Boivin G, Le Lannic J, Nénon JP (1999) Comparison of antennal sensilla of *Anaphes victus* and *A. listronoti* (Hymenoptera, Mymaridae), egg parasitoids of Curculionidae. Zoomorphology 119:1–8

Barlin MR, Vinson SB (1981) Multiporous plate sensilla in antennae of the Chalcidoidea (Hymenoptera). Int J Insect Morphol Embryol 10:29–42

Barlin MR, Vinson SB, Piper GL (1981) Ultrastructure of the antennal sensilla of the cockroach-egg parasitoid, *Tetrastichus hagenowii* (Hymenoptera: Eulophidae). J Morphol 168:97–108

Bartlet E, Isidoro N, Williams IH (1994) Antennal glands in *Psylliodes chrysocephala*, and their possible role in reproductive behaviour. Physiol Entomol 19:241–250

Basibuyuk HH, Quicke DLJ (1998) Gross morphology of multiporous plate sensilla in the Hymenoptera (Insecta). Zool Scr 28:51–67

Battaglia D, Isidoro N, Romani R, Bin F, Pennacchio F (2002) Mating behaviour of *Aphidius ervi* (Hymenoptera: Braconidae): the role of the antennae. Eur J Entomol 99:451–456

Billen J (1991) Ultrastructural organization of the exocrine glands in ants. Ethol Ecol Evol 1:67–73

Billen J, Morgan ED (1998) Pheromone communication in social insects: sources and secretions. In: van der Meer RK, Breed MD, Espelie KE, Winston ML (eds) Pheromone communication in social insects: ants, wasps, bees and termites. Westview Press, Boulder, pp 3–33

Bin F (1981) Definition of female antennal clava based on its plate sensilla in Hymenoptera Scelionidae Telenominae. Redia 64:245–261

Bin F, Vinson SB (1986) Morphology of the antennal sex-gland in male *Trissolcus basalis* (Woll) (Hymenoptera: Scelionidae), an egg parasitoid of the green stink bug, *Nezara viridula* (Hemiptera: Pentatomidae). Int J Insect Morphol Embryol 15:129–138

Bin F, Colazza S, Isidoro N, Solinas M, Vinson SB (1989) Antennal chemosensilla and glands, and their possible meaning in the reproductive behaviour of *Trissolcus basalis* (Woll.) (Hym.: Scelionidae). Entomologica 24:33–97

Bin F, Vinson SB, Strand MR, Colazza S, Jones WA (1993) Source of an egg kairomone for *Trissolcus basalis*, a parasitoid of *Nezara viridula*. Physiol Entomol 18:7–15

Bin F, Romani R, Isidoro N (1999a) Antennal structures of Hymenoptera: sensilla or glands? Atti Accad Naz Ital Entomol Rend XLVII:251–263

Bin F, Waeckers FL, Romani R, Isidoro N (1999b) Tyloids are release structures of male antennal glands involved in courtship behaviour (Hymenoptera: Ichneumonidae). Int J Insect Morphol Embryol 28:61–68

Cammaerts R (1974) Le système glandulaire tégumentaire du coléoptère myrmécophile *Claviger testaceus* Preyssler, 1790 (Pselaphidae). Z Morphol Tiere 77:187–219

Cave RD, Gaylor MJ (1987) Antennal sensilla of male and female *Telenomus reynoldsi* Gordh and Coker (Hymenoptera: Scelionidae). Int J Insect Morphol Embryol 16:27–39

Chiappini E, Solinas C, Solinas M (2001) Antennal sensilla of *Anagrus atomus* (L.) (Hymenoptera: Mymaridae) female and their possible behavioural significance. Entomologica 35:51–76

Cônsoli FL, Kitajima EW, Parra JRP (1999) Sensilla on the antenna and ovipositor of the parasitic wasps *Trichogramma galloi* Zucchi end *T. pretiosum* Riley (Hym., Trichogrammatidae). Microsc Res Tech 45:313–324

Dahms EC (1984) An interpretation of the structure and function of the antennal sense organs of *Melittobia australica* (Hymenoptera: Eulophidae) with the discovery of a large dermal gland in the male scape. Mem Queensl Mus 21:361–385

Diehl PA, Vlimant M, Guerenstein P, Guerin PM (2003) Ultrastructure and receptor cell responses of the antennal grooved peg sensilla of *Triatoma infestans* (Hemiptera: Reduviidae). Arthropod Struct Dev 31:271–285

Dickens JC, Callahan FE, Wergin WP, Erbe EF (1995) Olfaction in a hemimetabolous insect: antennal-specific protein in adult *Lygus lineolaris* (Heteroptera: Miridae). J Insect Physiol 41:857–867

Fresneau D (1979) Étude du role sensoriel de l'antenne dans l'ethogenese des soins aux cocons chez *Formica polyctena* Forst (Hymenoptera: Formicidae). Insectes Soc 26:170–195

Gaffal KP, Bassemir U (1974) Vergleichende untersuchung modifizierter cilienstrukturen in den dendriten mechano- und chemosensitiver rezeptorzellen der baumwollwanze *Dysdercus* und der Libelle *Agrion*. Protoplasma 82:177–202

Gaffal KP, Tichy H, Theiss J, Seelinger G (1975) Structural polarities in mechanosensitive sensilla and their influence on stimulus transmission (Arthropoda). Zoomorphologie 82:79–103

Giglio A, Ferrero EA, Zetto Brandmayr T (2005) Ultrastructural identification of the antennal gland complement in *Siagona europaea* Dejean 1826, a myrmecophagous carabid beetle. Acta Zool 86:195–203

Gnatzy W, Tautz J (1980) Ultrastructure and mechanical properties of an insect mechanoreceptor: Stimulus-transmitting structures and sensory apparatus of the cercal filiform hairs of *Gryllus*. Cell Tissue Res 213:441–463

Gnatzy W, Grünert U, Bender M (1987) Campaniform sensilla of *Calliphora vicina* (Insecta, Diptera). I. Topography. Zoomorphology 106:312–319

Goettler W, Kaltenpoth M, Herzner G, Strohm E (2007) Morphology and ultrastructure of a bacteria cultivation organ: the antennal glands of female European beewolves, Philanthus triangulum (Hymenoptera, Crabronidae). Arthropod Struct Dev 36:1–9

Grünert U, Gnatzy W (1987) Campaniform sensilla of *Calliphora vicina* (Insecta, Diptera). II. Typology. Zoomorphology 106:320–328

Guerrieri E, Pedata PA, Romani R, Isidoro N, Bin F (2001) Functional anatomy of male antennal glands in three species of Encyrtidae (Hymenoptera: Chalcidoidea). J Nat Hist 35:41–54

Hansson BS, Ochieng SA, Grosmaitre X, Anton S, Njagi PGN (1996) Physiological responses and central nervous projections of antennal olfactory receptor neurons in the adult desert locust, *Schistocerca gregaria* (Orthoptera: Acrididae). J Comp Physiol A – Sens Neural Behav Physiol 179:157–167

Hashimoto Y (1990) Unique features of sensilla on the antennae of Formicidae (Hymenoptera). Appl Entomol Zool 25:491–501

Henke K (1953) Über Zelldifferenzierungen im Integument der Insekten und ihre Bedingungen. J Embryol Exp Morphol 1:217–226

Hunger T, Steinbrecht RA (1998) Functional morphology of a double-walled multiporous olfactory sensillum: the sensillum coeloconicum of *Bombyx mori* (Insecta, Lepidoptera). Tissue Cell 30:14–29

Isidoro N (1991) I sensilli meccanorecettori delle antenne del *Trissolcus basalis* (Woll.) (Hymenoptera: Scelionidae). Atti XVI Congresso Nazionale Italiano di Entomologia, p 947

Isidoro N (1992) Fine structure of the sensillum coeloconicum in *Trissolcus basalis* (Woll.) (Hymenoptera, Scelionidae) antennae. Redia 75:169–178

Isidoro N, Bin F (1995) Male antennal gland of *Amitus spiniferus* (Brethes) (Hymenoptera: Platygastridae), likely involved in courtship behaviour. Int J Insect Morphol Embryol 24:365–373

Isidoro N, Bin F, Colazza S, Vinson SB (1996) Morphology of antennal gustatory sensilla and glands in some parasitoid Hymenoptera with hypothesis on their role in sex and host recognition. J Hymenopt Res 5:206–239

Isidoro N, Bin F, Romani R, Pujade-Villar J, Ros-Farré P (1999) Diversity and function of male antennal glands in Cynipoidea (Hymenoptera). Zool Scr 28:165–174

Isidoro N, Romani R, Velasquez D, Renthal R, Bin F, Vinson SB (2000) Antennal glands in queen and worker of the fire ant, *Solenopsis invicta* Buren: first report in female social Aculeata (Hymenoptera, Formicidae). Insectes Soc 47:236–240

Isidoro N, Romani R, Bin F (2001) Antennal multiporous sensilla: their gustatory features for host recognition in female parasitic wasps (Insecta, Hymenoptera: Platygastroidea). Microrsc Res Tech 55:350–358

Kaltenpoth M, Göttler W, Herzner G, Strohm E (2005) Symbiotic bacteria protect wasp larvae from fungal infestation. Curr Biol 15:475–479

Keil TA (1982) Contacts of pore tubules and sensory dendrites in antennal chemosensilla of a silkmoth: demonstration of a possible pathway for olfactory molecules. Tissue Cell 14:451–462

Keil TA (1997) Functional morphology of insect mechanoreceptors. Microsc Res Tech 39:506–531

Keil TA (1999) Morphology and development of the peripheral olfactory organs. In: Hansson BS (ed) Insect olfaction. Springer, Berlin, pp 5–47

Klein U (1987) Sensillum-lymph proteins from antennal olfactory hairs of the moth *Antheraea polyphemus* (Saturniidae). Insect Biochem 8:1193–1204

Lacher V (1964) Elektrophysiologische Untersuchungen an einzelnen Rezeptoren für Geruch, Kohlendioxyd, Luftfeuchtigkeit und Temperatur auf den Antennen der Arbeitsbiene und der Drohne (*Apis mellifera* L.). Z Vgl Physiol 48:587–623

Lawrence P (1966) Development and determination of hairs and bristles in the milkweed bug, *Oncopeltus fasciatus* (Lygeidae, Hemiptera). J Cell Sci 1:475–498

Locke M (1965) Permeability of insect cuticle to water and lipids. Science 147:295–298

Loudon C (2003) Antennae. In: Resh VH, Cardé RT (eds) Encyclopedia of insects. Academic, London, pp 26–28

McIver SB (1985) Mechanoreception. In: Kerkut GA, Gilbert LI (eds) Comprehensive insect physiology, biochemistry, and pharmacology, vol 6. Pergamon, Oxford, pp 71–132

McKenna MK, Hekmat-Scafe DS, Gaines P, Carlson JR (1994) Putative *Drosophila* pheromone-binding proteins expressed in a subregion of the olfactory system. J Biol Chem 23:16340–16347

Moran DT, Chapman KM, Ellis RA (1971) The fine structure of cockroach campaniform sensilla. J Cell Biol 48:155–173

Mou YC (1938) Morphologische und histologische Studien über Paussidendrüsen. Zool Jahrb Anat 64:287–346

Navasero RC, Elzen GW (1991) Sensilla on the antennae, foretarsi and palpi of *Microplitis croceipes* (Cresson) (Hymenoptera: Braconidae). Proc Entomol Soc Washington 93: 343–347

Noirot C, Quennedey A (1974) Fine structure of insect epidermal glands. Annu Rev Entomol 19:61–80

Olson DM, Andow DA (1993) Antennal sensilla of female *Trichogramma nubilale* (Ertle and Davis) (Hymenoptera: Trichogrammatidae) and comparisons with other parasitic Hymenoptera. Int J Insect Morphol Embryol 22:507–520

Paesen GC, Happ DM (1995) The beta proteins secreted by the tubular accessory sex glands of the male mealworm beetle, *Tenebrio molitor*, have sequence similarity to moth pheromone-binding proteins. Insect Biochem Mol Biol 25:401–408

Pophof B (1997) Olfactory responses recorded from sensilla coeloconica of the silkmoth *Bombyx mori*. Physiol Entomol 22:239–248

Pophof B, Stange G, Abrell L (2005) Volatile organic compounds as signals in a plant-herbivore system: Electrophysiological responses in olfactory sensilla of the moth *Cactoblastis cactorum*. Chem Senses 30:51–68

Pringle JWS (1961) Proprioception in arthropods. In: Ramsay JA, Wigglesworth VB (eds) The cell and the organism. Cambridge University Press, Cambridge, pp 256–282

Quennedey A (1998) Insect epidermal gland cells: ultrastructure and morphogenesis. In: Harrison FW, Locke M (eds) Microscopic Anatomy of invertebrates, vol 11A: Insecta. Wiley Liss, New York, pp 177–207

Romani R, Isidoro N, Bin F (1999) Further evidence of male antennal glands in Aphelinidae: the case of *Aphytis melinus* DeBach (Hymenoptera: Aphelinidae). J Hymenopt Res 8:109–115

Romani R, Isidoro N, Bin F (2002) Male antennal glands in *Bombus pascuorum* Scop.: morphology, possible function and comparison with other Hymenoptera Aculeata. Insect Soc Life 4:115–123

Romani R, Isidoro N, Riolo P, Bin F (2003) Antennal glands in male bees: structures for sexual communication by pheromones? Apidologie 34:603–610

Romani R, Isidoro N, Riolo P, Bin F, Fortunato A, Turillazzi S, Beani L (2005) A new role for antennation in paper wasps (Hymenoptera: Vespidae): antennal courtship and sex dimorphic glands in antennomeres. Insectes Soc 52:96–102

Romani R, Grasso DA, Mori A, Isidoro N, Le Moli F (2006) Antennal glands of the slave-making ant *Polyergus rufescens* and its slave species *Formica cunicularia* (Hymenoptera, Formicidae). Can J Zool 84:490–494

Romani R, Rosi MC, Isidoro N, Bin F (2008) The role of the antennae during courtship behaviour in the parasitic wasps *Trichopria drosophilae*. J Exp Biol 211:2486–2491

Schmidt JM, Smith JJB (1986) Correlations between body angles and substrate curvature in the parasitoid wasp *Trichogramma minutum*: a possible mechanism of host radius measurement. J Exp Biol 125:271–285

Schmidt JM, Smith JJB (1987a) The external sensory morphology of the legs and hairplate system of female *Trichogramma minutum* Riley (Hymenoptera: Trichogrammatidae). Proc R Soc Lond 232B:323–366

Schmidt JM, Smith JJB (1987b) Measurement of host curvature by the parasitoid was *Trichogramma minutum*, and its effect on host examination and progeny allocation. J Exp Biol 129:151–164

Schneider D (1964) Insect antennae. Annu Rev Entomol 9:103–122

Sen A, Raina R, Joseph M, Tungikar VB (2005) Response of *Trichogramma chilonis* to infochemicals: an SEM and electrophysiological investigation. BioControl 50:429–447

Steinbrecht RA (1970) Zur Morphometrie der Antenne des Seidenspinners, *Bombyx mori* L.: Zahl und Verteilung der Riechsensillen (Insecta: Lepidoptera). Zoologische Morphol Tiere 68:93–126

Steinbrecht RA (1997) Pore structures in insect olfactory sensilla: a review of data and concepts. Int J Insect Morphol Embryol 26:229–245

Steinbrecht RA, Müller B (1971) On the stimulus conducting structures in insect olfactory receptors. Z Zellforsch Mikrosk Anat 117:570–575

Thurm U (1964) Mechanoreceptors in the cuticle of the honeybee: fine structure and stimulus mechanism. Science 145:1063–1065

Thurm U (1970) Untersuchungen zur funktionellen organization sensorischer zellverbände. Verh Dtsch Zool Ges 64:79–88

Thurm U (1983) Mechano-electric transduction. In: Hoppe W, Lohmann W, Markl H, Ziegler H (eds) Biophysics. Springer, Berlin, pp 666–671

Tichy H (1979) Hygro-receptive and thermo-receptive triad in antennal sensillum of the stick insect, *Carausius morosus*. J Comp Physiol 132:149–152

Tuccini AR, Maida P, Rovero M, Mazza M, Pelosi P (1996) Putative odorant binding protein in antennae and legs of *Carausius morosus*. Insect Biochem Mol Biol 26:19–24

Voegelé J, Cals-Usciati J, Pihan JP, Daumal J (1975) Structure de l'antenne femelle des Trichogrammes. Entomophaga 20:161–169

Vogt RD, Riddiford LM (1981) Pheromone binding and inactivation by moth antennae. Nature 293:161–163

Vogt RG, Rybczynski R, Lerner MR (1990) The biochemistry of odorant reception and transduction. In: Schild D (ed) Chemosensory information processing, NATO ASI Ser. H, vol 39. Springer, New York, pp 33–76

Waldow U (1970) Electrophysiological investigations of moist, dry and cold receptors on antenna of migratory locust. Z Vgl Physiol 69:249–283

Wasmann E (1903) Zur näheren Kenntnis des echten Gastverhältnisses (Symphilie) bei den Ameisen- und Termitengästen. Biol Zentralbl 23:63–72, 232–248, 298–310

Weis A, Schönitzer K, Melzer RR (1999) Exocrine glands in the antennae of carabid beetle, *Platynus assimilis* (Paykull) 1970 (Coleoptera, Carabidae, Pterostichinae). Int J Insect Morphol Embryol 28:331–335

Yokohari F (1983) The coelocapitular sensillum, an antennal hygroreceptive and thermoreceptive sensillum of the honey bee, *Apis mellifera*. Cell Tissue Res 233:355–365

Yokohari F, Tominaga Y, Tateda H (1982) Antennal hygroreceptors of the honey bee, *Apis mellifera* L. Cell Tissue Res 226:63–73

Zacharuk RY (1985) Antennae and sensilla. In: Kerkut GA, Gilbert LI (eds) Comprehensive insect physiology, biochemistry, and pharmacology, vol 6. Pergamon, Oxford, pp 1–69

Ziegelberger G (1995) Redox-shift of the pheromone-binding protein in the silkmoth *Antheraea polyphemus*. Eur J Biochem 232:706–711

Zill SN, Moran DT (1981) The exoskeleton and insect proprioception. I. Responses of tibial campaniform sensilla to external and muscles-generated forces in the American cockroach, *Periplaneta americana*. J Exp Biol 91:1–24

Chapter 4
Host Searching by Egg Parasitoids: Exploitation of Host Chemical Cues

Stefano Colazza, Ezio Peri, Gianandrea Salerno, and Eric Conti

Contents

4.1 Introduction

Insect parasitoids are considered "keystone species" in many ecosystems in terms of biodiversity, ecological impact and economic importance (Vinson 1985, LaSalle and Gauld 1993, Hawkins et al. 1999). In the last decades, several reviews have been published on the relationships among plants, hosts and parasitoids, which

S. Colazza (✉)
Department of S.En.Fi.Mi.Zo., Section of Acarology, Entomology and Zoology,
University of Palermo, Viale delle Scienze 13, 90128, Palermo, Italy
e-mail: colazza@unipa.it

F.L. Consoli et al. (eds.), *Egg Parasitoids in Agroecosystems with Emphasis on Trichogramma*, Progress in Biological Control 9, DOI 10.1007/978-1-4020-9110-0_4,

reflect a strong interest in these insects both as models for behavioral ecologists and as important organisms for classical and applied biological control programs (Hawkins et al. 1999, Vet 1999, Bale et al. 2008). The majority of these studies have considered the larval parasitoids, besides the extensive use of egg parasitoids in biological control (Hawkins et al. 1999). Insect eggs can be parasitized by about 15 families of Hymenoptera parasitoids, among which several may have potential for biological control application, such as Aphelinidae, Encyrtidae, Eulophidae, Eupelmidae, Mymaridae, Platygastridae, Pteromalidae, Scelionidae, Tetracampidae and Trichogrammatidae (Bin 1994). Three families, Mymaridae, Scelionidae and Trichogrammatidae are exclusively composed of egg parasitoids, whereas the other families are represented by species developing in different host stages, and they also include egg-larval parasitoids, egg-prepupal parasitoids and egg predators (Bin 1994, Vinson 1994). In this chapter we will focus only on egg parasitoids. Successful parasitism of herbivores by insect parasitoids arises through several phases during host searching, which lead wasp females into the close vicinity/contact of their hosts (Vinson 1998). During the host location process, females encounter and explore a great variety of stimuli, among which the chemical cues, named semiochemicals or infochemicals, play a relevant role (Godfray 1994, Vet and Dicke 1992, Vinson 1998). Female parasitoids are under selection pressure to efficiently invest their limited time on the location and exploitation of host derived stimuli, so that the appropriateness and usability of semiochemicals could be influenced by their reliability in indicating host presence and by the degree to which stimuli can be detected, as explained by the reliability-detectability theory (Vet and Dicke 1992). In developing this theory, it was argued that the level of reliability and detectability of a particular stimulus is inversely correlated, e.g. cues from the hosts may be highly reliable, but are less detectable compared to volatiles from plants, which have a much larger biomass. To get through the reliability-detectability dilemma, wasp females can adopt three different strategies based on the exploitation of either: (1) cues originated from stages different from the one attacked (infochemical detour); (2) cues originated from the interaction of the plant and the herbivore (host-induced synomones); or (3) reliable but poorly detectable cues which were linked, through associative learning, with more detectable but unreliable cues (Vet and Dicke 1992).

In the specific case of egg parasitoids, hosts are generally available during a short time due to their rapid development, so that their quality has a tendency to decrease with time (Vinson 1998, see also Chapter 2). Furthemore, in term of biomass, host eggs represent a small component of a composite environment, so that, if they produce any information at all, it appears to be quantitatively negligible (Vinson 1994, see also Chapter 2). To find their hosts, egg parasitoids can use stimuli originated from the host eggs, i.e., direct host-related cues, or not, i.e., indirect host-related cues (Fig. 4.1). According to the reliability-detectability theory, the latter can provide an abundant supply of chemical information, but they are poorly reliable in indicating the host egg presence, while the former are more reliable, although they are produced in smaller amounts and, consequently, are more difficult to be detected. It is generally assumed that the infochemical detour strategy, based on the ability to

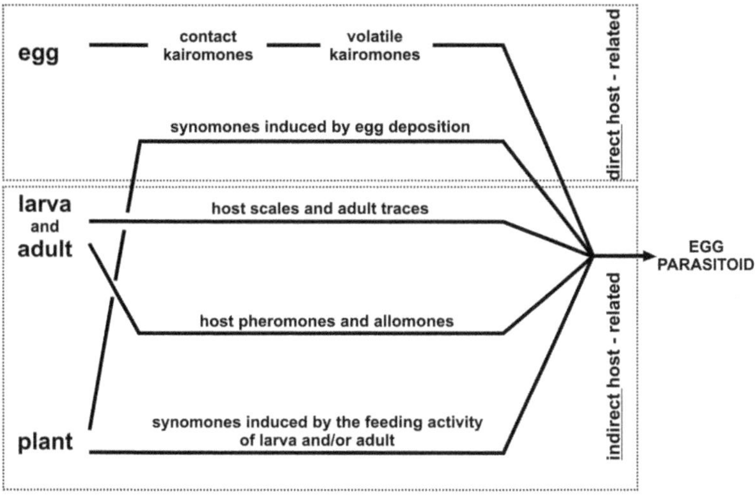

Fig. 4.1 Indirect and direct host-related chemical cues exploited by egg parasitoid females

eavesdrop indirect host-related cues, such as the host pheromones, is a main strategy for egg parasitoid females (Vinson 1998, Fatouros et al. 2008). Instead, evidences on the ability of egg parasitoids to exploit direct host-related cues are less numerous, so that for long time eggs have been considered an insect stage that has minimal interaction with the environment, since they do not feed nor release compounds as a consequence of their activity. However, novel data on the role of plant volatiles induced by egg deposition has opened a new scenario in the host location process by egg parasitoids.

In this chapter we will provide an overview on the different strategies and stimuli used by egg parasitoids to locate and recognize their hosts, discussing separately the less reliable indirect host-related cues from the more reliable direct host-related cues, which show different levels of detectability. The behavioral responses of egg parasitoids to chemical cues, the chemistry of such cues and the effect of learning on parasitoid responses will be also addressed.

4.2 Indirect Host-Related Chemical Cues

The infochemical detour strategy is based on the parasitoid's ability to spy on the chemical cues arising from the activities of stages different from the ones attacked (Vet and Dicke 1992, Vinson 1998). This set of cues is composed by chemical compounds with different volatile profiles, which can be perceived by parasitoid females through olfactory or gustatory sensilla (see Chapter 3 for detailed discussion). Indirect host-related cues do not provide information on egg location, but they lead females into the close vicinity of their potential host egg. Such cues are exploited at short or long range and act as arrestants or attractants, inducing peculiar searching patterns that will be discussed along the text. In this chapter, the following indirect host-related cues will be considered: (1) synomones induced by the feeding activity of the host herbivores (larvae and/or adults), (2) host pheromones and host allomones, and (3) kairomones from the scales of adult lepidopterous hosts and from the traces left behind by adult hosts while moving on the plant.

4.2.1 Synomones Induced by the Feeding Activity of the Host Herbivores

The attraction of egg parasitoid females to synomones induced by the feeding activity of immatures and adults has been explored in a few cases, involving Mymaridae and Scelionidae (Table 4.1). A clear example is represented by rice plants infested by nymphs and gravid females of *Nilaparvata lugens* (Stål), that attracted females of the egg parasitoid *Anagrus nilaparvatae* Pang and Wang, whereas undamaged plants, artificially damaged plants or each isolated *N. lugens* stage did not elicit any parasitoid response (Lou et al. 2005b). Similarly, two species of leguminous plants, soybean and pigeon pea, when infested by nymphs, males or females of *Euschistus heros* (F.) became attractive for females of the egg parasitoid *Telenomus podisi* (Ashmead) (Moraes et al. 2005). In another study, Manrique et al. (2005) reported that volatiles released by several herbaceous plants infested by adults of *Lygus hesperus* Knight attracted the egg parasitoid *Anaphes iole* Girault. However, Moraes et al. (2005) and Manrique et al. (2005) allowed the herbivore females both to feed and oviposit on the host plants, therefore synomone induction also as a consequence of oviposition cannot be excluded. In addition to volatile synomones, herbivory can induce synomones that are perceived by parasitoids only when they alight on damaged plants. In the system *Murgantia histrionica* (Hahn) – *Trissolcus brochymenae* (Ashmead), the parasitoid showed an intense antennation and increased residence time on plants damaged by feeding compared to healthy plants. However, in Y-tube olfactometer the wasp females were not attracted by host-damaged plants (Conti et al. 2006).

4.2.2 Host Pheromones and Allomones

The ability of egg parasitoid females to eavesdrop the pheromonal communication (sex and other attractant pheromones), and the allomonal defenses of their hosts is documented by an extensive literature (reviewed by Powell 1999 and Fatouros et al. 2008). There are reports for Trichogrammatidae, Scelionidae, Encyrtidae and Eulophidae (Table 4.2).

The volatile sex pheromone of several adult moths mediates the foraging behavior of egg parasitoid females belonging to Trichogrammatidae, Scelionidae and Encyrtidae. This behavior was first evidenced by Lewis et al. (1982), who demonstrated in field and greenhouse experiments that the sex pheromone of *Heliothis zea* (Boddie) attracts *Trichogramma pretiosum* (Riley). Similar results have been

Table 4.1 Responses to synomones induced by the herbivore feeding stages (larva and/or adult)

Taxon	Host	Plant(s)	• Induction by • Induction time	• Location of elicitor • Spatial scale	• Synomone composition • Stimulus perception • Stimulus duration	Referemces
Mymaridae						
Anaphes iole	*Lygus hesperus*	*Gossypium hirsutum* (Malvaceae), *Senecio vulgaris* (Asteraceae), *Medicago sativa* (Fabaceae), *Ambrosia artemisifolia* (Asteraceae), *Amaranthus retroflexus* (Amaranthaceae)	• adult feeding (egg deposition not excluded) • n.r.	• possibly saliva • n.r.	• qualitative and quantitative differences • olfaction • n.r..	Rodrigues-Saona et al. (2002), Manrique et al. (2005)
Anagrus nilaparvatae	*Nilaparvata lugens*	*Oriza sativa* (Graminaceae)	• nymph and adult feeding • n.r.	• possibly saliva • local and systemic	• n.r. • olfaction • n.r.	Lou et al. (2005 a, b)
Scelionidae						
Telenomus podisi	*Euschistus heros*	*Glycine max*, *Cajanus cajan* (Leguminosae)	• adult feeding • n.r.	• possibly saliva • n.r.	• quantitative differences • olfaction • n.r.	Moraes et al. (2005)
Trissolcus brochymenae	*Murgantia histrionica*	*Brassica oleracea* (Cruciferae)	• adult feeding • 18 h	• possibly saliva • local and systemic	• quantitative differences • taste/olfaction • < 100 h	Conti et al. (2006, 2008)

n.r. = not reported

Table 4.2 Responses to host pheromones (sex and attractant) and to host allomones

Taxon	Host(s)	Chemical(s)	Function(s)	Response	References
Encyrtidae					
Ooencyrtus nezarae	*Riptortus clavatus*	(*E*)-2-Hexenyl (Z)-3-hexenoate	Attractant pheromone	Attraction in the field	Leal et al. (1995), Mizutani et al. (1997)
O. pityocampae	*Thaumetopoea pityocampa*	(Z)-13-Hexadecen-11-yn-1-ol acetate	Sex pheromone	Attraction in the field	Battisti (1989)
Eulophidae					
Chrysonotomyia ruforum	*Diprion pini*	(2S,3R,7R)-3,7-Dimethyl-2-tridecanol	Sex pheromone	Arrestment	Hilker et al. (2000)
Dipriocampe diprioni	*Neodiprion sertifer*	(2S,3S,7S)-3,7-Dimethyl-2-pentadecyl acetate	Sex pheromone	Attraction in the field	Hilker et al. (2000), Kennedy (1979)
Entedon leucogramma (Ratzeburg)	*Scolytus multistriatus*	4-Methyl-heptan-3-ol, α-cubebene	Aggregation pheromone	Attraction in the field	Kennedy (1979), Kennedy (1984)
Scelionidae					
Gryon pennsylvanicus	*Leptoglossus australis*		Attractant pheromone	Attraction in the field	Yasuda and Tsurumachi (1995)
Telenomus busseolae	*Sesamia nonagrioides*	(Z)-11-Hexadecenyl acetate, (Z)-11-Hexadecenal, Dodecyl acetate	Sex pheromone	Attraction in Y-tube olfactometer	Colazza et al. (1997)
	Sesamia calamistis		Sex pheromone	Arrestment	Fiaboe et al. (2003)
Te. calvus	*Podisus neglectus*		Attractant pheromone		Aldrich (1995)
	Podisus maculiventris		Attractant pheromone	Phoresy	Orr et al. (1986), Aldrich et al. (1984)

Table 4.2 (continued)

Taxon	Host(s)	Chemical(s)	Function(s)	Response	References
		(*E*)-2-Hexenal, α-terpineol, benzyl alcohol		Attraction in the field	Bruni et al. (2000)
Te. euproctidis	*Euproctis taiwana*	(*Z*)-16-Methyl-9-heptadecenyl isobutyrate	Sex pheromone	Attraction in the field/phoresy	Wakamura (2006), Arakaki et al. (1996, 1997, 2000)
	Euproctis pseudoconspersa		Sex pheromone	Attraction in the field/phoresy	Wakamura (2006)
Te. isis	*Sesamia calamistis*		Sex pheromone	Arrestment	Fiaboe et al. (2003)
Te. podisi	*Euschistus heros*	Methyl 2,6,10-trimethyltridecanoate	Male sex pheromone	Choice in closed arena	Silva et al. (2006)
	Euschistus conspersus	Methyl (E,Z)-2,4-decadienoate	Male attractant pheromone	No parasitism increase in the field	Krupke and Brunner (2003)
Te. remus	*Spodoptera frugiperda*	(*Z*)-9-Dodecenyl acetate	Sex pheromone	Attraction	Nordlund et al. (1983)
Trissolcus basalis	*Nezara viridula*	(*E*)-2-Decenal	Defensive allomone	Attraction in Y-tube olfactometer	Mattiacci et al. (1993)
	Nezara viridula		Attractant pheromone		Colazza et al. (1999)
Tr. brochymenae	*Murgantia histrionica*		Pheromone from different stages and genders	Attraction in Y-tube olfactometer	Conti et al. (2003b)
Tr. euschisti (Ashmead)	*Euschistus conspersus*	Methyl (E,Z)-2,4-decadienoate	Male attractant pheromone	No parasitism increase in the field	Krupke and Brunner (2003)
Tr. grandis Thomson	*Eurygaster integriceps*		Male attractant pheromone and female sex pheromone	Attraction in the field	Buleza and Mikheev (1979)
Tr. simoni (Mayr)	*Eurydema ventrale*		Attractant pheromone	Attraction in Y-tube olfactometer	Colazza and Bin (1988)

Table 4.2 (continued)

Taxon	Host(s)	Chemical(s)	Function(s)	Response	References
Tr. utahensis (Ashmead)	*Euschistus conspersus*	Methyl (E,Z)-2,4-decadienoate	Sex pheromone	No parasitism increase in the field	Krupke and Brunner (2003)
Trichogrammatidae					
Lathromeris ovicida	*Sesamia calamistis*		Sex pheromone	Arrestment	Fiaboe et al. (2003)
Trichogramma brassicae	*Lobesia botrana*		Volatiles from virgin females	Parasitism increase and kinetic response	Garnier-Geoffroy et al. (1999)
	Pieris brassicae	2-Phenylacetonitrile	Anti aphrodisiac pheromone	Attraction on mated female	Fatouros et al. (2005b)
	Ostrinia nubilalis	Oleic acid	Sex pheromone	attraction in linear and four-arm Olfactometer	Frenoy et al. (1991, 1992), Kaiser et al. (1989b)
T. chilonis	*Helicoverpa assulta*	(*Z*)-11-Hexadecenyl acetate	Sex pheromone	Attraction in four-arm olfactometer	Boo and Yang (2000)
	Ostrinia furnacalis	(*E*)-12-Tetradecenyl acetate			
T. cordubensis Vargas and Cabello	*Heliothis armigera*		Volatiles from females	Attraction in olfactometer	Cabello Garcia and Vargas Piqueras (1985)
	Earias insulana				
T. evanescens	*Pectinophora gossypiella*		Two components of synthetic sex pheromone	Attraction and parasitism increase	Zaki (1985)
	Spodoptera littoralis			Poor attraction and no parasitism increase	

Table 4.2 (continued)

Taxon	Host(s)	Chemical(s)	Function(s)	Response	References
	Earias insulana			Slight attraction and no parasitism increase	
	Ephestia spp.	(*Z,E*)-9,12-Tetradecadenyl acetate	Sex-pheromone component	Attraction in four-arm olfactometer	Schöller and Prozell (2002)
	Plodia interpunctella				
T. evanescens = *T. maidis*	*Mamestra brassicae*		Volatiles from virgin females, sex pheromone	Attraction in wind tunnel	Noldus et al. (1991a,b), Noldus and van Lenteren (1983, 1985)
	Pieris brassicae		Volatiles from virgin females	Attraction in Y-tube olfactometer	
	Pieris rapae				
T. oleae (Voegele and Poitale)	*Palpita unionalis*	(*E*)-11-Hexadecenyl acetate	Sex pheromone component	Choice and parasitism increase	Abdelgader and Mazomenos (2002)
T. ostriniae	*Ostrinia nubilalis*		Synthetic sex pheromone	Attraction in Y-tube olfactometer and wind tunnel	Yong et al. (2007)
	Ostrinia furnacalis	(*E*)-12-Tetradecenyl acetate	Sex-pheromone	Attraction in four-arm olfactometer	Bai et al. (2004)
T. pretiosum	*Heliothis zea*	Hexadecanal, (*Z*)-7-Hexadecenal, (*Z*)-9-Hexadecenal, (*Z*)-11-Hexadecenal	Synthetic sex pheromone sex pheromone and virgin female	Parasitism increase in greenhouse attraction in wind tunnel and Y-tube olfactometer	Lewis et al. (1982), Noldus et al. (1988, 1990, 1991b)

Table 4.2 (continued)

Taxon	Host(s)	Chemical(s)	Function(s)	Response	References
	Plutella xylostella		Synthetic sex pheromone component	Parasitism increase	Klemm and Schmutterer (1993)
T. sp. near *buesi*	*Anagasta kuehniella*		Volatiles from females	Attraction in olfactometer	Cabello Garcia and Vargas Piqueras (1985)
T. sibericum Sorkina	*Rhodobota naevana*	(*Z*)-11-Tetradecen-1-ol, (*Z*)-11-Tetradecenyl acetate	Sex-pheromone	Increase of searching time	McGregor and Henderson (1998)
T. turkestanica Westwood	*Sesamia nonagrioides*		Sex-pheromone	No attraction in Y-tube olfactometer	Conti et al. (2003a)
Uscana lariophaga Steffan	*Callosobruchus maculatus*		Volatiles from virgin females	Attraction	van Huis et al. (1994)

observed for other Trichogrammatidae exposed to host virgin females or to synthetic sex pheromone blends (see refs. in Table 4.2). In addition, Fiaboe et al. (2003) reported that calling females of *Sesamia calamistis* (Hampson) attracted the trichogrammatid *Lathromeris ovicida* Risbec and the scelionids *Telenomus busseolae* (Gahan) and *Telenomus isis* (Polaszek). The response pattern for these *Telenomus* species was the same, while it was weaker for *L. ovicida*, reflecting the different host specificity of the three parasitoids. *Telenomus busseolae* females are also attracted by the synthetic sex pheromones of *Sesamia nonagrioides* Lefebvre (Colazza et al. 1997, Fiaboe et al. 2003). In the field, females of the encyrtid *Ooencyrtus pityocampae* (Mercet) were caught by traps baited with "pityolure", the synthetic sex pheromone of the host, *Traumatocampa* (=*Thaumetopoea*) *pityocampa* (Denis and Schiffermüller) (Battisti 1989).

Volatile attractant pheromones of Heteroptera are exploited by Scelionidae and oophagous Encyrtidae (Table 4.2). In field experiments, traps baited with male *Leptoglossus australis* F. placed next to host eggs increased the efficacy of *Gryon pennsylvanicum* (Ashmead) (Yasuda and Tsurumachi 1995). In addition, traps baited with the synthetic male attractant pheromone of *Riptortus clavatus* (Thunberg) captured females of the egg parasitoid *Ooencyrtus nezarae* Ishii (Leal et al. 1995). In laboratory experiments, *Trissolcus basalis* (Wollaston) and *Tr. brochymenae* females were attracted by volatiles released by immatures and adults of their hosts, although host gravid females elicited a higher response (Colazza et al. 1999, Conti et al. 2003b). Females of *Tr. basalis* are also attracted by host allomones like the compounds present in the defensive secretion of *Nezara viridula* (L.) adults (Mattiacci et al. 1993). It was speculated that host allomones and volatile male sex attractant pheromones can direct wasp females towards host aggregates, while volatile cues from mated host females may induce wasp females to shift from flying to walking behavior (Colazza et al. 1999).

The adaptation value for egg parasitoid females to use host pheromones and/or allomones is mainly related to the temporal and/or spatial vicinity of the calling host stages with the eggs and the oviposition sites. To overcome the temporal gap between host calling activity and deposited eggs, diurnal foraging females of *Trichogramma evanescens* Westwood detect the volatile host pheromone from their nocturnal host, *Mamestra brassicae* (L.), adsorbed by the epicuticular waxes of Brussels sprouts leaves for up to 24 h (Noldus et al. 1991a). Similarly, females of *Telenomus euproctidis* Wilcox detect the volatile sex pheromone of their nocturnal host, *Euproctis taiwana* (Shiraki), adsorbed on the scales covering the egg masses, thus extending the time for host location up to 48 h (Arakaki and Wakamura 2000). Another bridge-in-time (and bridge-in-space) strategy developed by egg parasitoids to reduce spatial and temporal discontinuity between the place where host mates and oviposits is the phoresy (Vinson 1998, see also Chapter 2). Phoresy is known to occur in Trichogrammatidae and Scelionidae, where the role of the sex pheromone in regulating such relationship between host and parasitoid has been clearly demonstrated. The phoretic egg parasitoid *Te. euproctidis* uses the host sex pheromone to find virgin females of *E. taiwana* (Arakaki et al. 1996), and females of *Telenomus calvus* Johnson use the attractant pheromone of *Podisus maculiventris*

(Say) to facilitate their phoretic behavior (Aldrich et al. 1984, Bruni et al. 2000). Recently, another strategy has been described by Fatouros et al. (2005b) for *Trichogramma brassicae* (Bezdenko), showing that wasp females can recognize and climb on mated *Pieris brassicae* L. females, based on the presence of the anti-aphrodisiac pheromone benzyl cyanide, which is transferred by the male during mating. Virgin host females still need to mate before ovipositing, therefore phoretic wasps that are able to discriminate between virgin and mated females can increase their chance to find host eggs.

4.2.3 Host Adult Scales and Traces

The indirect host-related kairomones can elicit arrestment (female remaing immobile with the antennae held in contact with the substrate) and prolonged searching behavior (female starting to walk slowly with many turns while drumming) in egg parasitoids, especially in Trichogrammatidae and Scelionidae (Table 4.3). Contact indirect host-related kairomones are generally perceived by taste and have been identified from lepidopteran scales and from footprints left on the substrate by true bugs. Other contact adult-derived chemical cues e.g. honeydew and feces, can also induce arrestment of egg parasitoid females, but their role has not been studied in details, so they will be only mentioned here.

(i) *Scales from Lepidoptera.* Prolonged searching behavior induced by scales shed by lepidopterans has been reported for Trichogrammatidae and Scelionidae (Table 4.3). According to the literature, this phenomenon appears quite common in *Trichogramma*, while very few cases are described for *Telenomus* (see references in Table 4.3). The role of scales as a contact cue for egg parasitoid females was reported for the first time by Laing (1937) in a pioneer paper which described that traces left by adults of the grain moth, *Sitotroga cerealella* (Olivier), induced an arrestment response in females of *T. evanescens*. Then, this subject was left unexplored up to the early 1970s, when Lewis and collaborators initiated in-depth studies on the role of scale kairomones on the behavioral ecology of *Trichogramma*. They demonstrated that wing scales were the kairomonal source, and that hexane extracts of scales sprayed over the plant could increase the likelihood of a host encounter and, consequently, the rate of parasitism (Lewis et al. 1972). The role of scale extracts on wasp's efficacy has been reported for several species of this genus, such as *Trichogramma nubilale* Ertle and Davis and *Trichogramma chilonis* Ischii (Shu and Jones 1989, Boo and Yang 2000). Noldus and van Lenteren (1985) reported that wasp females react to host scales only after contact, and Zaborski et al. (1987) pointed out that scales lack any physical cue that can elicit parasitoid arrestment. In the case of *Telenomus*, the response to lepidoteran scales has been studied for few systems, such as *Telenomus remus* Nixon – *Spodoptera frugiperda* (J.E. Smith) (Gazit et al. 1996) and *T. isis* – *Sesamia calamistis* (Chabi-Olaye et al. 2001). In another system, it was shown that parasitoid response to the host kairomone could be influenced by its geografical distribution. In fact, for example, two strains of *Te. busseolae*, one from Africa and

Table 4.3 Responses to scales and traces from host adults

Taxon	Host(s)	Stimulus	Bioassay	Response	Chemical(s)	• Gender discrimination ability • Host specificity	References
Encyrtidae							
Ooencyrtus kuvanae	*Lymantria dispar*	Scales and hairs on egg mass	In vivo				Lee et al. (1997)
Eulophidae							
Tetrastichus schoenobii	*Scirpophaga incertulas*	Hairs on egg mass	In vivo		proteins and amino acids		Ding et al. (1988)
Scelionidae							
Gryon boselli	*Gonocerus acuteangu-latus*	Footprints	In vivo in open arena	Arrestment and local search			Colazza et al. personal observation
Telenomus busseolae	*Sesamia nonagri-oides*	Scales	In vivo and solvent extracts in open arena	Arrestment and local search	scale hydrocarbons	• no discrimination • n.r.	Colazza and Rosi (2001)
Te. euproctidis	*Euproctis taiwana*	Scales on egg mass	In vivo in the field	Attraction and retention			Arakaki and Wakamura (2000), Wakamura (2006),
Te. isis	*Sesamia calamistis*	Host traces (most likely scales)	In vivo in open arena	Arrestment and local search		• discrimination • n.r.	Chabi-Olaye et al. (2001)
Te. podisi	*Euschistus heros*	Footprints	In vivo in open arena	Arrestment and local search			Borges et al. (2003)
Te. remus	*Spodoptera frugiperda*	Host body	Solvent adult body wash	Arrestment and local search			Gazit et al. (1996)

Table 4.3 (continued)

Taxon	Host(s)	Stimulus	Bioassay	Response	Chemical(s)	• Gender discrimination ability • Host specificity	References
Trissolcus basalis	*Nezara viridula*	footprints	In vivo and solvent extracts in open arena	Arrestment and local search	Cuticular hydrocarbons	• discrimination based on the presence or absence of *n*C19 • highly specific cue, allows to distinguish traces left by non-host pentatomids	Colazza et al. (1999, 2007), Salerno et al. (2006)
Tr. brochymenae	*Murgantia histrionica*	Footprints	In vivo and solvent extracts in open arena	Arrestment and local search	Cuticular hydrocarbons	• discrimination • highly specific cue, allows to distinguish traces left by non-host pentatomids	Conti et al. (2003b, 2004)
Tr. simoni	*Eurydema ventrale*	Footprints	In vivo in open arena	Arrestment and local search		• n.r. • poorly specific cue, does not allow to distinguish traces left by non-host pentatomids	Conti et al. (2004)
Trichogrammatidae							
Trichogramma achaeae Nagaraja and Nagarkatti	*Heliothis zea*	Scales	Petri dishes in the laboratory	Increased parasitism			Gross et al. (1976), Lewis et al. (1975a)

Table 4.3 (continued)

Taxon	Host(s)	Stimulus	Bioassay	Response	Chemical(s)	• Gender discrimination ability • Host specificity	References
T. brassicae	*Ostrinia nubilalis*	Scales	Solvent extracts	Increased parasitism in artificial host eggs			Grenier et al. (1993)
	Pieris brassicae	Scales	2-choice contact bioassay	Increased retention time			Fatouros et al. (2005a)
T. chilonis	*Heliothis armigera*	Scales	Solvent extracts	Increased parasitism	Saturated hydrocarbons		Ananthakrishnan et al. (1991)
	Helicoverpa assulta	Scales from males	Solvent extracts and chromato-graphic fractions in petri dish arena	Retention and increased parasitism	Saturated hydrocarbons		Boo and Yang (2000)
	Corcyra cephalonica	Scales	Solvent extract	Increased parasitism			Ananthakrishnan et al. (1991), Bakthavatsalam and Tandon (2006)
	Pectinophora gossypiella	Scales	Solvent extracts in open arena	Increased parasitism			Wang and Zong (1991)
T. cacoeciae Marchal	*Lobesia botrana*	Scales	Exposing the eggs of *Anagasta kuehniella* and *L. botrana*	Increased discovery rate			Barnay et al. (1999)
T. evanescens	*Mamestra brassicae*	Scales		Arrestment			Laing (1937)

Table 4.3 (continued)

Taxon	Host(s)	Stimulus	Bioassay	Response	Chemical(s)	• Gender discrimination ability • Host specificity	References
	Sitotroga cerealella						
	Plodia interpunctella	Scales	In vivo in experimental cage	Increased parasitism			Lewis et al. (1971)
	Heliothis zea	Scales	In vivo in experimental cage; field and greenhouse experiments	Increased parasitism	Docosane, tricosane, tetracosane, pentacosane		Jones et al. (1973), Lewis et al. (1971)
	Cadra cautella	Scales	Solvent extracts in laboratory, greenhouse and field tests	Increased parasitism			Lewis et al. (1972)
	Choristoneura fumiferana	Scales	Solvent extracts in open arena	Arrestment and local search – dosage-dependent orthokinetic response			Schmidt and Carter (1992)
T. evanescens = *T. maidis* (Noldus 1989)	*Pieris rapae*	Wing scales	Treated leaves	Increased searching, arrestment			Noldus and van Lenteren (1983, 1985)
	Pieris brassicae						

Table 4.3 (continued)

Taxon	Host(s)	Stimulus	Bioassay	Response	Chemical(s)	• Gender discrimination ability • Host specificity	References
	Mamestra brassicae	Wing scales	Treated leaves	Increased searching, arrestment			Noldus and van Lenteren (1983), Smits (1982)
T. exiguum Pinto et al.	*Heliothis zea*	Scales from females	In vivo in laboratory	Klinokinetic response			Thomson and Stinner (1988)
	Heliothis zea, Manduca sexta, Ostrinia nubilalis	Scales from females	In vivo in open arena	Arrestment and local search		• n.r. • low specificity cue	Thomson and Stinner (1990)
T. japonicum (Ashmead)	*Scirpophaga incertulas (Tryporyza incertulas)*	Egg "fur" and scales from females and males	Solvent extracts in open arena	Increased searching		• present • n.r.	Zou et al. (2002)
T. maltbyi Nagaraja and Nagarkatti	*Heliothis zea, Manduca sexta, Ostrinia nubilalis*	Scales from females	In vivo in open arena	Arrestment and local search		• n.r. • low specificity cue	Thomson and Stinner (1990)
T. minutum Riley	*Heliothis zea, Manduca sexta, Ostrinia nubilalis*	Scales from females	In vivo in open arena	Arrestment and local search		• n.r. • low specificity cue	Thomson and Stinner (1990)
	Choristoneura fumiferana	Scale extract	Solvent extracts in open arena	Increased parasitism		• n.r. • specificity cue	Zaborski et al. (1987)

Table 4.3 (continued)

Taxon	Host(s)	Stimulus	Bioassay	Response	Chemical(s)	• Gender discrimination ability • Host specificity	References
T. sp. nr. *pretiosum*	*Heliothis zea, Manduca sexta, Ostrinia nubilalis*	Scales from females	In vivo in open arena	Arrestment and local search		• n.r. • low specificity cue	Thomson and Stinner (1990)
	Heliothis virescens	Scales	In vivo in open arena	Klinokinetic response			Thomson and Stinner (1988)
T. nubilale	*Ostrinia nubilalis*	Scales	Solvent extract and synthesized compounds in open arena	Arrestment and orthokinetic response; increased parasitism	11,15-, 13,17-, and 15,19-Dimethylnonatriacontane		Shu and Jones (1989), Shu et al. (1990)
T. oleae	*Palpita unionalis*	Scale extract	Solvent extract in petri dishes	Local search and increased parasitism			Abdelgader and Mazomenos (2002)
T. ostriniae	*Ostrinia nubilalis*	Scales	Y-tube olfactometer				Yong et al. (2007)
T. pretiosum	*Heliothis zea*	Scales	Solvent extracts in laboratory, greenhouse and field tests	Increased parasitism and increased retention time			Beevers et al. (1981), Gross et al. (1984), Lewis et al. (1975a, 1979), Nordlund et al. (1976)

Table 4.3 (continued)

Taxon	Host(s)	Stimulus	Bioassay	Response	Chemical(s)	• Gender discrimination ability • Host specificity	References
			Solvent extracts in open arena	Did not increase parasitism	Hexanoic, heptanoic, octanoic, nonanoic, 2- (or 3-) furan carboxylic, phenylacetic		Gueldner et al. (1984)
T. turkestanica	*Sesamia nonagri-oides*	Scales	In vivo in open arena	Arrestment and orthokinetic response			Conti et al. (2003a)

n.r. = not reported

another from Turkey, were evaluated in comparative tests showing that the African strain searches longer on the scales of *S. nonagrioides* (Colazza and Rosi 2001). In addition to the above systems, hairs and scales used to cover the egg mass arrest parasitoids and give information on the presence of host eggs, as shown in the cases of *Ooencyrtus kuvanae* (Howard), *Tetrastichus schoenobii* Ferriere and *T. euproctidis* (Ding et al. 1988, Lee et al. 1997, Arakaki and Wakamura 2000).

(ii) *Traces (foot prints) left by true bugs.* Motivated searching induced by chemical residues left on a substrate by true bugs has been observed in various species of *Trissolcus* (Colazza et al. 1999, Conti et al. 2003b, 2004, Salerno et al. 2006), in *Te. podisi* (Borges et al. 2003), and in *Gryon boselli* Mineo and Szabo (Colazza, Lo Bue and Cusimano, personal observation). Egg parasitoid females respond to footprints left by immature and adult stages, but with a clear preference for those left by adult females (Colazza et al. 1999, Conti et al. 2003b). The capacity to distinguish and prefer footprints left by females or by mated females was reported for *T. basalis* and *T. brochymenae*, respectively (Colazza et al. 1999, 2007).

(iii) *Honeydew and feces.* Other detectable contact chemical cues produced by adults can induce arrestment response in egg parasitoid females. For example, females of *Oomyzus gallerucae* Fonscolombe were arrested when in contact with adult feces of *Xanthogaleruca luteola* (Muller) (Meiners and Hilker 1997), and females of *A. nilaparvatae* were arrested once in contact with honeydew produced by the host *Sogatella furcifera* (Horvath) (Lou and Cheng 2001). The relative contribution of each of these indirect host-related cues as kairomones for the egg parasitoids and how reliable they are as cue for egg presence remains to be elucidated.

4.3 Direct Host-Related Chemical Cues

Stimuli associated with host eggs can provide reliable information on the presence of suitable target hosts. To date, direct egg-related cues have been reported to play a role mainly in host recognition rather than in the host location process, like for example the contact kairomones from host eggs, which are fundamental in host recognition for many egg parasitoid species (see below). It has been shown recently that egg parasitoids are also able to detect volatile compounds induced by egg deposition (Meiners and Hilker 1997, 2000, Colazza et al. 2004a, Hilker and Meiners 2006). Such oviposition-induced synomones are as reliable as direct host kairomones, and due to the remarkable plant biomass, they have the advantage of being produced in large amounts providing highly detectable cues (see discussion in Section 4.3.1).

Direct host-related cues can be exploited at long range, orienting wasp females towards the source, or at close range, inducing wasp females to evaluate the host egg surface by antennation. Here, the following direct host-related cues will be discussed: (1) synomones induced by egg deposition, (2) volatile and (3) contact kairomones from host eggs.

4.3.1 Synomones Induced by Egg Deposition

An intrinsic characteristic of egg parasitoids is that they attack an immobile host stage, which, by itself, does not cause plant damage. In spite of that, recent data show that host egg deposition and associated behaviors of parental females may determine a change in plant emission of volatile compounds, which consequently may act as host-induced synomones for the egg parasitoids (Table 4.4). From an evolutionary point of view, both symbionts will have advantage from this "early alert" mechanism (*sensu* Hilker and Meiners 2006), as the egg parasitoids will use such highly detectable and reliable volatiles induced in plants soon after herbivore eggs are laid, whereas the plants would also increase their fitness by recruiting parasitoids which will attack the herbivore eggs before significant damage has occurred, i.e. before the herbivore egg hatches (Colazza et al. 2004a, Hilker and Meiners 2006).

Most oviposition-induced synomones known so far are perceived by the parasitoid as olfactory stimuli (volatile synomones) (Meiners and Hilker 1997, 2000, Hilker et al. 2002, 2005, Colazza et al. 2004a, b), although there are few exceptions of chemical cues which act after parasitoid's landing on the plant substrate, thus apparently being perceived through contact chemoreception (contact or short range synomones) (Fatouros et al. 2005a, 2007, Conti et al. 2006).

Volatile induced synomones are known to be exploited by Eulophidae, Scelionidae and Mymaridae in 5 tri-trophic systems, two of which based on arboreal, perennial plant species, and the others on herbaceous annual plants (Table 4.4). On the two perennial plants, the eulophids *O. gallerucae* and *Chrysonotomyia ruforum* Krausse are attracted to synomones emitted, respectively, by elm leaves as a consequence of oviposition by the elm leaf beetle *Xanthogaleruca luteola*, and by pine needles after oviposition by the pine sawfly *Diprion pini* L. (Meiners and Hilker 1997, 2000, Hilker et al. 2002, 2005). On annual plants, the scelionid *Tr. basalis* is attracted to volatiles emitted by bean and faba bean plants carrying egg masses deposited by the Southern green stink bug *N. viridula* (Colazza et al. 2004a), whereas the mymarid *A. iole* responds to volatiles from different plant species attacked by the plant bug *L. hesperus* (Rodriguez-Saona et al. 2002, Manrique et al. 2005).

In all of these systems, the emission of oviposition-induced volatiles allows parasitoids to locate plants that carry host eggs. Additionally, another scelionid wasp, *Tr. brochymenae*, responds from a short distance to chemicals emitted by cabbage plants as a consequence of oviposition by the harlequin bug, *M. histrionica*. However, such type of short distance response seems to be connected to the parasitoid's behavior on the host plant (see below) rather than to a true attractive response (Conti et al. 2006).

The induction of synomones perceived by female parasitoids after alighting on the host plant has been recently described for three tri-trophic systems (Fatouros et al. 2005a, 2007, Conti et al. 2006), *T. brassicae* and *T. evanescens* on cabbage plants, as consequences of oviposition by the large cabbage white butterfly *P. brassicae* (Fatouros et al. 2005a, 2007), *Tr. brochymenae* also on cabbage plants, after oviposition by the harlequin bug *M. histrionica* (Conti et al. 2006), and for *Te.*

Table 4.4 Responses to synomones induced by egg deposition

Taxon	Host	Plant(s)	• Induction activity • Induction time • Distribution	• Elicitor • Source of elicitor	• Synomone composition • Stimulus perception • Stimulus duration	References
Eulophidae						
Chrysonotomyia ruforum	*Diprion pini*	*Pinus sylvestris* (Pinaceae)	• oviposition + damaged leaf • 3 d • local and systemic	• protein/peptide • oviduct secretion coating the eggs	• quantitative differences (> (*E*)-β-farnesene) • olfaction • n.r.	Hilker et al. (2002, 2005)
Oomyzus gallerucae	*Xanthogaleruca luteola*	*Ulmus minor* (Ulmaceae)	• oviposition + damaged leaf • 3 h • local and systemic	• n.r. • oviduct secretion coating the eggs	• qualitative differences • olfaction • n.r.	Meiners et al. (1997, 2000)
Scelionidae						
Trissolcus basalis	*Nezara viridula*	*Vicia faba*, *Phaseolus vulgaris* (Leguminosae)	• oviposition + feeding • 3-4 d • local and systemic	• n.r. • n.r.	• quantitative differences (> (*E*)-β-caryophyllene) • olfaction • n.r.	Colazza et al. (2004a, b)
Trissolcus brochymenae	*Murgantia histrionica*	*Brassica oleracea* (Cruciferae)	• oviposition / oviposition + feeding • 18 h • local and systemic	• n.r. • egg mass	• quantitative differences • taste/olfaction • < 100 h	Conti et al. (2006, 2008)
Telenomus busseolae	*Sesamia nonagrioides*	*Zea mais* (Graminaceae)	• oviposition • 24 h • local and systemic	• n.r. • eggs (ovarian)	• n.r. • taste • n.r.	

Table 4.4 (continued)

Taxon	Host	Plant(s)	• Induction activity • Induction time • Distribution	• Elicitor • Source of elicitor	• Synomone composition • Stimulus perception • Stimulus duration	References
Trichogrammatidae						
Trichogramma brassicae	*Pieris brassicae*	*Brassica oleraceae* (Cruciferae)	• oviposition • 72 h • local and systemic	• n.r. • secretion of accessory glands coating the eggs	• n.r. • taste • > 96 h	Fatouros et al. (2005a)
Trichogramma evanescens			• oviposition • up to 24 h • n.r.	• n.r. • n.r.	• n.r. • taste • 72 -96 h	Fatouros et al. (2007)

n.r. = not reported

busseolae on corn with egg masses of the noctuid stemborer *S. nonagrioides* (Conti et al. 2006). Such oviposition-induced contact synomones appear to open a new scenario in the host location process by egg parasitoids.

With a single exception, i.e. that of *P. brassicae* on cabbage (Fatouros et al. 2005a), almost all the oviposition induced synomones known so far show a systemic distribution on the plant (Meiners and Hilker 1997, 2000, Hilker et al. 2002, 2005, Colazza et al. 2004a, Conti et al. 2006). This is important both for volatile and for contact synomones. In the case of the volatiles, by maximising the release surface, the plant is expected to emit high synomone amounts making them easily detectable by parasitoids. In the case of contact synomones, being distributed on the plant surface, parasitoids are informed of the presence of host eggs independently from their landing site. Synomone activity over time also seems to be finely tuned with the parasitoid behavior and biology, since the attraction fades when host eggs are closed to eclosion. The time necessary for the induction is also variable, ranging from few hours to some days (Colazza et al. 2004a, Fatouros et al. 2005a, Conti et al. 2006). Interestingly, when induction takes longer the parasitoid initially shows positive response to contact kairomones from the host female, and such response decreases when synomone emission starts (Fatouros et al. 2005a).

4.3.2 Volatile Kairomones from Host Eggs

In spite of the very limited biomass of host eggs, the volatiles they emit have been shown to elicit positive behavioral responses in several egg parasitoids, mostly belonging to the family Trichogrammatidae, and few belonging to Scelionidae, Eulophidae and Mymaridae (Table 4.5). Most of the active compounds identified are hydrocarbons, but other egg-derived compounds were also found to be active (Table 4.5).

Among the Trichogrammatidae, *T. brassicae* (Renou et al. 1989, 1992, Frenoy et al. 1992) and *T. ostriniae* Pang and Chen (Bai et al. 2004, Yong et al. 2007) were the most investigated. Females of *T. brassicae* responded in a linear olfactometer to *Ostrinia nubilalis* Hübner eggs, and to chemical extracts from eggs of *O. nubilalis* and *M. brassicae*. Females responded most readily to a mixture of five synthetic saturated hydrocarbons, although when tested isolated, they had a very low response only to pentacosane. Females of *T. brassicae* were also found to be responsive to palmitic acid and ethyl palmitate, indicating that different compounds on the host egg may be used for host location (Renou et al. 1992). Females of *T. ostriniae* were shown to respond to accessory glands from mated females besides the extracts and egg masses of the host *Ostrinia furnacalis* (Guenée) (Bai et al. 2004; Yong et al. 2007).

The olfactory response of Scelionidae, Eulophidae and Mymaridae has been poorly studied and, whenever egg volatiles were tested, they were found to be unattractive (Bloem and Yeargan 1982, Meiners and Hilker 1997, Meiners et al. 1997). However, there are two cases, one in Scelionidae and another in Eulophidae, that clearly show parasitoid attraction to the host eggs. In an open

Table 4.5 Responses to volatile kairomones from host eggs

Taxon	Host(s)	Source	Response	Chemical(s)	References
Eulophidae					
Aprostocetus hagenowii	*Periplaneta americana*	Ootheca	Attraction in Y-tube olfactometer	(*Z*, *Z*)- 6, 9-Heptacosadiene)	Suiter et al. (1996)
Oomyzus gallerucae	*Xanthogaleruca luteola*	Eggs	No response in four-arm olfactometer		Meiners and Hilker (1997)
O. galerucivorus (Hedqvits)	*Galeruca tanaceti*	Eggs	No response in four-arm olfactometer		Meiners and Hilker (1997)
Mymaridae					
Anagrus nilaparvatae	*Sogatella furcifera*	Eggs	Increased searching in petri dish		Lou and Cheng (2001)
Patasson lameerei Debauche	*Sitona hispidulus*	Eggs	No response in olfactometer		Bloem and Yeargan (1982)
Scelionidae					
Trissolcus brochymenae	*Murgantia histrionica*	Egg masses	Attraction in Y-tube olfactometer		Conti et al. (2003b)
Tr. basalis	*Nezara viridula*	Egg mass	No response in Y-tube olfactometer		Colazza personal observation
Trichogrammatidae					
Trichogramma brassicae (= *T. maidis*)	*Ostrinia nubilalis*	Eggs and solvent extracts from eggs	Attraction in linear olfactometer	Hydrocarbons blend, ethyl palmitate and palmitic acid	Frenoy et al. (1992), Renou et al. (1989, 1992)
T. chilonis	*Corcyra cephalonica*	Egg mass and hexane extracts	Attraction in four-arm olfactometer	Hexacosane, tricosane	Lu et al. (2006)
T. closterae Pang and Chen	*Corcyra cephalonica*	Egg mass and hexane extracts			Lu et al. (2006)
T. evanescens	*Sitotroga cerealella*		No response in static olfactometer		Laing (1937)

Table 4.5 (continued)

Taxon	Host(s)	Source	Response	Chemical(s)	References
T. maidis	*Ostrinia nubilalis*		No response in four-arm olfactomete		Kaiser et al. (1989b)
T. ostriniae	*Ostrinia nubilalis*	Egg mass	Attraction in Y-tube olfactometer		Yong et al. (2007)
	Ostrinia furnacalis	Egg mass	Attraction in four-arm olfactometer	(*E*)-12-Tetradecenyl acetate	Bai et al. (2004)
T. platneri Nagarkatti	*Boarmia selenaria*	Eggs	Attraction in olfactometer		Wysoki and de Jong (1989)
	Cryptoblabes gnidiella	Eggs			
T. raoi Nagaraja	*Corcyra cephalonica*	Egg mass and hexane extracts	Attraction in four-arm olfactometer		Lu et al. (2006)
T. spp.	*Anagasta kuehniella*	Eggs	Attraction in diffusion olfactometer		Ferreira et al. (1979)
T. spp.	*Mythimna unipuncta*	Eggs			Ferreira et al. (1979)
Trichogrammatoidaea bactrae Nagaraja	*Corcyra cephalonica*	Egg mass and hexane extracts	Attraction in four-arm olfactometer		Lu et al. (2006)
Uscana lariophaga	*Callosobruchus maculatus*	Egg mass	Attraction in diffusion olfactometer		van Alebeek et al. (1997)

arena, females of *T. brochymenae* showed an orientation response towards host egg clusters or to dummies with a chemical extract of host eggs. When the chemical egg extract was applied without dummies, it elicited the same response, whereas dummies without extract did not influence parasitoid behavior, indicating that visual factors are not necessary in mediating this last step of host location (Conti et al. 2003b). Concerning the Eulophidae, females of *Aprostocetus hagenowii* (Ratzeburg) responded in olfactometer to oothecae of *Periplaneta americana* L. (Suiter et al. 1996). Strictly, this cannot be considered as a response to host eggs because it is the ootheca that attracts the parasitoid. However, since the ootheca is produced by the host accessory glands (Bai et al. 2004), the kairomonal source may actually not differ from that of other systems where the kairomone source is the host adhesive colleterial secretions (see Tables 4.5 and 4.6). The active compound found in the *P. americana* oothecae was identified as (*Z,Z*)-6,9-heptacosadiene, which was also found in the frass and in adult females (Suiter et al. 1996).

4.3.3 Contact Kairomones from Host Eggs

Species of the three specialized families of egg parasitoids, Trichogrammatidae, Scelionidae and Mymaridae, as well as egg parasitoids from Encyrtidae and Eulophidae, use chemical cues from the host eggs as host recognition kairomones (Table 4.6). In most cases the kairomone source is the adhesive secretion from the ovipositing female, although there are cases of recognition kairomones originating from scales, hairs and oviposition deterrents in Lepidoptera, or from rectal secretions in Blattodea (Tables 4.3 and 4.6).

The first and most intensively studied families are Trichogrammatidae and Scelionidae on exposed eggs of Lepidoptera and Heteroptera (Table 4.6). Host recognition in trichogrammatids is generally less specific compared to that of other egg parasitoids, as they may accept even inorganic objects not treated with chemicals and mercury globules if they are of a suitable size (Salt 1935). Huang and Gordh (1998) suggested that females of *Trichogramma australicum* Girault do not use chemical cues to recognize their host as they only depend on physical cues like egg size and texture. However, although the presence of a host recognition kairomone may not be necessary for these generalists, the presence of chemicals, depending on the host and parasitoid species, would elicit or increase recognition in several species or, by contrast, inhibit acceptance through repellency (Nordlund et al. 1987, Schmidt 1994). Saturated long chain hydrocarbons from the host eggs or associated materials, such as host scales or hairs, have been shown to increase parasitism by *Trichogramma* species (Lewis et al. 1975b, Grenier et al. 1993, Paul et al. 2002, Zou et al. 2002, Paramasivan et al. 2004). However, these compounds seem to operate as searching stimulants, acting on parasitoid responsiveness to the host, rather than being host recognition kairomones.

The effects of chemicals on host recognition by scelionids seem much more specialized when compared to trichogrammatids. Nordlund et al. (1987), in a cross comparison of *Te. remus* and *T. pretiosum* towards their respective hosts,

Table 4.6 Responses to contact kairomones from host eggs. n.r. = not reported; n.e.= no effect

Taxon	Host(s)	Source(s)	Chemical(s)	• Visual cues • Physical cues	References
Encyrtidae					
Ooencyrtus kuvanae	*Lymantria dispar*	From accessory glands, scales and hairs		• background color effect • n.r.	Lee et al. (1997), Hofstaetter and Raffa (1998)
Eulophidae					
Edovum puttleri Grissell	*Leptinotarsa decemlineata*	Hexane extract of eggs		• n.r. • curve surface	Leonard et al. (1987)
Oomyzus gallerucae	*Xanthogaleruca luteola*	Dichloromethane extract of eggs			Meiners and Hilker (1997)
O. galerucivorus	*Galeruca tanaceti*	Dichloromethane extract of eggs		• n.e. • n.e.	Meiners and Hilker (1997)
Tetrastichus hagenowii	*Periplaneta feliginosa, P. japonica*	Rectal secretions			Dai (1992)
	Periplaneta americana	From salivary glands	Mucopolisaccharides		Vinson and Piper (1986)
		From accessory glands	calcium oxalate		
Mymaridae					
Anagrus nilaparvatae	*Sogatella furcifera*				Lou and Cheng (2001)
Anaphes iole	*Lygus hesperus*	Compounds from eggs		• n.r. • wounds; egg shape and position related to substrate	Conti et al. (1996)

Table 4.6 (continued)

Taxon	Host(s)	Source(s)	Chemical(s)	• Visual cues • Physical cues	References
		Compounds on eggs and/or from host adults			Takasu and Nordlund (2001)
A. victus (Huber)	*Listronotus oregonensis*	Egg plugs			Cournoyer and Boivin (2005)
Scelionidae					
Telenomus chloropus (Thomson)	*Eurygaster spp.*	Substances on the eggs			Buleza (1985)
Te. busseolae	*Sesamia nonagrioides*	From colleterial glands	In glicoconjugates	• n. r. • dimension	De Santis et al. (2008)
Te. heliothidis	*Heliothis virescens*	From accessory glands	300-k protein fraction (2)	• no colour effect • dimension and shape	Strand and Vinson (1982, 1983a, b)
Te. podisi	*Euschistus heros*	Acetone extracts of egg. major component of pheromonal blend.	2,6,10-Trimethyltridecanoate	• no colour effect • n.r.	Borges et al. (1999)
Te. remus	*Heliothis zea*	From accessory glands (rejection)	1100-k protein fraction	•	Nordlund et al. (1987)
	Spodoptera frugiperda	From accessory glands	700-k protein fraction		
		Hexane extracts of egg			Gazit et al. (1996)

Table 4.6 (continued)

Taxon	Host(s)	Source(s)	Chemical(s)	• Visual cues • Physical cues	References
Te. theophilae Wu and Chen	*Theophila mandarina*	From accessory glands		• n.r. • dimension	Gao and Hu (1995, 1997), Gao et al. (2002), Wei et al. (2005)
Trissolcus basalis	*Nezara viridula*	Follicular secretion	Mupolysaccharide-protein complex	• n.r. • spherical substrate	Bin and Vinson (1985), Bin et al. (1993)
		Dichloromethane extract of eggs			Sales et al. (1978), Sales (1985)
Tr. brochymenae	*Murgantia histrionica*	Follicular secretion		• n.e. • dimension	Conti et al. (2003b, pers. observation)
Tr. grandis	*Eurygaster* spp.	Compounds on the eggs			Buleza (1985)
Trichogrammatidae					
Trichogramma australicum	*Helicoverpa armigera*	No use of chemical cues			Huang and Gordh (1998)
T. brasiliensis (Ashmead)	*Corcyra cephalonica*		Pentacosane		Paul et al. (2002), Singh et al. (2002)
T. brassicae	*Ostrinia nubilalis*	Satured hydrocarbons from eggs			Grenier et al. (1993)
	Pieris brassicae	From accessory glands of mated females			Fatouros (2006)
T. buesi Voegele	*Pieris brassicae*	From accessory glands (rejection)			Pak and de Jong (1987)

Table 4.6 (continued)

Taxon	Host(s)	Source(s)	Chemical(s)	• Visual cues • Physical cues	References
	Mamestra brassicae	From accessory glands			
T. chilonis	*Chilo partellus*	Hexane extracts of eggs	(heneicosane, tricosane, hexacosane, octacosane)		Paramasivan et al. (2004)
T. evanescens	*Pieris brassicae*	From accessory glands, egg washes with oviposition deterrent for the butterflies			Noldus and van Lenteren (1985), Pak and de Jong (1987)
T. exiguum	*Corcyra cephalonica*		Pentacosane		Paul et al. (2002), Singh et al. (2002)
T. pretiosum	*Heliothis zea*	From accessory glands			Nordlund et al. (1987)

S. frugiperda and *H. zea*, showed that the presence of the host's accessory gland secretion applied on glass beads is necessary to elicit probing in *Te. remus*, while it would only increase probing by *T. pretiosum*. Host egg size and shape are also important in Scelionidae (Strand and Vinson 1983a, Bin et al. 1993), but chemicals from the host egg surface are the ultimate cues to elicit egg recognition (Table 4.6). For some *Telenomus* and/or *Trissolcus* species it was shown that kairomones are contained in the adhesive secretion from the colleterial glands of Lepidoptera (Strand and Vinson 1983b, Nordlund et al. 1987, Gao and Hu 1995, De Santis et al. 2008) or follicular cells of heteropteran hosts (Bin et al. 1993, Borges et al. 1999, Conti et al. 2003b), but their chemical nature has not been clarified yet, except that they are glycoconjugate complexes (Bin et al. 1993, De Santis et al. 2008) and that active fractions are represented by large proteins (Strand and Vinson 1983b, Nordlund et al. 1987).

Eulophid egg parasitoids also rely on chemicals from accessory glands or from a different origin. Host recognition of *P. americana* eggs by *Tetrastichus* (= *Aprostocetus*) *hagenowii* is mediated by calcium oxalate originated from the host accessory glands and found as a common component of cockroach oothecae, as well as mucopolisaccharides from salivary glands (Vinson and Piper 1986). Another kairomonal source for *Tt. hagenowii* was found to be the rectal secretions from *Periplaneta* spp. (Dai 1992). For species attacking chrysomelid eggs, chemical extracts of eggs were necessary to elicit host recognition (Leonard et al. 1987, Meiners et al. 1997, Meiners and Hilker 1997).

More complex appear to be the host recognition mechanisms in mymarids that attack embedded eggs of turnished plant bugs (Conti et al. 1996, Takasu and Nordlund 2001), plant hoppers (Moratorio 1990, Cronin and Strong 1990, Lou and Cheng 2001) and weevils (Cournoyer and Boivin 2005), since chemicals from different host sources as well as the plant surface may elicit probing behavior. In addition, physical factors such as the presence of a wound and/or of an extruding egg, and the different positions of eggs related to the plant surface, significantly influence probing when combined with chemical cues (Conti et al. 1996). This apparently intricate mechanism may indicate that the probing behavior by this mymarid is the final step of the host searching, rather than the recognition behavior (Conti et al. unpublished). This would explain why *A. iole* intensively probes artificial lesions in plant tissues and artificial parafilm/gelcarin substrates, which were not contaminated by the host. Then, chemicals from the host would act synergistically with substrate and shape by enhancing probing intensity and, eventually, leading to host egg drilling (Conti et al. 1996, 1997).

4.4 The Chemistry of Host-Related Cues

Chemical cues from first and second trophic level are the stimuli that play a major role in the host searching behavior of egg parasitoids. In the last decade the identity of these semiochemicals has been intensively studied. Egg parasitoid females can perceive them using olfactory and taste sensilla according to the nature and the chemical properties of the compounds (Vinson 1985, see also Chapter 3).

4.4.1 Synomones

The chemistry of synomones emitted in response to herbivory has been deeply investigated in the system represented by rice plants, the rice brown planthopper, *N. lugens,* and *A. nilaparvatae* (Lou et al. 2005a, b). The co-occurrence of all development stages in the same plant, at the same time, allowed females of *A. nilaparvatae* to efficiently associate synomones induced by damage inflicted by immature and adult stages with the presence of suitable host eggs. Changes in the plant's volatiles emission induced by *N. lugens* nymphs and gravid females consisted of an increase of aliphatic aldehydes, alcohols, and several mono- and sesquiterpens. Synomones released by rice is not limited to leaves infested by herbivores, but also involves undamaged leaves. In addition, infested plants are attractive for the wasp only after 6–24 h from the beginning of infestation, while early infested plants or plants infested for more than 48 h are unattractive. Finally, rice plants treated with jasmonic acid (JA) were more attractive for the wasps than control plants.

These aspects are analogous with those observed in the systems where egg deposition is crucial for the induction of plant synomones, especially in the induction of a systemic response, the time window for synomone release, and the role of jasmonate signaling pathway on synomone emission (Colazza et al. 2004b, Hilker and Meiners 2006). A first aspect of synomone induction is the type of relationship that occurs between the ovipositing female and the eggs on one side, and the plant substrate on the other side. For some herbivores, such as *X. luteola, D. pini* (Meiners and Hilker 1997, 2000, Hilker et al. 2002, 2005) and *L. hesperus* (Rodriguez-Saona et al. 2002 Manrique et al. 2005), oviposition is strictly associated with mechanical lesions made by the ovipositing female in the oviposition site. Instead, in the case of the pentatomid bugs *N. viridula* and *M. histrionica,* eggs are glued on the plant surface with the female's follicular secretion and are not strictly associated with wounding, although females generally feed on the plant before and/or after oviposition (Colazza et al. 2004a, b, Conti et al. 2006). Finally, the lepidopterans *P. brassicae* and *S. nonagrioides* use their colleterial gland secretions to glue the egg masses on the plant surface and do not cause apparent lesions at oviposition, although a possible presence of micro-lesions cannot be excluded (Fatouros et al. 2005a, Conti et al. 2006).

A main question that arises at this point is: what are the elicitors? In the case of *X. luteola* and *D. pini* it was found that these are part of the secretion from the oviduct associated glands, which must be in direct contact with the plant wound in order to be active (Meiners and Hilker 2000, Hilker et al. 2005). Conversely, the colleterial gland secretion of *S. nonagrioides* was found to be inactive, but ovarian eggs artificially applied on maize plants induced parasitoid response to plant synomones (Conti et al. 2006). In pentatomid bugs, whether the elicitor originates from the eggs or from the follicular secretion is unknown; however, the combined presence of feeding punctures is necessary for synomone induction by *N. viridula* and to enhance parasitoid response in the case of *M. histrionica* (Colazza et al. 2004a, Conti et al. 2006). A recent paper shows the presence of JA in the eggs of several lepidopterans, which may provide an explanation for the oviposition-induced plant resistance (Tooker and De Moraes 2005). However, the presence of this compound in the eggs of the herbivores considered here is unknown. A proteic fraction of the oviduct secretion appears to act as the elicitor of oviposition-induced synomones in *D. pini* (Hilker et al. 2005), but no elicitors have been identified from host genitalia, eggs or saliva for the other oviposition-induced synomones to date. In the case of piercing-sucking insects, Rodriguez-Saona et al. (2002) showed that the volatile blend induced by *L. hesperus* salivary glands is similar to that induced by volicitin, but chemical analyses of the salivary glands from *Lygus* species have so far shown no evidence of volicitin. Therefore, it is possible that other fatty acid-amino acid coniugates (FACs) are present. Overall, there is evidence that the mechanisms of volatile induction caused by *L. hesperus* are similar to those induced by chewing caterpillars, and that the plant response is also similar to responses induced by volicitin.

A second main question concerns the chemistry of the plant synomones and the biochemical pathways involved in their production. Qualitative and/or quantitative changes in the profile of volatile organic compounds (VOCs) characterize the response of some plants to herbivore oviposition. Specifically, oviposition of *X. luteola* on elm plants, *Ulmus minor* Miller, induces an increase of green leaf volatiles (GLVs) and terpenic compounds, as well as *de novo* synthesis of some homoterpenoids (Wegener et al. 2001, Wegener and Schulz 2002), whereas *D. pini* induces a higher emission of (*E*)-β-farnesene in pine plants, *Pinus sylvestris* L., which elicit parasitoid response only when bioassayed as a blend with the common volatiles of undamaged pine plants (Mumm and Hilker 2005). Combined oviposition with feeding punctures of *N. viridula* on *Vicia faba* L. and *Phaseolus vulgaris* L. induce an increase of (*E*)-β-cariophyllene as well as two sesquiterpenes (Colazza et al. 2004b).

In the case of the oviposition-induced synomones in corn and cabbage, parasitoids respond only after having contacted the plant or from a very short distance (Fatouros et al. 2005a, Conti et al. 2006). Therefore, it may be suspected that the active compounds have a very low volatility. In fact, non-volatile secondary metabolites can be deposited on the plant surface via diffusion and associated with the waxes, or they can be stored in glandular trichomes (Muller and Riederer 2005). Many studies have addressed the presence of secondary metabolites on the leaf surface, stems or reproductive tissue (seeds and fruits), where they act as allomones or kairomones (Muller and Riederer 2005). Alternatively, because feeding and/or oviposition causes different changes in the VOCs profile of plants, we cannot exclude possible absorption of volatile compounds by the plant epicuticular waxes (Conti et al. 2006, 2008).

4.4.2 Host Pheromones with Kairomonal Activity

Males of the bean bug *R. clavatus* emit an aggregation pheromone that attracts adults of both sexes and nymphs (Mizutani 2006). The aggregation pheromone consists of three components: (*E*)-2-hexenyl (*E*)-2-hexenoate, (*E*)-2-hexenyl (*Z*)-3-hexenoate, and tetradecyl isobutyrate. One of the three, (*E*)-2-hexenyl (*Z*)-3-hexenoate, attracts females of the egg parasitoid *O. nezarae*, but does not attract *R. clavatus*. This compound, sprayed in the crops, induce parasitoids to immigrate into the fields earlier than *R. clavatus*, and to remain at higher densities than in untreated ones. According to the data obtained from eggs artificially placed on soybean plants, field application of (*E*)-2-hexenyl (*Z*)-3-hexenoate resulted in higher parasitism as compared with the untreated fields.

The egg parasitoid *Te. busseolae*, a natural enemy of various noctuids belonging to the genera *Sesamia*, is attracted to the pheromone emitted by females of the Mediterranean stem borer, *S. nonagrioides*, and the pink stem borer, *Sesamia calamistis* (Hampson) (Colazza et al. 1997, Chabi-Olaye et al. 2001). In particular, in laboratory experiments, it was shown that three components of the synthetic sex pheromone of the Mediterranean stem borer, e.g. (*Z*)-11-hexadecenyl acetate (the main component), (*Z*)-11-hexadecenal and dodecyl acetate, attract wasp females, while a fourth compound, (*Z*)-11-hexadecenol, does not. Interestingly, these active compounds are also present in the sex pheromone blends of other *Sesamia* hosts, and in the sex pheromone blends of some non-host noctuid species. However, egg parasitoid female may use other compunds present in the sex pheromone of non host species to avoid wrong decisions, as in the case of (*Z*)-9-hexadecenal, present in the corn earworm sex pheromone, *Heliothis armigera* (Hübner), which in laboratory experiments prevents *Te. busseolae* females from responding to the sex pheromone of this non-host species (Peri et al. 2007).

4.4.3 Contact Kairomones

In addition to host pheromones, other host-related kairomonal cues are generally composed by slightly volatile chemicals that induce behavioral response only after wasp females have touched the kairomone with their antennae, involving, in this way, the gustatory sensilla (see Chapter 3).

Chemical analysis of lepidopteran scales revealed the presence of linear hydrocarbons and various organic acids. Jones et al. (1973) isolated dodecosane, tricosane, tetracosane and pentacosane from *H. zea* scales which were able to arrest *T. evanescens* females. Shu et al. (1990) found 3-dimethyl-nonatriacontane from scale extracts of the European corn borer that increase parasitism rates of *T. nubilale*. Instead a blend of various organic acids recovered from *O. nubilalis* scales, had a minor influence on the field performance of *T. pretiosum* (Gueldner et al. 1984). Altogether, these studies show that straight-chain hydrocarbons and related long-chain carboxylic acids are important factors for host location. However, it is possible that moth scales can adsorb and release volatile compounds, such as the sex pheromone emitted by virgin females (Arakaki and Wakamura 2000, Yong et al. 2007).

The chemistry of the traces (footprints) that induce arrestment response in egg parasitoids was investigated in the association *N. viridula-Tr. basalis*. Females of *T. basalis* are able to perceive the chemical traces left by host adults of *N. viridula* at the dose of about 7.0 ng/cm^2 of cuticular hydrocarbons applied on a filter paper arena (Colazza et al. 2007). However, how *Tr. basalis* females can discriminate the host cuticular hydrocarbons amid the waxy background of the plant cuticle remains to be elucitade. Chemical analysis of extracts of *N. viridula* cuticular lipids revealed the presence of linear alkanes with quantitative and qualitative differences between the sexes. The linear alkan *n*-nonadecane (nC_{19}) was recovered only from extracts of males and from solid-phase microextraction (SPME) of residues left by *N. viridula* males walking on a glass surface. This alkan added to the crude extracts of *N. viridula* females induced on *Tr. basalis* females a significant reduction of their residence time in the arena, similar to what occurs when female wasps are in the presence of hexane extracts of male hosts (Colazza et al. 2007). This evidence demonstrates that *Tr. basalis* females can discriminate between residues left by male or female hosts based on the presence/absence of *n*-nonadecane.

4.5 How Egg Parasitoids Exploit Host-Related Chemical Cues

The behavioral responses of egg parasitoids to kairomones from their hosts or to host-induced plant synomones often reflect the general behavior of hymenopterous parasitoids attacking other host stages, with peculiar characteristics associated with the fact that the host eggs are generally available for a short time and, obviously, immobile, and that the host oviposition sites can be exposed, embedded in plant tissues and/or protected by different materials (see Chapter 2). In addition, egg parasitoids are generally not strong fliers, and, thereby, directional searching to a source of volatiles from a distance is unlikely.

In the case of *Trichogramma* it has been hypothesized that they "float" in the wind until the presence of suitable cues will cause them to land on the substrate (Nordlund 1994), although short oriented flies are possible especially in the case of low wind (see Fatouros et al. 2008 and literature therein). Walking behavior of egg parasitoids has been extensively studied in olfactometers and other types of arenas. Air flows carrying volatile stimuli generally elicit arrestment behavior or odour-regulated positive anemotaxis and reduction of linear speed.

Trichogrammatidae and Scelionidae react in a similar way to indirect host-related contact cues. Once touched an area contaminated by moth scales or by true bug footprints, they delay the tendency of fly and start to adopt a searching behavior in and around the vicinity of the contaminated patches (called "motivated searching" by Vinson 1998). At the beginning, females remain for a relevant amount of time motionless rubbing the surface with the antennae (Laing 1937, Zaborsky et al. 1987, Colazza et al. 1999). Then, while drumming the surface with the antennae, they exhibit a searching behavior characterized by orthokinetic and klinokinetic responses (Laing 1937, Gardner and van Lenteren 1986, Schmidt 1994, Colazza et al. 1999). During such motivated searching behavior, wasp females remain in or around the contaminated area where host eggs are more likely to be found, and their response is reinforced by systematic returns to the stimuli after losing contact with them. However, wasp females which are not rewarded by successful oviposition within a certain amount of time gradually lose their motivated searching response and progressively move back to a more general host searching behavior (Peri et al. 2006).

Host eggs can be exposed on plant tissues or embedded (partially or completely), and as a consequence the way that eggs are laid can affect parasitoid response. When at a very short distance from hosts, parasitoids of exposed eggs may react with directional movements, elicited by visual and/or volatile stimuli. In experiments conducted on a filter paper open arena, females of *Tr. brochymenae* showed chemotaxic response to volatiles from host egg extracts, but no response to visual cues from host eggs (Conti et al. 2003b). This behavior is quite different when host eggs are embedded in plant tissues, as parasitoids often also respond to plant wounds (Conti et al. 1996, 1997, Takasu and Nordlund 2001). A combined and/or sinergistic effect of stimuli from host eggs and host oviposition incisions can be hypothesized in the case of *L. hesperus* (see above), although the type of behavioral response has not been elucidated yet.

When a female parasitoid encounters the host egg or egg mass, or an oviposition incision containing host eggs, the response is characterized by intense host antennation or probing. *Trichogramma* females carefully examine their exposed host by mounting and walking on it from side to side, showing a reduced walking speed and an intensified frequency of antennal drumming (Schmidt and Smith 1987, 1989). If the egg is rejected, the parasitoid walks away, while in case of acceptance, it assumes a probing posture and uses its ovipositor to drill the host chorion (Schmidt 1994). The chemical and physical cues perceived before and during host examination and after drilling are used by females to choose whether or not to oviposit in the host, the number of eggs to lay and the sex to allocate (see Chapter 2). Similar arrestment responses and intensification of antennal drumming are also observed in other egg parasitoid families (see references in Tables 4.3 and 4.6). Females of *Tr. basalis* first antennate the lateral side of their barrel-shaped host eggs, then they examine one or more eggs of the mass before mounting on it. Generally the females examine the entire edge of the egg mass before giving attention to a single egg. If the host eggs are recognized, they will finally probe them with the ovipositor by assuming a probing posture (Bin et al. 1993). Functional morphology evidences on *Tr. basalis* females suggested the contribution of the papillary sensilla on host recognition (Bin et al. 1989, Isidoro et al. 1996). These sensilla are closely associated with accessory glands that produce a secretion which may play a role as a solvent for contact kairomones (Isidoro et al. 1996).

Mymaridae that attack embedded eggs show typical differences in their host recognition behavior, as they may not directly contact hosts with their antennae. After encountering an egg that is partially exposed above the substrate, mymarids show arrestment behavior as earlier described, with strong decrease of linear speed and intense antennal drumming. However, when the host eggs are deeply embedded into plant tissues, the parasitoids may try to contact them by inserting the antennae in the lesion before probing with the ovipositor (Cronin and Strong 1990, Moratorio 1990, Conti et al. 1996, 1997).

4.6 How Learning Can Change the Response to Host-Related Chemical Cues

Several reports show that egg parasitoid females are innately attracted to synomones and/or kairomones (Vinson 1985, Lewis and Tumlinson 1988, Turlings et al. 1993, Godfray 1994, Vet et al. 1995). For example, naïve females of *C. ruforum*, *Dipriocampe diprioni* (Ferrière), *Tr. basalis*, and *Tr. brochymenae* respond to host egg-induced synomones (Meiners and Hilker 2000, Colazza et al. 2004a, Conti et al. 2006). Moreover, naïve females of *C. ruforum* and *D. diprioni* positively react to the sex pheromones of their hosts (Hilker et al. 2000). Such wasp's innate behavioral repertoires can be deeply modified by experiences acquired during the development and/or during the adult stage, although a clear separation between preadult and adult experiences is not always possible, or at least it could be the subject of controversy (Vet and Groennewold 1990). Studies indicate that egg parasitoids can become experienced during larval development inside the host or at eclosion while females antennate the host chorion. In addition, the more polyphagous an egg parasitoid is, the more likely pre-adult and/or adult learning can occur. For example, *Trichogrammatoidea bactrae fumata* Nagaraja shows a clear preference for *Epiphyas postvittana* (Walker) when it is reared on this host species (Stevens 1995). The polyphagous egg parasitoid *Trichogramma maidis* Pint. et Voeg. prefers *O. nubilalis* over *Anagasta kuehniella* (Zeller), but this preference decreases when wasps are reared on *A. kuehniella* (Kaiser et al. 1989a). Females of *T.* nr. *brassicae* (formerly *T.* nr. *ivelae*) would search longer on tomato leaf if emerged from hosts laid on a tomato plants (Bjorksten and Hoffmann 1998).

As stated earlier, the host location process of egg parasitoid females is regulated by indirect and direct host-related cues. It is generally assumed that direct host-related cues elicit fixed responses, whereas indirect host-related cues require various foraging decisions which could be influenced by the adult ability to learn. Current knowledge indicates that the responses of naïve egg parasitoids to indirect host-related cues are improved by rewarding, i.e. oviposition experiences accumulated while foraging. For example, the response of *T. evanescens* females to the host sex pheromone is increased markedly by oviposition experience in the presence of this indirect host-related cue (Schöller and Prozell 2002). In *T. maidis*, females can learn to associate the presence of host eggs with some olfactory cues, such as the food plant of the host and the host pheromone (Kaiser et al. 1989b). By contrast, other egg parasitoids seem unable to associate host-derived stimuli with successful oviposition, as in the cases the eulophids *C. ruforum* and *D. diprioni* (Hilker et al. 2002).

The recent discovery of the attractiveness of egg parasitoids for synomones induced by egg deposition opened a new perspectives about the possible roles played by adult experience on the response to direct host-related cues. Females of *T. brassicae* that had been previously exposed to plants with host eggs were attracted to the host-induced synomones, whereas naïve females were not responsive (Fatouros et al. 2005a). Similarly, naïve females of *C. ruforum* were not attracted to oviposition-induced pine volatiles (Mumm et al. 2005), while they showed attractiveness after experience with food source and host eggs (Schröder et al. 2008). In this case, the sex pheromone can direct wasp females towards the host population and closer to future host oviposition sites, whereas oviposition-induced pine volatiles reliably indicate the host egg presence, so that experienced wasps can improve host search and increase their chance to locate a host egg.

On the other hand, egg parasitoid females that have responded to a set of cues and were unsuccessful in locating a host can adjust their foraging behavior accordingly. For example, *Tr. basalis* change their innate searching behavior on host traces according to the experience gained during foraging. The response to host footprints became weaker when wasps encountered patches without oviposition rewards repeatedly, whereas females responded strongly on kairomone patches where oviposition had taken place (Peri et al. 2006). The time window needed by *Tr. basalis* females to 'forget' unrewarded experiences was estimated in about 3 days. After this latent period, female wasps that encountered a contaminated patch behaved similar to naïve females (Peri et al. 2006). Taken together, these learning and forgetting mechanisms are seen as having an adaptive meaning since they are most likely leading foraging parasitoid females to find their hosts and to attack them in a more efficient way (Dauphin et al. 2009). A spatially-explicit Monte Carlo simulation showed that females of *Tr. basalis* are able to learn the features of their foraging environment and to adjust accordingly the amount of time spent on the patches of kairomones they are visiting, depending on whether or not host eggs are found (Dauphin et al. 2009).

Different to Scelionidae, in *Trichogramma* the oviposition experience does not seem to be able to increase or restore the wasp's interest for moth scales (Gardner and van Lenteren 1986, Thomson and Stinner 1990, Nurindah et al. 1999). This divergence among Trichogrammatidae and Scelionidae in their ability to associate indirect host-related contact kairomones with reward experiences might be explained with the rank of correlation of the stimuli with the presence of host eggs. Adult chemical foot prints represent a reliable predictor of host eggs presence for Scelionidae as they allow the egg parasitoid females to discriminate between coevolved and non-coevolved host species (Salerno et al. 2006), and among coevolved hosts, to discriminate between host males and females (Colazza et al. 1999, Conti et al. 2003b, Colazza et al. 2007). On the contrary, the poor level of host-specificity evoked by adult scales might explain the small role played by host scales in inducing associative links with oviposition experience, so that for Trichogrammatidae females associating host scales with oviposition would be counter-adaptive (Thomson and Stinner 1990). Moreover, host scales cannot provide information that allow Trichogrammatidae females to discriminate between host genders.

4.7 Concluding Remarks

In this chapter we have shown how intense the study has been on searching behavior in egg parasitoids during the last seven decades. Several mechanisms of the tri-trophic interactions have been elucidated, revealing that, most likely due to the

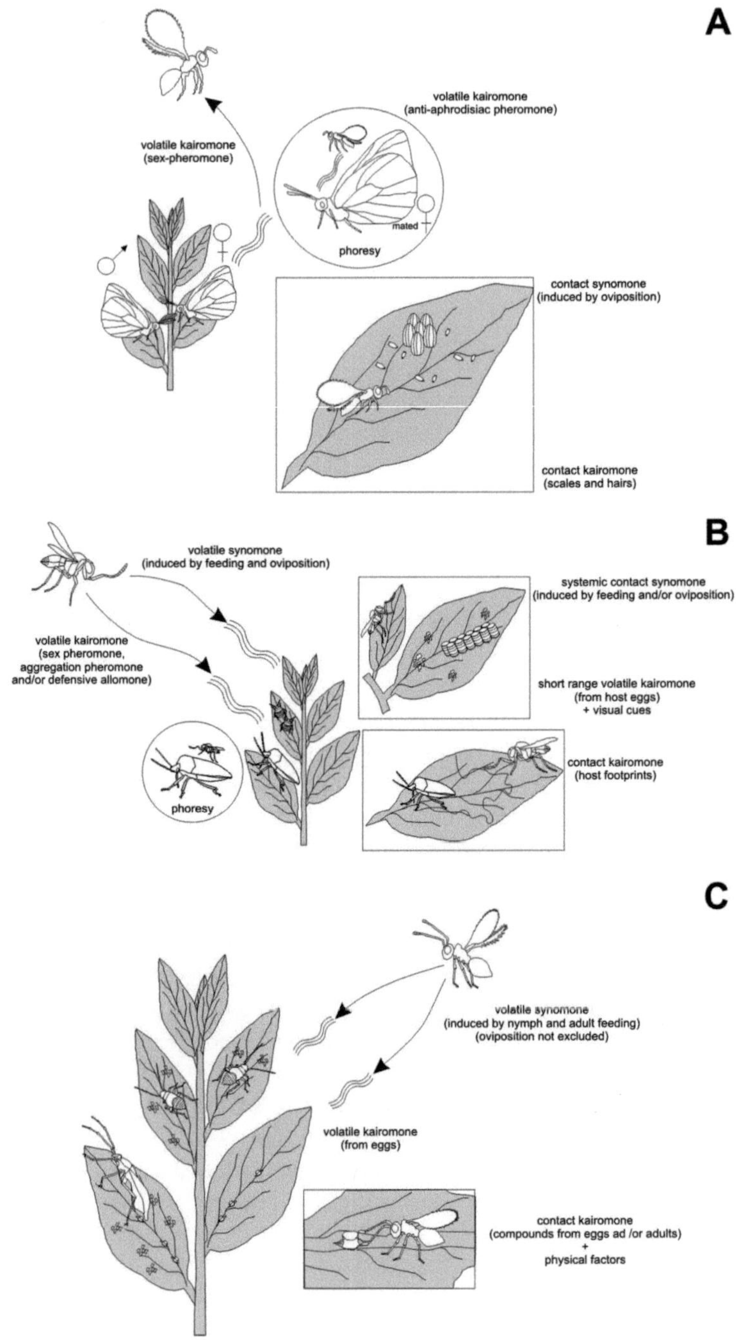

Fig. 4.2 Three examples of host searching strategies and relative chemical cues exploited by egg parasitoids: (**a**) *Trichogramma* spp. *vs. Pieris brassicae* on Brussels sprouts; (**b**) *Trissolcus* spp. *vs.* pentatomid bugs on different plant species; (**c**) *Anaphes iole vs. Lygus* spp. on different plant species (partly modified from Conti et al. 2006)

small biomass of the host eggs, volatile kairomones from host adults or oviposition-induced synomones from the plants often have a major importance (Fig. 4.2). However, because of the wide diversity of host-parasitoid associations, many aspects still need to be clarified. In particular, it is necessary to investigate the "host unit" for most of such associations, defined as the ensemble of the physical and semiochemical characters from the host and from associated macro-organisms (i.e., the plant) and microorganisms which elicit host selection in parasitoids (Conti et al. 2000, Vinson et al. 2002, see also Chapter 2).

In addition, in-depth studies would shed light on whether different physical and, especially, semiochemical stimuli used by egg parasitoids act hierarchically, sequentially, or even alternatively.

Some limitations to the acquisition of such knowledge probably depend on the fact that research has been conducted mainly in the laboratory and was focused especially on host-parasitoid associations that are economically relevant for agriculture. Because laboratory investigations clearly tend to simplify the systems, they are useful to understand single interactions, and the different aspects of more complex interactions. However, in order to understand what the real complexity of multitrophic interactions is, it would be important to carry out studies in more realistic conditions, i.e. semi-field and/or field.

Understanding the strategies of host selection by egg parasitoids and the cues involved would provide important potential to improve the efficacy of biological control, through behavioral manipulation of parasitoids in the field, and to develop *in vivo* and *in vitro* techniques for mass rearing of egg parasitoids (Cônsoli and Parra 1999). Interesting application prospects are offered by the new acquisitions on the oviposition-induced synomones in the development of plants characterized by high expression of indirect induced resistance. In addition, field spraying with resistance elicitors may also be considered (Dicke and Hilker 2003). In general, direct field treatments with synthetic analogues or natural synomones or kairomones, or pre-release conditioning of parasitoid behavior have also been proposed for these and other parasitoid groups (Lewis and Martin 1990, Papaj and Vet 1990, Dicke and Hilker 2003).

Finally, determining the host selection strategies in egg parasitoids would be important to define release methods and spreading (Lewis and Martin 1990), develop quality control procedures in biofactories (Lewis and Martin 1990, van Lenteren 2003), select egg parasitoids, evaluate their specificity and assess the risk of their introduction in classical biologiocal control (Wajnberg et al. 2001). More generally, these studies would be fundamental for a finely tuned planning of biocontrol and IPM programmes (van Lenteren 2006).

Acknowledgments The authors are grateful to Ferdinando (Nando) Bin for his useful suggestions and encouragement. This work was financially supported by Cofin/PRIN 2005 (Potential for biological control of *Sesamia* spp. using egg parasitoids), FISR 2005 (SIMBIO-VEG), and is part of the European Science Foundation (ESF) – Behavioural Ecology of Insect Parasitoids (BEPAR) scientific programme.

References

Abdelgader H, Mazomenos B (2002) Response of *Trichogramma oleae* (Hymenoptera: Trichogrammatidae), to host pheromones, frass and scales extracts. Egg Parasitoid News 14:16–17

Aldrich JR (1995) Chemical communication in true bugs and exploitation by parasitoids and commensals. In: Carde RT, Bell WJ (eds) Chemical ecology of insects II. Chapman & Hall, London, pp 318–363

Aldrich JR, Kochansky JP, Abrams CB (1984) Attractant for a beneficial insect and its parasitoids: pheromone of a predatory spined soldier bug, *Podisus maculiventris* (Hemiptera: Pentatomidae). Environ Entomol 13:1031–1036

Ananthakrishnan TN, Senrayan R, Murugesan S, Annadurai RS (1991) Kairomones of *Heliothis armigera* and *Corcyra cephalonica* and their influence on the parasitic potential of *Trichogramma chilonis* (Trichogrammatidae: Hymenoptera). J Biosci 16:11–119

Arakaki N, Wakamura S (2000) Bridge in time and space for an egg parasitoid kairomonal use of trace amount of sex pheromone adsorbed on egg mass scale hair of the Tussock moth, *Euproctis taiwana* (Shiraki) (Lepidoptera: Lymantriidae), by egg parasitoid, *Telenomus euproctidis* Wilcox (Hymenoptera: Scelionidae), for host location. Entomol Sci 3:25–31

Arakaki N, Wakamura S, Yasuda T (1996) Phoretic egg parasitoid, *Telenomus euproctidis* (Hymenoptera: Scelionidae), uses sex pheromone of tussock moth *Euproctis taiwana* (Lepidoptera: Lymantriidae) as a kairomone. J Chem Ecol 22:1079–1085

Arakaki N, Wakamura S, Yasuda T, Yamagishi K (1997) Two regional strains of a phoretic egg parasitoid, *Telenomus euproctidis* (Hymenoptera: Scelionidae), that use different sex pheromones of two allopatric tussock moth species as kairomones. J Chem Ecol 23:153–161

Bai SX, Wang ZY, He KL, Zhou DR (2004) Olfactory responses of *Trichogramma ostriniae* Pang et. Chen to kairomones from eggs and different stages of adult females of *Ostrinia furnacalis* (Guenee). Acta Entomol Sin 47:48–54

Bakthavatsalam N, Tandon PL (2006) Kairomones, their optimum concentrations, and application techniques to enhance the parasitization efficacy of *Trichogramma chilonis* Ishii (Hymenoptera: Trichogrammatidae). J Biol Control 20:169–174

Bale JS, van Lenteren JC, Bigler F (2008) Biological control and sustainable food production. Philos Trans R Soc Lond 363B:761–776

Barnay O, Pizzol J, Gertz C, Kienlen JC, Hommay G, Lapchin L (1999) Host density-dependence of discovery and exploitation rates of egg patches of *Lobesia botrana* (Lepidoptera: Tortricidae) and *Ephestia kuehniella*. J Econ Entomol 92:1311–1320

Battisti A (1989) Field studies on the behaviour of two egg parasitoids of the pine processionary moth *Thaumetopoea pityocampa*. Entomophaga 34:29–38

Beevers M, Lewis WJ, Gross HEJr, Nordlund DA (1981) Kairomones and their use for management of entomophagous insects: X. Laboratory studies on manipulation of host-finding behavior of *Trichogramma*. J Chem Ecol 7:635–648

Bin F (1994) Biological control with egg parasitoids other than *Trichogramma*. In: Wajnberg E, Hassan SA (eds) Biological control with egg parasitoids. CAB International, Oxford, pp 245–271

Bin F, Vinson SB (1985) Egg recognition kairomone of *Trissolcus basalis* (Woll.) parasitoid of the southern green stick bug (Hym., Scelionidae; Hem., Pentatomidae). Atti XIV Congresso Nazionale Italiano di Entomologia, pp 903–904

Bin F, Colazza S, Isidoro N, Solinas M, Vinson SB (1989) Antennal chemosensilla and glands, and their possible meaning in the reproductive behaviour of *Trissolcus basalis* (Woll.) (Hym.: Scelionidae). Entomologica 24:33–97

Bin F, Vinson SB, Strand MR, Colazza S, Jones WA (1993) Source of an egg kairomone for *Trissolcus basalis*, a parasitoid of *Nezara viridula*. Physiol Entomol 18:7–15

Bjorksten TA, Hoffmann AA (1998) Persistence of experience effects in the parasitoid *Trichogramma* nr *brassicae*. Ecol Entomol 23:110–117

Bloem KA, Yeargan KV (1982) Host-finding behaviour of *Patasson lameerei* (Hym., Chalcidoidea), a parasitoid of *Sitona* eggs (Col., Curculionidae). Entomophaga 27:93–97

Boo KS, Yang JP (2000) Kairomones used by *Trichogramma chilonis* to find *Helicoverpa assulta* eggs. J Chem Ecol 26:359–375

Borges M, Costa MLM, Sujii ER, Cavalcanti MDG, Redigolo GF, Resck IS, Vilela EF (1999) Semiochemical and physical stimuli involved in host recognition by *Telenomus podisi* (Hymenoptera: Scelionidae) toward *Euschistus heros* (Heteroptera: Pentatomidae). Physiol Entomol 24:227–233

Borges M, Colazza S, Ramirez-Lucas P, Chauhan KR, Kramer M, Moraes MCB, Aldrich JR (2003) Kairomonal effect of walking traces from *Euschistus heros* (Heteroptera: Pentatomidae) on two strains of *Telenomus podisi* (Hymenoptera: Scelionidade). Physiol Entomol 28:349–355

Bruni R, SantÁna J, Aldrich JR, Bin F (2000) Influence of host pheromone on egg parasitism by scelionid wasps: Comparison of phoretic and nonphoretic parasitoids. J Insect Behav 13: 165–173

Buleza VV (1985) Mechanisms of search and choice of host in egg parasites (Hymenoptera, Scelionidae). Zool Zhurnal 64:1309–1317

Buleza VV, Mikheev AV (1979) Localisation in the host′s body of substances stimulating host-seeking by the egg parasite *Trissolcus grandis* (Thom.). Khemoretseptsiya Nasekomykh 4: 95–100

Cabello Garcia T, Vargas Piqueras P (1985) Olfactometer studies of the influence of the plant and of the insect host in the searching activity of *Trichogramma cordubensis* Vargas, Cabello and *T. sp.* near *buesi* (Hym.: Trichogrammatidae). Boletin Servicio Defensa contra Plagas Inspeccion Fitopatologica 11:237–241

Chabi-Olaye A, Schulthess F, Poehling HM, Borgemeister C (2001) Host location and host discrimination behavior of *Telenomus isis*, an egg parasitoid of the African cereal stem borer *Sesamia calamistis*. J Chem Ecol 27:663–678

Colazza S, Bin F (1988) Risposta di *Trissolcus simoni* (Mayr) (Hym.: Scelionidae) ad *Eurydema ventrale* Klt. (Het.: Pentatomidae) in olfattometro. Atti XV Congresso Nazionale Italiano di Entomologia, L'Aquila, Italy, pp 833–840

Colazza S, Rosi MC (2001) Differences in the searching behaviour of two strains of the egg parasitoid *Telenomus busseolae* (Hymenoptera: Scelionidae). Eur J Entomol 98:47–52

Colazza S, Rosi CM, Clemente A (1997) Response of egg parasitoid *Telenomus busseolae* to sex pheromone of *Sesamia nonagrioides*. J Chem Ecol 23:2437–2444

Colazza S, Salerno G, Wajnberg E (1999) Volatile and contact chemicals released by *Nezara viridula* (Heteroptera: Pentatomidae) have a kairomonal effect on the egg parasitoid *Trissolcus basalis* (Hymenoptera: Scelionidae). Biol Control 16:310–317

Colazza S, Fucarino A, Peri E, Salerno G, Conti E, Bin F (2004a) Insect oviposition induces volatiles emission in herbaceous plant that attracts egg parasitoids. J Exp Biol 207:47–53

Colazza S, Mcelfresh JS, Millar JG (2004b) Identification of volatile synomones, induced by *Nezara viridula* feeding and oviposition on bean spp. that attract the egg parasitoid *Trissolcus basalis*. J Chem Ecol 5:939–958

Colazza S, Aquila G, De Pasquale C, Peri E, Millar J (2007) The egg parasitoid *Trissolcus basalis* uses n-nonadecane, a cuticular hydrocarbon from its stink bug host *Nezara viridula*, to discriminate between female and male hosts. J Chem Ecol 33:1405–1420

Cônsoli F, Parra JRP (1999) In vitro rearing of parasitoids: constraints and perspectives. Trends Entomol 2:19–32

Conti E, Bin F, Vinson SB (2000) Host range in egg parasitoids: a tentative approach through the analysis of the host unit. In: 7th European Workshop on Insect Parasitoids, Haarlem, The Netherlands, p 32

Conti E, Jones WA, Bin F, Vinson SB (1996) Physical and chemical factors involved in host recognition behavior of *Anaphes iole* Girault, an egg parasitoid of *Lygus hesperus* Knight (Hymenoptera: Mymaridae; Heteroptera: Miridae). Biol Control 7:10–16

Conti E, Jones WA, Bin F, Vinson SB (1997) Oviposition behavior of *Anaphes iole*, an egg parasitoid of *Lygus hesperus* Knight (Hymenoptera: Mymaridae; Heteroptera: Miridae). Ann Entomol Soc Am 90:91–101

Conti E, Salerno G, Bayram A, Bin F (2003a) Strategies involved in host location of *Telenomus busseolae* and *Trichogramma turkestanica*, egg parasitoids of *Sesamia nonagrioides*. XII International Entomophagous Workshop, Tucson, AZ 27–31 July 2003. J Insect Sci 3.33:6

Conti E, Salerno G, Bin F, Williams H, Vinson SB (2003b) Chemical cues from adults, nymphs and eggs of *Murgantia histrionica* eliciting host location and recognition in the egg parasitoid *Trissolcus brochymenae*. J Chem Ecol 29:115–130

Conti E, Salerno G, Bin F, Vinson SB (2004) The role of host semiochemicals in parasitoid specificity: a case study with *Trissolcus brochymenae* and *Trissolcus simoni* on pentatomid bugs. Biol Control 29:435–444

Conti E, Salerno G, De Santis F, Leombruni B, Bin F (2006) Difese indirette delle piante: i sinomoni per contatto indotti da ovideposizione. Atti Accad Naz Ital Entomol Rend 54:129–148

Conti E, Zadra C, Salerno G, Leombruni B, Volpe D, Frati F, Marucchini C, Bin F (2008) Changes in the volatile profile of *Brassica oleracea* due to feeding and oviposition by *Murgantia histrionica* (Heteroptera: Pentatomidae). Eur J Entomol 105:839–847

Cournoyer M, Boivin G (2005) Evidence for kairomones used by the egg parasitoid *Anaphes victus* (Hymenoptera: Mymaridae) when searching for its host. Can Entomol 137:230–232

Cronin JT, Strong DR (1990) Biology of *Anagrus delicatus* (Hymenoptera: Mymaridae), an egg parasitoid of *Prokelisia marginata* (Homoptera: Delphacidae). Ann Entomol Soc Am 83: 846–854

Dai XH (1992) Studies on the stimulating behaviour of rectal secretions of cockroach on the oviposition behaviour of *Tetrastichus hagenowii* (Hymenoptera, Eulophidae). Chin J Biol Control 8:13–15

Dauphin G, Coquillard P, Colazza S, Peri E, Wajnberg E (2009) Host kairomone learning and foraging success in an egg parasitoid: a simulation model. Ecol Entomol 34:193–293

De Santis F, Conti E, Romani R, Salerno G, Parillo F, Bin F (2008) Colleterial glands of *Sesamia nonagrioides* as a source of the host recognition kairomone for the egg parasitoid *Telenomus busseolae*. Physiol Entomol 33:7–16

Dicke M, Hilker M (2003) Induced plant defences: from molecular biology to evolutionary ecology. Basic Appl Ecol 4:3–14

Ding DC, Qiu HQ, Du JW (1988) Host recognition and host acceptance behaviour of *Tetrastichus schoenobii*. Colloques l′ INRA 43:173–180

Fatouros NE (2006) Parasitic wasps on butterfly expedition. Foraging strategies of egg and larval parasitoids exploiting infochemicals of Brussels sprouts and their *Pieris* hosts. Ph.D. Dissertation, Berlin (Germany), Freie Universität Berlin,187 pp

Fatouros NE, Bukovinszkine'Kiss G, Kalkers LA, Soler Gamborena R, Dicke M, Hilker M (2005a) Oviposition-induced plant cues: do they arrest *Trichogramma* wasps during host location? Entomol Exp Appl 115:207–215

Fatouros NE, Huigens ME, van Loon JJA, Dicke M, Hilker M (2005b) Chemical communication – Butterfly anti-aphrodisiac lures parasitic wasps. Nature 433:704

Fatouros NE, Bukovinszkine′Kiss G, Dicke M, Hilker M (2007) The response specificity of *Trichogramma* egg parasitoids towards infochemicals during host location. J Insect Behav 20:53–65

Fatouros NE, Dicke M, Mumm R, Meiners T, Hilker M (2008) Foraging behavior of egg parasitoids exploiting chemical information. Behav Ecol 19:677–689

Ferreira L, Pintureau B, Voegele J (1979) A new type of olfactometer. Application to the measurement of the ability to search for and locate attractant substances in the host in *Trichogramma* (Hym. Trichogrammatidae). Ann Zool, Ecol Anim 11:271–279

Fiaboe MK, Chabi-Olaye A, Gounou S, Smith H, Borgemeister C, Schulthess F (2003) *Sesamia calamistis* calling behavior and its role in host finding of egg parasitoids *Telenomus busseolae*, *Telenomus isis*, and *Lathromeris ovicida*. J Chem Ecol 29:921–929

Frenoy C, Farine JP, Hawlitzky N, Durier C (1991) Role of kairomones in the relations between *Ostrinia nubilalis* Hubner (Lep., Pyralidae) and *Trichogramma brassicae* Bezdenko (Hym., Trichogrammatidae). Redia 74:143–151

Frenoy C, Durier C, Hawlitzky N (1992) Effect of kairomones from egg and female adult stages of *Ostrinia nubilalis* (Hubner) (Lepidoptera, Pyralidae) on *Trichogramma brassicae* Bezdenko (Hymenoptera, Trichogrammatidae) female kinesis. J Chem Ecol 18:761–773

Gao QK, Hu C (1995) Source and characterization of an host egg recognition kairomone of *Telenomus theophilae* Wu et Chen. J Zhejiang Agr Univ 21:583–587

Gao QK, Hu C (1997) Factors affecting host recognition and acceptance in the egg parasitoid *Telenomus theophilae* Wu et Chen. J Zhejiang Agr Univ 23:631–634

Gao QK, Lou BG, Dong HT, Hu C (2002) The relationships between development, nucleic acids and kairomone proteins of *Bombyx mori* female accessory glands. Acta Entomol Sin 45: 583–587

Gardner SM, van Lenteren JC (1986) Characterization of the arrestment responses of *Trichogramma evanescens*. Oecologia 68:265–270

Garnier-Geoffroy F, Malosse C, Durier C, Hawlitzky N (1999) Behaviour of *Trichogramma brassicae* Bezdenko (Hym.: Trichogrammatidae) towards *Lobesia botrana* Denis, Schiffermuller (Lep.: Tortricidae). Ann Soc Entomol France 35:390–396

Gazit Y, Lewis WJ, Tumlinson JH (1996) Arrestment of *Telenomus remus* (Hymenoptera: Scelionidae) by a kairomone associated with eggs of its host, *Spodoptera frugiperda* (Lepidoptera: Noctuidae). Biol Control 6:283–290

Godfray HCJ (1994) Parasitoids. Behavioral and evolutionary ecology. Princeton University Press, Princeton

Grenier S, Veith V, Renou M (1993) Some factors stimulating oviposition by the oophagous parasitoid *Trichogramma brassicae* Bezd. (Hym., Trichogrammatidae) in artificial host eggs. J Appl Entomol 115:66–76

Gross HR Jr, Lewis WJ, Jones RL, Nordlund DA (1976) Kairomones and their use for management of entomophagous insects: III. Stimulation of *Trichogramma achaeae*, *T. pretiosum*, and *Microplitis croceipes* with host-seeking stimuli at time of release to improve their efficiency. J Chem Ecol 1:431–438

Gross HR Jr, Lewis WJ, Beevers M, Nordlund DA (1984) *Trichogramma pretiosum* (Hymenoptera: Trichogrammatidae): effects of augmented densities and distributions of *Heliothis zea* (Lepidoptera: Noctuidae) host eggs and kairomones on field performance. Environ Entomol 13:981–985

Gueldner RC, Nordlund DA, Lewis WJ, Thean JE, Wilson DM (1984) Kairomones and their use for management of entomophagous insects. XV. Identification of several acids in scales of *Heliothis zea* and comments on their possible role as kairomone for *Trichogramma pretiosum*. J Chem Ecol 10:245–251

Hawkins BA, Mills NJ, Jervis MA, Price PW (1999) Is the biological control of insects a natural phenomenon? Oikos 86:493–506

Hilker M, Meiners T (2006) Early herbivore alert: insect eggs induce plant defense. J Chem Ecol 32:1379–1397

Hilker M, Bläske V, Kobs C, Dippel C (2000) Kairomonal effects of sawfly sex pheromones on egg parasitoids. J Chem Ecol 26:2591–2601

Hilker M, Kobs C, Varama M, Schrank K (2002) Insect egg deposition induces *Pinus sylvestris* to attract egg parasitoids. J Exp Biol 205:455–461

Hilker M, Stein C, Schroder R, Varama M, Mumm R (2005) Insect egg deposition induces defence responses in *Pinus sylvestris*: characterisation of the elicitor. J Exp Biol 208:1849–1854

Hofstetter RW, Raffa KF (1998) Endogenous and external factors affecting parasitism of gypsy moth egg masses by *Ooencyrtus kuvanae* (Hymenoptera: Encyrtidae). Entomol Exp Appl 88:123–135

Huang K, Gordh G (1998) Does *Trichogramma australicum* Girault (Hymenoptera: Trichogrammatidae) use kairomones to recognise eggs of *Helicoverpa armigera* (Hubner) (Lepidoptera: Noctuidae)? Aust J Entomol 37:269–274

Isidoro N, Bin F, Colazza S, Vinson SB (1996) Morphology of antennal gustatory sensilla and glands in some parasitoid Hymenoptera with hypothesis on their role in sex and host recognition. J Hymenopt Res 5:206–239

Jones RL, Lewis WJ, Beroza BA, Sparks AN (1973) Host-seeking stimulants (kairomones) for the egg parasite, *Trichogramma evenescens*. Environ Entomol 2:593–596

Kaiser L, Pham-Delegue MH, Masson C (1989a) Behavioural study of plasticity in host preferences of *Trichogramma maidis* (Hym.: Trichogrammatidae). Physiol Entomol 14:53–60

Kaiser L, Pham-Delegue MH, Bakchine E, Masson C (1989b) Olfactory responses of *Trichogramma maidis* Pint. et Voeg.: effects of chemical cues and behavioral plasticity. J Insect Behav 2:701–712

Kennedy BH (1979) The effect of multilure on parasites of the European elm bark beetle, *Scolytus multistriatus*. Bull Entomol Soc Am 25:116–118

Kennedy BH (1984) Effect of multilure and its components on parasites of *Scolytus multistriatus* (Coleoptera: Scolytidae). J Chem Ecol 10:373–385

Klemm U, Schmutterer H (1993) Wirkungen von Niempräparaten auf die Kohlmotte *Plutella xylostella* L. und ihre natürlichen Feinde der Gattung Trichogramma. Z Pflanzenkrankh Pflanzenschutz 100:113–128

Krupke CH, Brunner JF (2003) Parasitoids of the consperse stink bug (Hemiptera: Pentatomidae) in North Central Washington and attractiveness of a host-produced pheromone component. J Entomol Sci 38:84–92

Laing J (1937) Host-finding by insect parasitoids. 1. Observations on the finding of hosts by *Alysia manducator*, *Mormoniella vitripennis* and *Trichogramma evanescens*. J Anim Ecol 6:298–317

LaSalle J, Gauld ID (1993) Hymenoptera & biodiversity. CAB International, Wallingford

Leal WS, Higushi H, Mizutani N, Nakamori H, Kadosawa T, Ono M (1995) Multifunctional communication in *Riptortus clavatus* (Heteroptera: Alydidae): conspecific nymphs and egg parasitoid *Ooencyrtus nezarae* use the same adult attractant pheromone as chemical cue. J Chem Ecol 211:973–985

Lee HP, Boo KS, Kim SO, Lee KS (1997) Gypsy moth kairomones affecting host acceptance behavior of the egg parasitoid *Ooencyrtus kuvanae* (Hymenoptera: Encyrtidae). Korean J Appl Entomol 36:88–95

Leonard DE, Wu ZX, Ferro DN (1987) Responses of parasite *Edovum puttleri* to kairomone from eggs of Colorado potato beetle, *Leptinotarsa decemlineata*. J Chem Ecol 13:335–344

Lewis WJ, Martin J (1990) Semiochemicals for use with parasitoids: status and future. J Chem Ecol 16:3067–3089

Lewis WJ, Tumlinson JH (1988) Host detection by chemically mediated associative learning in a parasitic wasp. Nature 331:257–259

Lewis WJ, Sparks AN, Redlinger LM (1971) Moth odor: a method of host-finding by *Trichogramma evanescens*. J Econ Entomol 64:557–558

Lewis WJ, Jones RL, Sparks AN (1972) A host-seeking stimulant for the egg parasite *Trichogramma evanescens*: its source and a demostration of its laboratory and field activity. Ann Entomol Soc Am 65:1087–1089

Lewis WJ, Jones RL, Nordlund DA, Gross HR (1975a) Kairomones and their use for management of entomophagous species: II. Mechanism causing increase in rate parasitization by *Trichogramma* spp. J Chem Ecol 1:349–360

Lewis WJ, Jones RL, Nordlund DA, Sparks AN (1975b) Kairomones and their use for management of entomophagous insects: I. Evaluation for increasing rates of parasitization by *Trichogramma* spp. in the field. J Chem Ecol 1:343–347

Lewis WJ, Beevers M, Nordlund DA, Gross HR Jr, Hagen KS (1979) Kairomones and their use for management of entomophagous insects. IX. Investigations of various kairomone-treatment patterns for *Trichogramma* spp. J Chem Ecol 5:673–679

Lewis WJ, Nordlund DA, Gueldner RC, Teal PEA, Tumlinson JH (1982) Kairomones and their use for management of entomophagous insects. XIII. Kairomonal activity for *Trichogramma* spp. of abdominal tips, excretion and a synthetic sex pheromone blend of *Heliothis zea* (Boddie) moths. J Chem Ecol 8:1323–1331

Lou YG, Cheng JA (2001) Host-recognition kairomone from *Sogatella furcifera* for the parasitoid *Anagrus nilaparvatae*. Entomol Exp Appl 101:59–67

Lou YG, Du MH, Turlings TCJ, Cheng JA, Shan WF (2005a) Exogenous application of jasmonic acid induces volatile emissions in rice and enhances parasitism of *Nilaparvata lugens* eggs by the parasitoid *Anagrus nilaparvatae*. J Chem Ecol 31:1985–2002

Lou YG, Ma B, Cheng JA (2005b) Attraction of the parasitoid *Anagrus nilaparvatae* to rice volatiles induced by the rice brown planthopper *Nilaparvata lugens*. J Chem Ecol 31: 2357–2372

Lu YQ, Chen KW, He YR, Tang C, Ye JS (2006) Olfactory response of four species of trichogrammatid to kairomones of *Corcyra cephalonica* (Stainton). J South China Agr Univ 27:14–17

Manrique V, Jones WA, Williams LH III, Bernal JS (2005) Olfactory responses of *Anaphes iole* (Hymenoptera: Mymaridae) to volatile signals derived from host habitats. J Insect Behav 18:89–104

Mattiacci L, Vinson SB, Williams HJ (1993) A long-range attractant kairomone for egg parasitoid *Trissolcus basalis*, isolated from defensive secretion of its host, *Nezara viridula*. J Chem Ecol 19:1167–1181

McGregor R, Henderson D (1998) The influence of oviposition experience on response to host pheromone in *Trichogramma sibericum* (Hymenoptera: Trichogrammatidae). J Insect Behav 11:621–632

Meiners T, Hilker M (1997) Host location in *Oomyzus gallerucae* (Hymenoptera: Eulophidae), an egg parasitoid of the elm leaf beetle *Xanthogaleruca luteola* (Coleoptera: Chrysomelidae). Oecologia 112:87–93

Meiners T, Hilker M (2000) Induction of plant synomones by oviposition of a phytophagous insect. J Chem Ecol 26:221–232

Meiners T, Kopf A, Stein C, Hilker M (1997) Chemical signals mediating interactions between *Galeruca tanaceti* L. (Coleoptera, Chrysomelidae) and its egg parasitoid *Oomyzus galerucivorus* (Hedqvits) (Hymenoptera, Eulophidae). J Insect Behav 10:523–539

Mizutani N (2006) Pheromones of male stink bugs and their attractiveness to their parasitoids. Jpn J Appl Entomol Zool 50:87–99

Mizutani N, Wada T, Higuchi H, Ono M, Leal WS (1997) A component of a synthetic aggregation pheromone of *Riptortus clavatus* (Thunberg) (Heteroptera: Alydidae), that attracts an egg parasitoid, *Ooencyrtus nezarae* Ishii (Hymenoptera: Encyrtidae). Appl Entomol Zool 32:504–507

Moraes MCB, Laumann R, Sujii ER, Pires C, Borges M (2005) Induced volatiles in soybean and pigeon pea plants artificially infested with the neotropical brown stink bug, *Euschistus heros*, and their effect on the egg parasitoid, *Telenomus podisi*. Entomol Exp Appl 115:227–237

Moratorio MS (1990) Host finding and oviposition behavior of *Anagrus mutans* and *A. silwoodensis* (Hymenoptera: Mymaridae). Environ Entomol 19:142–147

Müller C, Riederer M (2005) Plant surface properties in chemical ecology. J Chem Ecol 31: 2621–2651

Mumm R, Hilker M (2005) The significance of background odour for an egg parasitoid to detect plants with host eggs. Chem Senses 30:337–343

Mumm R, Tiemann T, Varama M, Hilker M (2005) Choosy egg parasitoids: specificity of oviposition-induced pine volatiles exploited by an egg parasitoid of pine sawflies. Entomol Exp Appl 115:217–225

Noldus LPJJ (1989) Semiochemicals, foraging behavior and quality of entomophagous insects for biological control. J Appl Entomol 108:425–451

Noldus LPJJ, van Lenteren JC (1983) Kairomonal effects on searching for eggs of *Pieris brassicae*, *Pieris rapae* and *Mamestra brassicae* of the parasite *Trichogramma evanescens* Westwood. Mededelingen Faculteit Landbouwwetenschappen, Rijksuniversiteit Gent 48:183–194

Noldus LPJJ, van Lenteren JC (1985) Kairomones for the egg parasite *Trichogramma evanescens* Westwood. II. Effect of contact chemicals produced by two of its hosts, *Pieris brassicae* L. and *Pieris rapae* L. J Chem Ecol 11:793–800

Noldus LPJJ, Lewis WJ, Tumlinson JH, van Lenteren JC (1988) Olfactometer and wind tunnel experiments on the role of sex pheromones of noctuid moths in the foraging behaviour of *Trichogramma* spp. Colloques l'INRA 43:223–238

Noldus LPJJ, Lewis WJ, Tumlinson JH (1990) Beneficial arthropode behavior mediated by airborne semiochemicals. IX. Differential response of *Trichogramma pretiosum*, an egg parasitoid of *Heliothis zea*, to various olfactory cues. J Chem Ecol 16:3531–3544

Noldus LPJJ, Potting RPJ, Barendregt HE (1991a) Moth sex-pheromone adsorption to leaf surface-bridge in time for chemical spies. Physiol Entomol 16:329–344

Noldus LPJJ, van Lenteren JC, Lewis WJ (1991b) How *Trichogramma* parasitoids use moth sex pheromones as kairomones: orientation behaviour in a wind tunnel. Physiol Entomol 16: 313–327

Nordlund DA (1994) Habitat location by *Trichogramma*. In: Wajnberg E, Hassan SA (eds) Biological control with egg parasitoids. CAB International, Oxford, pp 155–163

Nordlund DA, Lewis WJ, Jones RL, Gross HRJr (1976) Kairomones and their use for management of entomophagous insects. IV. Effect of kairomones on productivity and longevity of *Trichogramma pretiosum* Riley (Hymenoptera: Trichogrammatidae). J Chem Ecol 2:67–72

Nordlund DA, Lewis WJ, Gueldner RC (1983) Kairomones and their use for management of entomophagous insects XIV. Response of *Telenomus remus* to abdominal tips of *Spodoptera frugiperda*, (Z)-9-tetradecene-1-ol acetate and (Z)-9-dodecene-1-ol acetate. J Chem Ecol 9:695–701

Nordlund DA, Strand MR, Lewis WJ, Vinson SB (1987) Role of kairomones from host accessory gland secretion in host recognition by *Telenomus remus* and *Trichogramma pretiosum*, with partial characterization. Entomol Exp Appl 44:37–44

Nurindah, Cribb BW, Gordh G (1999) Experience acquisition by *Trichogramma australicum* Girault (Hyemnoptera: Trichogrammatidae). Aust J Entomol 38:115–119

Orr DB, Russin JS, Boethel DJ (1986) Reproductive biology and behavior of *Telenomus calvus* (Hymenoptera: Scelionidae), a phoretic egg parasitoid of *Podisus maculiventris* (Hemiptera: Pentatomidae). Can Entomol 118:1063–1072

Pak GA, de Jong EJ (1987) Behavioural variations among strains of *Trichogramma* spp.: host recognition. Neth J Zool 37:137–166

Papaj R, Vet LEM (1990) Odor learning and foraging success in the parasitoid, *Leptopilina heterotoma*. J Chem Ecol 16:3137–3150

Paramasivan A, Paul AVN, Dureja P (2004) Kairomones of *Chilo partellus* (Swinhoe) and their impact on the egg parasitoid *Trichogramma chilonis* Ishii. Indian J Entomol 66:78–84

Paul AVN, Singh S, Singh AK (2002) Kairomonal effect of some saturated hydrocarbons on the egg parasitoids, *Trichogramma brasiliensis* (Ashmead) and *Trichogramma exiguum* Pinto, Platner and Oatman (Hym., Trichogrammatidae). J Appl Entomol 126:409–416

Peri E, Sole MA, Wajnberg E, Colazza S (2006) Effect of host kairomones and oviposition experience on the arrestment behavior of an egg parasitoid. J Exp Biol 209:3629–3635

Peri E, Guarino S, Lo Bue P, Cork A, Colazza S (2007) Host specificity in the egg parasitoid *Telenomus busseolae* is mediated by sex pheromone compounds. J Insect Sci 7:16–17

Powell W (1999) Parasitoid hosts. In: Hardie J, Minks AK (eds) Pheromones of non-lepidopteran insects associated with agricultural plants. CABI Publishing, Wallingford, pp 405–427

Renou M, Hawlitzky N, Berthier A, Malosse C, Ramiandrasoa F (1989) Evidence of a kairomonal activity from eggs of the European corn borer on females of *Trichogramma maidis*. Entomophaga 34:569–580

Renou M, Nagnan P, Berthier A, Durier C (1992) Identification of compounds from the eggs of *Ostrinia nubilalis* and *Mamestra brassicae* having kairomone activity on *Trichogramma brassicae*. Entomol Exp Appl 63:291–303

Rodriguez-Saona C, Crafts-Brandner SJ, Paré PW (2002) *Lygus hesperus* feeding and salivary gland extracts induce volatile emissions in plants. J Chem Ecol 28:1733–1747

Salerno G, Conti E, Peri E, Colazza S, Bin F (2006) Kairomone involvement in the host specificity of the egg parasitoid *Trissolcus basalis* (Hymenoptera: Scelionidae). Eur J Entomol 103: 311–318

Sales FM (1985) Normal reactions of females of the parasite *Trissolcus basalis* (Wollaston) (Hym.: Scelionidae) to the kairomonal extract of the eggs of the host, *Nezara viridula* (L.) (Hem.: Pentatomidae). Fitossanidade 6/9:109–110

Sales FM, Tumlinson JH, McLaughlin JR, Sailer R (1978) Behaviour of the parasitoid *Trissolcus basalis* (Wollaston) in response to kairomones produced by the host, *Nezara viridula* (L.). Fitossanidade 2:80–83

Salt G (1935) Experimental studies in insect parasitism. III. Host selection. Proc R Soc Lond 117B:413–435

Schmidt JM (1994) Host recognition and acceptance by *Trichogramma*. In: Wajnberg E, Hassan SA (eds) Biological control with egg parasitoids. CAB International, Oxford, pp 165–200

Schmidt JM, Carter MH (1992) The locomotory response of the egg parasitoid *Trichogramma evanescens* Westwood to hexane extracts of eastern spruce budworm scales (Choristoneura fumiferana (Clemens)). Can J Zool 70:941–949

Schmidt JM, Smith JJ (1987) Measurement of host curvature by the parasitoid wasp *Trichogramma minutum*, and its effects on host examination and progeny allocation. J Exp Biol 129: 151–164

Schmidt JM, Smith JJ (1989) Host examination walk and oviposition site selection of *Trichogramma minutum*: studies on spherical hosts. J Insect Behav 2:143–171

Schöller M, Prozell S (2002) Response of *Trichogramma evanescens* to the main sex pheromone component of *Ephestia* spp. and *Plodia interpunctella*, (Z,E)-9,12-tetra-decadenyl acetate (ZETA). J Stored Prod Res 38:177–184

Schröder R, Wurm L, Varama M, Meiners T, Hilker M (2008) Unusual mechanisms involved in learning of oviposition-induced host plant odours in an egg parasitoid? Anim Behav 75: 1423–1430

Shu SQ, Jones RL (1989) Kinetic effects of a kairomone in moth scales of the European corn borer on *Trichogramma nubilale* Ertle, Davis (Hymenoptera: Trichogrammatidae). J Insect Behav 2:123–131

Shu SQ, Swedenborg PD, Jones RL (1990) A kairomone for *Trichogramma nubilale* (Hymenoptera: Trichogrammatidae) isolation, identification, and synthesis. J Chem Ecol 16:521–529

Silva CC, Moraes MCB, Laumann RA, Borges M (2006) Sensory response of the egg parasitoid *Telenomus podisi* to stimuli from the bug *Euschistus heros*. Pesqui Agropecu Bras 41: 1093–1098

Singh S, Paul AVN, Dureja P, Singh AK (2002) Kairomones of two host insects and their impact on the egg parasitoids, *Trichogramma brasiliensis* (Ashmead) and *T. exiguum* Pinto, Platner and Oatman. Indian J Entomol 64:96–106

Smits PH (1982) The influence of kairomones of *Mamestra brassicae* L. on the searching behavour of *Trichogramma evanescens* Westwood. Colloques l'INRA 9:139–150

Stevens PS (1995) Host preferences of *Trichogrammatoidea bactrae fumata* (Hym.: Trichogrammatidae) an egg parasitoid of leafrollers (Lep.:Tortricidae). Entomophaga 40: 379–385

Strand MR, Vinson SB (1982) Stimulation of oviposition and successful rearing of *Telenomus heliothidis* (Hym.: Scelionidae) on nonhosts by use of a host-recognition kairomone. Entomophaga 27:365–370

Strand MR, Vinson SB (1983a) Factors affecting host recognition and acceptance in the egg parasitoid *Telenomus heliothidis* (Hymenoptera: Scelionidae). Environ Entomol 12: 1114–1119

Strand MR, Vinson SB (1983b) Analyses of an egg recognition kairomone of *Telenomus heliothidis* (Hymenoptera; Scelionidae): isolation and host function. J Chem Ecol 9:423–432

Suiter DR, Carlson DA, Patterson RS, Koehler PG (1996) Host-location kairomone from *Periplaneta americana* (L.) for parasitoid *Aprostocetus hagenowii* (Ratzeburg). J Chem Ecol 22:637–651

Takasu K, Nordlund DA (2001) Host recognition kairomones for *Anaphes iole* Girault, an egg parasitoid of the western tarnished plant bug. Biol Control 22:60–65

Thomson MS, Stinner RE (1988) Comparative responses of feral and laboratory *Trichogramma* spp. (Hymenoptera: Trichogrammatidae) to *Heliothis* spp. (Lepidoptera: Noctuidae) moth scales and inert particles. J Entomol Sci 23:245–250

Thomson MS, Stinner RE (1990) The scale response of *Trichogramma* [Hymenoptera: Trichogrammatidae]: variation among species in host specificity and the effect of conditioning. Entomophaga 35:7–21

Tooker JF, De Moraes CM (2005) Jasmonate in Lepidoptera eggs and neonates. J Chem Ecol 31:2753–2759

Turlings TCJ, Wäckers FL, Vet LEM, Lewis WJ, Tumlinson JH (1993) Learning of host-finding cues by hymenopterous parasitoids. In: Papaj DR, Lewis AC (eds) Insect learning: ecological and evolutionary perspectives. Chapman & Hall, New York, pp 51–78

van Alebeek FAN, van Huis A (1997) Host location in stored cowpea by the egg parasitoid *Uscana lariophaga* Steffan (Hym., Trichogrammatidae). J Appl Entomol 121:399–405

van Huis A, Schütte C, Cools MH, Fanget P, van der Hoek H, Piquet SP (1994) The role of semiochemicals in host location by *Uscana lariophaga*, egg parasitoid of *Callosobruchus maculatus*. Proceedings of the 6th International Working Conference on Stored-Product Protection, vol 2. pp 1158–1164

van Lenteren JC (ed) (2003) Quality control and production of biological control agents. Theory and testing procedures. CABI Publishing, UK

van Lenteren JC (2006) How not to evaluate augmentative biological control. Biol Control 39: 115–118

Vet LEM (1999) From chemical to population ecology: Infochemical use in an evolutionary context. J Chem Ecol 25:31–49

Vet LEM, Dicke M (1992) Ecology of infochemical use by natural enemies in a tritrophic context. Annu Rev Entomol 37:141–172

Vet LEM, Groenewold AW (1990) Semiochemicals and learning in parasitoids. J Chem Ecol 16:3119–3135

Vet LEM, Lewis WJ, Cardè R (1995) Parasitoid foraging and learning. In: Cardè R, Bell WJ (eds) Chemical ecology of insects. Chapman & Hall, New York, pp 65–101

Vinson SB (1985) The behaviour of parasitoids. In: Kerkut GA, Gilbert LI (eds) Comprehensive insect physiology biochemistry and pharmacology. Pergamon, New York, pp 417–469

Vinson SB (1994) Physiological interactions between egg parasitoids and their hosts. In: Wajnberg E, Hassan SA (eds) Biological control with egg parasitoids. CAB International, Oxford, pp 245–271

Vinson SB (1998) The general host selection behavior of parasitoid Hymenoptera and a comparison of initial strategies utilized by larvaphagous and oophagous species. Biol Control 11: 79–97

Vinson SB, Piper GL (1986) Source and characterization of host recognition kairomones of *Tetrastichus hagenowii*, a parasitoid of cockroach eggs. Physiol Entomol 11:459–468

Vinson SB, Conti E, Salerno G, Bin F (2002) The "Egg-Host Unit", what is it and its role in parasitoid host location and acceptance, Egg Parasitoids, 6th International Symposium, Perugia, p 67

Wajnberg E, Scott JK, Quimby PC (eds.) (2001) Evaluating indirect ecological effects of biological control. CABI Publishing, UK

Wakamura S (2006) Behavioral response to female sex pheromone by insects. Aroma Res 7: 228–233

Wang JJ, Zong LB (1991) A study on host-seeking kairomone for *Trichogramma confusum* Viggiani. Colloques l'INRA 56:93–96

Wegener R, Schulz S (2002) Identification and synthesis of homoterpenoids emitted from elm leaves after elicitation by beetle eggs. Tetrahedron 58:315–319

Wegener R, Schulz S, Meiners T, Hadwich K, Hilker M (2001) Analysis of volatiles induced by oviposition of elm leaf beetle *Xanthogaleruca luteola* on *Ulmus minor*. J Chem Ecol 27: 499–515

Wei GD, Rong NH, Ye GY, Gao QK (2005) A new host based on the host-recognition kairomone of *Telenomus theophila* (Hymenoptera: Scelionidae) – *Dendrolimus punctatus*. Acta Agric Zhejiangensis 17:69–73

Wysoki M, de Jong M (1989) Attraction of *Trichogramma platneri* to eggs of some lepidopterous pests of avocado. Phytoparasitica 17:315–318

Yasuda K, Tsurumachi M (1995) Influence of male-adults of the leaf-footed plant bug, *Leptoglossus australis* (Fabricius) (Heteroptera, Coreidae), on host-searching of the egg parasitoid, *Gryon pennsylvanicum* (Ashmead) (Hymenoptera, Scelionidae). Appl Entomol Zool 30:139–144

Yong TH, Pitcher S, Gardner J, Hoffmann MP (2007) Odor specificity testing in the assessment of efficacy and non-target risk for *Trichogramma ostriniae* (Hymenoptera: Trichogrammatidae). Biocontrol Sci Technol 17:135–153

Zaborski E, Teal PEA, Laing JE (1987) Kairomone-mediated host finding by the spruce budworm egg parasite, *Trichogramma minutum*. J Chem Ecol 13:113–122

Zaki FN (1985) Reactions of the egg parasitoid *Trichogramma evanescens* Westw. to certain insect sex pheromones. Z Angew Entomol 99:448–453

Zou WH, Lei CL, Zhang F (2002) Effect of the host kairomone on the host selection *Trichogramma japonicum*. Entomol Knowl 39:370–373

Chapter 5
Genetics of the Behavioral Ecology of Egg Parasitoids

Eric Wajnberg

Contents

5.1 Introduction

The release of insect parasitoids for biological control programs of phytophagous pests has regularly increased over the last decades all over the world to protect crops in open fields (e.g., cereals), greenhouses (e.g., vegetables) and even in forestry (Boller et al. 2006). Important successes were obtained in different

E. Wajnberg (✉)
INRA – 400, Route des Chappes, BP 167 – 06903, Sophia Antipolis, Cedex, France
e-mail: wajnberg@sophia.inra.fr

F.L. Consoli et al. (eds.), *Egg Parasitoids in Agroecosystems with Emphasis on Trichogramma*, Progress in Biological Control 9, DOI 10.1007/978-1-4020-9110-0_5,

agro-ecosystems, and egg parasitoids (also called "oophagous") are most probably the most important amongst the different parasitoids species used as biocontrol agents, leading sometimes to spectacular pest control efficacy (Wajnberg and Hassan 1994). Egg parasitoids, and especially those belonging to the Trichogrammatidae, the Scelionidae and the Mymaridae families, present interesting features that elect them as real good candidates to control several pests attacking a variety of crops. They are usually easy and cheap to rear in the laboratory and in large-scale mass-rearing facilities, since they can be produced on factitious hosts that are easier to handle than the natural hosts. Since they are attacking the egg stage, they are also preventing hatching of the larvae which are the damaging stages for the crop to be protected (Wajnberg and Hassan 1994). These are most likely the main reasons why egg parasitoid species are the most intensively produced animals for biological control programs worldwide (Wajnberg and Hassan 1994).

However, biological control programs with insect parasitoids and especially with egg parasitoids, even if sometimes they lead to spectacular pest control efficacy, still need to be improved, at least for reducing the cost and/or increasing the success of several pest control programs or sometimes just to enable the biological control on new-coming pests that was not initially feasible. For this, several authors over the last 15–20 years progressively proposed to develop a more scientifically-based, formal approach of biological control that is then not only based on an usual trial-and-error method (e.g., Waage 1990, Wajnberg 2004, Roitberg 2007, Wajnberg et al. 2008). The general idea is to use the concept and development of behavioral ecology, a growing scientific discipline that aims at understanding the behavior animals should adopt in different environmental situations to maximize their long-term offspring production (their so-called fitness) (Stephens and Krebs 1986, Godfray 1994). Any optimal behavioral strategy leading parasitoid females to maximize the total number of progeny produced will directly maximize the number of hosts attacked and killed, and will thus maximize the pest control efficacy when parasitoids are released in biological control programs (Mills and Wajnberg 2008).

As pointed out by Wajnberg (2004), considering optimal a behavioral strategy that maximizes long-term offspring production supposes that the corresponding behaviors have been settled progressively by natural selection, and thus that there is/was, in the parasitoid populations, genetic variability upon which natural selection could act. Without such genetic variability, transmissible over generations, no adaptation and evolution toward optimal behavioral strategies would be possible.

The aim of this chapter is thus to discuss the genetics of behavioral ecology of insect parasitoids, and especially egg parasitoids. After presenting the general concepts of behavioral ecology of insect parasitoids, I will give a rapid summary of the main methods that are available to quantify the genetic variation in quantitative behavioral traits in parasitoid populations. I will then present an exhaustive list of what has been done already on the quantification of the genetic variability in behavioral traits in insect parasitoids. Finally, I will discuss on what is the necessary knowledge that is currently missing in this field of research leading to the proposal of several future research directions that should hopefully be developed in the near future.

5.2 Different Behavioral Ecology Approaches on Insect Parasitoids

As briefly mentioned above, the aim of behavioral ecology is to study and to understand the behavior of animals within an ecological and evolutionary framework. The general idea is thus to understand the roles of animals' behavioral decisions in enabling them to adapt to their biotic and abiotic environment (Krebs and Davies 1997). There are several ways in which behaviors can be studied according to such a framework. Among others, looking for the so-called ultimate causation tries to identify the function of the behaviors studied to see how they can contribute to both animal survival and the number of offspring produced. Usually, optimization theory by means of theoretical models is used for this. On the other hand, the aim of the so-called proximate approach is to accurately identify the behavioral mechanisms animals are adopting to give rise to the observed behaviors. Obviously, studies of function and mechanisms must go hand to hand (Krebs and Davies 1997).

During the last 15 years or so, such a scientific approach was developed on a large variety of different animals, but a recent broad survey of the literature (i.e., more than a 1000 papers) published during a five-years period in the two main scientific international journals "Behavioral Ecology" and "Behavioral Ecology and Sociobiology" showed that only 1.7% of them were dealing with parasitoids. The remaining papers were on (1) birds: 31.8%, (2) insects, mites, etc. (except parasitoids): 25.6%, (3) fishes, reptiles, etc.: 19.3%, (4) mammals: 17.8%, (5) shellfishes, etc.: 3.1%, (6) miscellaneous: 0.7%. There is, thus, a huge gap on insect parasitoids to be exploited, which is rather surprising. Parasitoids share several ecological and biological features that are of utmost importance to understand how animal behavioral decision can be shaped by natural selection (see below). Besides, studies on these animals can be directly translated in terms of field application for biological control purposes (Godfray 1994, Wajnberg et al. 2008).

In this section, I am providing some details of the different specific behavioral features of insect parasitoids that are within a behavioral ecology framework. All parasitoid species are concerned but, for egg parasitoids, important results were recently published on Scelionidae and Trichogrammatidae.

5.2.1 Optimal Residence Time on Host Patches

Hosts of most parasitoid species are distributed in the environment in patches that are distant from each other. For egg parasitoids, this is for example the case for egg masses of the European corn borer *Ostrinia nubilalis* (Hübner) or the Southern Green Stinkbug *Nezara viridula* (L.). Furthermore, parasitoids are usually drastically short-lived animals and are thus usually so-called time-limited in the sense that they usually die before they can deposit all of their eggs (Driessen and Hemerik 1992, Seventer et al. 1998, Rosenheim 1999). Hence, parasitoids should likely maximize their offspring production per time unit, and have thus been likely selected to optimize their time allocation on every patch of hosts they visit (Wajnberg 2006). Actually, understanding how female parasitoids are managing the time they are allocating to different patches of their hosts has arguably been the most studied problem of behavioral ecology over the past few decades (van Alphen et al. 2003, Wajnberg 2006, van Alphen and Bernstein 2008).

There is now an important body of theoretical models enabling to know what should be the optimal time females should remain on each host patch under different environmental conditions. An important number of experimental work demonstrated that female parasitoids usually behave according to the corresponding theoretical predictions (see Wajnberg 2006 for a recent review). There are also now powerful statistical methods to identify from experimental data the mechanistic proximate behavioral patch-leaving decision rules females are using to reach such theoretical predictions (van Alphen et al. 2003, Wajnberg 2006). Some specific statistical methods have also been used to compare patch-leaving decision mechanisms adopted by different species of Trichogrammatidae taking into account their phylogenetic relationships through a so-called comparative analysis (Wajnberg et al. 2003, see Harvey and Pagel 1991 and Martins 1996 for details on the statistical methods used). It is clear that such a comparative approach should lead to a better description of egg parasitoid species as potential biocontrol agents, and should optimize the choice of the proper species to control a given pest, on a given crop and in a given environment.

Finally, the time parasitoid females allocate to host patches they exploit should be directly linked to their dispersion ability (i.e., the longer the time invested on each visited patch, the lower the dispersion). Therefore, an accurate understanding of the decision-making processes involved in patch-time allocation in female parasitoids should provide relevant information on spatial population dynamics, which would be especially useful for their release for biological control purposes.

5.2.2 Optimal Clutch Size

A parasitoid female encountering a host to attack has to decide how many eggs she should lay in it. Such behavioral decision is known to depend mainly on the size of the host (Vinson 1976, 1985, Waage 1986), and this is especially true for semi-gregarious species, including most egg parasitoids that attack eggs of different host species which display egg size variability (Waage and Ng 1984, Schmidt 1994). Sometimes, hosts to be attacked have been already parasitized, and the decision of the wasp female to lay additional eggs is called "superparasitism". In self-superparasitism, female attacks a host she already attacked before, while in conspecific-superparasitism, female attacks a host that has been previously attacked by a conspecific (Waage 1986, van Dijken and Waage 1987).

Several theoretical models have been developed over the years to understand what should be the optimal number of progeny females should deposit in each attacked host (Godfray 1994). Some models suggest that females have been selected to lay a clutch size that maximizes the total number of adult progeny emerging from the hosts attacked (e.g., Lack 1947), some others are assuming that this is rather the rate, per time unit, at which progeny are produced that matter (e.g., Charnov and Skinner 1984, 1985, Parker and Courtney 1984). The predictions of these models were verified on different parasitoid species (see Godfray 1994). Interestingly, these different models are also able to explain how many eggs parasitoid females should lay in a situation of superparasitism and, here again, a number of studies demonstrated that the fundamental assumptions of such theoretical approaches are well supported by experimental data.

In most cases, offspring fitness declines with increasing clutch size (Godfray 1994). This is especially true with superparasitism, leading both to progeny with lower fecundity, longevity, host searching rate, etc. (Chacko 1969, Waage 1986), and to the emergence of a population with a male-biased sex ratio (see below) (Waage 1986). This can be a real problem for designing an efficient biological control program with egg parasitoids, especially at a large-scale mass production step (Wajnberg et al. 1989). Thus, an accurate understanding of the behavioral proximate mechanisms used by parasitoid females to adjust the number of progeny laid in each host should provide us with some means to improve the quality of mass-produced parasitoids used in biological control programs.

5.2.3 Optimal Sex Ratio

Another important feature of insect parasitoids that has been intensively studied both theoretically and experimentally within a behavioral ecology framework concerns sex allocation (Godfray 1994, Ode and Hunter 2002). There are three main reasons for that. The first one is that sex allocation in most parasitoid species is usually extremely labile and is often directly linked to fitness (Ode and Hardy 2008). Moreover, since only female parasitoids are attacking hosts, releasing a wasp population with female-biased sex ratio has been assumed repeatedly to produce better pest control efficacy (e.g., Waage 1982, Heimpel and Lundgren 2000), and the study of sex allocation strategy in parasitic wasps can thus produce results that can be directly used to improve pest control strategies though biological control programs (Hardy and Ode 2007). Finally, insect parasitoids appear to be an excellent model for studying sex allocation since most species are arrhenotokous (i.e., haplodiploid), meaning that mated females can lay either unfertilized haploid males or fertilized diploid females. Hence, mated females are – at least theoretically – able to accurately control the proportion of sons and daughters in their progeny, and have been most likely selected to use such sex determination system to optimize their offspring production (Charnov 1982, Cook 1993).

Several well-known theoretical models are predicting what should be the optimal proportion of sons and daughters animals should produce to maximize their fitness. The conceptual and historical foundation of sex ratio theory is represented by the model of Fisher (1930) explaining that animal should invest equally in sons and daughters if individuals are finding mates randomly (i.e., panmixis). When sons and daughters are equally costly to produce and the sex ratio in a population is unbiased, each son will on average mate with one female and a mother will indeed realize equal fitness gains from investing in a son or in a daughter.

For most insect parasitoids, however, mating is not occurring randomly. As we have seen above, females are attacking hosts that are usually aggregated into patches, and their progeny are usually mating together before dispersing to forage for oviposition opportunities. Since only a limited number of so-called "foundress" females are contributing to offspring in a patch, mating between progeny are then among relatives, and the optimal sex ratio should be female-biased and should depend on the number of foundresses attacking the host patches (Ode and Hardy 2008). This is the Local Mate Competition theory (Hamilton 1967, Taylor and Bulmer 1980) and the predictions have been observed in several parasitoid species, including egg parasitoids (Godfray 1994).

Different proximate mechanistic rules have been identified leading parasitoid females to adjust their sex ratio according to the number of foundresses on host patches. Females were shown to respond to the frequency of physical contacts with other females (Wylie 1976) or to the perception of the chemical traces they left on host patches (Viktorov and Kochetova 1973). Females were also shown to use the frequency of encounters with already attacked hosts on the patch as a cue to the number of competing females (Wylie 1973). However, the most efficient and simple mechanism was originally observed by Waage and Lane (1984) and consists for the females in laying their son and daughter eggs in a particular order, usually with males first (Wajnberg 1993, Colazza and Wajnberg 1998, Bayram et al. 2004). The efficiency of such mechanism was quantified with a simulation model by Wajnberg (1994).

Finally, it has to be noted that parasitoid females are not only responding to the number of other conspecifics on patches of their hosts to adjust their own sex ratio, as host quality (e.g., size) was also shown to be a determinant factor (Charnov et al. 1981, Werren 1984). Parasitoid sex ratio can also be influenced by so-called sex ratio distorters (see Chapter 6, Werren 1997, Stouthamer et al. 2002, Ode and Hardy 2008), with the most well-known being the symbiotic bacteria *Wolbachia* that are especially common in egg parasitoids of the genus *Trichogramma* (Stouthamer et al. 1990, see Chapter 6). Theoretical models leading to understand accurately the evolutionary meaning of such particular sex determination system are still drastically needed.

5.2.4 Optimal Marking Strategy

In order to be able to optimize the number of eggs laid in each host encountered and attacked (see Section 5.2.2) and what sex those eggs should be (see Section 5.2.3), parasitoid females should be able to assess, in most cases, whether hosts have been already previously attacked either by themselves or by conspecifics. In most species, such ability is usually mediated through the use of marking pheromones that are chemical substances deposited by egg-laying females and that serve as messages conveying information (Roitberg and Mangel 1988). From an ecological point of view, such marking pheromones are usually complex signals that (1) can be used by the marking females to recognize where are the hosts they already attacked, (2) can be used to prevent females to attack hosts already parasitized, or (3) are used at the population level to spread more efficiently parasitoid offspring across all available hosts (Roitberg and Mangel 1988).

Marking pheromones can also have negative effects when they are used as kairomones by hyperparasitoids to locate and attack parasitoid offspring. Moreover, marking takes time and is necessarily associated with metabolic costs for producing the marking compounds (Godfray 1994). Finally, after being deposited, signal efficiency might suffer a decay with time though, e.g., evaporation (Hoffmeister and Roitberg 1998). There is thus a benefit/cost issue in such marking behavior and a growing body of theory is progressively developed to understand under what circumstance a marking behavior can evolve (Roitberg and Mangel 1988). Briefly, the general theory suggests that the production and use of signals can only evolve when there is a net benefit to the sender (Krebs and Dawkins 1984, Harper 1991). More accurately, under strong resource competition for limited resources, an increase in signal intensity and duration is expected if the animals issuing the signals obtain a disproportionate part of the resources (Hoffmeister and Roitberg 1998).

Understanding optimal marking strategies in insect parasitoids, and especially in egg parasitoids, is important if we want to use these insects in biological control programs more efficiently. However, although the theory is becoming progressively more accurate, there is currently a clear lack of experimental data to understand the associated evolutionary meaning and what can be the corresponding applications in terms of biological control (Roitberg 2007).

5.2.5 Optimal Diet Choice

As we have seen above, females of several parasitoid species are encountering and attacking different host instars or species showing different biological and/or ecological features. Especially egg parasitoids, that are mostly polyphages, can attack host eggs from different species, differing in e.g., size and/or shape, etc. Furthermore, in this respect, and as discussed before (see section 5.2.2), parasitoid females are sometimes encountering either unparasitized or already attacked hosts. In all these situations, the environment offers to the wasp females a set of different resource items (i.e., hosts) which might vary both in profitability (i.e., number of progeny that can be produced divided by handling time) and abundance and there is a growing theory, so-called optimal diet selection models, that predicts what resource type a foraging animal should accept in order to maximize its rate of progeny production (Stephens and Krebs 1986, Godfray 1994, Sih and Christensen 2001). Briefly, resource items (i.e., hosts) vary both in the number of progeny a female can produce attacking them, and in handling time (i.e., the time needed in host handling and parasitization). In this case, the theory predicts that a foraging female should accept and thus attack a host type if the progeny she will produce is worth the handling time invested (Charnov 1976, Houston and McNamara

1999, Hamelin et al. 2007). In other words, if the different host types are ranked according to their profitability, it can be shown that the foraging animal will maximize its rate of progeny production if it accepts all hosts with profitability above a given threshold and reject all hosts beneath this cutoff (Krebs and Davies 1987, Godfray 1994). The threshold level is known to depend on the number and quality of the hosts available in the environment. In rich environments with many good-quality hosts, females should optimally accept a narrow range of hosts, but should attack progressively poorer hosts if good-quality hosts are becoming less frequent.

Although the corresponding theory is still under development, there is clearly a lack of experimental data for parasitoids, and especially for egg parasitoids, trying to test the predictions of host acceptance models within the theory of optimal diet selection. What seems to be currently well established however is that hosts of better quality are usually those that are more readily accepted for oviposition, but this has been mainly demonstrated on *Drosophila* Fallén parasitoids (e.g., van Alphen and Vet 1986, Driessen et al. 1991) or on aphid parasitoids (e.g., Liu et al. 1984, Liu 1985). Trichogrammatidae egg parasitoids usually prefer young hosts, which yield higher survival and reproductive success (see Strand 1986 for a review). On these parasitoids, this problem was also approached by the accurate quantification of host handling time and the corresponding variation (e.g., Wajnberg 1989).

Like for the other behavioral decisions previously discussed, understanding why and how a host should be accepted for oviposition by a parasitoid female appears to be important for optimizing both the mass-rearing system used to produce the parasitoids and their efficiency when they are released for biological control programs.

5.3 Main Methods Available for Quantifying the Genetic Variation in Behavioral Traits

We have seen that a behavioral ecology approach on insect parasitoids can provide a list of important behaviors of parasitoid females that are most likely determining their efficiency to control pests when they are produced and released in the field for biological control programs. Quantifying the genetic variation in these behavioral traits in different parasitoid species and populations regularly appears to be an important task for different reasons. Briefly, and among others, it has been regularly admitted that an accurate quantification of the intra-population genetic variation in biological attributes of insect parasitoids, and especially egg parasitoids, can lead (1) to better estimate the ability of biocontrol agents to evolve in response to the environmental characteristics of the system used in their rearing and production (Mackauer 1976, Boller 1979), (2) to better estimate the parasitoid survival potential after field release (Hopper et al. 1993), and (3) to start a breeding selection program in order to improve the efficacy of the released animals to control the target pests (Hoy 1976, Mackauer 1976, Wajnberg 1991, 2004).

There are several review articles and book chapters presenting the main experimental and statistical methods that can be used to estimate the intra-population level of genetic variation in quantitative biological traits in animals, and particularly insects (e.g., Boller 1979, Collins 1984, Joslyn 1984, Falconer 1989, Roush 1990, Hopper et al. 1993, Margolies and Cox 1993, Wajnberg 2004, Beukeboom and Zwaan 2005). Here, I will give just a brief overview of there different methods and invite the reader to look at these review articles to understand more in details how these methods can be implemented.

5.3.1 Basic Concepts

Phenotypic variation among individuals is a common feature of all biological studies (Lewis et al. 1990), especially for behavioral traits (Roitberg 1990), and such variation is known to be the result of interactions between the genotype of each animal and the environment it lives in (Collins 1984). All methods available to quantify the genetic variation in any biological trait aim at estimating the effect of these two sources of variation, genetic and environment, and to see whether the genetic part is significantly contributing to the total phenotypic variation observed (Falconer 1989).

Variations among individuals in a population are quantified with variances and, using both standardized experimental methods and statistical models, the total phenotypic variance can be divided into the variance due to the environment and the variance due to genotypes. In turn, the genotypic variance can further be divided into (1) the variance due to additive genetic effects (i.e., the genetic variance associated with the additive effect of each allele involved in the phenotypic trait studied), (2) the variance due to dominance effect between alleles, and (3) the variance due to interaction between the loci involved in the trait under consideration (see, e.g., Falconer 1989 for a detailed description of this). Of the three types of genetic variance, the variance due to additive genetic effects is considered to be the most important since selection – natural or artificial – will essentially act on it. Hence, the methods available to quantify the genetic variation of phenotypic traits are built to estimate this particular type of variance, either directly or indirectly.

5.3.2 Parent–Offspring Regression

One of the first methods that have been proposed is based on the fact that, if a phenotypic trait is genetically determined, then offspring should resemble their parents. Then, the idea is to quantify the trait in a set of parents (mothers, fathers or both) and also on their progenies. A regression analysis is then performed with the offspring's values on the Y-axis and the parents' values on the X-axis and the slope of the regression is statistically tested. Since most of the behavioral traits discussed in the previous sections of this chapter are females' attributes, and more generally since only females are useful in most cases to control target pests in biological control programs, parent-offspring regressions are in most of the cases mother-daughter regressions. On Trichogrammatidae egg parasitoids, this method has been used successfully, for example, by Wajnberg (1989, 1993) and Bruins et al. (1994).

5.3.3 Sib Analysis

Another method consists of mating a number of males with a number of females. Each male has to mate with more than one female, but each female is mated with only one male. The trait is analyzed at the offspring generation, and an analysis of variance is then used to quantify the variation between males (i.e., fathers), females (i.e., mothers) and within the progeny of each female. In turn, the estimated parameters of the analysis of variance are used to test the significance of the genetic variation in the population studied for the quantified trait. As far as I know, this method was never used to quantify the genetic variation in biological traits of parasitic wasps.

5.3.4 Family Analysis

The most used method in the literature is the so-called isofemale strains method, also called isofemale lines or family analysis (Parsons 1980, Hoffmann and Parsons 1988). In this case, several families are founded, each from a single mated female taken at random in the population, and the trait is quantified in several offspring in each family at the F1 generation. Then, a simple one-way analysis of variance is used to test for a significant difference among average values of the different families compared. Such a significant difference will indeed indicate a significant genetic variation in the trait studied in the population. On Trichogrammatidae, this method was successfully used by Wajnberg (1994), Bruins et al. (1994) and Wajnberg and Colazza (1998), among others, and on Scelionidae by Wajnberg et al. (1999, 2004).

5.3.5 Breeding Selection

A modification in the average value of a quantitative trait through several generations of breeding selection proves that the phenotypic variation observed in this trait in a population is under significant genetic control (Roush 1990). This is the reason why a few number of studies used breeding selection as a mean to demonstrate significant genetic variation in biological attributes of parasitic wasps. In this case, the basic method is to quantify the trait under study on several individuals at a given generation and the individuals that are used to found the next generation are chosen according to they own phenotype. On egg parasitoids, this method was successfully used on Trichogrammatidae by Urquijo (1950), Ashley et al. (1974), Schmidt (1991) and Fleury et al. (1993). Since most, if not all, behavioral traits listed in previous sections of this chapter are most likely implicated in the efficacy of insect parasitoids to control pests in biological control programs, a significant change in the average value of such traits through breeding selection not only proves that traits have a significant genetic variation in the population studied, but also represents the starting point for improving the pest control efficacy of a biocontrol agent.

5.3.6 Common Features and Generalization

All methods briefly described previously can be used to estimate the so-called heritability of the traits studied, which can be defined as the ratio of the additive genetic variance to the total phenotypic variance (Hoffmann and Parsons 1988, Falconer 1989). Such a genetic parameter can be used to estimate the expected response to selection in the course of several breeding generations (Falconer 1989).

Also, with appropriate experimental set-ups, these different methods can sometimes be combined together. For example, a mother-daughter regression analysis can be planed over two successive generations, with several offspring measured for each mother. Then, at the F1 generation, the daughters are treated like different isofemale lines that are compared by means of a family analysis. On Trichogrammatidae egg parasitoids, such a combination of methods was successfully used by Chassain and Boulétreau (1991), Mimouni (1991), Fleury et al. (1993), Bruins et al. (1994) and Pompanon et al. (1994).

Finally, since behaviors are usually biological traits that can be difficult to measure and analyze (Martin and Bateson 1994), there is sometimes the need for the quantification of single behaviors, each with several values. In this case, all methods earlier described can be generalized using multidimensional statistical methods. For example, canonical correlation analysis can be used to perform a multidimensional parent-offspring regression, as this was done on egg parasitoids by Wajnberg (1993), or a discriminant analysis can be used to compare different isofemale lines in a multidimensional family analysis.

5.4 What Has Been Done and What Remains to Be Done?

Following Wajnberg (2004), an exhaustive survey was done over the main scientific database to find all references describing intra-population genetic variation in biological traits of egg parasitoids. Only 20 references were found, covering 11 species (since *Trichogramma brassicae* Bezdenko and *Trichogramma maidis* Pintureau and Voegelé are two synonymous names of the same species), among which nine are Trichogrammatidae and two are Scelionidae (see Table 5.1).

Among these 20 references, two did not discussed about genetic variation in behavioral traits (Ashley et al. 1974, Bennett and Hoffmann 1998), and the distribution of the 18 remaining references in the different behavioral features of egg parasitoids is the following one: optimal patch time allocation: 2; optimal clutch size: 2; optimal sex ratio: 3; optimal marking strategy: 0; optimal diet choice: 1; miscellaneous: 10. The last category regroups all behavioral traits that have not been explicitly analyzed within a behavioral ecology framework (e.g., females' walking activity).

Needless to say that this literature survey represents a very small number of published data, and thus clearly demonstrates a drastic lack of studies on the quantification of the genetic variation in behavioral traits of egg parasitoids. Important information is currently missing regarding the existence of genetic variation in

Table 5.1 List of all egg parasitoid species in which intra-population genetic variation in quantitative traits was studied

Family *Species*	References
Trichogrammatidae	
Trichogramma brassicae Bezdenko	Chassain and Boulétreau (1991), Fleury et al. (1993), Wajnberg (1993), Bruins et al. (1994), Pompanon et al. (1994), Pompanon et al. (1999), Wajnberg (1994), Wajnberg and Colazza (1998)
Trichogramma cacoeciae Marchal	Chassain and Boulétreau (1991), Pompanon et al. (1994)
Trichogramma carverae Oatman and Pinto	Bennett and Hoffmann (1998)
Trichogramma dendrolimi Matsumara	Limburg and Pak (1991), Schmidt (1991)
Trichogramma evanescens Westwood	Limburg and Pak (1991), Schmidt (1991)
Trichogramma maidis Pintureau and Voegelé	Chassain and Boulétreau (1987), Wajnberg (1989), Wajnberg et al. (1989)
Trichogramma minutum Riley	Urquijo (1950), Liu and Smith (2000)
Trichogramma pretiosum Riley	Ashley et al. (1974)
Trichogramma semifumatum (Perkins)	Ashley et al. (1974)
Trichogramma voegelei Pintureau	Mimouni (1991)
Scelionidae	
Telenomus busseolae Gahan	Wajnberg et al. (1999)
Trissolcus basalis (Wollaston)	Wajnberg et al. (2004)

behavioral decision rules adopted by egg parasitoids. In turn, this prevents, for example, starting mass breeding selection in order to improve pest control efficacy of oophagous species used in biological control programs.

5.5 Concluding Remarks and Future Directions

After decades of biological control programs designed using a trial-and-error method, behavioral ecology, by identifying the key behavioral traits implicated in the pest control efficiency of insect parasitoids and especially egg parasitoids, seems now to provide a modern, scientifically-based and formal approach leading hopefully to define optimized and more efficient pest control programs (Waage 1990, Roitberg 2007, Wajnberg et al. 2008). The information collected following such a scientific approach can help us in optimizing the choice of the correct species and population to control an indentified pest in a given environment (Wajnberg 2004, Mills and Wajnberg 2008). Moreover, quantifying the genetic variation in the key behavioral components of parasitoid reproduction strategies could, among others, provide us with the means to improve through breeding selection the efficacy of biocontrol agents released in biological control programs. This is what this chapter was about, and we have seen that, although quantifying genetic variability in insect parasitoids has been continuously presented as an important task, there is currently a drastic deficit of studies done to achieve this goal, and especially for egg parasitoids.

In this chapter, we have seen that the most important behavioral traits of parasitic wasps are implicated in (1) the way females optimally allocated their time to patches of their hosts, (2) the optimal number of progeny they should lay in each host attacked, (3) the optimal proportion of males and females they should produce, (4) the optimal marking strategy they should adopt, and (5) the optimal host acceptance decision they should develop when they are facing a choice of hosts of different quality. Actually, all these different points still remain to be analyzed in details in egg parasitoids following a behavioral ecology framework. The genetic variation in the corresponding behavioral traits still remains to be quantified and our ability to improve the pest control efficacy of oophagous insects still remains to be tested. The aim of this chapter is to stimulate research in this direction.

References

Ashley TR, Gonzalez D, Leigh TF (1974) Selection and hybridization of *Trichogramma*. Environ Entomol 3:43–48

Bayram A, Salerno G, Conti E, Wajnberg E, Bin F, Kornosor S (2004) Sex allocation in *Telenomus busseolae*, a solitary parasitoid of concealed eggs: the influence of host patch size. Entomol Exp Appl 111:141–149

Bennett DM, Hoffmann AA (1998) Effects of size and fluctuating asymmetry on field fitness of the parasitoid *Trichogramma carverae* (Hymenoptera: Trichogrammatidae). J Anim Ecol 67:580–591

Beukeboom LW, Zwaan BJ (2005) Genetics. In: Jervis MA (ed) Insects as natural enemies: a practical perspective. Springer, Dordrecht, pp 167–218

Boller EF (1979) Ecological genetics and quality control. In: Hoy MA, McKelvey JJ Jr (eds) Genetics in relation to insect management. A Rockefeller Foundation Conference, Bellagio, Italy, pp 153–160

Boller EF, van Lenteren JC, Delucchi V (2006) International Organization for Biological Control of Noxious Animals and Plants – History of the first 50 years (1956–2006). IOBC

Bruins EBAW, Wajnberg E, Pak GA (1994) Genetic variability in the reactive distance in *Trichogramma brassicae* after automatic tracking of the walking path. Entomol Exp Appl 72:297–303

Chacko MJ (1969) The phenomenon of superparasitism in *Trichogramma evanescens minutum* Riley I. Beiträge Entomol 19:617–635

Charnov E (1976) Optimal foraging: attack strategy of a mantid. Am Nat 110:141–151

Charnov EL (1982) The theory of sex allocation. Princeton University Press, New Jersey

Charnov EL, Skinner SW (1984) Evolution of host selection and clutch size in parasitoid wasps. Florida Entomol 67:5–21

Charnov EL, Skinner SW (1985) Complementary approaches to the understanding of parasitoid oviposition decisions. Environ Entomol 14:383–391

Charnov EL, los-den Hartogh RL, Jones WT, van den Assem J (1981) Sex ratio evolution in a variable environment. Nature 289:27–33

Chassain C, Boulétreau M (1987) Genetic variability in the egg-laying behaviour of *Trichogramma maidis*. Entomophaga 32:149–157

Chassain C, Boulétreau M (1991) Genetic variability in quantitative traits of host exploitation in *Trichogramma* (Hymenoptera: Trichogrammatidae). Genetica 83:195–202

Colazza S, Wajnberg E (1998) Effect of host egg-mass size on sex ratio and oviposition sequence of *Trissolcus basalis* (Hym.: Scelionidae). Environ Entomol 27:329–336

Collins AM (1984) Artificial selection of desired characteristics in insects. In: King EG, Leppla NC (eds) Advances and challenges in insect rearing. USDA/ARS, New Orleans, pp 9–19

Cook JM (1993) Sex determination in the hymenoptera: a review of models and evidence. Heredity 71:421–435

Driessen G, Hemerik L (1992) The time and egg budget of *Leptopilina clavipes*, a parasitoid of larval *Drosophila*. Ecol Entomol 17:17–27

Driessen G, Hemerik L, Boonstra B (1991) Host selection behaviour in relation to survival in hosts of *Leptopilina clavipes*, a parasitoid of larval *Drosophila*. Neth J Zool 41:99–111

Falconer DS (1989) Introduction to quantitative genetics, 3rd edn. Longman, New York

Fisher RA (1930) The genetical theory of natural selection. Oxford University Press, Oxford

Fleury F, Chassain C, Fouillet P, Boulétreau M (1993) La dispersion spatiale de la pontes des trichogrammes (Hymenoptera : Trichogrammatidae): Bases génétiques et épigénétiques de la variabilité. Bull Soc Zool France 118:149–157

Godfray HCJ (1994) Parasitoids. Behavioral and evolutionary ecology. Princeton University Press, Princeton

Hamelin F, Bernhard P, Shaiju AJ, Wajnberg E (2007) Diet selection as a differential foraging game. SIAM J Control Optim 46:1539–1561

Hamilton WD (1967) Extraordinary sex ratios. Science 156:477–488

Hardy ICW, Ode PJ (2007) Why biocontrol practitioners should be more interested in parasitoid sex ratios. Biocontrol News Inf 28:49–51

Harper DGC (1991) Communication. In: Krebs JR, Davis NB (eds) Behavioural ecology: an evolutionary approach. 3rd edition. Blackwell Scientific Publications, Oxford, pp 374–397

Harvey PH, Pagel MD (1991) The comparative method in evolutionary biology. Oxford Series in Ecology and Evolution. Oxford University Press, Oxford

Heimpel GE, Lundgren JG (2000) Sex ratios of commercially reared biological control agents. Biol Control 19:77–93

Hoffmann AA, Parsons PA (1988) The analysis of quantitative variation in natural populations with isofemale strains. Genet Select Evol 20:87–98

Hoffmeister TS, Roitberg BD (1998) Evolution of signal persistence under predator exploitation. Ecoscience 5:312–320

Hopper KR, Roush RT, Powell W (1993) Management of genetics of biological-control introductions. Annu Rev Entomol 38:27–51

Houston A, McNamara J (1999) Models of adaptive behavior: an approach based on state. Cambridge University Press, Cambridge

Hoy MA (1976) Genetic improvement of insects: fact or fantasy. Environ Entomol 5:833–839

Joslyn DJ (1984) Maintenance of genetic variability in reared insects. In: King EG, Leppla NC (eds) Advances and Challenges in Insect Rearing. USDA/ARS, New Orleans, pp 20–29

Krebs JR, Davies NB (1987) An introduction to behavioural ecology. Blackwell, Oxford

Krebs JR, Davies NB (1997) The evolution of behavioural ecology. In: Krebs JR, Davies NB (eds) Behavioural ecology: an evolutionary approach, 4th edn. Blackwell, Oxford, pp 3–12

Krebs JR, Dawkins R (1984) Animal signals: mind-reading and manipulation. In: Krebs JR, Davis NB (eds) Behavioural ecology: an evolutionary approach. 2nd edition. Blackwell Scientific Publications, Oxford, pp 380–402

Lack D (1947). The significance of clutch size. Ibis 89:309–352

Lewis WJ, Vet LEM, Tumlinson JH, van Lenteren JC, Papaj DR (1990) Variations in parasitoid foraging behavior: essential element of a sound biological control theory. Environ Entomol 19:1183–1193

Limburg H, Pak GA (1991) Genetic variation in the walking behaviour of the egg parasite *Trichogramma*. In: Bigler F (ed) Proceedings of the 5th Workshop on Quality Control of mass-reared arthropods. Wageningen (NL), pp 47–55

Liu SS (1985) Development, adult size and fecundity of *Aphidius sonchi* reared in two instars of aphid hosts, *Hyperomyzus lactucae*. Entomol Exp Appl 37:41–48

Liu FH, Smith SM (2000) Measurement and selection of parasitoid quality for mass-reared *Trichogramma minutum* Riley used in inundative release. Biocontrol Sci Technol 10: 3–13

Liu SS, Morton R, Hughes RD (1984) Oviposition preferences of a hymenopterous parasite for certain instars of its aphid host. Entomol Exp Appl 35:249–254

Mackauer M (1976) Genetic problems in the production of biological control agents. Annu Rev Entomol 21:369–385

Margolies DC, Cox TS (1993) Quantitative genetics applied to haplodiploid insects and mites. In: Wrensch DL, Ebbert ME (eds) Evolution and diversity of sex ratio. Chapman & Hall, New York, pp 548–559

Martin P, Bateson P (1994) Measuring behaviour. An introduction guide, 2nd edn. Cambridge University Press, Cambridge

Martins EP (1996) Phylogenies and the comparative method in animal behaviour. Oxford University Press, Oxford

Mills NJ, Wajnberg E (2008) Optimal foraging behavior and efficient biological control methods. In: Wajnberg E, Bernstein C, van Alphen J (eds) Behavioural ecology of insect parasitoids – from theoretical approaches to field applications. Blackwell, Oxford, pp 3–30

Mimouni F (1991) Genetic variations in host infestation efficiency in two *Trichogramma* species from Morocco. Redia 74:393–400

Ode PJ, Hardy ICW (2008) Parasitoid sex ratios and biological control. In: Wajnberg E, Bernstein C, van Alphen J (eds) Behavioural ecology of insect parasitoids – from theoretical approaches to field applications. Blackwell, Oxford, pp 253–291

Ode PJ, Hunter MS (2002) Sex ratios of parasitic Hymenoptera with unusual life-histories. In: Hardy ICW (ed), Sex ratios: concepts and research methods. Cambridge University Press, Cambridge, pp 218–234

Parker GA, Courtney SP (1984) Models of clutch size in insect oviposition. Theor Popul Biol 26:21–48

Parsons PA (1980) Isofemale strains and evolutionary strategies in natural populations. In: Hecht M, Steere W, Wallace B (eds) Evolutionary biology, vol 13. Plenum, New York, pp 175–217

Pompanon F, Fouillet P, Boulétreau M (1994) Locomotion behaviour in females of two *Trichogramma* species. Description and genetic variability. Norw J Agric Sci 16:185–190

Pompanon F, Fouillet P, Boulétreau M (1999) Physiological and genetic factors as sources of variation in locomotion and activity rhythm in a parasitoid wasp (*Trichogramma brassicae*). Physiol Entomol 24:346–357

Roitberg BD (1990) Variation in behaviour of individual parasitic insects: Bane or Boon? In: Mackauer M, Ehler LE, Roland E (eds) Critical issues in biological control. Intercept, Andover, Hants, pp 25–39

Roitberg BD (2007) Why pest management needs behavioral ecology and vice versa. Entomol Res 37:14–18

Roitberg BD, Mangel M (1988) On the evolutionary ecology of marking pheromones. Evol Ecol 2:289–315

Rosenheim JA (1999) The relative contributions of time and eggs to the cost of reproduction. Evolution 53:376–385

Roush RT (1990) Genetic variation in natural enemies: Critical issues for colonization in biological control. In: Mackauer M, Ehler LE, Roland J (eds) Critical issues in biological control. Intercept, Andover, Hants, pp 263–288

Schmidt JM (1991) The inheritance of clutch size regulation in *Trichogramma* species (Hymenoptera: Chalcidoidea: Trichogrammatidae). In: Bigler F (ed), Proceedings of the 5th Workshop on Quality Control of mass-reared arthropods, Wageningen (NL), pp 26–37

Schmidt JM (1994) Host recognition and acceptance by *Trichogramma*. In: Wajnberg E, Hassan SA (eds) Biological control with egg parasitoids. CAB International, Wallingford, pp 165–200

Seventer JG, Ellers J, Driessen G (1998) An evolutionary argument for time limitation. Evolution 52:1241–1244

Sih A, Christensen B (2001) Optimal diet theory: when does it work, and when and why does it fail? Anim Behav 61:379–390

Stephens DW, Krebs JR (1986) Foraging theory. Princeton University Press, Princeton

Stouthamer R, Hurst GDD, Breeuwer JAJ (2002) Sex ratio distorters and their detection. In: Hardy ICW (ed) Sex ratios: concepts and research methods. Cambridge University Press, Cambridge, pp 195–215
Stouthamer R, Pinto JD, Platner GR, Luck RF (1990). Taxonomic status of thelytokous forms of *Trichogramma* (Hymenoptera: Trichogrammatidae). Ann Entomol Soc Am 83:475–481
Strand MR (1986) The physiological interactions of parasitoids with their hosts and their influence on reproductive strategies. In: Waage J, Greathead D (eds) Insect parasitoids. Academic, London, pp 97–136
Taylor PD, Bulmer MG (1980) Local mate competition and the sex ratio. J Theor Biol 86:409–419
Urquijo P (1950) Aplicacion de le genetica al aumento de la eficacia des *Trichogramma minutum* en la lucha biologica. Boletin Patologia Vegetal Entomol Agric 18:1–12
van Alphen JJM, Bernstein C (2008) Information acquisition, information processing and patch time allocation in insect parasitoids. In: Wajnberg E, Bernstein C, van Alphen J (eds) Behavioural ecology of insect parasitoids – from theoretical approaches to field applications. Blackwell, Oxford, pp 172–192
van Alphen JJM, Vet LEM (1986) An evolutionary approach to host finding and selection. In: Waage J, Greathead D (eds) Insect parasitoids. Academic, London, pp 23–61
van Alphen JJM, Bernstein C, Driessen G (2003) Information acquisition and time allocation in insect parasitoids. Trends Ecol Evol 18:81–87
van Dijken MJ, Waage JK (1987) Self and conspecific superparasitism by the egg parasitoid *Trichogramma evanescens*. Entomol Exp Appl 43:183–192
Viktorov GA, Kochetova NI (1973) The role of trace pheromones in regulating the sex ratio in *Trissolcus grandis* (Hymenoptera: Scelionidae). Zhurnal Obshchei Biologii 34:559–562
Vinson SB (1976) Host selection by insect parasitoids. Annu Rev Entomol 21:109–133
Vinson SB (1985) The behaviour of parasitoids. In: Kerkut GA, Gilbert LI (eds) Comprehensive insect physiology, biochemistry and pharmacology, vol 9. Pergamon, Oxford, pp 417–469
Waage JK (1982) Sex ratio and population dynamics of natural enemies – some possible interactions. Ann Appl Biol 101:159–164
Waage JK (1986) Family planning in parasitoids: adaptive patterns of progeny and sex allocation. In: Waage JK, Greathead DJ (eds) Insect parasitoids. Academic, London, pp 63–95
Waage JK (1990) Ecological theory and the selection of biological control agents. In: Mackauer M, Ehler LE, Roland E (eds) Critical issues in biological control. Intercept, Andover, Hants, pp 135–157
Waage JK, Lane JA (1984). The reproductive strategy of a parasitic wasp. II. Sex allocation and local mate competition in *Trichogramma evanescens*. J Anim Ecol 53:417–426
Waage JK, Ng SM (1984) The reproductive strategy of a parasitic wasp. I. Optimal progeny and sex allocation in *Trichogramma evanescens*. J Anim Ecol 53:401–416
Wajnberg E (1989) Analysis of variations of handling time in *Trichogramma maidis*. Entomophaga 34:397–407
Wajnberg E (1991) Quality control of mass-reared arthropods: a genetical and statistical approach. In: Bigler F (ed) Proceedings of the 5th Workshop on Quality Control of mass-reared arthropods. Wageningen (NL), pp 15–25
Wajnberg E (1993) Genetic variation in sex allocation in a parasitic wasp. Variation in sex pattern within sequences of oviposition. Entomol Exp Appl 69:221–229
Wajnberg E (1994) Intra-population genetic variation in *Trichogramma*. In: Wajnberg E, Hassan SA (eds) Biological control with egg parasitoids. CAB International, Wallingford, pp 245–271
Wajnberg E (2004) Measuring genetic variation in natural enemies used for biological control: why and how? In: Ehler L, Sforza R, Mateille T (eds) Genetics, evolution and biological control. CAB International, Wallingford, pp 19–37
Wajnberg E (2006) Time-allocation strategies in insect parasitoids: from ultimate predictions to proximate behavioural mechanisms. Behav Ecol Sociobiol 60:589–611
Wajnberg E, Colazza S (1998) Genetic variability in the area searched by a parasitic wasp. Analysis from automatic video tracking of the walking path. J Insect Physiol 44:437–444

Wajnberg E, Hassan SA (1994) Biological control with egg parasitoids. CAB International, Wallingford
Wajnberg E, Pizzol J, Babault B (1989) Genetic variation in progeny allocation in *Trichogramma maidis*. Entomol Exp Appl 53:177–187
Wajnberg E, Rosi MC, Colazza S (1999) Genetic variation in patch-time allocation in a parasitic wasp. J Anim Ecol 68:121–133
Wajnberg E, Gonsard PA, Tabone C, Curty E, Lezcano N, Colazza S (2003) A comparative analysis of patch-leaving decision rules in a parasitoid family. J Anim Ecol 72:618–626
Wajnberg E, Curty C, Colazza S (2004) Genetic variation in the mechanisms of direct mutual interference in a parasitic wasp: consequences in terms of patch-time allocation. J Anim Ecol 73:1179–1189
Wajnberg E, Bernstein C, van Alphen J (2008) Behavioural ecology of insect parasitoids – from theoretical approaches to field applications. Blackwell, Oxford
Werren JH (1984) Brood size and sex ratio regulation in the parasitic wasp *Nasonia vitripennis* (Walker) (Hymenoptera: Pteromalidae). Neth J Zool 34:123–143
Werren JH (1997) Biology of *Wolbachia*. Annu Rev Entomol 42:587–609
Wylie HG (1973) Control of egg fertilization by *Nasonia vitripennis* (Walk.) (Hymenoptera: Pteromalidae) reared from super-parasitized housefly pupae. Can Entomol 98:645–653
Wylie HG (1976) Interference among females of *Nasonia vitripennis* (Hymenoptera: Pteromalidae) and its effect on the sex ratio of their progeny. Can Entomol 108:655–661

Chapter 6
Sex Ratio Modulators of Egg Parasitoids

James E. Russell and Richard Stouthamer

Contents

6.1 Introduction

The importance of sex ratios in the biology and biocontrol application of egg parasitoids can be summarized as the cost of producing males. For both the biocontrol practitioner and the biocontrol agent a surplus of healthy fecund females is highly desirable. The sex ratio investment advice of a biological banker might be to maximize the total number of reproductive females and minimize the number of males. In this chapter we will describe two inherited genetic elements found in several *Trichogramma* species that can be considered either excellent or terrible investment options. These non-Mendelian genetic elements effectively distort offspring sex ratios in a manner that can be called selfish- since both promote their own transmission at the expense of host nuclear genes. One, a bacterium called *Wolbachia* can convert haplodiploid females into fully functional completely parthenogenetic organisms (Stouthamer et al. 1993). In fact, several *Wolbachia* infected species are only known from parthenogenetic females- few or no males are found (Huigens and Stouthamer 2003). The other, a nuclear "extra" chromosome called PSR (Paternal

J.E. Russell (✉)
Department of Biology, University of California, Riverside, CA 92521, USA
e-mail: jeruss002@gmail.com

F.L. Consoli et al. (eds.), *Egg Parasitoids in Agroecosystems with Emphasis on Trichogramma*, Progress in Biological Control 9, DOI 10.1007/978-1-4020-9110-0_6,

Sex Ratio) found only in male haplodiploids behaves in a manner that can only be described as extremely selfish. The PSR chromosome turns eggs destined to develop as females into males that carry the PSR chromosome (Werren and Stouthamer 2003). These two sex ratio distorters have opposite but sometimes complimentary consequences in parasitoid populations. We will discuss the biology and applied potential of both these sex ratio distorters in *Trichogramma* species.

From a biological control perspective it is clear to see how bacteria that create a continuous source of females without need of males can be beneficial. Female parasitoids are the agents of action in biocontrol practice. The relatively recent discovery that *Wolbachia* induced parthenogenesis can be infectious (Huigens et al. 2000), i.e. horizontally transferred rather than inherited, means that sexual populations or cultures can be converted to parthenogenetic reproduction. We will discuss this potential boon for insectaries and biocontrol practitioners and suggest that short and long-term fitness effects of *Wolbachia* infection still require careful analysis.

PSR chromosomes, on the other hand, would seem to spell disaster for biocontrol purposes. An inherited parasite that converts female offspring to male can literally destroy populations. With this in mind we discuss detection techniques for PSR chromosomes and their population biology. Applied strategies that incorporate PSR chromosomes appear to be confined to the destructive potential of male biased sex ratio distortion. But, the genetic behavior of PSR chromosomes are unique in that they are extra chromosomes and do not recombine in meiosis. We discuss a potential for biocontrol application that takes advantage of the unique genetic behavior of PSR chromosomes.

Sex ratio distortion, whether it is female or male-biased distortion, is the subject of this chapter. And, as alluded to earlier, the sex ratio is the domain of the host organism: bacteria and "extra" chromosomes are sexless and have no inherent sex ratio. Specifically, the sex ratio is the domain of the host nuclear genes. Sex ratio distortion imposed by non-Mendelian genetic elements is nothing short of a genetic war over control of male production. To understand the forces involved in this war we will summarize the rather large body of theoretical and empirical research that can be classified under the title "sex ratio theory". Sex ratio theory is considered one of the most successful areas of evolutionary biology and we rely on its findings to emphasize both the short and long-term consequences of sex ratio distortion. We begin with an explanation of sex ratio theory and how it applies to egg parasitoids. Then we discuss the prevalence and impact of non-Mendelian genetic elements that selfishly distort sex ratios. We end by describing two important sex ratio distorters found in *Trichogramma* species and how they can be detected in other egg parasitoids.

6.2 Sex Ratio Theory

Thus a tendency towards equalization of the sexes would be brought about... (Darwin 1871)

Darwin observed that in a population composed of excess males a mother who produces mostly female offspring will produce more descendents than mothers who produce even or male-biased offspring sex ratios. Carl Düsing (Düsing 1884) formalized the concept in a book titled *The Regulation of the Sex Ratio (Die Regulierung des Geschlechtsverhaltnisses)*. Düsing showed that indeed the rarer sex was more valuable genetically and deviations from a balanced sex ratio would be self-correcting. Taking Darwin's hypothetical scenario as an example: because all males (nx) together produce as many offspring (z) as all females (x) together; and if the population is male-biased ($nx > x$), then the reproductive output of each male is z/nx and for each female is z/x. From this it is clear to see the inequality in reproductive output, $z/x > z/nx$, making females more valuable (Segar and Stubblefield 2002). Thus selection acting on sex ratios would favor those females that produced more daughters. The reciprocal situation would be resolved using the same logic and the overall sex ratio would tend toward a balanced 1:1 state.

RA Fisher's explanation for the influence of natural selection on sex ratios (Fisher 1930) was essentially the same as Düsing's, except he framed it in terms of parental expenditure. Fisher did not claim that the sex ratio was necessarily balanced, but that parental expenditure was balanced, "...the sex ratio will so adjust itself under the influence of Natural Selection that the total parental expenditure incurred in respect of children of each sex, shall be equal...." (1930, p.142). This means that if daughters are more expensive than sons, the offspring sex ratio will be male biased, resulting in a balanced expenditure for the parent. It is not easy to imagine a situation where the cost of sons and daughters is intrinsically different in Hymenoptera. The production of honeybee queens is extremely expensive compared to male production in honeybee colonies and the corresponding sex ratio conforms to Fisher's prediction: many more males are produced (Queller 2006). For heteronomous parasitoids (parasitoids which lay female and male eggs in separate hosts) the cost of locating and handling different hosts may result in sex ratios biased toward the "less expensive" sex (Godfray and Waage 1990).

Perhaps a better way to imagine the sex ratio equation is in terms of the greatest return per investment. W. D. Hamilton's *Extraordinary Sex Ratios* (Hamilton 1967) considered the sex ratio question from an investment strategy perspective. The best return for reproductive investment, according to Hamilton, was determined by factors such as the unequal reproductive value of the sexes, unisexual modes of inheritance, and population structure. His argument began with the observation that balanced sex ratios are not as common in nature as Fisher's theory would predict. He then suggested that observed unbalanced sex ratios may be the result of selection acting in two unique ways. In the first section Hamilton showed that driving genetic elements linked to sex determination could spread through populations, rapidly distorting sex ratios in the process.

The second section dealt with a phenomenon Hamilton termed Local Mate Competition (LMC), which specifically dealt with population mating structure and parental control of offspring sex ratios. Both Düsing and Fisher assumed populations were unstructured and mating was panmictic. Hamilton suggested that populations were divided into subgroups in which mating took place before females dispersed to form the new next generation subgroups. The population structure Hamilton had in mind when he formulated his theory, was that of the parasitoid wasp *Nasonia vitripennis* (Walker), where their fly pupal hosts can occur in different patches. Often single pupae are found, but in some cases large groups of fly pupae are found together. Females in this species are winged while males have short wings and cannot fly. It was well known that very female biased offspring sex ratios were produced by single females parasitizing patches containing one or only a few pupae, whereas in large aggregations with many females the offspring sex ratio tended towards equality.

Population structures such as these, where mating among siblings (sib-mating) takes place, are well known among parasitoids (Hamilton 1967, and references therein). Hamilton reasoned that in population subgroups mating was not random and mothers could maximize their fitness by controlling mate competition amongst their offspring. With this in mind the "unbeatable sex ratio" was derived: sex ratio [M/(M+F)] = $(n-1)/2n$; where n = the number of foundresses establishing a local mating group. Foundresses are females that create the next generation of females and males among which mating will occur. In practical terms the unbeatable sex ratio meant that if only 2 female parasitoids parasitize a host upon which their offspring will mate, the unbeatable offspring sex ratio for both females would be 0.25, or three daughters for every son. If 20 females parasitize such a host the offspring sex ratio would be 0.475, or very close to 1 male for every female. As the number of foundresses increases and future mating opportunities approaches panmixia, the unbeatable sex ratio becomes the balanced sex ratio of Düsing and Fisher.

Female-biased sex ratios are not uncommon in many parasitoid species and are now often referred to as Hamiltonian, or LMC, sex ratios. While the exact causal mechanisms explaining the evolution of biased sex ratios have been debated (Colwell 1981, Werren 1983, Nunney 1985), the optimality of biased sex ratios and Hamilton's model have been repeatedly confirmed with laboratory tests using parasitoids (Werren 1983, Waage and Lane 1984, King and Skinner 1991).

Of the 26 species mentioned in Hamilton's original article as examples of biased sex ratios, 19 were parasitic wasps (one a fig wasp, 7 egg parasitoids, and 11 larval or pupal parasitoids) and all were haplodiploid. Haplodiploidy is characteristic of the order Hymenoptera and is described as the direct development of haploid males from unfertilized eggs and diploid female development from fertilized eggs (this is more appropriately called arrhenotoky). Diplodiploid sex chromosomal systems, such as XY and ZW where the presence or absence of a chromosome determines sex, are genetically constrained when it comes to sex ratio adjustments. Mendelian rules of segregation ensure half of the offspring of an XY male will carry his X chromosome and thus develop as females while the other half will carry his Y chromosome and develop as males. Haplodiploids, on the other hand, display a fascinating array of sex ratios largely because no genetic constraint exists; sex is determined by whether a female chooses to fertilize an egg or not. Thus, the sex of the hymenopteran egg is subject to the behavioral decisions of an ovipositing female, and numerous experiments have shown that these decisions are influenced by the local mating environment of her future offspring (Werren 1983, Waage and Lane 1984, Herre 1985, Vanwelzen and Waage 1987, Strand 1988, Godfray 1994, Shuker et al. 2005).

For gregarious parasitoids that mate in local patches where sibmating is prevalent, the benefits of female biased offspring sex ratios are (1) less male competition for mates, (2) an optimal maximum number of daughters means high reproductive return for the few males produced, and (3) in haplodiploid species, females are more related to their mothers and thus better transporters of her genes. Though the exact predictions given the "unbeatable sex ratio" are rarely observed (the theory is purely a mathematical explanation of optimal nuclear genetic strategy, in the real world environmental uncertainties influence actual reproductive decision-making), the consistent trend in field observations and experimental results follow the predicted pattern.

The work of Waage and colleagues (Waage 1982, Waage and Lane 1984, Waage and Ming 1984, Vanwelzen and Waage 1987) showed that for several egg parasitoids adaptive sex ratio changes not only conformed to the LMC models of Hamilton, but were also a function of the sex allocation strategy of individual ovipositing females. Waage and Ming (1984) suggested that in *Trichogramma evanescens* Westwood single male eggs were always laid early in a given oviposition bout, along with one or several female eggs. A similar allocation strategy has also been observed in laboratory studies of *Trichogramma chilonis* (Ishii) (Suzuki et al. 1984). Laboratory and field tests of sex allocation in *Trichogramma pretiosum* Riley characterized a similar oviposition pattern as a precise allocation strategy (Luck et al. 2001). Precise sex allocation is when variance in male offspring number is lower than binomial variance would predict. Luck et al. (2001) found that male eggs were consistently placed last in a host, preceded by one or more female eggs. This pattern suggested sex allocation was a product of behavioral decisions. Field observations and laboratory experiments have shown that in restricted locally mating environments parasitoid wasps are able to optimize offspring sex ratios to meet local conditions.

Optimal sex ratio theory was extended to include differential fitness effects of the sexes as a function of resource quality (Charnov et al. 1981). Sex allocation theory, as it is called, suggests that sex ratios vary according to the resources available as a consequence of parental control of sex ratios. Typically, this means that if one sex benefits more from high quality resources (*eg* host size, food availability or food quality), then parental control of offspring sex ratios will shift the balance toward that sex in high quality environments (Charnov 1982). Confirmation of sex allocation theory comes from parasitic wasps that have been shown to allocate resources differently to offspring sexes (Charnov et al. 1981).

Optimal offspring sex allocation is based on the assumption of parental control over offspring sex ratios. However, there are many examples of factors affecting offspring sex ratios over which parents have no control. Deviations from optimal offspring sex ratios can occur due to variation in ecological variables such as host size, diet, and temperature (Luck et al. 2001, Santolamazza-Carbone et al. 2007). Life history traits can also interact with local conditions to distort optimal sex ratios. Complimentary sex determination systems which rely on heterozygosity at sex determining loci for female development can produce male-biased offspring sex ratios under inbreeding conditions (Whiting 1943, Cook 1993). In polyembryonic species like *Copidosoma floridanum* (Ashmead), where hundreds of individual larvae develop from a single egg, differential mortality of the sexes can distort offspring sex ratios (Strand 1989a, b, Hardy et al. 1993). In mixed sex broods of *C. floridanum*, i.e. at least one female and one male egg oviposited in a single host, female larvae will consume male larvae (Grbic et al. 1992), distorting the offspring sex ratio.

6.3 Sex Ratio Distortion

In contrast to the ecological and life history traits that indirectly alter optimal sex ratios, the focus of this chapter is on non-Mendelian sex ratio distorters that directly derail optimal offspring sex ratios. To understand the nature of this effect it is important to recognize the essential conflict at its core. Optimal offspring sex ratios are considered the domain of the nuclear genes, and since all sexually produced offspring are the product of genes transmitted from one father and one mother, a balanced sex ratio is expected (with the caveat of Hamiltonian (LMC) sex ratios explained above). But, there are some sex specific nuclear and cytoplasmic genes whose transmission and consequent optimal sex allocation strategies are quite different. For example, cytoplasmic genes that are solely transmitted by females and suffer zero fitness in males are known to kill and feminize males in some species, resulting in higher female frequencies (Hurst et al. 1997, Rigaud 1997). Certain nuclear genes associated more with one sex than the other will also convert or destroy the non-transmitting sex. Examples include driving nuclear sex chromosomes in *Drosophila* spp. (Jaenike 2001), and nuclear B chromosomes ("extra" chromosomes) in some Hymenopteran species (Werren and Stouthamer 2003). These sex ratio distorters are genetic elements that effectively parasitize sex determination systems. In all cases sex ratio distorting elements exploit the inherent reproductive asymmetries present in sexual species- uniparental inheritance of cytoplasmic and sex chromosome genes.

Most sex ratio distorters are cytoplasmically inherited, and as a consequence tend to shift the sex ratio toward the transmitting sex – females. There are at least four groups of cytoplasmically inherited bacteria in insects that distort offspring sex ratios through various means (Table 6.1). Skinner (1982) also identified an unknown maternally inherited factor (MSR) in *Nasonia vitripennis* (Walker) that caused nearly 100% fertilization (resulting in 100% female offspring) in mated affected females. The list of infected species is rapidly growing with the application of molecular techniques that simplify the detection process of these fastidious symbionts.

Female biased sex ratios are common in parasitoids and as mentioned previously may be the result of life history, adaptation, or ecological conditions. For detection of cytoplasmic sex ratio distortion a combined experimental approach is recommended. Ideally, results showing sex ratio distortion from mating and rearing experiments on isofemale lines should be followed by a nested PCR design. Once sex ratio distortion is suspected a mating design that includes antibiotic-treated and untreated mated and virgin females will distinguish between the various forms of

Table 6.1 Cytoplasmic sex ratio distorters found in insects (partial list of infected insects)

Name	Phenotype	Host
Arsenophonus nasoniae	Male-killing	*Nasonia vitripennis*
Cardinium hertigii	Cytoplasmic incompatibility, Parthenogenesis	*Encarsia pergandiella* Howard
Rickettsia sp.	Male-killing	*Adalia bipuntata* (L.)
	Parthenogenesis	*Neochrysocharis formosa* (Westwood)
Wolbachia pipientis	Cytoplasmic incompatibility	*Drosophila simulans* Sturtevant
	Feminization	*Ostrinia furnacalis* (Guenée)
	Male-killing	*Acraea encedon* (L.)
	Parthenogenesis	*Trichogramma pretiosum*

symbiont-induced sex ratio distortion (Fig. 6.1). PCR based techniques have been developed for the cytoplasmic sex ratio distorters listed in Table 6.1 (Gherna et al. 1991, Werren et al. 1994, Wu et al. 2004, Zchori-Fein et al. 2004, Hagimori et al. 2006). The PCR protocol should include a general bacterial primer that will amplify a conserved region of all bacterial genomes, followed by specific primers that can

	♂♀ Mated, untreated	♂♀ Mated, antibiotic treated	♀ Unmated, untreated	♀ Unmated, treated
Male-killing	All female offspring, reduced offspring #	Female and male offspring, no reduction in offspring #	No or few offspring	All male offspring, no reduction in offspring #
Parthenogenesis	All female	Male and female, or all male	All female	All male
Reproductive incompatibility (symbiont-induced)	Some normal sex ratio broods, Some male-biased broods or no offspring	Normal sex ratio broods and offspring #	All male	All male

Fig. 6.1 A mating scheme designed to distinguish sex ratio distortion in haplodiploid organisms is shown. The top row represents the treatment, the leftmost column lists the type of sex ratio distortion distinguished, and the intersecting boxes describe the expected results for each treatment and type of sex ratio distortion

diagnose the bacterial taxon. The advantage of this approach is the clear laboratory results demonstrating sex ratio distortion followed by diagnostic PCR describing the sex ratio distorting agent. Antibiotic treatment of lines and the reversal of sex ratio distortion, which persists after the treatment has ended is an indication of bacterial involvement. The nested PCR design allows for the detection of previously undiscovered bacterial sex ratio distorters. The general bacterial primers amplify sequences from all bacteria present in the insect including previously undiscovered sex ratio distorting bacteria.

For the most part, biparental inheritance of nuclear genes limits their potential for sex ratio distortion, since both sexes transmit them equally. However, those nuclear genes that are not transmitted equally between the sexes have displayed extraordinary sex ratio distortion capabilities. We will discuss one such nuclear sex ratio distorter found in *Trichogramma* called PSR (Paternal Sex Ratio) that has been referred to as the most extreme selfish genetic element (Werren and Stouthamer 2003). But first we discuss a unique reproductive parasite found in several *Trichogramma* species and other egg parasitoids that not only distorts offspring sex ratios, but also introduces a new form of reproduction to infected individuals, populations, and species.

6.3.1 Wolbachia: Parthenogenesis Induction

Thelytokous parthenogenesis, the development of diploid females from unfertilized haploid eggs, is not uncommon in parasitoid species, with more than 270 species reported (Luck et al. 1992). Stouthamer et al. (1990) showed that in some species of *Trichogramma*, thelytokous parthenogenesis could be "cured" with antibiotic treatment, and thelytokous females previously fed a honey/antibiotic mixture began producing male offspring. Based on these experiments bacterial involvement was suspected and further research confirmed the presence of a cytoplasmic parasite called *Wolbachia pipientis* (Stouthamer et al. 1993), a bacterium known to cause various reproductive alterations in several arthropod species (Yen and Barr 1973, Binnington and Hoffmann 1989, Breeuwer and Werren 1990, Rousset et al. 1992) (Table 6.1). The ability to induce parthenogenesis marked a unique development in the evolution of reproductive parasitism. Rather than inducing reproductive incompatibility, killing males, or feminizing genetic males, as had been shown previously with maternally-inherited reproductive parasites, the parthenogenesis-inducing (PI) form of *Wolbachia* showed that it was capable of creating a new form of reproduction. PI *Wolbachia* can be regarded as the ultimate cytoplasmic solution. Rather than simply removing or penalizing males (genetic dead-ends for cytoplasmic symbionts), PI *Wolbachia* removes the *need* for males.

Evidence of *Wolbachia* involvement in thelytokous parthenogenesis has been found in at least 70 haplodiploid species (Huigens and Stouthamer 2003), 18 of which are egg parasitoids (Table 6.2). The exclusive occurrence of *Wolbachia*-induced thelytoky in haplodiploid species may be a product of the different developmental processes in haplodiploid and diplodiploid species. The initiation

Table 6.2 Evidence for symbiont induced parthenogenesis in egg parasitoid species

Species	Reference
Anagrus sophiae	Rosenheim, pers. comm.
Ooencyrtus submetallicus	Wilson (1962)
Ooencyrtus fecundus	Laraichi (1978)
Telonomus nawai	Arakaki et al. (2000)
Trichogramma brevicapillum	Stouthamer et al. (1990)
T. chilonis	Schilthuizen and Stouthamer (1997)
T. cordubensis	Silva and Stouthamer (1996)
T. deion	Stouthamer et al. (1993)
T. embryophagum	Stouthamer et al. (1990)
T. evanescens	Stouthamer et al. (1990)
T. kaykai	Stouthamer and Kazmer (1994)
T. nubilale	van Meer et al. (1999)
T. oleae	Roussett et al. (1992)
T. platneri	Schilthuizen and Stouthamer (1997)
T. pretiosum	Stouthamer et al. (1990)
T. sibericum	van Meer et al. (1999)
T. atopovirilla	Ciociola et al. (2001)
T. semblidis	Pintereau et al. (2000)

of developmental processes in haplodiploid species does not rely on egg penetration by sperm as in diplodiploids; rather, the deformation of the eggs during oviposition is sufficient to initiate development (Stouthamer 1997). Therefore, in haplodiploids the only change needed to induce thelytokous parthenogenesis is in the ploidy level of eggs, whereas for diplodiploids a ploidy and developmental change is needed. This may explain why *Wolbachia*-induced thelytoky has only been observed in haplodiploid species. However, the cytogenetic mechanism by which symbiont-induced thelytokous parthenogenesis generates diploid females does not necessarily preclude parthenogenesis in diplodiploid species (Weeks and Breeuwer 2001).

Several detailed cytogenetic studies have been done on *Wolbachia*-induced parthenogenesis (Stille and Davring 1980, Gottlieb et al. 2002, Pannebakker et al. 2004). The only detailed cytogenetic study to date of symbiont-induced parthenogenesis in egg parasitoids comes from the analysis of *Wolbachia* in *Trichogramma* spp. (Stouthamer and Kazmer 1994). Using molecular markers (allozymes), they discovered two important features of the parthenogenesis induction mechanism: (1) in unfertilized eggs, a failure of the first mitotic division in *Wolbachia* infected eggs after normal meiosis and fusion of the two mitotic products restored diploidy and converted eggs destined to develop as haploid males to develop as diploid female and, (2) in fertilized eggs, *Wolbachia* did not interfere with fertilization and normal sexual development of female offspring. The mechanism by which unfertilized eggs are converted to functional *Wolbachia* infected females is called gamete duplication and results in complete homozygosity for infected offspring developing from unfertilized eggs. How fertilization suppresses *Wolbachia* induced gamete duplication is currently an open question. Genetically determined thelytoky, i.e. not induced by cytoplasmic symbionts, contrasts with the findings of Stouthamer and Kazmer (1994) in that the restoration of diploidy is accomplished by various modifications during the meiotic process (Suomalainen et al. 1987, Vavre et al. 2004).

It appears parthenogenesis via gamete duplication is not a cost-free mode of reproduction. Fitness effects reported for PI *Wolbachia* infection in a number of *Trichogramma* species suggest infected females pay a fitness penalty. Stouthamer and Luck (1993) compared infected and cured females derived from the same isofemale lines. In all cases studied they found that the total number of offspring produced by the cured females was significantly higher than that produced by the infected females. In several of the species the number of daughters produced by the cured females was also higher, with the exception of the *T. pretiosum* line from Hawaii, where they produced similar numbers of daughters. Stouthamer and Luck (1993) also tested to see if this difference in offspring production was influenced by the fact that the infected females used in these experiments were virgins, while the cured females had mated. This was tested for one species *Trichogramma deion* Pinto and Oatman and no significant difference in offspring production was found between mated and virgin infected females. This contrasted with the findings of Tagami et al. (2002) who compared mated and virgin infected *T. deion* and *Trichogramma kaykai* Pinto, Platner and Stouthamer females and found that virgin females show between 25 and 45% reductions in fitness as measured by development from egg to adult (Tagami et al. 2002). This suggests that sexual reproduction is more efficient than gamete duplication, and virgin infected females whose only reproductive option is gamete duplication pay a significant fitness cost. Tagami et al. (2002) show that the eggs that did not develop had their development interrupted in the early mitotic divisions. Later experiments by Huigens et al. (2004b) also found a reduction in the number of offspring produced by virgin infected females in several *T. kaykai* lines, although the reduction was much less than that found by Tagami et al. (2002). A significant cost of *Wolbachia* infection has also been reported in several studies comparing infected and uninfected *Trichogramma* under superparasitization conditions (Stouthamer and Luck 1993, Hohmann and Luck 2000, Huigens et al. 2004b). Delayed development of infected larvae (Hohmann and Luck 2000) was implicated in the near 50% reduction in survival for infected larvae in superparasitized hosts (Huigens et al. 2004b). In most cases, infected *Trichogramma* lines show reductions in offspring production when compared to antibiotic cured versions of the same line (Stouthamer and Luck 1993, Huigens et al. 2004b). PI *Wolbachia* infection, at least in the *Trichogramma* species studied in the laboratory, does not result in an increase in female production as might be expected; rather, it reduces the fitness of infected females compared to their uninfected sexual counterparts.

Stouthamer and Luck (1993) argued that these disadvantages are only apparent when these traits are measured in the laboratory where the fitness of the wasps is often expressed as the total number of offspring produced under unlimited host access. In the field, the conditions may be quite different and individual females may only be able to parasitize a few host eggs in their lifetime. When the total offspring production was compared under host limitation, infected females produced more daughters than uninfected females.

Under laboratory conditions it is also clear that infected females, again under conditions of unlimited access to hosts, are not capable of producing only daughters. Often after two or three days of offspring production, the infected females start producing an increasing number of male offspring. The possible explanation for this is that the replication of *Wolbachia* in the females is not capable of keeping up with egg production, such that an insufficient number of *Wolbachia* per wasp egg are present to allow for successful gamete duplication.

With the fitness costs and the inefficient transmission in mind it is not a foregone conclusion that a PI *Wolbachia* infection will spread through a population of sexually reproducing females. Yet some populations and species are fixed for PI *Wolbachia* infection (Stouthamer 1997). Several studies have reported females from PI *Wolbachia* fixed populations that fail to use sperm from males to which they were paired, even though the males were fully functional (Pijls et al. 1996, Arakaki et al. 2001, Weeks and Breeuwer 2001, Jeong and Stouthamer 2005, Pannebakker et al. 2005). In mixed populations with low to intermediate infection frequencies, infected females readily fertilize eggs when mated to conspecific males (Stouthamer et al. 1990). Since *Wolbachia* itself does not interfere with the fertilization process, the reported data suggests that the population infection frequency somehow selects for the loss of female sexual function.

The sex ratio hypothesis (Huigens and Stouthamer 2003) attempts to explain this phenomenon through sex ratio theory (Fisher 1930, Hamilton 1967). When PI *Wolbachia* enters a population, it immediately introduces a more female-biased sex ratio. As the infection spreads and the population sex ratio becomes more female biased, the value of outcrossing males increases. Uninfected females that no longer fertilize their eggs, hence producing all male offspring, gain the sex ratio benefit of more valuable sons. The female biased sex ratio distortion brought with the spreading *Wolbachia* infection selects for those females in the uninfected portion of the population that no longer fertilize their eggs. Assuming a genetic basis for non-fertilization, selection for a "non-fertilization" allele will result in uninfected females becoming rarer since fewer females are willing to fertilize their eggs. The ultimate result, described in population models (Stouthamer et al. in prep.) is the replacement of uninfected sexual females with PI *Wolbachia*-infected females that no longer fertilize their eggs, as observed in the field.

Testing this hypothesis is difficult due to the initial rarity of a non-fertilization mutation in an infected population and the predicted rapid spread to fixation of such an allele. However, theoretical models show that a non-fertilizing genotype can spread through a PI *Wolbachia*-infected population taking PI *Wolbachia* to fixation with it (Stouthamer et al. in prep.). Crossing experiments between allopatric infected and uninfected populations of two parasitoids, *Telenomus nawai* Ashmead and *Leptopilina clavipes* (Hartig), have shown the non-fertilization "virginity" mutation to be heritable (Jeong and Stouthamer 2005, Pannebakker et al. 2005). In these experiments, females from populations fixed for PI *Wolbachia* infection were found to be incapable of fertilizing eggs when mated to functional males (males derived from antibiotic curing of the infected population and from the uninfected sexual population). The antibiotic-derived males from these populations however are sexually functional and capable of fertilizing eggs of females from the uninfected populations. However, crossing experiments showed that as the genome from the infected population was introgressed into females from the uninfected population, egg fertilization rates dropped to levels indistinguishable from those of the sexually dysfunctional infected females, indicating a genetic basis to functional "virginity" as the sex ratio hypothesis predicts.

An exception to the theoretical spread and observed fixation of PI *Wolbachia* infection in numerous parasitoid species is the intermediate infection frequencies found in some *Trichogramma* species. In the Mojave Desert, years of sampling *T. kaykai* and *T. deion* populations have shown that infection frequencies reach no higher than 30% (Stouthamer et al. 2001, Huigens et al. 2004a). Likewise, *Wolbachia* infection in the *T. chilonis* population of Hawaii has yet to spread to fixation (Stouthamer et al. 1993). Field surveys have also found low infection frequencies in *Trichogramma* species in Portuguese tomato fields (Gonçalves et al. 2006). These intermediate *Wolbachia* infection frequencies are curious given the predicted and observed fixation of PI *Wolbachia* in many species.

PI *Wolbachia* plays an important role in the biology of *Trichogramma* species and other infected parasitoids. The form of sex ratio distortion imposed by PI *Wolbachia*, at first glance, does not appear to negatively impact infected populations since sex is no longer necessary for reproduction. But this derived form of reproduction brought about by PI *Wolbachia*, is not without cost in the short and long-term. The detailed work of the immediate fitness consequences of *Wolbachia* infection in *Trichogramma* should now be followed with tests on the longer-term consequences of PI *Wolbachia* infection. Sampling and genetic analysis of infected and uninfected populations would shed light on the long-term genetic consequences of sex ratio distortion imposed by PI *Wolbachia* infection, the obligate asexuality brought about by such sex ratio distortion, and the factors influencing the frequency of infection.

6.3.2 PSR: Paternal Sex Ratio

A unique discovery in 2001 found that the factor influencing the intermediate *Wolbachia* infection frequency in the *T. kaykai* population of the Mojave Desert was actually another sex ratio distorting element. This element was paternally inherited, unique to haplodiploids, and caused fertilized infected and uninfected eggs to develop into males (Stouthamer et al. 2001). Haploid males produce gametes through essentially a mitotic process that distinguishes them from the normal meiotic process common to diploid organisms, including diploid females of haplodiploid species. In at least two species, *N. vitripennis* and *T. kaykai*, this process is parasitized by an "extra" or B chromosome called paternal sex ratio (PSR) (Werren and Stouthamer 2003). When sperm from a PSR-carrying male enters an egg, the entire paternal chromosomal compliment, aside from the PSR chromosome, forms a condensed mass at the first mitotic division and is eventually lost after successive mitotic divisions (van Vugt et al. 2003). The result is a "haploidized" fertilized egg carrying the maternal chromosomal compliment and the PSR chromosome. Phenotypically this means that an egg destined to develop as a diploid female is converted to one developing as a haploid male carrying PSR. To put it in another way, all fertilized eggs of females mated to a PSR male will be PSR-carrying males, and all unfertilized eggs will be normal males. PSR chromosomes simply use males and the nuclear genes they possess for transportation to the next generation. In this manner PSR moves through generations eliminating chromosomal compliments on its way; hence the phrase "extreme selfish genetic element".

Mitotic spermatogenesis appears to be an essential part of PSR chromosome ecology since PSR chromosomes, and paternally inherited PSR effects (Hunter et al. 1993), have only been found in male Hymenoptera. Research into B chromosomes shows that they have very poor transmission capability in diploid females (Beukeboom and Werren 1993, Perrot-Minnot and Werren 2001). Meiosis may inhibit PSR transmission due to the absence of a homologous chromosome with which to pair during the meiotic process.

Cytological and genetic analysis of the two known PSR chromosomes shows they are between 9 MB (*T. kaykai*) and 21 MB (*N. vitripennis*) in size, representing 3.9% and 5.7% of the respective haploid genomes and contain repeat sequences associated with closely related species (van Vugt et al. 2005). In *T. kaykai*, van Vugt et al. (2005) found 45S ribosomal DNA on the PSR chromosome that appears to have originated from the related species *Trichogramma oleae* Voegelé and Pointel. In *Nasonia*, repeat sequences found on the PSR chromosome were closely related to transposable element sequences found in *Nasonia* and *Trichomalopsis* species (Eickbush et al. 1992). These findings suggest the origin of PSR chromosomes may be associated with hybridization events.

The unique mode of PSR action makes PSR transfer across species boundaries through hybridization more likely since the incompatible genome is effectively eliminated prior to the first mitotic event. Indeed, Jeong and Stouthamer (2006) were able to pass the PSR chromosome from *T. kaykai* into a *T. deion* genomic background through induced hybridization. Normally, *T. kaykai* and *T. deion* are reproductively incompatible. Similar results have been found within the *Nasonia* species complex (Dobson and Tanouye 1998, Beukeboom and Werren 2000). With such a prolific transmission capability, one may be tempted to ask why the PSR chromosome has not been found in more species. Two possible reasons for the apparent rarity of the PSR chromosome are the ecological parameters under which PSR chromosomes are likely to exist and the ability to detect low frequency PSR chromosomes in populations.

Extensive population modeling has shown that PSR chromosomes are unlikely to evolve in haplodiploid populations that mate randomly where the fertilization frequency is 50% (i.e. an observed sex ratio of approximately 1:1) (Skinner 1987, Werren and Beukeboom 1993, Stouthamer et al. 2001, Werren and Stouthamer 2003). Only haplodiploid populations that normally produce female biased sex ratios, i.e. fertilization rates above 50%, are suitable environments for PSR invasion. Female-biased sex ratios are not uncommon in many parasitoids, including egg parasitoids, and have been explained mainly as a consequence of local mating competition (Hamilton 1967), the differential cost of sexes in relation to host quality (Charnov 1982), constrained sex ratios (Godfray 1990), and infection with female-biasing sex ratio distorters (Stouthamer et al. 1999).

For the two PSR chromosomes found to date, the presence of female-biasing sex ratio distorters appears to play a critical role in their maintenance. The *Wolbachia* infection in *T. kaykai* is the perfect complement to PSR, providing a source of virgin females for the PSR males produced in all male broods of PSR mated females. Emerging PSR males find themselves on patches of sibling males; PI *Wolbachia* females find themselves on patches of sibling females. Mating opportunities for each, to a certain extent, rely on the presence of the other. When a PSR male mates with a PI *Wolbachia* infected female, all fertilized eggs develop as PSR males and all unfertilized eggs develop as PI *Wolbachia* infected females – haplodiploid sex determination is literally reversed. In *N. vitripennis*, a cytoplasmically inherited factor of unknown origin called MSR (maternal sex ratio) induces near complete fertilization in MSR females and appears to be important in the maintenance of PSR in that species (Beukeboom and Werren 1993, Werren and Beukeboom 1993).

The necessary conditions for the spread of PSR chromosomes summarized by Werren and Stouthamer (2003), show there is a tradeoff between population mating structure and PSR invasion and maintenance. Structured populations that experience local mate competition and populations that experience partial inbreeding (or sib-mating) select for the female-biased sex ratios seen in many parasitoid species (Hamilton 1967, Suzuki and Iwasa 1980, Taylor and Bulmer 1980). In a subdivided population where mating takes place in local demes and offspring sex ratios are governed by LMC selection pressure, PSR chromosomes are likely to exist in only a limited range of deme sizes (Werren and Beukeboom 1993). The limiting factor in these situations is competition for mates at the local deme level.

In populations that experience partial sib-mating, mating opportunities for PSR males are restricted to the portion of the females that do not sib-mate and mate at large in the population. Assuming the population is not subdivided into local demes, and optimal offspring sex ratios are under parental control (Suzuki and Iwasa 1980), PSR chromosomes can exist at low to intermediate equilibrium frequencies under a wide range of sib-mating frequencies. As in the case of *T. kaykai*, where 70% sib-mating has been estimated (Huigens 2003), a large fraction of the population is unavailable for, and protected from, PSR males. However, broods from the PI *Wolbachia* infected portion of the population are all-female and consequently cannot mate a sibling. Therefore mating among the infected portion of the *T. kaykai* population occurs with males as they occur in the environment. A 10% PSR infection frequency would result in 10% of the *Wolbachia* infected population mating with PSR males. This is more than twice the rate estimated for uninfected females, where 70% are already mated (to siblings) and the remaining 30% mate with males as they occur in the environment, resulting in only 3% PSR matings. The higher mating rate of *Wolbachia*-infected *T. kaykai* females with PSR males keeps the *Wolbachia* infection frequency at intermediate levels. Indeed, only with the inclusion of realistic levels of sib-mating in population models can equililibrium frequencies like those observed in the field for the two sex ratio distorters, PSR and *Wolbachia*, be found for *T. kaykai* (Stouthamer et al. 2001).

Detection of PSR chromosomes is made difficult by the predicted low equilibrium frequencies of PSR chromosomes in populations, by the fact that different PSR chromosomes appear not to share DNA sequences and by constrained sex ratios (Godfray 1990), when unmated haplodiploid females produce all-male broods indistinguishable from females mated with PSR males. Therefore population surveys that show all male, or male-biased offspring sex ratios, should be followed with controlled matings between males from such broods and virgin females (Fig. 6.2). The detection of PSR like factors requires at least two generations of backcrossing. A PSR brood will consist of two different types of males: PSR males from the fertilized eggs and normal males from the unfertilized eggs. The most efficient method of testing for PSR is to take several males from an all-male brood and mate each of these males with several virgin females. For PSR males in this brood we expect the offspring of the females they have mated with to be all male, while for the normal males we expect the females they have mated with to produce offspring with normal sex ratios. Next, we take the male offspring of fathers that have produced only male offspring and again mate them each with several females. We again expect part of the males to father only male broods and other males to father broods with normal sex ratios. Once this pattern has been found, the chromosome behavior in fertilized eggs can be studied, later followed by karyotyping to confirm the presence of PSR chromosomes. Molecular screening for PSR chromosomes is confounded by the fact that the two PSR chromosomes known thus far have completely independent origins. Sequence analysis of the two PSR chromosomes (Reed et al. 1994, van Vugt et al. 2005) failed to find common sequences that can be used for

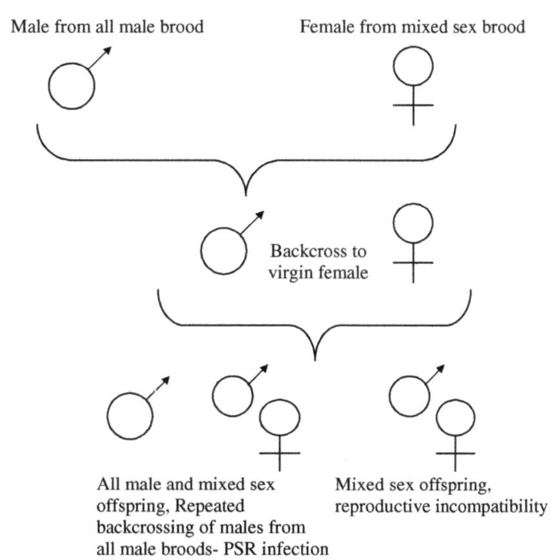

Fig. 6.2 For the detection of PSR chromosomes in haplodiploid populations a recommended mating scheme is presented. The all male results of the F1 generation can be either a result of a PSR infection or reproductive incompatibility. The results presented should be treated as preliminary; to be followed by detection methods mentioned in text

molecular detection. Although, the apparent hybrid origins, PSR chromosomes make them less tractable from a molecular standpoint, it also suggests PSR chromosomes may be quite common in sympatric populations of related haplodiploid organisms. Only a single study has reported on efforts to find PSR-like factors in a species that has a mating structure conductive to PSR invasion (Henter 2004). Despite extensive sampling no case of a PSR like factor was found.

6.4 Sex Ratio Distortion and Biological Control

The prevalence of sex ratio distorting genetic elements in biocontrol agents such as *Trichogramma* has potential benefits and applications that should be considered. For biological control agents in insectaries and laboratories care should be taken to monitor the infection status and offspring sex ratios of breeding stocks. PSR infection can threaten the health and productivity of such stocks. PI *Wolbachia* infection, on the other hand, represents a potential biological control boon for several reasons (Stouthamer 1993). First, since female egg parasitoids are the agents of action in biocontrol practice, a self-sustaining mechanism of continuous female production such as PI *Wolbachia* infection is highly desirable. The clear advantage of PI *Wolbachia* sex ratio distortion in *Trichogramma* and other egg parasitoids used in biocontrol is the relative increase in female production. An important consideration in this regard is the fitness of *Wolbachia* infected parasitoids versus uninfected or cured parasitoids. The apparent benefit of female biased sex ratio distortion should be weighed against the short and long term fitness consequences of infection with sex ratio distorting symbionts like *Wolbachia*.

There are several scenarios in which a choice between infected and uninfected biocontrol agents has relevance (Stouthamer 2003). Work with several *Trichogramma* species has shown the cured arrhenotokous versions of infected lines display higher fecundity and parasitization capabilities (Stouthamer and Luck 1993, Silva et al. 2000, Tagami et al. 2001, Huigens et al. 2004b). However, host density appears to play an important role in the fitness differences observed. Stouthamer and Luck (1993) found fitness penalties for infected *Trichogramma* lines only at high host densities; at low host density infected females produced more female offspring. Therefore, a biocontrol strategy involving the use of infected and uninfected natural enemies should incorporate host density expectations. For example, at high host density releasing uninfected arrhenotokous females will have a greater impact on the target population due to their overall higher fitness. However, as host density decreases, the supplemental release of infected thelytokous females would serve two useful purposes: (1) prevent a biocontrol agent population crash due to scattered reproductive resources (Allee effect – difficulty of finding mates in a scattered low density population) and (2) provide mating opportunities for any locally adapted males produced in the uninfected arrhenotokous portion of the population (Stouthamer 1993). Mating between locally adapted males and infected females has the potential to create superior genotypes which will be amplified in the thelytokous portion of the population by gamete duplication.

One of the unique features of PI *Wolbachia* infection and gamete duplication in *Trichogramma* is the ability to rearrange and fix genotypes. Inbred *Wolbachia* infected females are homozygous for all loci. When such females are mated to males of a different genotype, all F1 offspring emerging from fertilized eggs will be heterozygous females and *Wolbachia*-infected. These genotypically identical female offspring, if unmated, will produce every conceivable genotypic combination in the F2 generation (Fig. 6.3). This is due to the unique gamete duplication mechanism in which meiosis is followed by a diploidization of the haploid eggs. In this manner new beneficial gene combinations can be created and selected. Whether in the uncontrolled setting of a biocontrol release or in a controlled laboratory, this type of genotypic selection is a unique function of *Wolbachia* infection and the gamete duplication mechanism (Stouthamer 2003, 2004).

The ability to horizontally transfer *Wolbachia* within and between species (Huigens et al. 2000, 2004a) suggests a possible means of parthenogenesis transfer to uninfected biocontrol agents. Indeed, *Wolbachia* transfection techniques have already been tested in population suppression experiments involving the

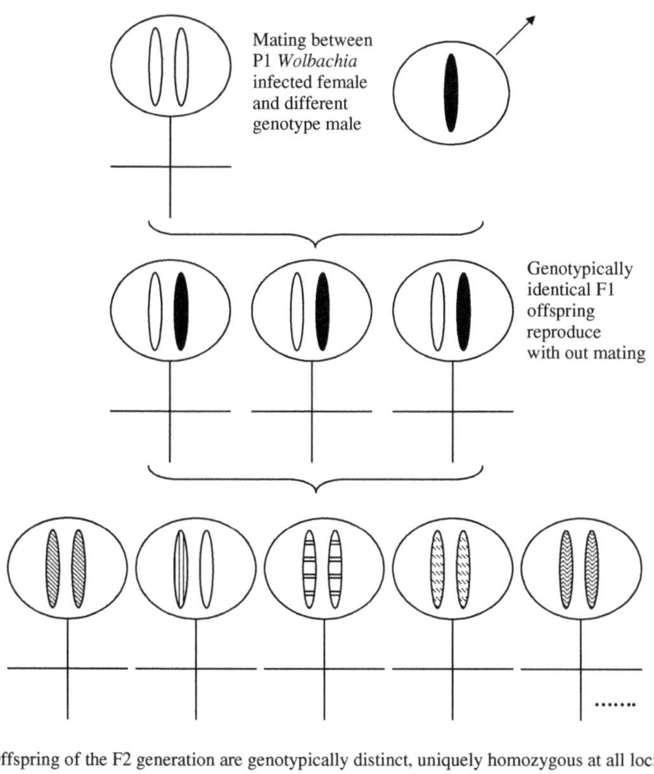

Fig. 6.3 A mating scenario designed to take advantage of the unique gamete duplication mechanism in *Trichogramma* is diagrammed. The F2 generation theoretically represents every conceivable gene combination in homozygous form

medfly *Ceratitis capitata* Wiedmann (Zabalou et al. 2004), though these experiments involved a reproductive incompatibility effect rather than parthenogenesis. But, placing a cytoplasmically inherited symbiont like *Wolbachia* in a novel nuclear background may have serious fitness consequences. Recent experiments have shown significant fitness variation in PI *Wolbachia* transfected lines of *T. kaykai* (Russell et al. in prep.). The results of these experiments showed both potential advantages and disadvantages of horizontal transfer. While most transfected lines show a decrease in fitness values associated with fecundity, a few show significant increases in all fitness values. The results from these experiments indicate an interaction between *Wolbachia* and nuclear genes that should be considered before wholesale transfection of biocontrol agents takes place.

PSR chromosomes in *Trichogramma* or any other biocontrol agent are clearly undesirable due to their male-biasing sex ratio distortion effect. Though PSR may have potential as a genetic control devise designed to depress pest populations (Werren and Stouthamer 2003), such effects are unwanted in egg parasitoids used in biocontrol. One positive application would be to introduce a PSR chromosome into populations of hyperparasitoids. An interesting use of the PSR chromosome may involve its ability to transfer across species boundaries. One of the obstacles in effective genetic control is the ability to transform an organism with a stable genetic construct that will then drive itself through the target population. A PSR chromosome, genetically modified so it carries a gene(s) of choice, is perhaps the most stable of all constructs (no homologous chromosomes with which to recombine) with proven drive capability (Braig and Yan 2002). The disadvantage of the PSR chromosome is the exclusive occurrence in male haplodiploids and the continuous destruction of the paternal chromosomes.

6.5 Concluding Remarks

Understanding the immediate and enduring impact of sex ratio distortion in *Trichogramma* and other organisms has been the emphasis and recommendation of this chapter. The relatively recent discovery of the biological basis of sex ratio distortion, PI *Wolbachia* and PSR chromosomes, has several implications. First, research has really just begun on these "selfish" genetic elements. More research into the physiology, ecology, and evolutionary biology of each (and other genetic elements that distort sex ratios) will gradually fill in the biological portrait of sex ratio distortion. Questions relating to physiological mechanisms responsible for sex ratio distortion, the population dynamics, and origins of these non-Mendelian genetic elements are just the tip of the "future research iceberg". Second, the future discovery of more sex ratio distorting elements seems extremely likely, if not certain. Sex ratio distorting bacteria unrelated to *Wolbachia* have recently been found (Zchori-Fein et al. 2004, Hagimori et al. 2006), and there is no reason to believe more will not be found in the future. It appears sex ratio distortion is a frequently exploited niche for bacteria and other genetic parasites. Sex ratio surveys and experiments with other organisms will surely reveal new host-sex ratio distorter associations. Third, the breadth of infection for both PI *Wolbachia* and PSR is unknown. Future sampling of haplodiploid species is likely to expand the host range for both. As mentioned above, care should be taken in conducting diagnostic screening procedures since there are numerous non-genetic factors that can induce sex ratio distortion. Several *Wolbachia* sequences attributed to parasitoids available in online databases are actually *Wolbachia* sequences from the host used to culture the parasitoids. With such fastidious organisms as intracellular bacteria and such common procedures as PCR it is easy to misdiagnose an infection. This is why we strongly recommend laboratory mating and rearing protocols.

The sex ratio niche is so interesting because it is so prevalent; every sexually reproducing organism expresses a sex ratio. The current molecular revolution has only recently allowed us to identify genetic parasites that distort sex ratios. But that is not to say these parasites have not been around for long. Indeed, sex ratio distorting elements have been mentioned as causative agents for everything from hybrid sterility to meiotic recombination (Hurst and Werren 2001). It appears sex ratio distorting elements have a long evolutionary history.

The long-term consequences of infection with PI *Wolbachia* are particularly intriguing since the sex ratio distorting effect is accompanied by a derived mode of reproduction, parthenogenesis. By most accounts, the loss of sexual reproduction for a species is disastrous; the end result being extinction (Barton and Charlesworth 1998). Are the PI *Wolbachia* infections that have reached fixation examples of parasite-mediated death of a species? The loss of sexual reproduction promoted by PI *Wolbachia* brings with it the accumulation of deleterious mutations, the Muller's rachet (Muller 1932). However, the gamete duplication mechanism used by *Wolbachia* quickly filters out those deleterious mutations by making them homozygous. The disadvantage of asexuality is compensated by the slow accumulation of a genetic load. The long-term fitness consequences of PI *Wolbachia* infection are unknown and deserve research attention. A fitness comparison across a range of PI *Wolbachia* infection frequencies, i.e. from fixation/asexual to uninfected/sexual would be useful. The predicted spread of a PI *Wolbachia* infection should bring with it competition among clonal lines. Population genetic analysis of the same populations will shed light on the clonal variation present before, during, and after PI *Wolbachia* has spread to fixation and sexual reproduction is lost.

Is sex really lost in populations fixed for *Wolbachia* infection? Perhaps rare sexual events occur in these populations bringing genetic variation with it. Phylogenetic analysis of so-called ancient asexual lineages have indicated sexual reproduction likely occurs (Judson and Normark 1996). As mentioned previously, sexual reproduction combined with PI *Wolbachia* infection and gamete duplication can combine to form new genetic combinations in ways and quantities not capable with other forms of reproduction. How would rare sexual events in PI *Wolbachia* infected populations affect population genetic variation and subsequent clonal selection?

The potential benefits of parthenogenetic reproduction for biocontrol agents have been noted for almost a century (Timberlake and Clausen 1924). Attempts to transfer PI *Wolbachia* across species boundaries within *Trichogramma* have shown that although the bacteria can be transferred, parthenogenesis is not maintained (Huigens et al. 2004a). Within species, the rate of horizontal transfer in the laboratory is no higher than 40% (Huigens et al. 2000), but it only takes a single transfer event to start a colony. The horizontal transfer of parthenogenesis using symbionts like *Wolbachia* may achieve the goal of improved efficiency in biological control. Future research should take advantage of intraspecies horizontal transfer to look into the role of nuclear-cytoplasmic interactions in parthenogenesis induction and other fitness related traits. Perhaps nuclear-cytoplasmic interactions involving *Wolbachia* and *Trichogramma* explain why PI *Wolbachia* appears to be species specific in some cases. Even within species, the effects of horizontal transfer and consequent novel nuclear-*Wolbachia* combination may have important fitness consequences.

The outlook for non-Mendelian sex ratio distorting elements in biocontrol practices is promising. They are proven genetic drivers – the rapid spread of *Wolbachia* through populations provides evidence of drive. Perhaps future genetic transformation of these sex ratio distorting elements will allow pest populations to be replaced or, at least controlled. But the most important finding in sex ratio distortion research is the rapid rise in discovery of infections. Clearly sex ratio distorting genetic elements influence arthropod communities whether we know it or not. The immediate future for sex ratio distortion in egg parasitoids and other arthropods should include comprehensive surveys of species and populations, and experimental fitness tests associated with infection.

References

Arakaki N, Miyoshi T, Noda H (2001) *Wolbachia*-mediated parthenogenesis in the predatory thrips *Franklinothrips vespiformis* (Thysanoptera: Insecta). Proc R Soc Biol Sci 268B:1011–1016

Barton NH, Charlesworth B (1998) Why sex and recombination? Science 281:1986–1990

Beukeboom LW, Werren JH (1993) Transmission and expression of the parasitic paternal sex ratio (PSR) chromosome. Heredity 70:437–443

Beukeboom LW, Werren JH (2000) The paternal-sex-ratio (PSR) chromosome in natural populations of *Nasonia* (Hymenoptera: Chalcidoidea). J Evol Biol 13:967–975

Binnington KC, Hoffmann AA (1989) *Wolbachia*-like organisms and cytoplasmic incompatibility in *Drosophila simulans*. J Invertebr Pathol 54:344–352

Braig HR, Yan G (2002) The spread of genetic constructs in natural insect populations. In: Letourneau DK, Burrows BE (eds) Genetically engineered organisms: assessing environmental and human health effects. CRC, Boca Raton, pp 251–314

Breeuwer JAJ, Werren JH (1990) Microorganisms associated with chromosome destruction and reproductive isolation between two insect species. Nature 346:558–560

Charnov EL (1982) The theory of sex allocation. Princeton University Press, Princeton

Charnov EL, Los-den Hartogh RL, Jones WT, van den Assem J (1981) Sex ratio evolution in a variable environment. Nature 289:27–33

Colwell RK (1981) Group selection is implicated in the evolution of female-biased sex-ratios. Nature 290:401–404

Cook JM (1993) Sex determination in the Hymenoptera – a review of models and evidence. Heredity 71:421–435

Darwin C (1871) The descent of man and selection in relation to sex. John Murray, London

Dobson SL, Tanouye MA (1998) Interspecific movement of the paternal sex ratio chromosome. Heredity 81:261–269

Düsing C (1884) Die Regulierung des Geschlectsverhaltnisses bei der Vermehrung der Menschen, Tiere und Pflanzen. Gustav Fischer Verlag, Jena
Eickbush DG, Eickbush TH, Werren JH (1992) Molecular characterization of repetitive DNA sequences from a B chromosome. Chromosoma 101:575–583
Fisher RA (1930) The genetical theory of natural selection. Oxford University Press, Oxford
Gherna RL, Werren JH, Weisburg W, Cote R, Woese CR, Mandelco L, Brenner DJ (1991) *Arsenophonus nasoniae* gen-nov, sp-nov, the causative agent of the son-killer trait in the parasitic wasp *Nasonia vitripennis*. Int J Syst Bacteriol 41:563–565
Godfray HCJ (1990) The causes and consequences of constrained sex allocation in haplodiploid animals. J Evol Biol 3:3–17
Godfray HCJ (1994) Parasitoids: behavioral and evolutionary ecology. Princeton University Press, Princeton
Godfray HCJ, Waage JK (1990) The evolution of highly skewed sex-ratios in aphelinid wasps. Am Nat 136:715–721
Gonçalves CI, Huigens ME, Verbaarschot P, Duarte S, Mexia A, Tavares J (2006) Natural occurrence of *Wolbachia*-infected and uninfected *Trichogramma* species in tomato fields in Portugal. Biol Control 37:375–381
Gottlieb Y, Zchori-Fein E, Werren JH, Karr TL (2002) Diploidy restoration in *Wolbachia*-infected *Muscidifurax uniraptor* (Hymenoptera:Pteromalidae). J Invertebr Pathol 81: 166–174
Grbic M, Ode PJ, Strand MR (1992) Sibling rivalry and brood sex-ratios in polyembryonic wasps. Nature 360:254–256
Hagimori T, Abe Y, Date S, Miura K (2006) The first finding of a *Rickettsia* bacterium associated with parthenogenesis induction among insects. Curr Microbiol 52:97–101
Hamilton WD (1967) Extraordinary sex ratios. Science 156:477–488
Hardy ICW, Ode PJ, Strand MR (1993) Factors influencing brood sex-ratios in polyembryonic Hymenoptera. Oecologia 93:343–348
Henter HJ (2004) Constrained sex allocation in a parasitoid due to variation in male quality. J Evol Biol 17:886–896
Herre EA (1985) Sex-ratio adjustment in fig wasps. Science 228:896–898
Hohmann CL, Luck RF (2000) Effect of temperature on the development and thermal requirements of *Wolbachia*-infected and antibiotically cured *Trichogramma kaykai* Pinto and Stouthamer (Hymenoptera:Trichogrammatidae). Ann Soc Entomol Brasil 29:497–505
Huigens ME (2003) On the evolution of *Wolbachia*-induced parthenogenesis in *Trichogramma* wasps. Entomology. Wageningen University, Netherlands, Wageningen, 183p
Huigens ME, Stouthamer R (2003) Parthenogenesis associated with *Wolbachia*. In: Bourtzis K, Miller TA (eds) Insect symbiosis. CRC, Boca Raton, pp 247–266
Huigens ME, de Almeida RP, Boons PAH, Luck RF, Stouthamer R (2004a) Natural interspecific and intraspecific horizontal transfer of parthenogenesis-inducing *Wolbachia* in *Trichogramma* wasps. Proc R Soc Lond 271B:509–515
Huigens ME, Hohmann CL, Luck RF, Gort G, Stouthamer R (2004b) Reduced competitive ability due to *Wolbachia* infection in the parasitoid wasp *Trichogramma kaykai*. Entomol Exp Appl 110:115–123
Huigens ME, Luck RF, Klaassen RHG, Maas MFPM, Timmermans MJTN, Stouthamer R (2000) Infectious parthenogenesis. Nature 405:178–179
Hunter MS, Nur U, Werren JH (1993) Origin of males by genome loss in an autoparasitoid wasp. Heredity 70:162–171
Hurst GDD, Werren JH (2001) The role of selfish genetic elements in eukaryotic evolution. Nat Rev Genet 2:597–606
Hurst GDD, Hurst LD, Majerus MEN (1997) Cytoplasmic sex ratio distorters. In: O'Neill S, Hoffman AA, Werren JH (eds) Influential passengers: inherited microorganisms and arthropod reproduction. Oxford University Press, Oxford, pp 125–154
Jaenike J (2001) Sex chromosome meiotic drive. Annu Rev Ecol Systemat 32:25–49

Jeong G, Stouthamer R (2005) Genetics of female functional virginity in the parthenogenesis *Wolbachia*-infected parasitoid wasp *Telenomus nawai* (Hymenoptera: Scelionidae). Heredity 94:402–407

Jeong G, Stouthamer R (2006) Influence of postzygotic reproductive isolation on the interspecific transmission of the paternal sex ratio chromosome in *Trichogramma*. Entomol Exp Appl 120:33–40

Judson OP, Normark BB (1996) Ancient asexual scandals. Trends Ecol Evol 11:A41–A46

King BH, Skinner SW (1991) Sex-ratio in a new species of *Nasonia* with fully-winged males. Evolution 45:225–228

Luck RF, Stouthamer R, Nunney L (1992) Sex determination and sex ratio patterns in parasitic Hymenoptera. In: Wrench NDL, Ebbert MA (eds), Evolution and diversity of sex ratio in haplodiploid insects and mites. Chapman & Hall, New York, pp 442–476

Luck RF, Janssen JAM, Pinto JD, Oatman ER (2001) Precise sex allocation, local mate competition, and sex ratio shifts in the parasitoid wasp *Trichogramma pretiosum*. Behav Ecol Sociobiol 49:311–321

Muller HJ (1932) Some genetic aspects of sex. Am Nat 66:118–138

Nunney L (1985) Female-biased sex-ratios – individual or group selection. Evolution 39:349–361

Pannebakker BA, Pijnacker LP, Zwaan BJ, Beukeboom LW (2004) Cytology of *Wolbachia*-induced parthenogenesis in *Leptopilina clavipes* (Hymenoptera: Figitidae). Genome 47: 299–303

Pannebakker BA, Schidlo NS, Boskamp GJF, Dekker L, van Dooren TJM, Beukeboom LW, Zwaan BJ, Brakefield PM, van Alphen JJM (2005) Sexual functionality of *Leptopilina clavipes* (Hymenoptera:Figitidae) after reversing *Wolbachia*-induced parthenogenesis. J Evol Biol 18:1019–1028

Perrot-Minnot MJ, Werren JH (2001) Meiotic and mitotic instability of two EMS-produced centric fragments in the haplodiploid wasp *Nasonia vitripennis*. Heredity 87:8–16

Pijls JWAM, van Steenbergen HJ, van Alphen JJM (1996) Asexuality cured: the relations and differences between sexual and asexual *Apoanagyrus diversicornis*. Heredity 76:506–513

Queller DC (2006) Sex ratios and social evolution. Curr Biol 16:R664–R668

Reed KM, Beukeboom LW, Eickbush DH, Werren JH (1994) Junctions between repetitive DNAs on the PSR chromosome of *Nasonia vitripennis*: Association of palindromes with recombination. J Mol Evol 38:352–362

Rigaud T (1997) Inherited microorganisms and sex determination of arthropod hosts. In: O'Neill S, Hoffman AA, Werren JH (eds) Influential passengers: inherited microorganisms and arthropod reproduction. Oxford University Press, Oxford, pp 81–101

Rousset F, Vautrin D, Solignac M (1992) Molecular identification of *Wolbachia*, the agent of cytoplasmic incompatibility in *Drosophila simulans* and variability in relation with host mitochondrial types. Proc R Soc Lond 247B:163–168

Santolamazza-Carbone S, Nieto MP, Rivera AC (2007) Maternal size and age affect offspring sex ratio in the solitary egg parasitoid *Anaphes nitens*. Entomol Exp Appl 125:23–32

Segar J, Stubblefield JW (2002) Models of sex ratio evolution. In: Hardy ICW (ed) Sex ratios: concepts and research methods. Cambridge University Press, Cambridge, pp 2–25

Shuker DM, Pen I, Duncan AB, Reece SE, West SA (2005) Sex ratios under asymmetrical local mate competition: theory and a test with parasitoid wasps. Am Nat 166:301–316

Silva IMMS, van Meer MMM, Roskam MM, Hoogenboom A, Gort G, Stouthamer R (2000) Biological control potential of *Wolbachia*-infected versus uninfected wasps: laboratory and greenhouse evaluation of *Trichogramma cordubensis* and *T. deion* strains. Biocontrol Sci Technol 10:223–238

Skinner SW (1982) Maternally inherited sex-ratio in the parasitoid wasp *Nasonia vitripennis*. Science 215:1133–1134

Skinner SW (1987) Paternal transmission of an extrachromosomal factor in a wasp – evolutionary implications. Heredity 59:47–53

Stille B, Davring I (1980) Meiosis and reproductive strategy in the parthenogenetic gall wasp *Diplolepis rosae* (L) (Hymenoptera, Cynipidae). Hereditas 92:353–362

Stouthamer R (1993) The use of sexual versus asexual wasps in biological control. Entomophaga 38:3–6

Stouthamer R (1997) *Wolbachia*-induced parthenogenesis. In: O'Neill S, Hoffman AA, Werren JH (eds) Influential passengers: inherited microorganisms and arthropod reproduction. Oxford University Press, Oxford, pp 102–124

Stouthamer R (2003) The use of unisexual wasps in biological control. In: van Lenteren JC (ed) Quality control and production of biological control agents: theory and testing procedures. CAB International, London, pp 93–113

Stouthamer R (2004) Sex-ratio distorters and other selfish genetic elements: Implications for biological control. In: Ehler RSL, Mateille T (eds) Genetics, evolution and biological control. CABI Publishing, Wallingford, UK, pp 235–252

Stouthamer R, Kazmer DJ (1994) Cytogenetics of microbe-associated parthenogenesis and its consequences for gene flow in *Trichogramma* wasps. Heredity 73:317–327

Stouthamer R, Luck RF (1993) Influence of microbe-associated parthenogenesis on the fecundity of *Trichogramma deion* and *T. pretiosum*. Entomol Exp Appl 67:183–192

Stouthamer R, Luck RF, Hamilton WD (1990) Antibiotics cause parthenogenetic *Trichogramma* (Hymenoptera, Trichogrammatidae) to revert to sex. Proc Natl Acad Sci USA 87:2424–2427

Stouthamer R, Breeuwer JAJ, Hurst GDD (1999) *Wolbachia pipientis*: microbial manipulator of arthropod reproduction. Annu Rev Microbiol 53:71–102

Stouthamer R, van Tilborg M, de Jong JH, Nunney L, Luck RF (2001) Selfish element maintains sex in natural populations of a parasitoid wasp. Proc R Soc 268B:617–622

Strand MR (1988) Variable sex-ratio strategy of *Telenomus heliothidis* (Hymenoptera, Scelionidae) – adaptation to host and conspecific density. Oecologia 77:219–224

Strand MR (1989a) Clutch size, sex-ratio and mating by the polyembryonic encyrtid *Copidosoma floridanum* (Hymenoptera, Encyrtidae). Florida Entomol 72:32–42

Strand MR (1989b) Development of the polyembryonic parasitoid *Copidosoma floridanum* in *Trichoplusia ni*. Entomol Exp Appl 50:37–46

Suomalainen E, Saura A, Lokki J (1987) Cytology and evolution in parthenogenesis. CRC, Boca Raton

Suzuki Y, Iwasa Y (1980) A sex ratio theory of gregarious parasitoids. Res Popul Biol 22:366–382

Suzuki Y, Tsuji H, Sasakawa M (1984) Sex allocation and effects of superparasitism on secondary sex-ratios in the gregarious parasitoid, *Trichogramma chilonis* (Hymenoptera, Trichogrammatidae). Anim Behav 32:478–484

Tagami Y, Miura K, Stouthamer R (2001) How does infection with parthenogenesis-inducing *Wolbachia* reduce the fitness of *Trichogramma*? J Invertebr Pathol 78:267–271

Tagami Y, Miura K, Stouthamer R (2002) Positive effect of fertilization on the survival rate of immature stages in a *Wolbachia*-associated thelytokous line of *Trichogramma deion* and *T. kaykai*. Entomol Exp Appl 105:165–167

Taylor PD, Bulmer MG (1980) Local mate competition and the sex-ratio. J Theor Biol 86:409–419

Timberlake PH, Clausen CP (1924) The parasites of *Pseudococcus maritimus* in California. Univ Calif Publ Tech Bull Entomol 3:223–292

van Vugt JFA, Salverda M, de Jong JH, Stouthamer R (2003) The paternal sex ratio chromosome in the parasitic wasp *Trichogramma kaykai* condenses the paternal chromosomes into a dense chromatin mass. Genome 46:580–587

van Vugt JJFA, de Nooijer S, Stouthamer R, de Jong J (2005) NOR activity and repeat sequences of the paternal sex ratio chromosome of the parasitoid wasp *Trichogramma kaykai*. Chromosoma 114:410–419

Vanwelzen CRL, Waage JK (1987) Adaptive responses to local mate competition by the parasitoid, *Telenomus remus*. Behav Ecol Sociobiol 21:359–365

Vavre F, de Jong JH, Stouthamer R (2004) Cytogenetic mechanism and genetic consequences of thelytoky in the wasp *Trichogramma cacoeciae*. Heredity 93:592–596

Waage JK (1982) Sib-mating and sex-ratio strategies in scelionid wasps. Ecol Entomol 7:103–112

Waage JK, Lane JA (1984) The reproductive strategy of a parasitic wasp. 2. Sex allocation and local mate competition in *Trichogramma evanescens*. J Anim Ecol 53:417–426

Waage JK, Ming NS (1984) The reproductive strategy of a parasitic wasp. 1. Optimal progeny and sex allocation in *Trichogramma evanescens*. J Anim Ecol 53:401–415
Weeks AR, Breeuwer JAJ (2001) *Wolbachia*-induced parthenogenesis in a genus of phytophagous mites. Proc R Soc 268B:2245–2251
Werren JH (1983) Sex-ratio evolution under local mate competition in a parasitic wasp. Evolution 37:116–124
Werren JH, Beukeboom LW (1993) Population genetics of a parasitic chromosome: theoretical analysis of PSR in subdivided populations. Am Nat 142:224–241
Werren JH, Stouthamer R (2003) PSR (paternal sex ratio) chromosomes: the ultimate selfish genetic elements. Genetica 117:85–101
Werren JH, Hurst GDD, Zhang W, Breeuwer JAJ, Stouthamer R, Majerus MEN (1994) Rickettsial relative associated with male killing in the ladybird beetle (*Adalia bipunctata*). J Bacteriol 176:388–394
Whiting PW (1943) Multiple alleles complementary sex determination of *Habrobracon*. Genetics 28:365–382
Wu M, Sun LV, Vamathevan J, Riegler M, Deboy R, Brownlie JC, McGraw EA, Martin W, Esser C, Ahmadinejad N, Wiegand C, Madupu R, Beanan MJ, Brinkac LM, Daugherty SC, Durkin AS, Kolonay JF, Nelson WC, Mohamoud Y, Lee P, Berry K, Young MB, Utterback T, Weidman J, Nierman WC, Paulsen IT, Nelson KE, Tettelin H, O'Neill SL, Eisen JA (2004) Phylogenomics of the reproductive parasite *Wolbachia pipientis* wMel: A streamlined genome overrun by mobile genetic elements. PLoS Biol 2:327–341
Yen JH, Barr AR (1973) The etiological agent of cytoplasmic incompatibility in *Culex pipiens*. J Invertebr Pathol 22:242–250
Zabalou S, Riegler M, Theodorakopoulou M, Stauffer C, Savakis C, Bourtzis K (2004) *Wolbachia*-induced cytoplasmic incompatibility as a means for insect pest population control. Proc Natl Acad Sci USA 101:15042–15045
Zchori-Fein E, Perlman SJ, Kelly SE, Katzir N, Hunter MS (2004) Characterization of a '*Bacteroidetes*' symbiont in *Encarsia* wasps (Hymenoptera:Aphelinidae): proposal of '*Candidatus* Cardinium hertigii'. Int J Syst Evol Microbiol 54:961–968

Chapter 7
Systematics of the Trichogrammatidae (Hymenoptera: Chalcidoidea) with a Focus on the Genera Attacking Lepidoptera

Ranyse B. Querino, Roberto A. Zucchi, and John D. Pinto

Contents

7.1 Introduction

Hymenoptera are placed among the mega-diverse orders (Whitfield 1998), it is estimated that at least 75% of Parasitic Hymenoptera species remain undescribed (LaSalle and Gauld 1992). Moreover, hymenopterous parasitoids represent an important component of the neotropical fauna through their role in regulating populations of other insects (Goulet and Huber 1993).

The superfamily Chalcidoidea, divided into 19 families, presents considerable biological, ecological and morphological diversity. There are approximately

R.B. Querino (✉)
Embrapa Meio-Norte, Av. Duque de Caxias 5650, 64006-220 Teresina, Piauí, Brazil
e-mail: ranyse@cpamn.embrapa.br

F.L. Consoli et al. (eds.), *Egg Parasitoids in Agroecosystems with Emphasis on Trichogramma*, Progress in Biological Control 9, DOI 10.1007/978-1-4020-9110-0_7,

22,000 described chalcidoid species. Most of these measure less than 3 mm in length (an average of ca. 1.5 mm), the smallest being about 0.1 mm (Mymaridae: *Dicopomorpha echmepterygis* Mockford, male). This small size makes the collection and study of these Hymenoptera difficult, which is why they have received comparatively little attention from taxonomists (Noyes 2003).

The Trichogrammatidae are egg parasitoids and one of the most poorly known groups of Chalcidoidea. This is largely due to the small size and fragility of these insects, in addition to problems associated with their collection and subsequent care (Pinto and Stouthamer, 1994). *Trichogramma*, with over 200 described species, is the best known genus in the family due to its use in the biological control of agricultural pests.

7.2 Systematics

Representatives of the Trichogrammatidae are relatively homogeneous morphologically and characterized by the 3-segmented tarsi, reduced number of antennal segments, compact body structure and their development as egg parasitoids. The trimerous tarsi, previously considered unique to trichogrammatids, have now been found in a few species in other families, namely in Aphelinidae (*Pteroptrix*) (Kim and Triapitsyn 2004), Eulophidae (*Trisecodes*) (Delvare and LaSalle 2000) and Mymaridae (*Kikiki*) (Huber and Beardsley 2000).

The placement of Trichogrammatidae within the Chalcidoidea remains ambiguous, although the family is usually considered closest to the Aphelinidae and Eulophidae (Fig. 7.1). Its affinity to the Eulophidae is suggested by the reduced number of tarsomeres and antennal segments and the short, straight tibial spur. Other

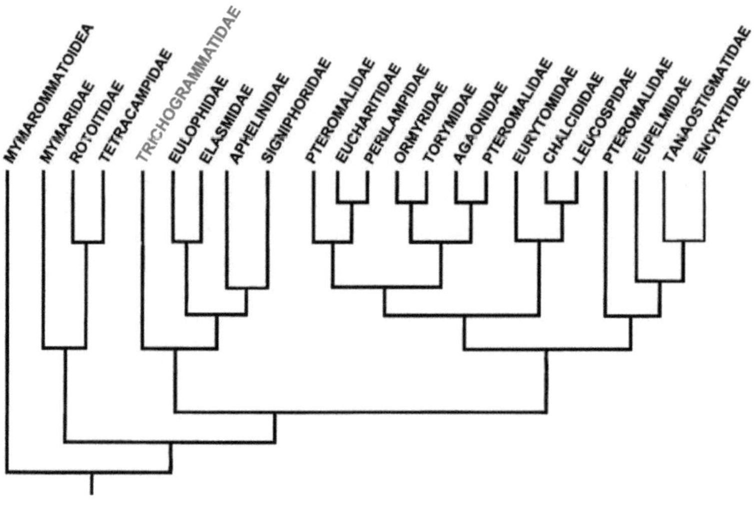

Fig. 7.1 Dendogram showing the relationship of Trichogrammatidae within the Chalcidoidea (adapted from Heraty et al. 1997)

morphological characteristics including tarsal segmentation and male genitalia suggest the Aphelinidae as a potential sister-group of Trichogrammatidae (Heraty et al. 1997, Gibson et al. 1999, Pinto 2006, Owen et al. 2007). The presence of the straight tibial spur has suggested proximity to *Cales* (Hayat 1994, 1998). Although the family-level placement of *Cales* remains questionable, it has traditionally been assigned to the Aphelinidae (Woolley 1997).

The most complete phylogenetic study of the Trichogrammatidae was presented by Owen et al. (2007). Analyses were based on molecular data and placed in the context of morphological features. Their results propose a new classification which includes a narrower definition of tribes, and the removal of the Paracentrobiini from the Trichogrammatinae and its placement together with the Chaetostrichini and Oligositini in the Oligositinae.

7.3 Classification

One of the most important taxonomic studies of Trichogrammatidae is that of Doutt and Viggiani (1968). It includes the only key to world genera. Other important publications include a list of all genera (Yousuf and Shafee 1986a), a list of the species of the world (Yousuf and Shafee 1986b) and catalogues of species from the USA (Burks 1979), Central Europe (Peck et al. 1964), the Oriental Region (Hayat and Viggiani 1984) and Latin America (De Santis 1979, 1980, 1981), a list of genera and species (Lin 1994), and a key to Nearctic genera (Pinto 1997b). The review of New World genera by Pinto (2006) and the phylogenetic study by Owen et al. (2007) are the most recent treatments of family classification.

The family Trichogrammatidae (Pinto 2006, Owen et al. 2007) is divided into the subfamilies Trichogrammatinae (paraphyletic) and Oligositinae (monophyletic). In the Trichogrammatinae, the genitalia possess two separate components, aedeagus

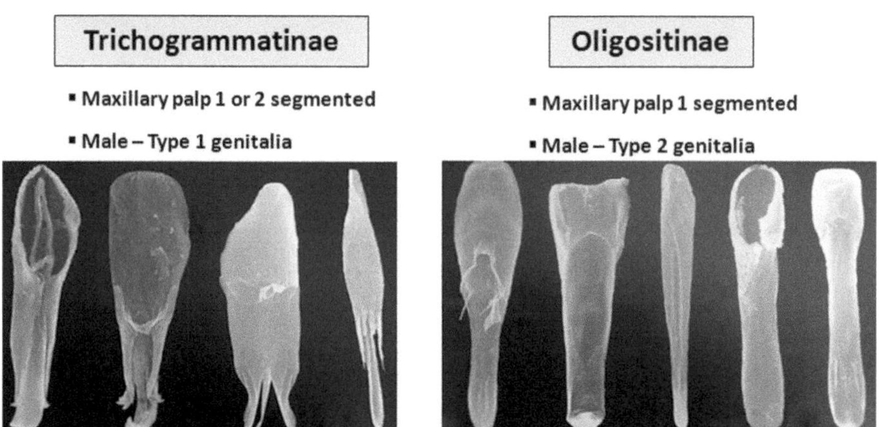

Fig. 7.2 Characters employed for separation of subfamilies of Trichogrammatidae (adapted from Pinto 2006)

Table 7.1 Classification of the New World Trichogrammatidae (based on Pinto 2006 and Owen et al. 2007)

Classification	Host	Distribution[b]
Subfamily: Trichogrammatinae		
Tribe: Trichogrammatini		
Thanatogramma Pinto	**Unknown**	NH
Trichogramma Westwood	Coleoptera, Lepidoptera, Diptera, Hymenoptera, Neuroptera	Cosmopolitan
Trichogrammatoidea Girault	Lepidoptera	CA, WI, SA (AR, BR, CH, CO, VE). Extralimital.
Xenufens Girault	Lepidoptera (Nymphalidae, Hesperiidae)	NA, CA, WI, SA (AR, BR, CO, EC, PE). Extralimital.
Incertae sedis within Trichogrammatinae		
Brachyufens Viggiani	Coleoptera (Curculionidae)	NA, WI
Ceratogramma De Santis	Coleoptera (Curculionidae)	NA, CA, WI, SA (AR, BR, BO, CO, EC, FrG, PE). Extramilital.
Haeckeliania Girault	Coleoptera (Curculionidae)	NA, CA, WI, SA (AR, BR, BO, CO, EC, PE, VE). EXtralimital.
Hydrophylita Ghenquière	Odonata (Coenagrionidae, Lestidae)	NA, CA, WI, SA (AR, BR, BO, CO, EC, VE). Extralimital.
Mirufens Girault	Hemiptera (Membracidade, Cicadellidae)	NA, CA, SA (CH). Extralimital.
Pachamama Owen & Pinto	**Unknown**	CA, SA (EC).
Paratrichogramma Girault	Lepidoptera (Gracillariidae, Noctuidae, Lycaenidae)	NA, CA, SA (AR). Extralimital.
Poropoea Forster	Coleoptera (Atellabidae)	NA, CA, WI, SA (BR, EC). Extralimital.
Pterandrophysalis Nowicki	Coleoptera (Curculionidae)	NA. Extralimital.
Soikiella Nowicki	Diptera (Asilidae)	NA. Extralimital.
Trichogrammatella Girault[a]	Hemiptera (Membracidae)	CA, WI, SA (BO, BR, EC)
Trichogrammatomyia Girault	Lepidoptera (Tortricidae)	NA, CA, WI, SA (AR, BR, CH, EC, VE). Extralimital.
Viggianiella Pinto[a]	**Unknown**	SA (BR, CO)

Table 7.1 (continued)

Classification	Host	Distribution[b]
Subfamily: Oligositinae		
Tribe: Chaetostrichini		
Adryas Pinto & Owen	Unknown	CA, SA (BO, CO, EC, PE, VE)
Brachista Walker	Coleoptera (Chrysomelidae), Hemiptera (Cicadidae), Diptera (Asilidae)	NA, CA, WI, SA (BR)
Burksiella De Santis	Coleptera (Chrysomelidae), Hemiptera (Cicadelldae), Orthoptera (Tettigoniidae)	NA, CA, WI, SA (AR, BO, BR, CH, CO, EC). Extralimital.
Chaetostricha Walker	Hemiptera (Membracidae, Miridae)	NA, WI. Extralimital
Lathromeroidea Girault	Hemiptera (Gerridae)	NA, CA, WI, SA (AR, BO, BR, CO, EC, PA, PE, VE). Extralimital.
Prouscana Viggiani & Velasquez	Unknown	SA (VE)
Pseuduscana Pinto	Unknown	NA, CA, WI, SA Extralimital.
Uscana Girault	Coleoptera (Bruchidae)	NA, CA, SA (AR, BR). Extralimital.
Uscanoidea Girault	Hemiptera (Cercopidae, Cicadellidae, Membracidae)	NA, CA, WI, SA (AR, BR, CO, EC, PE). Extralimital.
Zega Girault	Unknown	NA, CA, WI, SA (AR, BR, EC). Extralimital.
Zegella Girault	Hemiptera (Cicadellidae)	NA, SA (AR, BR, UR, VE).
Tribe: Oligositini		
Doirania Waterson	Orthoptera (Tettigoniidae)	NA, SA(?). Extralimital.
Epoligosita Girault	Hemiptera (Cicadellidae, Tingidae)	NA, CA, WI, SA, (AR, EC). Extralimital.
Megaphragma Timberlake	Thysanoptera	NA, CA, WI, SA (AR). Extralimital.
Oligosita Walker	Hemiptera (Cicadellidae)	NA, CA, WI, SA (AR, BO, BR, CH, CO, EC). Extralimital.
Prestwichia Lubbock[a]	Coleoptera, Hemiptera, Odonata	NA. Extralimital.
Pseudoligosita Girault	Hemiptera, Orthoptera, Coleoptera	NA, CA, WI, SA (AR, BO, BR, CO, EC, UR, VE). Extralimital.
Sinepalpigramma Viggiani & Pinto[a]	Unknown	NA, CA, SA (AR, BR, CO, EC, VE).
Tribe: Paracentrobiini		
Ittys Girault	Hemiptera (Auchenorrhyncha and Heteropter)	NA, CA, WI, SA (BR). Extralimital.
Ittysella Pinto & Viggiani	Hemiptera (Cicadellidae)	NA.

Table 7.1 (continued)

Classification	Host	Distribution[b]
Paracentrobia Howard	Hemiptera (Auchenorrhyncha and Heteroptera), Lepidoptera (?)	NA, CA, WI, SA (AR, BO, BR, CO, EC, PE, VE). Extralimital.
Incertae sedis within Oligositinae		
Adelogramma Pinto[a]	Unknown	NA, CA, SA (CO, EC).
Aphelinoidea Girault	Hemiptera (Cicadelllidae, Fulgoridae)	NA, CA, WI, SA (AR, BR, BO, CH, EC, UR). Extralimital.
Bloodiella Nowicky[a]	Unknown	SA (UR). Extralimital.
Brachygrammatella Girault[a]	Hemiptera (Cicadellidae, Membracidae, Miridae)	NA. Extralimital.
Centrobiopsis Girault[a]	Odonata (Lestidae)	NA.
Chaetogramma Doutt	Unknown	NA, CA, WI, SA (AR, CH). Extralimital.
Lathromeris Forster	Diptera (Cecidomyiidae), Lepidoptera (Noctuidae, Pyralidae)	NA, CA, SA (AR). Extralimital.
Monorthochaeta Blood	Coleoptera (Chrysomelidae)	NA. Extralimital.
Nicolavespa Pinto	Unknown	NA, CA, WI.
Pintoa Viggiani	Unknown	NA, CA, WI, SA (AR, CO, EC).
Pteranomalogramma Viggiani & Velàsquez[a]	Unknown	SA (VE).
Pterygogramma Perkins	Hemiptera (Aetalionidae, Cicadellidae)	NA, CA, WI, SA (AR, BO, BR, EC). Extralimital.
Tumidiclava Girault	Coleoptera (Curculionidae), Lepidoptera (Cossidae, Pyralidae, Noctuida	NA, CA, WI, SA (AR, CH, EC). Extralimital.
Tumidifemur Girault[a]	Hemiptera (Membracidae)	CA, WI, SA (CO, EC, VE).
Ufens Girault	Hemiptera (Cicadellidae), Orthoptera (Tettigoniidae)	NA, CA, WI. Extralimital.
Uscanella Girault[a]	Hemiptera (Membracidae)	WI.
Uscanopsis Girault[a]	Hemiptera (Membracidae)	WI.
Xiphogramma Nowicki	Unknown	NA. Extralimital.

[a]Genera not represented in molecular dataset; placement inferred from morphology, particularly male genitalia

[b]NH: North America, CA: Central America, WI: West Indies, SA: South America, AR: Argentina, BR: Brasil, BO: Bolivia, CH: Chile, CO: Colombia, EC: Ecuador, FrG: French Guyana; PA: Paraguay, PE: Peru, UR: Uruguay,VE: Venezuela

and genital capsule, the latter generally with well-developed parameres and volsellae. The subfamily includes only the tribe Trichogrammatini. The Oligositinae, with genitalia consisting of a single component and a varying degree of consolidation of the genital capsule and aedeagus (Fig. 7.2), are divided into three monophyletic tribes, Chaetostrichini, Oligositini and Paracentrobiini (Oligositinae) (Pinto 2006). Several genera are considered *incertae sedis* within Oligositinae (Table 7.1).

7.4 Distribution

The Trichogrammatidae are represented by 89 genera and more than 800 species distributed worldwide, both in terrestrial and aquatic habitats.

Fifty-six genera are recorded in the New World. 80% of these are found in North America (Fig. 7.3, Table 7.2) with nine unknown south of there (*Brachygrammatella*, *Centrobiopsis*, *Ittysella*, *Monorthochaeta*, *Prestwichia*, *Pterandrophysalis*, *Soikiella*, *Thanatogramma* and *Xiphogramma*).

The family remains poorly known in Central and South America, with 37 and 41 genera, respectively (Fig. 7.3, Table 7.1). Twenty-eight genera occur in Brasil (Table 7.2). Three genera are unique to South America (*Prouscana, Pteranomalogramma* and *Viggianiella*). The occurrence of the European genus *Bloodiella* in South America remains questionable (Pinto 2006). Most South American records of the family concern *Trichogramma* and *Trichogrammatoidea*, whose species are commonly used in biological control. However, there is no recent treatment of Trichogrammatidae in the Neotropical region.

The importance of the family is due almost entirely to *Trichogramma*, the largest genus, whose species parasitize numerous pest Lepidoptera. Due to the ease of breeding in the laboratory, species have been used extensively in biological control programmes. Of the 232 described New World trichogrammatids, ca. 40% belong to *Trichogramma*. Other relatively large genera include *Oligosita*, *Pseudoligosita* and

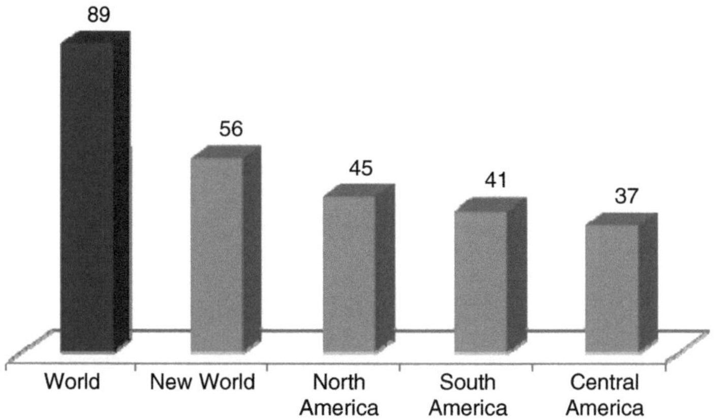

Fig. 7.3 Number of genera of Trichogrammatidae worldwide and in the New World

Table 7.2 Genera of Trichogrammatidae from South America, focusing on Brazil

Genera	South America	Brazil
Adelogramma	■	
Adryas	■	■
Aphelinoidea	■	■
Bloodiella	■	
Brachista	■	■
Brachygrammatella		
Brachyufens		
Burksiella	■	■
Centrobiopsis		
Ceratrogramma	■	■
Chaetogramma	■	
Chaetostricha		
Doirania	?	
Epoligosita	■	■
Haeckeliania	■	■
Hydrophylita	■	■
Ittys	■	■
Ittysella		
Lathromeris	■	
Lathromeroidea	■	■
Megaphragma	■	
Mirufens	■	
Monorthochaeta		
Nicolavespa	■	
Oligosita	■	■
Pachamama	■	
Paracentrobia	■	■
Paratrichogramma	■	
Pintoa	■	■
Poropoea	■	■
Prouscana	■	
Prestwichia		
Pseudoligosita	■	■
Pseuduscana	■	■
Pterandrophysalis		
Pteranomalogramma	■	
Pterygogramma	■	■
Sinepalpigramma	■	■
Solkiella		
Thanatogramma		
Trichogramma	■	■
Trichogrammatella	■	■
Trichogrammatoidea	■	■
Trichogrammatomyia	■	■
Tumidiclava	■	
Tumidifemur	■	
Ufens		
Uscana	■	■
Uscanella		
Uscanoidea	■	■
Uscanopsis		
Viggianiella	■	■
Xenufens	■	■
Xiphogramma		
Zaga	■	■
Zagella	■	■

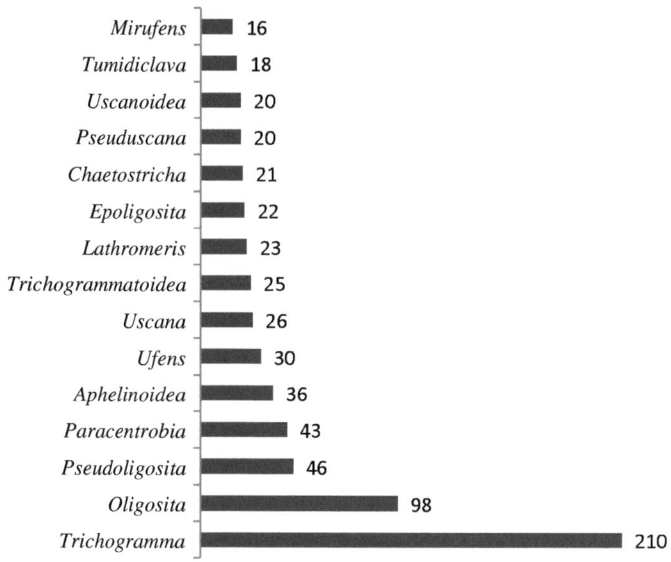

Fig. 7.4 Number of species in the larger genera of the Trichogrammatidae

Paracentrobia (Fig. 7.4). Genera such as *Burksiella* and *Lathromeroidea* also are represented by numerous species but only a few have been described thus far (Pinto 2006).

7.5 Biology

Members of the Trichogrammatidae are primary endoparasitoids (solitary or gregarious) of insect eggs. The host associations of representatives of the family are varied, although the recent phylogenetic study of Owen et al. (2007) suggests that the family initially parasitized Coleoptera, more recently spreading to other insect orders, such as Lepidoptera and Hemiptera.

Ten orders of insects are known hosts for Trichogrammatidae in the New World. The Hemiptera has the highest number of associated genera, followed by Coleoptera and Lepidoptera (Fig. 7.5). In the New World, Lepidoptera species are attacked by eight genera (Fig. 7.6). Of these, only *Trichogrammatoidea* and *Trichogrammatomyia* are exclusively associated with Lepidoptera. *Trichogramma* has the highest number of records on Lepidoptera. Lists of *Trichogramma* hosts in North and South America were published by Pinto (1999) and Zucchi and Monteiro (1997) (see Chapter 8).

Certain genera are known to parasitize eggs of several orders of insects (e.g. *Trichogramma* on Coleoptera, Diptera, Hemiptera, Hymenoptera and Neuroptera). Others are apparently restricted to a single order (Pinto 1997a). For example, *Oligosita* is known only from Hemiptera, *Uscana* only from Coleoptera, *Megaphragma* is restricted to Thysanoptera and *Hydrophylita* is recorded only from Odonata.

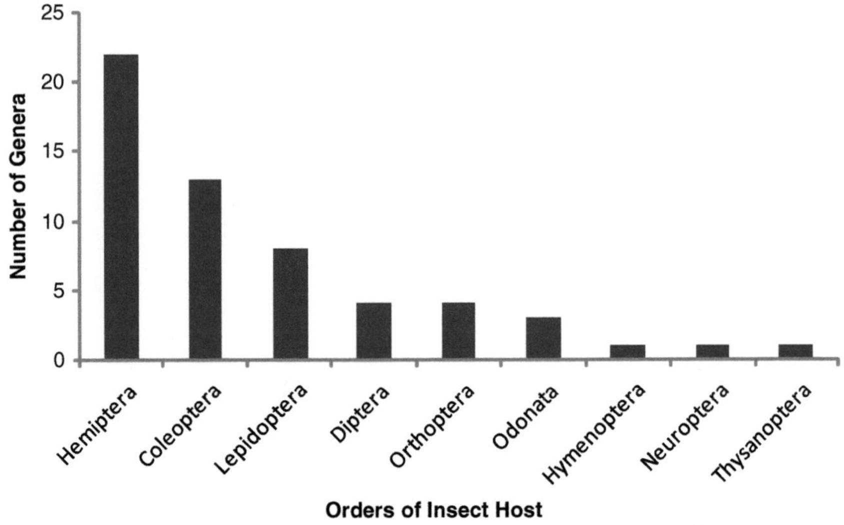

Fig. 7.5 Number of genera of Trichogrammatidae associated with orders of Insecta

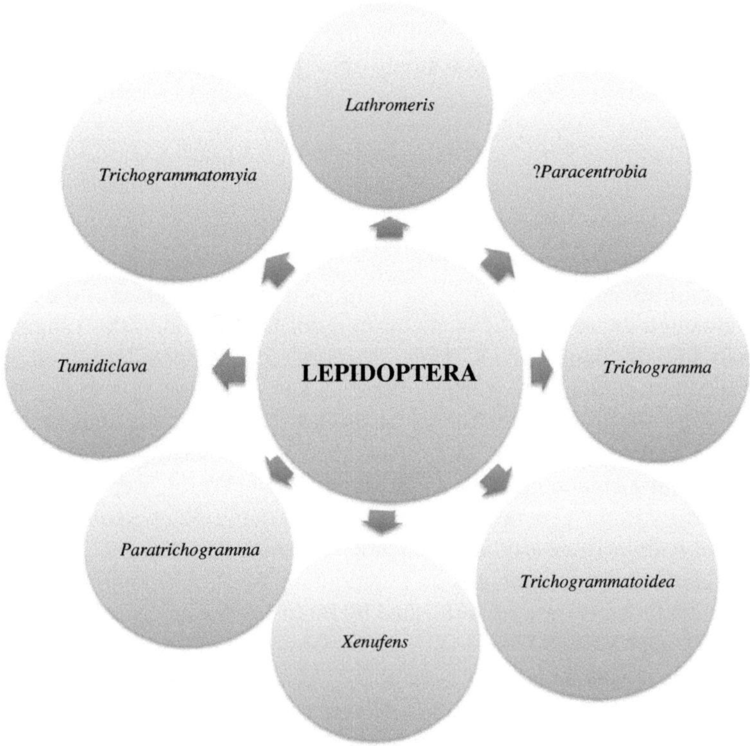

Fig. 7.6 Genera of New World Trichogrammatidae with species parasitizing eggs of Lepidoptera

It is likely that for some genera affinity to microhabitat is stronger than to host taxon. Thus, members of *Lathromeroidea* have been associated with eggs of Odonata and Hemiptera in aquatic environments, and certain species of *Trichogramma* parasitize eggs of Lepidoptera, Neuroptera and Hymenoptera placed on the same plant (Pinto and Stouthamer 1994).

The hosts of 15 genera of Trichogrammatidae remain unknown (*Adryas* Pinto and Owen, *Adelogramma* Pinto, *Bloodiella* Nowicky, *Chaetogramma* Doutt, *Nicolavespa* Pinto, *Pachamama* Owen and Pinto, *Pintoa* Viggiani, *Prouscana* Viggiani and Velàsquez, *Pseuduscana* Pinto, *Pteranomalogramma* Viggiani and Velàsquez, *Sinepalpigramma* Viggiani and Pinto, *Thanatogramma* Pinto, *Viggianiella* Pinto, *Zaga* Girault and *Xiphogramma* Nowicki). Unfortunately, little work has been done to clarify biological associations, with the exception of species of *Trichogramma* and *Trichogrammatoidea*, because of their importance in biological control.

Several new host records have become available for South American taxa. Those specific to Brasil include the following: *Pseudoligosita longifrangiata* (Viggiani) on *Argia insipida* Hagen in Selys (Odonata: Coenagrionidae) (Querino and Hamada 2009), *Epoligosita mexicana* Viggiani on Tingidae (Santos 2007), *Hydrophylita neusae* Querino and Pinto on Coenagrionidae (Odonata) (Querino and Pinto 2007); *H. backmanni* De Santis on Odonata (Ramalheira et al. 2006); *Trichogrammatomyia tortricis* on *Hypsipyla grandella* (Ohashi et al. 2005); *Zaga* sp. on *Enchenopa gracilis* (Hemiptera: Membracidae) (Aguiar-Menezes et al. 2005).

7.6 Identification

7.6.1 Diagnosis

The family Trichogrammatidae is distinguished from almost all other Chalcidoidea by the 3-segmented tarsi. Other important characters are: compact body without a constriction between the mesosoma and metasoma; non-metallic coloring; faintly sculptured cuticle; flagellum with a reduced number of segments (2–9, normally 3–7); forewing often with setae arranged in distinct rows and post-marginal vein absent or greatly reduced.

7.6.2 The Separation of Trichogrammatidae from Other Families That Parasitize Eggs

Among Chalcidoidea, Mymaridade is another family with members that parasitize insect eggs. Nevertheless mymarids rarely attack Lepidoptera, and thus are not likely to be confused with, for example, *Trichogramma* and *Trichogrammatoidea*, based on this fact alone. However, they are common parasites of eggs of Hemiptera and Coleoptera, groups also associated with a large number of Trichogrammatidae (Pinto and Stouthamer 1994, Owen et al. 2007). Mymarids are easily separated from trichogrammatids by the 4-5-segmented tarsi, by the position of the toruli (closer to the eyes than to one another) and by the distinctly pedunculate hind wing. Except for the probability of the poorly known Mymarommatidae (Huber et al. 2008), no other family of Chalcidoidea consists predominantly of egg parasitoids. The few taxa of Aphelinidae and Eulophidae that do attack insect eggs are easily separated by having more than three tarsomeres.

Two families of proctotrupoid Hymenoptera, Scelionidae and Platygastridae, are also important groups of insect egg parasitoids. The Scelionidae attack a wide range of insects. Representatives of the subfamily Telenominae, which includes the important genus *Telenomus*, are frequently associated with Hemiptera and Lepidoptera. The Platygastridae of the subfamily Sceliotrachelinae are endoparasites of eggs of Hemiptera (Auchenorryncha) and Coleoptera (Masner 1995), orders which also include important hosts of Trichogrammatidae. Differentiating characters include the 5-segmented tarsi, a distinct separation between the mesosoma and metasoma, and usually a well-developed post-marginal vein. Furthermore, as in all Proctotrupoidea, the pronotum reaches the tegula at the base of the forewing. In the Trichogrammatidae, as in all Chalcidoidea, the pronotum does not reach the tegula (Goulet and Huber 1993).

7.6.3 Key to the New World Genera of Trichogrammatidae Parasitizing Lepidoptera Eggs

1. Antenna with only 1 elongate postannelar segment...*Trichogramma* (*Trichogramma*) ♂
1. Antenna with more than 1 postannelar segment..2
2. Antenna with both a 2-segmented funicle and 1-segmented club.........................3
2. Antenna with funicle present or not and club at least 2-segmented......................6
3. Forewing with abbreviated venation, stigmal vein (SV) reduced to a small appendix at apex of marginal vein (MV). Antenna of male with second funicular segment (F2) noticeably narrowed at apex, bottle-shaped...*Paratrichogramma*
3'. Forewing venation not so abbreviated, SV variable but clearly diverging from MV, never reduced to a small appendix. Antenna with F2 not narrowed at apex, not bottle- shaped...4
4. Forewing venation sinuate, with marginal vein (MV) gradually curving away from forewing margin onto stigmal vein (SV). Mesophragma entire apically..........5
4'. Forewing venation not sinuate. MV ending abruptly distally, not curving gradually onto SV. Mesophragma apically notched..*Trichogrammatomyia* ♀
5. Forewing with RS1 track present behind stigma vein (SV); pre-marginal vein (PM) with 2 setae. Hindwing usually with at least 2 setal tracks (a complete middle-track and a complete or partial hind-track)......................................*Trichogramma* ♀

5'. Forewing without an RS1 track behind SV; PM with 1 seta. Hindwing with only a mid-setal track which is incomplete to wing apex.. *Trichogrammatoidea* ♀
6. Antenna with funicle present...7
6'. Antenna without a funicle..11
7. Forewing venation sinuate, greatest curvature distally where marginal vein (MV) gradually curves away from anterior wing margin onto stigmal vein (SV)...............8
7'. Forewing venation not sinuate, MV ending abruptly distally, not gradually curving onto SV..10
8. Antennal-club distinctly 3 segmented; funicular segments sub-quadrate or only slightly transverse. Forewing with marginal vein (MV) attaining anterior margin of wing..9
8'. Antennal-club with only 2 complete segments, a third (if present) incompletely separated from second; funicular segments strongly transverse. Forewing with MV placed slightly behind anterior margin of wing..*Xenufens*
9. Forewing with RS1 present; pre-marginal vein (PM) with 2 setae. Male genitalia with a dorsal lamina (DLA)...*Trichogramma* (*Vanlisus*) ♂
9'. Forewing without an RS1. PM with only 1 seta. Male genitalia without a DLA.. *Trichogrammatoidea* ♂
10. Antenna with F1 shorter or equal in length to F2, club with 2–3 segments. Forewing with long fringe setae, their maximum length at least half width of wing (FWW)... *Trichogrammatomyia* ♂, ♀ (in part)
10'. Antenna with F2 distinctly shorter than F1, club 3 segmented. Forewing usually with fringe-setae distinctly shorter than half FWW............................ *Paracentrobia*
11. Female antenna with a 2–3 segmented club. Forewing with stigmal vein (SV) broad, very short, indistinct, sessile to marginal vein (MV).................. *Tumidiclava*
11. Female antenna with a 5-segmented club. Forewing with SV distinct.... .... *Lathromeris*

7.6.4 A Brief Synopsis of the New World Genera of Trichogrammatidae Associated with Lepidoptera

1. *Lathromeris* Forster (23 species). This genus is known from all continents (worldwide). In the New World, species occur in the USA, Costa Rica, Canada and Argentina. Only *L. argentina* De Santis and *L. hesperis* Pinto are recorded from the New World. *Lathromeris* appears to be uncommon in the Neotropical region (Pinto 2006). It is associated with Lepidoptera, Noctuidae and Pyralidae (Polaszek et al. 1998).
2. *Paracentrobia* Howard (43 species). This genus is known from all continents (worldwide). Eight species occur in the New World where it has been recorded in Argentina, Bermuda, Belize, Bolivia, Brasil, Canada, Costa Rica, Colombia, Equador, Honduras, the USA, Peru, Venezuela and the West Indies (George 2003, Pinto 2006). An association with Lepidoptera remains questionable (Pinto 2006).

3. *Paratrichogramma* Girault (8 species). This genus is known from Australia, Africa, Israel and India, as well as the New World (USA, Costa Rica and Argentina). Only *P. californica* Doutt has been recorded in the Americas (Pinto 2006). It is associated with Lepidoptera, Gracillariidae (Hayat and Viggiani 1984), Noctuidae (Viggiani 1976) and Lycaenidae (Pinto 2006).
4. *Trichogramma* Westwood (210 species). This is the largest genus of Trichogrammatidae. It occurs on all continents. Ninety eight species have been recorded in the New World (Pinto 2006).
5. *Trichogrammatoidea* Girault (25 species). The number of species from the New World is not clear due to introductions for biological control. Of the 10 species recorded, only 5 are native, all from South America: *T. annulata* De Santis; *T. bennetti* Nagaraja; *T. brasiliensis* (Ashmead); *T. hypsipylae* Nagaraja; and *T. signiphoroides* Bréthes. Those introduced are: *T. armigera* Manjunath; *T. bactrae* Nagaraja; *T. eldanae* Viggiani; *T. nana* (Zehntner); and *T. robusta* Nagaraja (Querino and Zucchi 2004b, Pinto 2006). It is exclusively associated with Lepidoptera (several families) (Nagaraja 1978, De Santis 1981, Querino and Zucchi 2004b).
6. *Trichogrammatomyia* Girault (1 species). *Trichogrammatomyia tortricis* Girault was described from North America. It occurs from Canada to Argentina and also is known from the Asia Pacific region. There are at least two or three undescribed species in the New World (Pinto 2006). It was recorded from Lepidoptera, Tortricidae (Girault 1916) and on *Hypsipyla grandella* (Lepidoptera, Pyralidae) (Ohashi et al. 2005).
7. *Tumidiclava* Girault (18 species). Distribution is worldwide. Only two described species occur in the New World, *T. pulchrinotum* Girault in North America and *T. pampeana* De Santis in Argentina; several species remain to be described. It is uncommon in the tropics (Pinto 2006).
8. *Xenufens* Girault (2 species). This genus is known from the New World and the Asia Pacific region. *X. ruskini* Girault is widely distributed, whereas *X. forsythi* Yoshimoto (previously in *Pseudoxenufens*) occurs only in Brasil, Bolivia, Equador, Colombia and Costa Rica (Yoshimoto 1976, Pinto 2006). It was recorded from Lepidoptera, Nymphalidae (Malo 1961, Yoshimoto 1976) and Hesperiidae (Girault 1916, Pinto 2006).

7.6.5 *Trichogramma*

The chaotic taxonomic history of *Trichogramma* in North America summarized by Pinto (1999) can be applied to the entire New World. In general, interest in using species for biological control predated the development of taxonomic expertise by several decades. Consequently, inappropriate and often intraspecifically variable characters such as color and wing setation were used for identification. This resulted in virtually all of the pre-1970 literature at the species level being based on misidentifications. These misidentifications compromise the utility of earlier catalogues depending on this literature such as that of De Santis (1981) of South American

Chalcidoidea, which includes the 13 native and introduced species of *Trichogramma* recognized at that time. The discovery that the male genitalia are a rich source of characters (Nagarkatti and Nagaraja 1968, 1971) finally allowed potential discrimination of species. The first South American species description based on male genitalia was *T. rojasi* from Chile described by Nagaraja and Nagarkatti (1973). The use of reproductive compatibility studies (e.g. Pinto et al. 1983, 1986) and, more recently, the incorporation of molecular methods (e.g. Stouthamer et al. 1999) as tools for identification has continued to refine *Trichogramma* species taxonomy.

In spite of recent advances, the local fauna of many regions remain poorly understood, although this knowledge is important for furthering the use of *Trichogramma* in biological control programmes. In almost all habitats native species adapted to local conditions are present, suggesting that natural parasitism of pest species by *Trichogramma* could be high (Pinto 1999). Surveys of native species prior to the introduction of exotic ones in mass release programmes are rare.

Taxonomic knowledge is required before programmes for the biological control of pests can be optimally instituted. Only after species identification has been clarified, can the remaining biological areas be developed (Zucchi 2004). Thus, an investigation of *Trichogramma* diversity present in ecosystems (agricultural, forest and native) is fundamental for the success of biological control.

7.6.5.1 Classification

Trichogramma, with 210 species worldwide, is the largest genus of Trichogrammatidae (Pinto 2006). It is divided into three subgenera: the nominate, *Trichogramma* Westwood – Type-species *Trichogramma evanescens* Westwood – is cosmopolitan and contains the vast majority of species; *Trichogrammanza* Carver– type-species *Trichogramma* (*Trichogrammanza*) *funiculatum* Carver – is a small subgenus restricted to Australia and New Zealand, with only three species (*T. carverae* Oatman and Pinto, *T. funiculatum* and *T. tenebrosum* Oatman and Pinto); and the subgenus *Vanlisus* Pinto – type-species *Trichogramma lachesis* Pinto – occurs in Australia and in the New World; it includes four species, three in the *lachesis* group (*T. atropos* Pinto, *T. clotho* Pinto, *T. lachesis* Pinto) and one in the *primaevum* group (*T. primaevum* Pinto). The subgenera are based on male characters and may be separated as follows (Fig. 7.7):

(a) Subgenus *Trichogramma* Westwood: Segments of the antennal flagellum joined to form a 1-segmented club;
(b) Subgenus *Trichogrammanza* Carver: Antennal flagellum with two funicular segments and a 1-segmented club;
(c) Subgenus *Vanlisus* Pinto: Antennal flagellum with two funicular segments and a 3-segmented club.

There has been relatively little taxonomic work on the *Trichogramma* of Latin America and much of it has been recent. Most of what is available is regional in

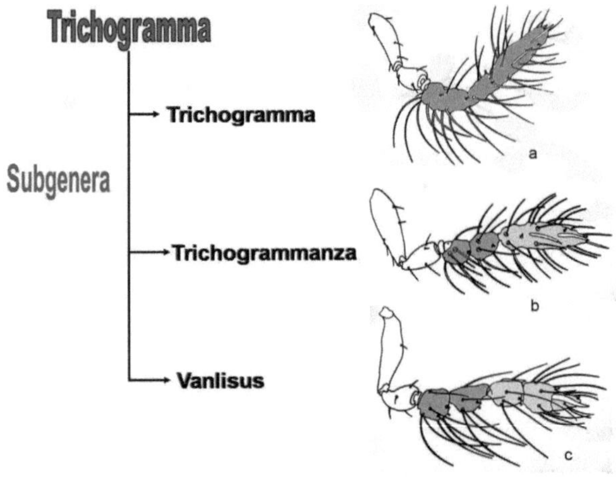

Fig. 7.7 Scheme indicating differences in the male antenna of *Trichogramma* subgenera

character. Ruiz and Korytkowski (1979) treated the Peruvian fauna and included a key to the species of Peru as well as a species catalogue for the Neotropics; Velásquez De Rios and Teran (1995, 2003) published keys to the Venezuelan species. There have also been descriptions of species from Chile (Pintureau et al. 1999), and Uruguay (Basso et al. 1999). In Brasil, starting in the 1980s, there were various sequential descriptions and records of species associated with *Diatraea saccharalis* (F.) (Zucchi 1985, 1988). In the 90's, there was an increase in research related to biology and applied biological control, with new records of host association and the distribution of species (e.g. Zucchi and Monteiro 1997). Additional morphologic characterization of species of *Trichogramma* recorded in the South America and new species were added by Querino and Zucchi (2003a–c) and a key to the Brazilian *Trichogramma* was published recently (Querino and Zucchi 2005). The revision of North American *Trichogramma* by Pinto (1999) also treats some species occurring in Central and South America. South American species have also been the focus of studies utilizing molecular identification (e.g. Ciociola et al. 2001, Almeida 2004).

The taxonomic confusion associated with *Trichogramma* species identification in the New World has already been noted. The identity of several species was clarified by Pinto et al. (1978, 1983). They pointed out that literature records of *T. fasciatum* (Perkins), *T. perkinsi* Girault, *T. retorridum* (Girault) and *T. semifumatum* (Perkins) actually corresponded to *T. fuentesi* Torre, *T. exiguum* Pinto and Platner, *T. brevicapillum* Pinto and Platner, and *T. pretiosum* Riley, respectively. Another example of misidentification, this one at the generic level, involves *T. brasiliense* Ashmead. This species was purportedly recorded from several South American countries. Nevertheless, *T. brasiliense* does not belong to the genus *Trichogramma*, but to *Trichogrammatoidea*. This name had been incorrectly attributed to specimens raised in the laboratory that probably refer to *T. fuentesi*. The name *T. brasiliense* may have also been incorrectly applied to other species (Pinto 1997c).

7.6.5.2 Distribution

Due to their small size (less than 1 mm), species of *Trichogramma* can potentially be transported by wind and intentionally or unintentionally, by man from one country to another in a short period of time. For these reasons it is often difficult to determine the natural region of distribution (Nagarkatti and Nagaraja 1977).

Trichogramma occurs in all vegetated terrestrial habitats that have been sampled. This includes areas in the tundra of the Arctic Circle, deserts, islands, mountain habitats, tropical areas and, at the most southerly point, on islands south of New Zealand (Pinto and Stouthamer 1994, Pinto and Oatman 1996). Thus, they are present in all six biogeographic regions: Palearctic, Oriental, Nearctic, Neotropical, Afrotropical and Australian. The number of species known from these regions is highly correlated with the intensity of collecting and their use in agriculture (Pinto and Stouthamer 1994).

In the New World, species distributions are best known in North America. Because early introductions for biological control preceeded a study of natives it is difficult to determine the original distribution of certain species. Some, with wide distributions, such as *T. pretiosum* in Australia (and perhaps in South America as well) and *T. chilonis* Ishii in Hawaii (Oatman et al. 1982, Pinto et al. 1993) could have been disseminated by man.

Species from North America and their respective hosts were discussed by Pinto (1999) and for South America by Zucchi and Monteiro (1997) and Querino (2002) (see Chapter 8).

7.6.5.3 Identification

Females of *Trichogramma* are separated from other trichogrammatids by the following characters: forewing with sigmoid venation and an RS_1 setal track; antenna with two funicular segments and a 1-segmented club. Males are also characterized by alar venation and genitalia with a dorsal lamina. The closest genus, *Trichogrammatoidea*, lacks both the RS_1 track and the dorsal lamina. The male antenna in most species is also of help in identification. In about 95% of known *Trichogramma* species, the segments of the funicle and club are fused into one unique elongate structure. In *Trichogrammatoidea*, the male antenna is composed of two funicular segments and three segments in the club. Nevertheless, the latter arrangement is also found in a few species of *Trichogramma* (see Classification) whose identification depends on characteristics of the genitalia and wings.

In spite of headway made in morphological studies, it is not always possible to identify species using morphology alone. This is due to the occurrence of cryptic species. In some cases reproductive compatibility studies and molecular identification are required (e.g. Pinto et al. 2003, Borghuis et al. 2004).

7.6.5.4 Morphological Characters

The primary structures used in identification are illustrated in Fig. 7.8. Species identification is principally based on the male antennae and male genitalia (see

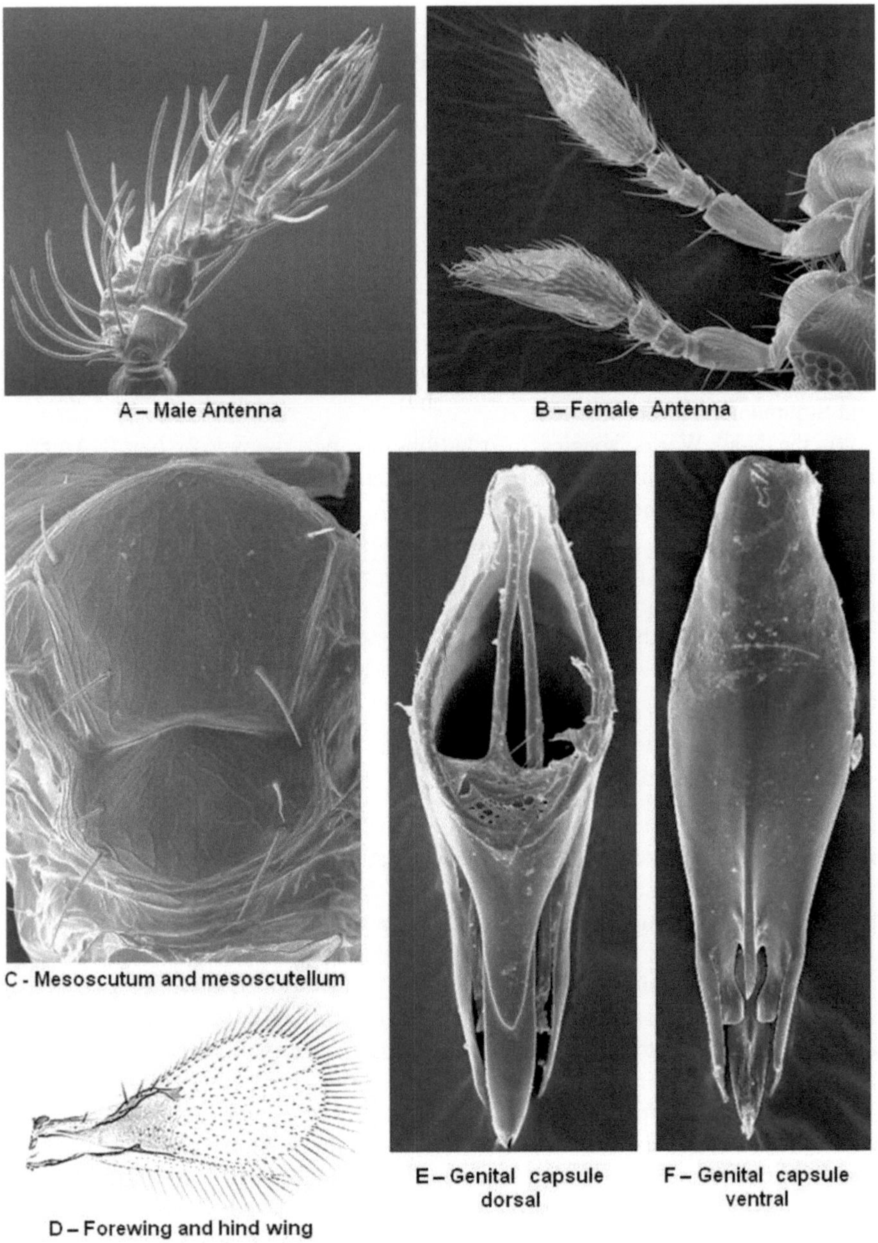

Fig. 7.8 General aspects of structures used for *Trichogramma* identification

Pinto 1997a, 1999, Querino and Zucchi 2003a). The female antennae and ovipositor structure may also be taxonomically useful but clearly are of secondary importance (Pinto 1999).

The fundamental characters of these structures are:

Antenna (Fig. 7.8a, b):

- Flagelliform setae: long setae on the flagellar surface.
- Unsocketed setae: very short and less conspicuous than flagelliform setae. They are stiff, apically directed hair-like structures which clearly lack a basal socket.
- Basiconic peg sensilla: small, stalked, bulbous structures on the flagellum of both sexes. Their shape, size and number vary interspecifically in males; they can occur at six positions along the flagellum, numbered 1–6 from base to apex, with a maximum of one or two sensilla at each, depending on position [e.g. 1-2-2-0-1-1, 1-2(1)-2-0-1-1]. In the latter formula the number in parentheses, indicates an alternate and less common condition at that position.
- Placoid sensillae: elongate, narrow structures, present in the apical half of the male flagellum; in *Trichogramma* they are adnate to the antennal surface for most of their length. Two occur at the very apex of the flagellum and the degree of overlap beyond the apex is sometimes useful in taxonomy. The number of sensilla is 3 in almost all species, although it varies in the subgenus *Vanlisus* (Pinto 1999).

Scutellum (Fig. 7.8c): There are two pair of setae on the scutellum of *Trichogramma*. The length of the anterior pair relative to the posterior pair varies in length and can be useful in identification of certain species.

Forewing (Fig. 7.8d):

- Fringe setae: arranged along the edge of the wing; the longest are found posterolaterally and are often measured and compared to total wing width.
- Setal density: estimated by number of setae between the fourth and fifth linear setal tracks on the wing-membrane.

Hind wing (Fig. 7.8d):

- Membrane contains three longitudinal rows of setae (anterior, median and posterior). The length of the anterior and posterior rows can be useful in identification. The median row always reaches the wing apex.

Genitalia: composed of several dorsal and ventral structures (Fig. 7.8e, f)

- Dorsal lamina: a structure that extends posteriorly, in general narrowing apically.
- Intervolsellar process: a ventral structure (usually pointed or subtriangular), positioned between the volsellae.
- Volsellae: two typically digitiform structures lateral to the intervolsellar process.
- Ventral ridge: a median carina of varying length, that extends forwards from the base of the intervolsellar process.

- Dorsal ridge: a median line of varying length that extends backwards from the base of the genital capsule; it is not present in all species.
- Ventral processes: two small structures situated beside the ventral ridge, generally close to the base of the intervolsellar process.
- Parameres: the most lateral structures at the apex of the genital capsule. Their length coincides with the apical distance of the genital capsule.
- Aedeagus: Occurs between the dorsal lamina and the ventral region of the genital capsule.
- Apodemes of the aedeagus: two rigid structures of varying length comprising the basal section of the aedeagus.

7.6.5.5 Intra-specific Variation

Intraspecific variation in morphological characters can be genetic or environmental (ecophenotypic) in origin. Genetic variation undoubtedly affects all characters to some degree. Ecophenotypic plasticity can be extensive in certain features such as forewings but fortunately, in most species, appears to have minimal impact on the male genitalia, the most important source of characters for identification of *Trichogramma* species (Pinto et al. 1989). Nevertheless, genitalic variation, especially of the dorsal lamina, cannot be completely discounted. Thus, in the case of *T. bruni*, a common species in South America, host and habitat, in addition to individual differences, are known to cause profound variation in the shape of the

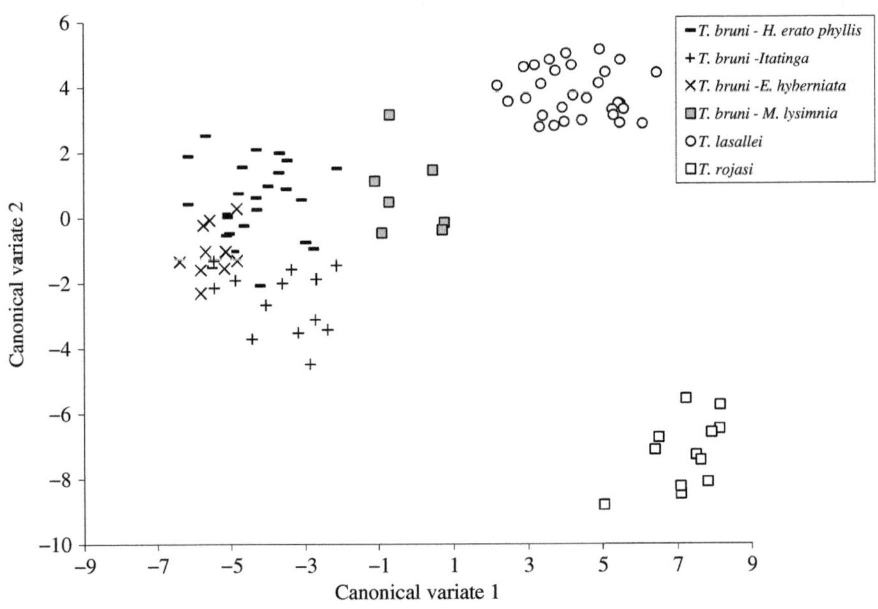

Fig. 7.9 Arrangement of *T. bruni* from different hosts, and of two related species (*T. lasallei* and *T. rojasi*), based on the analysis of morphometric variables (Querino and Zucchi 2002a)

dorsal lamina (Querino and Zucchi 2002a) (Fig. 7.9). This variation needs to be taken into consideration in the identification of *T. bruni*. The extent to which such extensive plasticity of the dorsal lamina occurs in other species is unknown. At present it appears to be an exceptional case.

7.6.5.6 Morphometry

When allied to morphology, morphometric methods can be used to detect subtle differences among populations (Reyment et al. 1981), and thus are important tools for specific identification. The analysis of main components, canonical variation and discriminate function are extremely useful in arranging morphometric data, thus allowing biological parameters underlying morphological relations between individuals or groups to be more easily detected and interpreted (Blackith and Reyment 1971, Reis 1988).

Multivariate morphometry has been little used in the study of *Trichogramma*. Factorial discriminative analysis has lead to knowing the structures and reasons responsible for the separation of five species from Cuba (Galán and Rodriguez 1993). Based on an analysis of principal components, Velásquez de Rios and Colmenares (1999) concluded that variables related to the male genitalia were those that most contributed to the separation of *T. pretiosum* and *T. atopovirilia* Oatman and Platner. Furthermore, by the analysis of principal components, morphological variation in *T. acacioi* Brun, Moraes and Soares, *T. atopovirilia*, *T. demoraesi* Nagaraja, *T. distinctum* Zucchi, *T. galloi* Zucchi and *T. bruni* Nagaraja (e.g. Querino and Zucchi 2004a) were studied. Morphometric methods also allowed separation of *T. minutum* Riley and *T. platneri* Nagarkatti, two species impossible to separate with conventional morphology (Burks and Heraty 2002).

The introduction of geometric morphometry in the study of shape caused a revolution in biometric analyses (Bookstein 1996, Rohlf and Marcus 1993). These methods were developed to analyze differences in shape between organisms based on anatomical points as defined by Cartesian co-ordinates (x and y). These co-ordinates are compared after removal of effects unrelated to shape (size, position and orientation). Thus, the shape of the object to be studied is contained in geometric properties which do not alter with changes in scale, translation or rotation (Rohlf 1996).

Exploratory results demonstrated subtle differences in the shape of the genital capsule in *T. pretiosum* specimens from different hosts, detected by means of geometric morphometry (Querino and Zucchi 2002b) (Fig. 7.10).

7.6.5.7 Reproductive Methods

Reproductive compatibility or crossing studies have frequently been used to complement morphology in solving taxonomic problems in *Trichogramma* (Nagarkatti and Nagaraja 1968, Pinto et al. 1978, Nagaraja 1987). In some cases, as with *T. deion* Pinto and Oatman and *T. pretiosum*, reproductive data have helped verify species

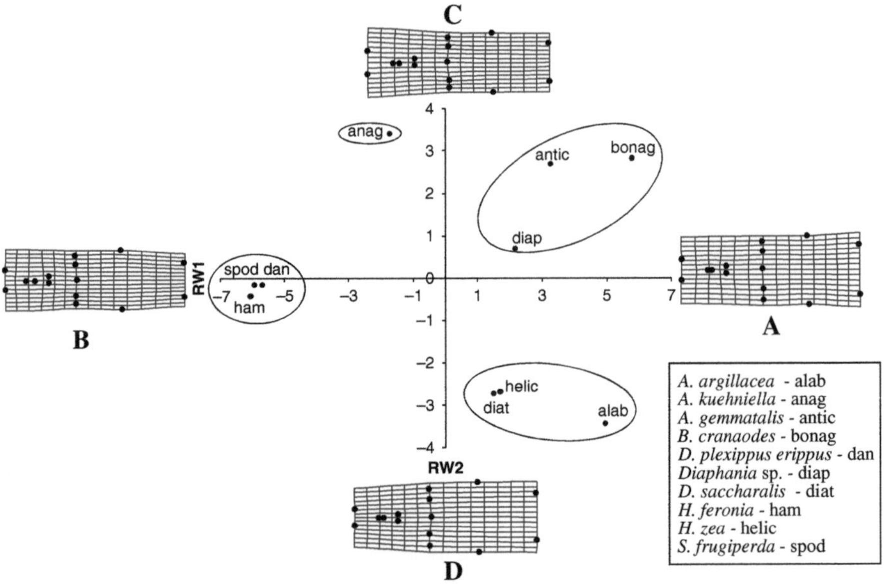

Fig. 7.10 Relative Warps (RW) for values of = 0 of *T. pretiosum* specimens obtained from 10 different Lepidoptera hosts. (**a**) variation in the shape of the genital capsule foreseen as positive deviations of the mean in the axis of relative warps 1. (**b**) variation in the shape of the genital capsule foreseen as negative deviations of the mean in the axis of relative warps 1. (**c**) variation in the shape of the genital capsule foreseen as positive deviations of the mean in the axis of the relative warps 2. (**d**) variation in the shape of the genital capsule foreseen as negative deviations of the mean in the axis of relative warps 2 (Querino and Zucchi 2002b)

originally suspected by minor morphological differences (e.g. Pinto et al. 1983). In others, as with *T. platneri* and *T. minutum*, it was only after morphologically identical cultures failed to cross were species even suspected (Nagarkatti 1975). Similar studies were also useful in separating two species from South America (*T. bruni* and *T. lasallei* Pinto) (Querino and Zucchi 2002a) (Table 7.3). Reproductive compatibility does not always clearly coincide with species limits, however, in certain species, such as *T. deion*, varying levels of hybridization occur ranging from complete compatibility to complete incompatibility with certain cultures displaying partial incompatibility (in one or both directions) (Pinto et al. 1991). In this case, the variation observed was not clearly correlated with geographic distribution.

The most common means of breeding in *Trichogramma* is arrhenotokous, but thelytoky or complete parthenogenesis also occurs. In *Trichogramma*, there are two forms of thelytoky, reversible (associated with microbial infection) and non-reversible or genetic (Stouthamer et al. 1990). Evidence points to the complete co-relationship between the presence of bacteria of the genus *Wolbachia* (α-Proteobacteria, Rickettsia) in eggs and the incidence of reversible parthenogenesis (thelytoky) in *Trichogramma* (Stouthamer et al. 1993). These bacteria lead to cytoplasmic incompatibility, thelytoky, feminization, or an increase of fertility in their

Table 7.3 Mean number of male and female descendants and total number of parasitized eggs obtained upon crossing of *T. lasallei* (L) and *T. bruni* (B) at 25°C (Querino and Zucchi 2002a)

Crossings	N	No. Male	No. Female	Total	Parasitized eggs
Heterogamic					
1 Male L × Female B	18	30.17 ±8.40	0	30.17 ±8.40	31.05 ±8.99
2 Male B × Female L	16	17.87 ±9.44	0	17.87 ±9.44	18.50 ±10.22
Homogamic					
3 Male L × Female L	09	14.22 ±4.38	11.33 ±6.18	25.55 ±7.52	27.55 ±7.21
4 Male B × Female B	10	15.50 ±7.11	12.10 ±5.72	27.60 ±9.14	31.60 ±9.87
Mode of reproduction					
Virgin Female L	10	23.9	0	23.9	23.5
Virgin Female B	11	35.5	0	35.5	36.73

hosts depending on the taxon infected (Van Meer 1999). Pinto and Stouthamer (1994) indicated that *Wolbachia* was known to induce parthenogenesis in more than 10 species of *Trichogramma*. The first cases of complete parthenogenetic reproduction induced by *Wolbachia* in South America (Brasil) were recorded by Almeida et al. (2001) in *T. pretiosum* and Ciociola Jr. et al. (2001) in *T. atopovirilia*.

Thelytokous females occurring in an arrhenotokous (non-infected) population can receive and use male sperm. There is no genetic isolation. These females if treated with antibiotic or high temperatures will transform into normal arrhenotokous individuals. For this reason, it is important to determine the type of thelytoky involved before considering unisexual populations as distinct species or subspecies (Pinto 1997a).

A modification of breeding could interfere with the dynamics of the host population, giving rise to consequences in the specifying process (Breeuweer and Werren 1990). The advantages and disadvantages for biological control in the use of thelytokous versus arrhenotokous females were discussed by Stouthamer (1993). Van Meer (1999) studied the ecological implications of using thelytokous females for biological control.

7.6.5.8 Molecular

Several studies have demonstrated the usefulness of techniques such as RAPD, RFLP, micro-satellites and sequencing of ITS1 and ITS2 in DNA, in the identification of species of *Trichogramma* (Orrego and Agudelo-Silva 1993, Silva et al. 1995, Pinto et al.1997a, Silva et al. 1995).

Stouthamer et al. (1999) demonstrated the utility of the ITS2 (Internal transcribed spacer 2) region of r-DNA as a new technique in the identification of *Trichogramma*. For South America, Ciociola Jr. et al. (2001) produced a molecular key for seven Brasilian species of *Trichogramma* using ITS2 region sequences and restriction analyses. ITS2 sequences were used to distinguish 17 species of native and introduced *Trichogramma* from South America (Almeida 2004).

Results obtained with molecular methods are promising and very useful in taxonomy, but should be combined with morphological and breeding procedures. The main advantages of molecular identification are: (a) the possibility of identifying females to species; to date, with few exceptions, identification has been based almost entirely on males; (b) identification of intra-specific variants, which are generally morphologically indistinguishable but potentially important, in biological control (Pinto 1999).

References

Aguiar-Menezes EL, Querino RB, Resende ALS, Paixão FHM, Guerra JGM, Almeida DL de (2005) Parasitismo de ovos de *Enchenopa gracilis* (Hemiptera: Membracidae) infestando guandu em sistema de produção agroecológico. In: 9o. Simpósio de Controle Biológico (9 SICONBIOL), 2005, Recife. Anais do 9o. Simpósio de Controle Biológico (9 SICONBIOL). Recife: J. Luiz Vasconcelos, p 142

Almeida RP de (2004) *Trichogramma* and its relationship with *Wolbachia*: Identification of *Trichogramma* species, phylogeny, transfer and costs of *Wolbachia* symbionts. PhD Thesis, Wagenignen University, The Netherlands, 142p

Almeida RP de, Ciociola AI, Stouthamer R (2001) *Wolbachia*-induced parthenogenesis: the first report in a Brazilian *Trichogramma pretiosum* population. Proc Exp Appl Entomol 12:41–44

Basso C, Pintureau B, Grille G (1999) Taxonomic study of two *Trichogramma* species from Uruguay (Hym.: Trichogrammatidae). Boletín Sanidad Vegetal Plagas 25:373–382

Blackith RE, Reyment RA (1971) Multivariate morphometrics. Academic Press, London

Bookstein FL (1996) Biometrics, biomathematics and the morphometric synthesis. Bull Math Biol 58:313–365

Borghuis A, Pinto JD, Platner GR, Stouthamer R (2004) Partial cytochrome oxidase II sequences distinguish the sibling species *Trichogramma minutum* Riley and *Trichogramma platneri* Nagarkatti. Biol Control 30:90–94

Breeuweer JAJ, Werren JH (1990) Microorganisms associated with chromosome destruction and reproductive isolation between two insect species. Nature 346:558–560

Burks BD (1979) Chalcidoidea (part) and Cynipoidea. In: Krombein KV, Hurd PD, Smith DR, Burks BD. Catalog of Hymenoptera in America North of Mexico, vol 1, pp 768–1107

Burks RA, Heraty JM (2002) Morphometric analysis of four species of *Trichogramma* Westwood (Hymenoptera: Chalcidoidea) attacking codling moth and other tortricid pests in North America. J Hymenopt Res 11:167–187

Ciociola Júnior AI, Zucchi RA, Stouthamer R (2001) Molecular key to seven brazilian species of *Trichogramma* (Hymenoptera: Trichogrammatidae) using sequences of the ITS2 region and restriction analysis. Neotrop Entomol 30:259–262

Delvare G, LaSalle J (2000) *Trisecodes* gen. n. (Hymenoptera, Eulophidae, Entedoninae) the first eulophid with three tarsal segments. J Hymenopt Res 9:305–312

De Santis L (1979) Catálogo de los himenopteros calcidoideos de América do Sur de los Estados Unidos. La Plata: Comisión de Investigaciones Científicas de la Provincia de Buenos Aires. 488p (Publicación Special)

De Santis L (1980) Catálogo de los himenopteros brasileños de la serie parasitica incluindo Bethyloidea. UFPR, Curitiba, 395p

De Santis L (1981) Catálogo de Los himenopteros calcidoideos de América do sur de los Estados Unidos – Primer Suplemento. Rev Peru Entomol 24:1–38

Doutt RL, Viggiani G (1968) The classification of the Trichogrammatidae (Hymenoptera: Chalcidoidea). Proc Calif Acad Sci 35:477–586

Galán M, Rodriguez J (1993) Caracterizacion morfometrica de Trichogramma Weswood (Hymenoptera: Trichogrammatidae). Rev Prot Veg 8:157–165

George J (2003) A review of the Paracentrobiini (Hymenoptera: Trichogrammatidae) and a review of the United States species of *Ittys* and *Ittysella*. MSc Thesis, University of California, Riverside, 124p

Gibson GAP, Heraty JM, Woolley JB (1999) Phylogenetics and classification of Chalcidoidea and Mymarommatoidea – a review of current concepts (Hymenoptera, Apocrita). Zool Scr 28:87–124

Girault AA (1916) Description of and observation on some chalcidoid Hymenoptera – II. Can Entomol 48:265–268

Goulet H, Huber JT (1993) Hymenoptera of the world: an identification guide to families. Ottawa, Agriculture Canada Publication, 668p

Hayat M (1994) Notes on some genera of the Aphelinidae (Hymenoptera: Chalcidoidea), with comments on the classification of the family. Orient Insects 28:81–96

Hayat M (1998) Aphelinidae of India (Hymenoptera: Chalcidoidea): a taxonomic revision. Mem Entomol, Int 13:1–416

Hayat M, Viggiani G (1984) A preliminary catalogue of the Oriental Trichogrammatidae (Hymenoptera, Chalcidoidea). Boll Laboratorio Entomol Agraria 'Filippo Silvestri' Portici 41:23–52

Heraty JM, Woolley JB, Darling DC (1997) Phylogenetic implications of the mesofurca in Chalcidoidea (Hymenoptera), with emphasis on Aphelinidae. Syst Entomol 22:45–65

Huber JT, Beardsley JW (2000). A new genus of fairyfly, *Kikiki*, from the Hawaiian Islands (Hymenoptera: Mymaridae). Proc Hawaiian Entomol Soc 34:65–70

Huber JT, Gibson GAP, Bauer LS, Liu H, Gates M (2008) The genus *Mymaromella* (Hymenoptera: Mymarommatidae) in North America, with a key to described extant species. J Hymenopt Res 17:175–194

Kim JW, Triapitsyn SV (2004) A new species of *Pteroptrix* (Hymenoptera: Aphelinidae) from Argentina, the first known aphelinid with three-segmented tarsi. Entomol News 114:10–17

LaSalle J, Gauld ID (1992) Parasitic Hymenoptera and the biodiversity crisis. Redia 74:315–334

Lin N (1994) Systematic studies of Chinese Trichogrammatidae. Fuzhou, Fujian

Malo F (1961) Phoresy in *Xenufens* (Hymenoptera: Trichogrammatidae), a parasite of *Caligo eurilochus* (Lepidoptera: Nymphalidae). J Econ Entomol 54:465–466

Masner L (1995) The proctotrupoid families, pp 209–245. In: Hanson P, Gauld ID (eds) Hymenoptera of Costa Rica. Oxford University Press, Oxford. 893p

Meer MMM van (1999). Phylogeny and host-symbiont interactions of thelytoky inducing *Wolbachia* in Hymenoptera. PhD Thesis, Wageningen Agricultural University, Wageningen, 118p

Nagaraja H (1978) Studies on *Trichogrammatoidea* (Hymenoptera: Trichogrammatidae). Orient Insects 12:489–530

Nagaraja H (1987) Recent advances in biosystematics of *Trichogramma* and *Trichogrammatoidea* (Hymenoptera: Trichogrammatidae). Proc Indian Acad Sci 96:469–477

Nagaraja H, Nagarkatti S (1973) A key to some New World species of *Trichogramma* (Hymenoptera: Trichogrammatidae) with descriptions of four new species. Proc Entomol Soc Washington 75:288–297

Nagarkatti S (1975) Two new species of *Trichogramma* (Hymenoptera: Trichogrammatidae) from the USA. Entomophaga 20:245–248

Nagarkatti S, Nagaraja H (1968) Biosystematics studies on *Trichogramma* species: experimental hybridization between *Trichogramma australicum* Girault, *T. evanescens* Westwood and *T. minutum* Riley. CIBC Tech Bull 10:81–96

Nagarkatti S, Nagaraja H (1971) Redescriptions of some known species of *Trichogramma* (Hymenoptera: Trichogrammatidae) showing the importance of the male genitalia as a diagnostic character. Bull Entomol Res 61:13–31

Nagarkatti S, Nagaraja H (1977) Biosystematics of *Trichogramma* and *Trichogrammatoidea* species. Annu Rev Entomol 22:157–176

Noyes JS (2003) Universal Chalcidoidea Database. Disponível em http://www.nhm.ac.uk/entomology/chalcidoids/index.html. Accessed on 20 Jan 2006

Oatman ER, Pinto JD, Platner GR (1982) *Trichogramma* (Hymenoptera: Trichogrammatidae) of Hawaii. Pacific Insects 24:1–24

Ohashi OS, Silva-Júnior ML, Lameira OA, Silva JNM, Leão NVM, Terezo EF, Batista TFC, Hidaka DZL, Almeida GB, Bittencourt PR da, Gomes F da S, Neves GA de M (2005) Danos e controle da broca *Hypsipyla grandella* em plantio de mogno *Swietenia macrophylla* no Estado do Pará. In: Poltronieri LS, Trindade DR, Santos IP dos (eds) Pragas e doenças de cultivos amazônicos. Embrapa Amazônia Oriental, Belém

Orrego C, Agudelo-Silva F (1993) Genetic variation in the parasitoid wasp *Trichogramma* (Hymenoptera: Trichogrammatidae) revealed by DNA amplification of a section of the nuclear ribosomal repeat. Florida Entomol 76:519–524

Owen AK, George J, Pinto JD, Heraty JM (2007) A molecular phylogeny of the Trichogrammatidae (Hymenoptera: Chalcidoidea), with an evaluation of the utility of their male genitalia for higher level classification. Syst Entomol 32:227–251

Peck O, Boucek Z, Hoffe A (1964) Keys to the Chalcidoidea of Czechoslovakia (Insecta: Hymenoptera). Mem Entomol Soc Can 34:170p

Pinto JD (1997a) Taxonomia de Trichogrammatidae (Hymenoptera) com ênfase nos gêneros que parasitam Lepidoptera, pp 13–40. In: Parra JRP, Zucchi RA (eds) *Trichogramma* e o controle biológico aplicado. FEALQ, Piracicaba, 324p

Pinto JD (1997b) Trichogrammatidae, pp 726–752. In: Gibson GAP, Huber JT, Woolley JB (eds), Annotated keys to the genera of Nearctic Chalcidoidea (Hymenoptera). NRC Research Press, Ottawa, 778p

Pinto JD (1997c) *Trichogrammatoidea brasiliensis* (Ashmead) – new combination for a species historically placed in *Trichogramma* (Hymenoptera: Trichogrammatidae). Proc Entomol Soc Washington 99:593–596

Pinto JD (1999) Systematics of the North American species of *Trichogramma* Westwood (Hymenoptera: Trichogrammatidae). Mem Entomol Soc Washington 22:1–287

Pinto JD (2006) A review of the New World genera of Trichogrammatidae (Hymenoptera). J Hymenopt Res 15:38–163

Pinto JD, Oatman ER (1996) Description of three new *Trichogramma* (Hymenoptera: Trichogrammatidae) from New Zealand and their relationship to New World species. Proc Entomol Soc Washington 98:396–406

Pinto JD, Stouthamer R (1994) Systematics of the Trichogrammatidae with emphasis on *Trichogramma*, pp 1–36. In: Wajnberg E, Hassan SA (eds) Biological control with egg parasitoids. CAB International, IOBC, Wallingford, 304p

Pinto JD, Platner GR, Oatman ER (1978) Clarification of the identity of several common species of North American *Trichogramma* (Hymenoptera: Trichogrammatidae). Ann Entomol Soc Am 71:169–180

Pinto JD, Oatman ER, Platner GR (1983) The identity of two closely related and frequently encountered species of New World *Trichogramma* (Hymenoptera: Trichogrammatidae). Proc Entomol Soc Washington 85:588–593

Pinto JD, Oatman ER, Platner GR (1986) *Trichogramma pretiosum* and a new cryptic species occurring sympatrically in Southwestern North America (Hymenoptera: Trichogrammatidae). Ann Entomol Soc Am 79:1019–1028

Pinto JD, Velten RK, Platner GR, Oatman ER (1989) Phenotypic plasticity and taxonomic character in *Trichogramma* (Hymenoptera: Trichogrammatidae). Ann Entomol Soc Am 82: 414–425

Pinto JD, Stouthamer R, Platner GR, Oatman ER (1991) Variation in reproductive compatibility in *Trichogramma* and its taxonomic significance (Hymenoptera: Trichogrammatidae). Ann Entomol Soc Am 84:37–46

Pinto JD, Platner GR, Sassaman CA (1993) An electrophoretic study of two closely related species of North American *Trichogramma*, *T. pretiosum* and *T. deion* (Hymenoptera: Trichogrammatidae). Ann Entomol Soc Am 86:702–709

Pinto JD, Platner GR, Stouthamer R (2003) The systematics of the *Trichogramma minutum* Complex (Hymenoptera: Trichogrammatidae), a group of important North American biological control agents: the evidence from reproductive compatibility and allozymes. Biol Control 27:167–180

Pintureau B, Gerding M, Cisternas E (1999) Description of three new species of Trichogrammatidae (Hymenoptera) from Chile. Can Entomol 131:53–63

Polaszek A, LaSalle J, Jongema Y (1998) Chalcidoidea, pp 191–203. In: Polaszek A (ed) African cereal stem borers: Economic importance, taxonomy, natural enemies and control. CAB International, Wallingford, Oxon, 530p

Querino RB (2002) Taxonomia do gênero *Trichogramma* (Hymenoptera: Trichogrammatidae) na América do Sul. Tese de doutorado, ESALQ/USP, Piracicaba, 214p

Querino RB, Hamada N (2009) An aquatic microhymenopterous egg-parasitoid of *Argia insipida* Hagen in Selys (Odonata: Coenagrionidae) and biological observations in the Central Amazon, Brasil. Neotrop Entomol 38:346–351

Querino RB, Pinto JD (2007) A new *Hydrophylita* (Hymenoptera: Trichogrammatidae) from the Neotropics, with a key to species. Zootaxa 1437:47–54

Querino RB, Zucchi RA (2002a) Intraspecific variation in *Trichogramma bruni* Nagarja, 1833 (Hymenoptera, Trichogrammatidae) associated with different hosts. Braz J Biol 62:665–679

Querino RB, Zucchi RA (2002b) Relative warps analysis to study morphological variations in the genital capsule of *Trichogramma pretiosum* Riley (Hymenoptera: Trichogrammatidae). Neotrop Entomol 31:217–224

Querino RB, Zucchi RA (2003a) Caracterização morfológica de dez espécies de *Trichogramma* (Hymenoptera: Trichogrammatidae) registradas na América do Sul. Neotrop Entomol 32:597–613

Querino RB, Zucchi RA (2003b) Six new species of *Trichogramma* Westwood (Hymenoptera: Trichogrammatidae) from a Brasilian forest reserve. Zootaxa 134:1–11

Querino RB, Zucchi RA (2003c) New species of *Trichogramma* Westwood (Hymenoptera: Trichogrammatidae) associated with lepidopterous eggs in Brasil. Zootaxa 163:1–10

Querino RB, Zucchi RA (2004a) Análise morfométrica em espécies de *Trichogramma* (Hymenoptera: Trichogrammatidae). Neotrop Entomol 33:583–588

Querino RB, Zucchi RA (2004b) Redescription of *Trichogrammatoidea annulata* De Santis (Hymenoptera: Trichogrammatidae). Zootaxa 677:1–6

Querino RB, Zucchi RA (2005) An illustrated key to the species of *Trichogramma* (Hymenoptera: Trichogrammatidae) of Brasil. Zootaxa 1073:37–60

Ramalheira CS, Feitosa MCB, Querino RB, Hamada N (2006) Primeiro registro do microimenóptero *Hydrophylita bachmanni* DeSantis (Hymenoptera: Trichogrammatidae) parasitando ovos de Odonata no Brasil. In: XXI Congresso Brasileiro de Entomologia, 2006, Recife. XXI Congresso Brasileiro de Entomologia

Reis SF dos (1988) Morfometria e estatística multivariada em biologia evolutiva. Rev Brasil Zool 5:571–580

Reyment RA, Blackith RE, Campbell NA (1981) Multivariate morphometrics. Academic, New York

Rohlf FJ (1996) Morphometric spaces, shape components, and the effects of linear transformations, pp 117–129. In: Marcus LF, Corti M, Loy A, Naylor GJP, Slice DE (eds) Advances in morphometrics. NATO ASI, New York (Ser. A life Science, 284), 580p

Rohlf FJ, Marcus LF (1993) A revolution in morphometrics. Trends Ecol Evol 8:129–132

Ruiz ER, Korytkowski CA (1979) Contribucion al conocimiento de los Trichogrammatidae (Hymenoptera: Chalcidoidea) del Peru. Rev Peru Entomol 22:1–8

Santos RS (2007) Parasitismo em ovos *de Letptopharsa heveae* Drake and Poor, 1935 (Hemiptera: Tingidae) em seringueira (*Hevea brasiliensis* Müell. Arg.) no Estado do Mato Grosso. Tese de Doutorado, Universidade Estadual Paulista Júlio de Mesquita Filho, UNESP, Brasil

Silva IMMS, Van Kan FJPM, Van Lenteren JC, Stouthamer R (1995) Analysis of Portuguese *Trichogramma* spp. (Hym., Trichogrammatidae) using ITS-rDNA and RAPDs. Colloques l'INRA 73:37–39

Stouthamer R (1993) The use of sexual versus asexual wasps in biological control. Entomophaga 38:3–6

Stouthamer R, Pinto JD, Platner GR, Luck RF (1990) Taxonomic status of thelytokous forms of *Trichogramma* (Hymenoptera: Trichogrammatidae). Ann Entomol Soc Am 83:475–481

Stouthamer R, Breeuwer JAJ, Luck RF, Werren JH (1993) Molecular identification of microorganisms associated with parthenogenesis. Nature 361:66–68

Stouthamer R, Hu J, Kan F van, Platner GR, Pinto JD (1999) The utility of internally transcribed spacer 2 DNA sequences of the nuclear ribosomal gene for distinguishing sibling species of *Trichogramma* (Hymenoptera: Trichogrammatidae). BioControl 43:421–440

Velásquez de Rios M, Terán J (1995) Description of the species of the *Trichogramma* genus (Hymenoptera: Trichogrammatidae) in Venezuela. Colloques l'INRA 73:41–46

Velásquez de Rios M, Colmenares O. (1999) Análisis morfométrico de dos espécies de *Trichogramma* (Hymenoptera: Trichogrammatidae) utilizando la metodología de componentes principales. Boletín Entomología Venezolana 14:91–200

Velásquez de Rios M, Terán J (2003) Los *Trichogramma* (Hymenoptera: Trichogrammatidae) de La région noroccidental del estado Guárico, Venezuela. Entomotropica 18:127–145

Viggiani G (1976) Ricerche sugli Hymenoptera Chalcidoidea XLIX. *Trichogramma confusum* n. sp. per *T. australicum* Nagarkatti e Nagaraja (1968), nec Girault (1912), con note su *Trichogrammatoidea* Girault e descrizione di *Paratrichogramma heliothidis* n.sp. Boll Laboratorio Entomol Agraria 'Filippo Silvestri' de Portici 33:182–187

Whitfield JB (1998) Phylogeny and evolution host-parasitoid interactions in Hymenoptera. Annu Rev Entomol 43:129–151

Woolley JB (1997) Aphelinidae, pp 134–150. In: Gibson G, Huber J, Woolley J (eds) Annotated keys to the genera of Nearctic Chalcidoidea (Hymenoptera). NRC Research Press, Ottawa, Ontario, 778p

Yoshimoto C (1976) *Pseudoxenufens forsythi* a new genus and species of Trichogrammatidae (Hymenoptera: Chalcidoidea) from western Ecuador. Can Entomol 108:419–422

Yousuf M, Shafee SA (1986a) Catalogue of genus-group names of world Trichogrammatidae (Hymenoptera). Indian J Syst Entomol 3:13–27

Yousuf M, Shafee SA (1986b) Check list of species and bibliography of the world Trichogrammatidae (Hymenoptera). Indian J Syst Entomol 3:29–82

Zucchi RA (1985) Taxonomia de espécies de *Trichogramma* (Hymenoptera: Trichogrammatidae) associadas a algumas pragas (Lepidoptera) no Brasil. Tese (Livre-Docência) – Escola Superior de Agricultura "Luiz de Queiroz", Universidade de São Paulo, Piracicaba, 77p

Zucchi RA (1988) New species of *Trichogramma* (Hymenoptera: Trichogrammatidae) associated with sugar cane borer *Diatraea saccharalis* (F.) (Lepidoptera: Pyralidae) in Brasil. Colloques l'INRA 43:133–140

Zucchi RA (2004) Taxanomia e o controle biológico de pragas, pp 17–27. In: Parra JRP, Botelho PSM, Corrêa-Ferreira BS, Bento JMS (eds) Controle Biológico no Brasil: parasitóide e predadores. Manole, São Paulo, 609p

Zucchi RA, Monteiro RC (1997) O gênero *Trichogramma* na América do Sul, pp 41–66. In: Parra JRP, Zucchi RA (eds) *Trichogramma* e o controle biológico aplicado, FEALQ, Piracicaba, 324p

Chapter 8
Diversity and Hosts of *Trichogramma* in the New World, with Emphasis in South America

Roberto A. Zucchi, Ranyse B. Querino, and Renata C. Monteiro

Contents

8.1 Introduction

Trichogramma diversity in the New World is better known in North America, which was highlighted in a revision of the genus by Pinto (1999). In Central America, there have been few studies, but significant advances on the knowledge of this genus over latter decades in some South American countries has been achieved.

Many surveys on *Trichogramma* species were done concerning to biological control programs; therefore knowledge on insect hosts and the distribution is highly correlated with crops and their application in agriculture. Consequently, diversity is better known in countries with active research groups on the applied use or on the development of rearing systems of *Trichogramma* species. On the other hand, collections in natural habitats have been neglected throughout South America, but these habitats can harbor species which are not associated to agroecosystems. As an example, six new species were described from a survey carried out in a natural habitat in the State of São Paulo (Querino and Zucchi 2003a). Therefore, collections in undisturbed areas should considerably increase the number of known species in the Americas.

R.A. Zucchi (✉)
Departamento Entomologia e Acarologia, ESALQ/USP, Av. Pádua Dias 11, 13418-900 Piracicaba, São Paulo, Brazil
e-mail: razucchi@esalq.usp.br

F.L. Consoli et al. (eds.), *Egg Parasitoids in Agroecosystems with Emphasis on Trichogramma*, Progress in Biological Control 9, DOI 10.1007/978-1-4020-9110-0_8,

In this chapter, a list for the 94 *Trichogramma* species in the American continent (ca. 45% of worldwide species) and their habitats was compiled. Also, an update of the list of South American species (Zucchi and Monteiro 1997) with their respective host insects, distribution and associated plants is presented.

8.2 Geographical Distribution

Around 210 *Trichogramma* species are known worldwide (Pinto 2006), distributed among the six biogeographical regions. In the New World, 60 species are recorded in North America, 22 in Central America and 41 in South America (Table 8.1). Considering only the native species in South America, Brazil has the largest number of known species (26), followed by Venezuela (13), Colombia (9) and Peru (7) (Fig. 8.1). Records of *Trichogramma* are unknown only for the French Guyana and Suriname. There is a record of *Trichogramma* in Guyana (Thompson 1958) in Noyes (2003), but it must be confirmed.

It is difficult to determine with accuracy the *Trichogramma* distribution in South America due to innumerous cases of introduction of species (official or accidental) during the last 100 years. In Peru, for example, there are records of six native and eight introduced species (Whu and Valdivieso 1999). Among the latter, only *T. pintoi* has been recovered, whereas others had been incorrectly identified or introduced under erroneous specific names. The report of the first occurrence of *T. cacoeciae* in Peru and its distribution in South America were discussed by Almeida (2004). It is still doubtful whether *T. cacoeciae* is a native or an introduced species. Among fifteen insect pests mentioned as hosts of *Trichogramma*, only *Cydia pomonella* (L.), which is commonly associated to *T. cacoeciae*, does not occurr in Peru (Whu and Valdivieso 1999). Only the introduction of *T. cacoeciae* in Argentina and Cuba (De Santis and Fidalgo 1994) has been reported for Latin America.

Another example is *T. minutum*, which has been traditionally recorded by several agricultural entomologists in Brazil. Until the 1960's, it was the only species of *Trichogramma* recorded in Brazil parasitizing lepidopteran eggs such as *Alabama argillacea* (Hueb.), *Calpodes ethlius* (Stoll), *Diatraea saccharalis* (Fabr.), *Erinnyis ello* (L.), *Helicoverpa zea* (Bod.), *Neoleucinodes elegantalis* (Guenée) and *Sitotroga cerealella* (Oliv.) (Silva et al.1968). However, these records of *T. minutum* in Brazil were not subjected to taxonomic identification. In fact, they were based on the use of an available name in the literature, without appropriate species identification. Nevertheless, surveys on *Trichogramma* species conducted in the past 20 years at several Brazilian locations, including the Luiz de Queiroz campus, in Piracicaba, SP, where, in the 1980's, specimens of *T. minutum* from Antibes (France) were reared, were unable to detect *T. minutum* in the field (Querino and Zucchi 2007a). Also, the record of *T. minutum* for Venezuela is a misidentification of *T. fuentesi* (Velásquez and Terán 2003). Therefore, records of *T. minutum* in South America must be reevaluated.

Some species are widely distributed in the American continent, such as *T. pretiosum*, *T. exiguum*, *T. galloi*, *T. lasallei* and *T. atopovirilia* (Fig. 8.2). *Trichogramma pretiosum* occurs in all South American countries. Some species have specific hosts

Table 8.1 *Trichogramma* species and their habitats in the New World

***Trichogramma* species**	**Distribution**			**Habitats**			
	North America (60 species)	**Central America (22 species)**	**South America (41 species)**	**Annual Crops**	**Fruit Trees**	**Forestry**	**Natural Reserve**
T. alpha Pinto	■						■
T. acacioi Brun, Moraes and Soares			■	■		■	■
T. acuminatum Querino and Zucchi			■				■
T. acutovirilia Pinto	■						■
T. alloeovirilia Pinto			■				■
T. arcanum Pinto	■						■
T. atopovirilia Oatman and Platner	■	■	■	■			■
T. atropos Pinto			■				■
T. aurosum Sugonjaev and Sorokina	■			■	■		
T. ballmeri Pinto	■						■
T. bellaunionensis Basso and Pintureau			■	■			
T. bennetti Nagaraja and Nagarkatti	■	■	■				■
T. bertii Zucchi and Querino			■			■	
T. bispinosum Pinto	■						■
T. brassicae Voegelé				■			
T. brevicapillum Pinto and Platner	■			■			■
T. browningi Pinto and Oatman	■			■	■		
T. bruni Nagaraja	■	■	■		■	■	■
T. cacoeciae Marchal	■				■		■
T. californicum Nagaraja and Nagarkatti	■				■		■
T. canadense Pinto[a]	■						
T. clotho Pinto		■	■				■

Table 8.1 (continued)

T. colombiensis Velásquez and Terán							
T. deion Pinto and Oatman							
T. demoraesi Nagaraja							
T. dianae Pinto							
T. diazi Velasquez and Terán							
T. dissimilis Zucchi							
T. distinctum Zucchi							
T. drepanophorum Pinto and Oatman							
T. erebus Pinto							
T. esalqueanum Querino and Zucchi							
T. exiguum Pinto and Platner							
T. fasciatum (Perkins)							
T. fuentesi Torre							
T. funestum Pinto and Oatman							
T. gabrielino Pinto							
T. galloi Zucchi							
T. gordhi Pinto							
T. huberi Pinto							
T. infelix Pinto							
T. interius Pinto							
T. inyoense Pinto and Oatman							
T. iracildae Querino and Zucchi							
T. jalmirezi Zucchi							
T. japonicum Ashmead[b]							
T. julianoi Platner and Oatman[b]							

Table 8.1 (continued)

T. kaykai Pinto and Stouthamer							
T. lachesis Pinto							
T. lasallei Pinto							
T. leviculum Pinto							
T. lopezandinensis Sarmiento							
T. maltbyi Nagaraja and Nagarkatti							
T. manicobai Brun, Moraes and Soares							
T. marandobai Brun, Moraes and Soares							
T. marthae Goodpasture							
T. marylandense Thorpe							
T. maxacalii Voegelé and Pointel							
T. meteorum Vincent							
T. minutum Riley							
T. mullensi Pinto							
T. nemesis Pinto[b]							
T. nerudai Pintureau and Gerding							
T. nomlaki Pinto and Oatman							
T. nubilale Ertle and Davis							
T. obscurum Pinto							
T. offella Pinto and Platner							
T. panamense Pinto[a]							
T. parkeri Nagarkatti							
T. parrai Querino and Zucchi							
T. parvum Pinto							
T. pintoi Voegelé							

Table 8.1 (continued)

T. pintureaui Rodríguez and Galán							
T. platneri Nagarkatti							
T. pluto Pinto							
T. pratissolii Querino and Zucchi							
T. pratti Pinto							
T. pretiosum Riley							
T. pussillus Querino and Zucchi							
T. retorridum (Girault)							
T. rojasi Nagaraja and Nagarkatti							
T. satarosae Pinto							
T. sathon Pinto							
T. semblidis (Aurivillius)[b]							
T. sibericum Sorokina[b]							
T. sinuosum Pinto							
T. stampae Vincent							
T. suorangelica Pinto							
T. terani Velásquez and Terán							
T. thalense Pinto and Oatman							
T. tupiense Querino and Zucchi							
T. viggianii Pinto							
T. zeta Pinto							
T. zucchii Querino							

[a] *Habitat unknown*
[b] *Aquatic habitats*

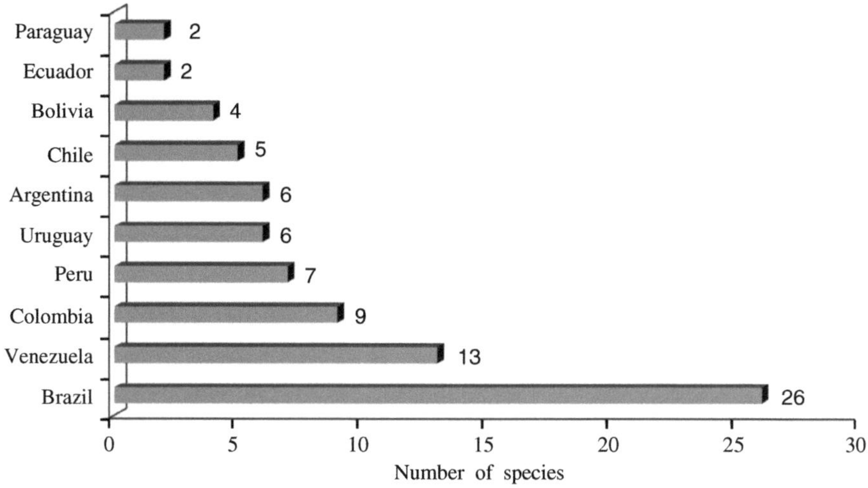

Fig. 8.1 Number of *Trichogramma* species in South America

and habitats, such as *T. galloi*, which is recorded in countries where sugarcane is cultivated. Moreover, more than half of the *Trichogramma* species have a narrow distribution, with 17 of them been recorded exclusively in Brazil (Table 8.2). In fact, studies on the taxonomy of *Trichogramma* are incipient in South American; however, in countries where these eggs parasitoids have been reared for biological control purposes, more information on species, hosts and associated plant are available.

Surveys have been addressed to areas in Brazil with the highest investments in biological control programs for economically important crops. Consequently, the largest number of species records is found in the Brazilian Southeastern region.

Species which predominantly occur in agroecosystems have been more frequently associated to their hosts in these systems, as *T. pretiosum* with about 240 records (Pinto 1999), which is not normally found in natural habitats. In sugarcane, seven *Trichogramma* species are associated to *Diatraea saccharalis* (Fabricius), with four of them known only to this host (Zucchi et al. 1996). However, little is known on the *Trichogramma* species associated to hosts from forests or natural reserves, as *T. bruni*.

Even though not widely studied, levels of natural parasitism by *Trichogramma* in agroecosystems are relatively high (Pinto 1999). In some cases, native species are more efficient than the introduced ones (Monje 1995). Therefore, besides being important for the conservation of native species, information on the local diversity can affect the success of the implementation of biological control initiatives. On the other hand, many exotic parasitoid species may be as efficient as the native ones, which are usually disregarded. Besides, the exotic species not always become established in the area where they have been introduced (Pinto and Stouthamer 1994). Preference for habitat (agroecosystems, fruit trees, forests, etc.) as shown by *Trichogramma* species is still another important aspect in the choice for the species

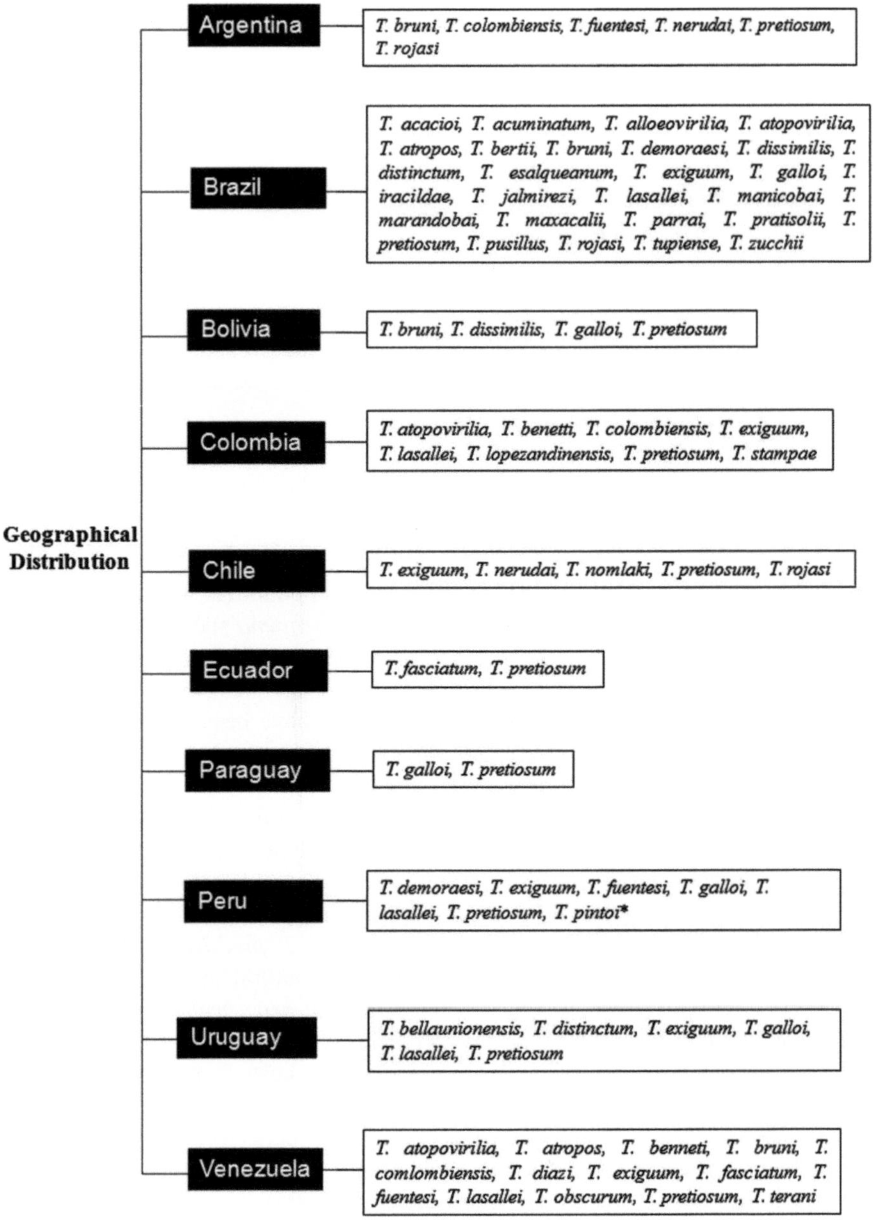

Fig. 8.2 *Trichogramma* species distribution in South American countries (*introduced and recovered species)

Table 8.2 *Trichogramma* species reported only to South America

Countries	Species numbers	Species
Brazil	17	*T. acacioi, T. acuminatum, T. bertii, T. dissimilis, T. distinctum, T. esalqueanum, T. galloi, T. iracildae, T. jalmirezi, T. manicobai, T. marandobai, T. maxacalii, T. parrai, T. pratisolii, T. pusillus, T. tupiense, T. zucchii*
Venezuela	02	*T. diazi, T. terani*
Uruguay	01	*T. bellaunionensis*
Colombia	01	*T. lopezandinensis*

to be used in biological control (Hassan 1994). Therefore, studies in these habitats must be stressed.

8.3 Hosts and Habitats

Trichogramma species are mainly associated to economically important lepidopterans (Fig. 8.3; Table 8.5). Most lepidopterans parasitized by *Trichogramma* belong to Noctuidae, including the major agricultural pests, as *Alabama argillacea* in cotton, *Anticarsia gemmatalis* in soybean and *Helicoverpa zea* and *Spodoptera frugiperda* in corn. Nevertheless, several non-economically important butterflies, such as *Heliconius erato phyllis* (Fabricius), *Hamadryas feronia* (Linnaeus) and *Mechanitis lysimnia* (Fabricius) are mainly attacked by *T. bruni* and *T. acacioi*, which occur in forest habitats (Table 8.5).

Some *Trichogramma* species occur in several different habitats (natural reserve, orchards, annual crops and forest), whereas others are restricted to agricultural areas (Table 8.2). In Brazil, several species have been recorded only in forest (*T. pusillum, T. acuminatum, T. esalqueanum, T. tupiense, T. alloeovirilia, T. maxacalii, T. zucchii, T. bertii, T. parrai*), while others that occur in agroecosystems can also parasitize forest pests (*T. acacioi, T. bruni, T. maxacalii*) (Querino and Zucchi 2003a, b, c) (Table 8.1).

Distribution of *Trichogramma* species varies according to the altitude. Only 12 *Trichogramma* species in the New World are found at altitudes higher than 2,000 m, and *T. fasciatum*, in Costa Rica, was recorded above 3,000 m (Table 8.3). Most *Trichogramma* species occur in terrestrial environments, but a few may also parasitize eggs of aquatic insects (Table 8.4) and others are found only in aquatic environments. Although some *Trichogramma* species can swim by using their legs and remain under the water for up to five days, most aquatic species will undoubtedly be reared from the terrestrial eggs of aquatic insect species. So far, no *Trichogramma* species has been reported as being a parasitoid of submerged host eggs (Bennett 2008).

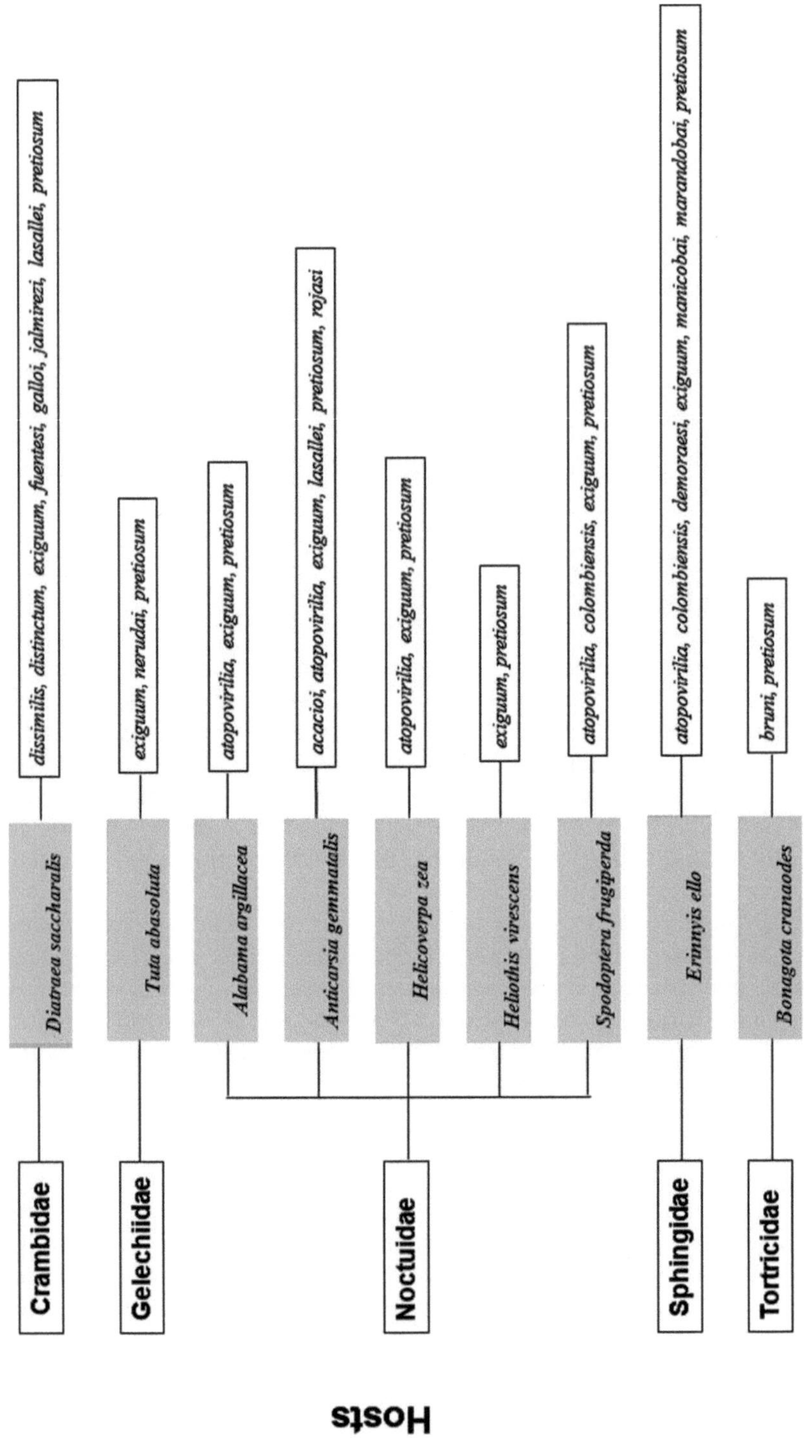

Fig. 8.3 Lepidopteran host pests of *Trichogramma* species

Table 8.3 *Trichogramma* species distribution in several altitudes

Species	500–1,000	1,001–2,000	2,001–3,000	>3,000
T. atropos	BR		VE	
T. arcanum	US			
T. suorangelica		ME		
T. atopovirilia	HO, ME	ME	GU, ME	
T. bruni	CR			
T. fasciatum		ME	CR, ME, VE	CR
T. gordhii			CR	
T. infelix		CR		
T. leviculum			ME	
T. maltbyi	US			
T. nomlaki	CA			
T. obscurum	CR, ME			
T. panamense		PA		
T. pluto	ME, CR, GU			
T. santarosae	CR	CR		
T. viggianii	ME			
T. zeta	RD			
T. acutovirilia	US			
T. alpha	CA			
T. aurosum	CA, US	US		
T. califomicum	US	US		
T. deion	US	US	US	
T. erebus	ME			
T. exiguum	HO, US	GU		
T. funestum	US			
T. inyoense	CA		US	
T. marylandense	US			
T. minutum	US	US		
T. parvum	US			
T. platneri		US		
T. pretiosum	CR, GU, HO, ME, US	GU, ME, US	ME	
T. sathon		US		
T. stampae	ME		ME	
T. parkeri	US			
T. pintoi	CA, US	US	US	
T. ballmeri		CA, US		
T. brevicapillum	US	US	US	
T. mullensi		US		
T. retorridum	US	US		
T. thalense		US	US	

*BR = Brazil; CA = Canada; CR = Costa Rica; HO = Honduras; MR = Mexico; GU = Guatemala; RD = Dominican Republic; US = United States; VE = Venezuela

8.4 A List for *Trichogramma* in South America

The list was based on the *Trichogramma* records compiled by Zucchi and Monteiro (1997), whose data were updated (Table 8.5).

Table 8.4 *Trichogramma* species associated with aquatic habitats and their hosts

Species	Hosts		Countries	References
T. nemesis	Stratiomyidae	*Odontomyia* sp.	USA	Pinto 1999
		Hedriodiscus sp.	USA	Juliano 1981
		Stratiomys sp.	USA	
T. semblidis	Sciomyzidae	*Sepedon fuscipennis*		
		Tetanocera sp.		
		Elgiva sp.	USA	Juliano 1981
	Tabanidae	*Chrysops* sp.	Canada	Pinto 1999
		Chrysops aestuans		
	Sialidae	*Sialis* sp.		
		Sialis californica	USA	Azam and Anderson 1969
		Sialis rotunda	USA	Azam and Anderson 1969
	Corydalidae	*Corydalus* sp.	USA	Bright 2007
		Chauliodes sp.	USA	Bennett 2008
T. japonicum	unknown	Unknown	Canada	Pinto 1999
T. julianoi	Sciomyzidae	*Sepedon fuscipennis*	USA	
		Eligiva sundewalli	USA	Juliano 1981
	Stratiomyidae	Unknown	USA	
Trichogramma sp.	Acari	*Hydracarina*	USA	Bright 2007

After more than a decade, some records were added to the *Trichogramma* 1997 list as follows: (i) 16 new *Trichogramma* species records, totaling 40 species; (ii) first record of *Trichogramma* in Ecuador; (iii) 18 new host insect records for *T. pretiosum*, totaling 44; (iv) 19 new host insects, totaling 61 species and (v) 9 plant associated records, totaling 38 species. A database for most of the species and their hosts in Brazil was compiled by Querino and Zucchi (2007b).

Table 8.5 *Trichogramma* species, insect hosts and respective associated plants in South America

1. *Trichogramma acacioi* Brun, Moraes and Soares
Anticarsia gemmatalis/*Glycine max*/Brazil
Dione juno juno/*Passiflora* sp./Brasil
Euselasia sp./*Eucalyptus* sp./Brazil
Hamadryas feronia/*Dalechampia* sp./Brazil
Nipteria panacea/*Persea americana*/Brazil
Noctuidae/*Phaseolus vulgaris*/Brazil
Psorocampa denticulata/*Eucalyptus* sp./Brazil

2. *Trichogramma acuminatum* Querino and Zucchi
unknown host/forest reserve/Brazil

3. *Trichogramma alloeovirilia* Querino and Zucchi
unknown host/forest reserve/Brazil

4. *Trichogramma atopovirilia* Oatman and Platner
Alabama argillacea/*Gossypium* sp. and *Solanum lycopersicum*/Venezuela
Anomis sp./unknown plant/Venezuela

Table 8.5 (continued)

Anticarsia gemmatalis/*Glycine max*/Brazil
Erinnyis ello/*Manihot esculenta*/Brazil
Helicoverpa zea/*Zea mays*/Brazil
Heliothis spp./*Gossypium* sp. and *Solanum lycopersicum*/Venezuela
Sitotroga cerealella trap eggs/Pepper/Brazil
Spodoptera frugiperda/*Zea mays*/Brazil
Spodoptera frugiperda/*Zea mays*/Venezuela
Spodoptera frugiperda/*Malachra* spp./Venezuela
unknown host/unknown plant/Colombia

5. *Trichogramma atropos* Pinto
unknown host/unknown plant/Brazil and Venezuela

6. *Trichogramma bellaunionensis* Basso and Pintureau
Diatraea saccharalis/*Oryza sativa*/Uruguai

7. *Trichogramma bennetti* Nagaraja and Nagarkatti
Anomis spp./*Malachra* sp./Venezuela
Heliothis sp./unknown plant/Colombia
Lepidoptera/*Spiracantha cornifolia*/Venezuela

8. *Trichogramma bertii* Zucchi and Querino
Glena sp./unknown plant/Brazil
Melanolophia sp./unknown plant/Brazil

9. *Trichogramma bruni* Nagaraja
Anomis spp./*Malachra* sp./Venezuela
Anomis spp./*Spiracantha cornifolia*/Venezuela
Anticarsia gemmatalis/*Glycine max*/Argentina
Bonagota cranaodes/*Lepidium meyenii*/ Brazil
Diatraea rufescens/*Saccharum* sp./Bolivia
Diatraea sp./*Saccharum* sp./Bolivia
Erosina hyberniata/*Acnistus arborescens*/Brazil
Hamadryas feronia/*Dalechampia* sp./Brazil
Heliconius erato phyllis sp./*Passiflora* sp. Brazil
Mechanistis lysimnia/*Solanum agrarium*/Brazil
Melanolophia sp./*Eucalyptus* sp./Brazil
Noctuidae/*Spiracantha cornifolia*/Venezuela

10. *Trichogramma colombiensis* Velásquez and Terán
Erinnyis ello/*Manihot esculenta*/Colombia and Venezuela
Spodoptera frugiperda/*Zea mays*/Colombia
Unknown host/unknown plant/Argentina

11. *Trichogramma clotho* Pinto
unknown host/unknown plant/Colombia

12. *Trichogramma demoraesi* Nagaraja
Erinnyis ello/*Manihot esculenta*/Brazil
Glena bipennaria/*Eucalyptus* sp./Brazil
unknown host/unknown plant/Peru

13. *Trichogramma diazi* Velásquez and Terán
Noctuidae/*Malachra* spp./Venezuela

Table 8.5 (continued)

14. *Trichogramma dissimilis* Zucchi
Diatraea saccharalis/*Saccharum* sp./Brazil
unknown host/unknown plant/Bolivia

15. *Trichogramma distinctum* Zucchi
Diatraea saccharalis/*Saccharum* sp./Brazil

16. *Trichogramma erebus* Pinto
Hesperiidae/*Desmodium* sp./Colombia

17. *Trichogramma esalqueanum* Querino and Zucchi
unknown host/forest reserve/Brazil

18. *Trichogramma exiguum* Pinto and Platner
Alabama argillacea/*Gossypium* sp./Colombia
Argyrotaenia sphaleropa/*Vitis* sp./Uruguay
Argyrotaenia sphaleropa/unknown plant/Peru
Bonagota cranaode/*Vitis* sp./Uruguay
Diatraea indigenella/*Saccharum* sp./Colombia
Diatraea rosa/*Saccharum* sp./Colombia
Diatraea saccharalis/*Saccharum* sp./Colombia and Peru
Diatraea spp./*Saccharum* sp./Venezuela
Dichomeris sp./*Sorghum* sp./Colombia
Dione Juno/*Passiflora* sp./Peru
Erinnyis ello/*Manihot esculenta*/Colombia
Erinnyis ello/*Manihot esculenta*/Peru
Helicoveerpa zea/*Zea mays*/Peru
Heliothis virescens Gossypium sp./Peru
Heliothis virescens/*Phaseolus vulgaris*/Colombia
unknown host/traps on *Pinus* sp./Chile
Neoleucinodes elegantalis/*Solanum lycopersicum*/Colombia
Palpita persimilis/ unknown plant/Peru
Pococera atramentalis/*Sorghum* sp./Colombia
Sacadodes pyralis/*Gossypium* sp./Colombia
Spodoptera frugiperda/*Zea mays*/Venezuela
Tuta absoluta/*Solanum lycopersicum*/Colombia

19. *Trichogramma fasciatum* (Perkins)
Peridroma saucia/*Agave sisalana*/Ecuador
unknown host/unknown plant/Venezuela
Noctuidae/*Zea mays*/Ecuador

20. *Trichogramma fuentesi* Torre
Anomis texana/*Gossypium* sp./Peru
Diatraea saccharalis/unknown plant/Peru
Helicoverpa zea/*Zea mays*/ Venezuela
Helicoveerpa zea/ Zea mays/Peru
Heliothis virescens/unknown plant/Peru
unknown host/unknown plant/Argentina

21. *Trichogramma galloi* Zucchi
Diatraea rufescens/*Saccharum* sp./Bolivia
Diatraea saccharalis/*Saccharum* sp./Bolivia, Brazil, Paraguay and Uruguay

Table 8.5 (continued)

Diatraea saccharalis/*Saccharum* sp./Peru
22. *Trichogramma iracildae* Querino and Zucchi
Calpodes ethlius/*Canna* spp./Brazil
23. *Trichogramma jalmirezi* Zucchi
Diatraea saccharalis/*Saccharum* sp./Brazil
24. *Trichogramma lasallei* Pinto
Anomis sp./*Malachra* spp./Venezuela
Anticarsia gemmatalis/*Glycine max*/Brazil
Diatraea saccharalis/*Saccharum* sp./Uruguay
Diatraea spp./*Oryza sativa*/Venezuela
Quinta cannae/*Canna* sp./Peru
25. *Trichogramma lopezandinensis* Sarmiento
Colias dimera/*Dasyphyllum diacanthoides* ?/Colombia
Copitarsia consueta/*Solanum tuberosum*/Colombia
26. ***Trichogramma manicobai* Brun, Moraes and Soares**
Erinnyis ello/*Hevea braziliensis* and *Manihot esculenta*/Brazil
27. *Trichogramma marandobai* Brun, Moraes and Soares
Erinnyis ello/*Manihot esculenta*/Brazil
28. *Trichogramma maxacalii* Voegelé and Pointel
Euselasia euploea eucerus/*Eucalyptus* sp./Brazil
Euselasia hygenius occulta/*Eucalyptus* sp./Brazil
Euselasia sp./*Eucalyptus* sp./Brazil
29. *Trichogramma nerudai* Pintureau and Gerding
Rhyacionia buoliana/*Pinus radiata*/Chile
Tuta absoluta/*Solanum lycopersicum*/Chile
30. *Trichogramma nomlaki* Pinto and Oatman
unknown host/*Pinus* sp. and "faia"/Chile
31. *Trichogramma obscurum* Pinto
Dione juno/*Passiflora edulis*/Venezuela
32. *Trichogramma parrai* Querino and Zucchi
unknown host/forest reserve/Brazil
33. *Trichogramma pratisolii* Querino and Zucchi
Anagasta kuehniella trap eggs/*Persea americana*/Brazil
34. *Trichogramma pretiosum* Riley
Agraulis vanillae/*Passiflora* sp./Brazil
Agrotis sp./*Pisum sativum*/Brazil
Alabama argillacea/*Gossypium* sp./Argentina, Bolivia, Brazil, Colombia, Peru, Uruguay and Venezuela
Anomis spp./*Malachra* spp./Venezuela
Anomis texana/*Gossypium* sp./Peru
Anticarsia gemmatalis/*Glycine max*/Argentina and Brazil
Bonagota cranaodes/*Lepidium meyenii*/ Brazil
Chrysoperla sp./*Gossypium* sp./Brazil
Chrysoperla sp./Zea mays/Brazil

Table 8.5 (continued)

Colias lesbia/*Medicago sativa* and *Glycine max*/Argentina
Danaus plexippus erippus/*Asclepias curassavia*/Brazil
Diatraea indigenella/*Saccharum* sp./Colombia
Diatraea saccharalis/*Saccharum* sp. and *Oryza sativa*/Brazil, Colombia and Uruguay
Diatraea saccharalis/*Saccharum* sp./Peru
Diatraea sp./*Saccharum* sp./Bolivia
Dione juno juno/*Passiflora* sp./Brazil
Ecdytolopha aurantiana/*Citrus* sp./Brazil
Erinnyis ello/*Manihot esculenta*/Brazil
Hamadryas feronia/*Dalechampia* sp./ Brazil
Helicoverpa gelotopoeon/*Gossypium* sp./Argentina
Helicoverpa zea/*Solanum lycopersicum*/Brazil and Venezuela
Helicoverpa zea/*Zea mays*/Brazil, Paraguay, Peru and Uruguay
Helicoverpa sp./*Gossypium* sp./ Venezuela
Heliothis spp./*Gossypium* sp. and *Solanum lycopersicum*/Argentina and Venezuela
Heliothis virescens/*Gossypium* sp./Brazil and Colombia
Heliothis virescens/unknown plant/Peru
Heraclides thoas/*Citrus* sp./ Peru
Hipsypila grandella/*Cedrella* sp./Peru
Mocis spp./*Digitaria* sp./Venezuela
Neoleucinodes elegantalis/*Solanum lycopersicum*/Brazil, Colombia and Venezuela
Phthorimaea operculella/stored potatoes/Venezuela
Plutella xylostella/*Brassica oleracea*/Brazil
Rachiplusia nu/*Glycine max*/Argentina
Sacadodes pyralis/*Gossypium* sp./Colombia
Sphingidae/*Solanum melongena*/Brazil
Spodoptera frugiperda/*Gossypium* sp./Brazil
Stenoma catenifer/*Persea americana*/Brazil
Trichoplusia oxygramma/*Bacharis* sp./Brazil
Tuta absoluta/*Solanum lycopersicum*/Chile

35. *Trichogramma pussilus* Querino and Zucchi
unknown host/forest reserve/Brazil

36. *Trichogramma rojasi* Nagaraja and Nagarkatti
Anticarsia gemmatalis/*Glycine max*/Argentina and Brazil
Colias lesbia/*Glycine max*/Argentina
Rachiplusia nu/*Glycine max*/Argentina
Tatochila sp./*Dasyphyllum diacanthoides*?/Chile

37. *Trichogramma stampae* Vincent
Chlosyne saundersii/*Helianthus annuus*/Colombia

38. *Trichogramma terani* Velásquez and Terán
Noctuidae/*Sida* sp./ Venezuela

39. *Trichogramma tupiense* Querino and Zucchi
unknown host/forest reserve/Brazil

40. *Trichogramma zucchii* Querino
unknown host/forest reserve/Brazil

Note: *Trichogramma pintoi* was not listed, although it was introduced in and recovered once from Peru.

References

Almeida RP de (2004) *Trichogramma* and its relationship with *Wolbachia*: identification of *Trichogramma* species, phylogeny, transfer and costs of *Wolbachia* symbionts. PhD Thesis, Wagenignen University, The Netherlands, 142p

Azam KM, Anderson NH (1969) Life history and habits of *Sialis rotunda* and *S. californica* in Western Oregon. Ann Entomol Soc Am 62:549–558

Bennett AMR (2008) Aquatic Hymenoptera. In: Merritt RW, Cummins KW (eds) An introduction to the aquatic insects of North America, 4th edn, pp 673–686

Bright E (2007) Aquatic Hymenoptera (Wasps) of Michigan. Available: http://insects.ummz.lsa.umich.edu/~ethanbr/aim/sp/Hymenoptera/sp_hymenom.html. Accessed: 07 jul 2008

De Santis L, Fidalgo P (1994) Catalogo de Los Himenopteros Calcidoideos. Serie de La Academia Nacional de Agronomia y Veterinaria, number 13, 154p

Hassan SA (1994) Strategies to select *Trichogramma* species for use in biological control, pp 55–71. In: Wajnberg E, Hassan SA (eds) Biological control with egg parasitoids. CAB International Wallingford, 304p

Juliano SA (1981) *Trichogramma* sp. (Hymenoptera: Trichogrammatidae) as egg parasitoids of *Sepedon fuscipennis* (Diptera: Sciomyzidae) and other aquatic Diptera. Can Entomol 113:271–279

Monje JC (1995) Present significance of *Trichogramma* spp. (Hymenoptera: Trichogrammatidae) for the control of sugarcane borers in Americas. Mitteilungen der Deutschen Gesellschaft für allgemeine und angewandte Entomologie 27:287–290

Noyes JS (2003) Universal Chalcidoidea Database. Avaliable em http://www.nhm.ac.uk/research. Accessed on 18 May 2009

Pinto JD (1999) Systematics of the North American species of *Trichogramma* Westwood (Hymenoptera: Trichogrammatidae). Mem Entomol Soc Washington 22:287

Pinto JD (2006) A review of the New World genera of Trichogrammatidae (Hymenoptera). J Hymenopt Res 15:38–163

Pinto JD, Stouthamer R (1994) Systematics of the Trichogrammatidae with emphasis on *Trichogramma*, pp 1–36. In: Wajnberg E, Hassan SA (eds) Biological control with egg parasitoids. CAB International, IOBC, Wallingford, 304p

Querino RB, Zucchi RA (2003a) Six new species of *Trichogramma* Westwood (Hymenoptera: Trichogrammatidae) from a Brasilian forest reserve. Zootaxa 134:1–11

Querino RB, Zucchi RA (2003b) New species of *Trichogramma* Westwood (Hymenoptera: Trichogrammatidae) associated with lepidopterous eggs in Brasil. Zootaxa 163:1–10

Querino RB, Zucchi RA (2003c) Caracterização morfológica de dez espécies de *Trichogramma* (Hymenoptera: Trichogrammatidae) registradas na América do Sul. Neotropic Entomol 32:597–613

Querino RB, Zucchi RA (2007a) Do *T. minutum* Riley and *T. bennetti* Nagaraja and Nagarkatti (Hymenoptera: T.tidae) Occur in Brasil? Neotropic Entomol 36:145–146

Querino RB, Zucchi RA (2007b). Species of *Trichogramma* – Collection of ESALQ-USP http://www.lef.esalq.usp.br/tricho/. Accessed on 30 April 2008

Silva AGd'A, Gonçalves CR, Galvão DM, Gonçalves AJL, Gomes J, Silva MN, Simoni L (1968) Quarto catálogo dos insetos que vivem nas plantas do Brasil. Seus parasitos e predadores. Lab. Central Patologia Vegetal, MA ed., 1°.Tomo, Parte II

Thompson WR (1958) A catalogue of the parasites and predators of insect pests. Section 2. Host parasite catalogue, Part 5. Commonwealth Agricultural Bureaux, Commonwealth Institute of Biological Control, Ottawa, Ontario, Canada, 669p

Velásquez de Rios M, Terán J (2003) Los *Trichogramma* (Hymenoptera: Trichogrammatidae) de La région noroccidental del estado Guárico, Venezuela. Entomotropica 18:127–145

Whu M, Valdivieso L (1999) Distribución y comportamiento de ocho especies de *Trichogramma* y Trichogrammatoidea (Hymenoptera: Trichogrammatidae) en el Perú. Rev Peru Entomol 41:61–68

Zucchi RA, Monteiro RC (1997) O gênero *Trichogramma* na América do Sul, pp 41–66. In: Parra JRP, Zucchi RA (eds) *Trichogramma* e o controle biológico aplicado, FEALQ, Piracicaba, 324p
Zucchi RA, Pinto JD, Monteiro RC (1996) Some records on the *Trichogramma* species associated with *Diatraea* in the New World (Hym., Trichogrammatidae – Lep., Pyralidae). In: International Congress of Entomology, 20, Firenze, Resumos. Firenze: s.ed., p 639

Chapter 9
Species Diversity and Host Associations of *Trichogramma* in Eurasia

Andrew Polaszek

Contents

9.1 Introduction

The first *Trichogramma* species described was *T. evanescens* Westwood, from Chelsea Physic Garden in London, England, 177 years ago (Westwood 1833). Unfortunately, the true identity of *T. evanescens* has long remained a mystery, leading to taxonomic confusion surrounding this and many other important species in the genus (Fursov 2000, Noyes et al. 2000). To a large extent, this taxonomic confusion is characteristic of much of the taxonomy of Old World *Trichogramma* species. The daunting task of using morphological studies for species characterisation of *Trichogramma* species was undertaken by several workers in the 20th century, notably in the former Soviet Union (Sorokina 1991, 1993) and India (Nagarkatti and Nagaraja 1971, 1977). The crucial importance of many *Trichogramma* species for biological and natural pest control was the incentive for these painstaking studies, in many cases supplemented by complex crossing experiments (Nagarkatti and Nagaraja 1977). The reliance on male genitalia as an extremely useful character (Nagarkatti and Nagaraja 1971) still meant that females were often impossible to identify. The relatively recent discovery of the utility for *Trichogramma* species identification of the ribosomal Internal Transcribed Spacer ITS-2 region (Stouthamer et al. 1998) is another significant advance in the taxonomy of this complex genus. The increasing use of ITS-2 to characterize known and new species

A. Polaszek (✉)
Natural History Museum, London, UK
e-mail: ap@nhm.ac.uk

F.L. Consoli et al. (eds.), *Egg Parasitoids in Agroecosystems with Emphasis on Trichogramma*, Progress in Biological Control 9, DOI 10.1007/978-1-4020-9110-0_9,

is both revolutionising *Trichogramma* systematics and has the potential to bring hitherto unknown and much-needed taxonomic stability to this important genus. The following account summarises current knowledge of the species diversity and host associations of *Trichogramma* species in Europe and Asia. Apart from certain species known to be widespread across Eurasia, particularly those occurring in both European and Asian Russia, the Palaearctic and Oriental *Trichogramma* faunas have tended to be treated separately. It seems probable that some hitherto undiscovered synonymy between species in these respective faunas will come to light when further molecular studies are completed.

Distribution and host records in the following account are drawn from two main sources: John Noyes' Universal Chalcidoidea Database ("UCD" – http://www.nhm.ac.uk/jdsml/research-curation/research/projects/chalcidoids/index.dsml") and the recently published account of European *Trichogramma* species by Pintureau (2008). Lin's (1994) account of Chinese Trichogrammatidae and Pinto's (1999) detailed treatment of the North American species have been the other main sources of information for this review. Because authoritative checking of these data is in many cases impossible, especially for records published pre-1970, a large proportion of the records drawn from the UCD are inevitably unverified repeats of earlier misidentifications. Where Pinto (1999, and personal communication) and/or Pintureau (2008) shed light on some of these records, they have been modified or commented upon accordingly.

9.2 Problematic Species

The literature on *Trichogramma*, especially records published pre-1970, contains a very large proportion of misidentifications. It was not until Nagarkatti and Nagaraja's pioneering work using male genitalia for species identifications during the 1970's that published records started to become more reliable. Pinto (personal communication) considers the following species *nomina dubia*, and thus best left treated as such, without perpetuating unreliable data associated with them: *T. flava* (=*flavum*) Ashmead, *T. carina* (Walker) and *T. erosicornis* (Westwood). Noyes (UCD) treats these species as valid. European and Asian records of *T. fasciatum*, *T. minutum*, and *T. pretiosum* are doubtful, as are North American records of *achaeae*, *evanescens* and *semifumatum*. With regard to *T. semifumatum*, Pinto *et al* (1978) and Pinto (personal communication) consider that all records except the original description from Hawaii represent misidentifications – and that how the name became applied to some common species is one of entomology's unsolved mysteries. *T. californicum* Nagaraja and Nagarkatti is known from Mexico and U.S.A. (Pinto 1999) but is recorded in error by Noyes (UCD) as occuring in India. *T. kaykai* is another North American species recorded in the UCD as occuring in Japan. *T. australicum* (Girault) is another name with a problematic history; this name was long applied in error to *T. chilonis*, and thus the majority of published records referring to *T. australicum* actually concern *T. chilonis*. According to Pinto et al. (1982), *T. australicum* is almost certainly restrcited to Australia. Records in the UCD of *T. stampae* Vincent from outside the Americas should also be treated as doubtful.

9.3 Taxonomy and Species Accounts

The following account summarises the taxonomic, distributional and host data for all the known Eurasian species of *Trichogramma*. Species are treated in alphabetical order, irrespective of their taxonomic affinity to any particular "section" or "species group" according to the taxonomic arrangements of Pinto (1999) or Pintureau (2008). As stated above, the account follows Noyes' online account, with modifications from Pinto (1999 and personal communication) or Pintureau (2008) commented on below where appropriate.

Trichogramma acantholydae Pintureau and Kenis

Identification: Pintureau et al. (2000), Pintureau (2008).
Distribution: **Europe**: Italy.
Hosts: **HYMENOPTERA**: **Pamphiliidae**: *Acantholyda erythrocephala*; *A. posticallis*.

Trichogramma achaeae Nagaraja and Nagarkatti

Synonymy: *Trichogramma achaea* Nagaraja and Nagarkatti.
Identification: Nagaraja and Nagarkatti (1970), Lin (1994).
Distribution: **Asia**: China, India, Russia; **Europe**: France (introduced); Portugal (Azores); Russia; Spain (introduced to mainland – see Cabello et al. 2009); **Africa**: Cape Verde; **Americas**: Argentina, Barbados, Chile, Trinidad and Tobago.
Hosts: **DIPTERA**: **Anthomyiidae**: *Atherigona soccata*; **LEPIDOPTERA**: **Gelechiidae**: *Pectinophora gossypiella*; *Sitotroga cerealella*; *Tuta absoluta*; **Geometridae**: *Boarmia variegata*; **Noctuidae**: *Achaea janata*; *Anticarsia gemmatalis*; *Chrysodeixis chalcites*; *Earias* sp.; *E. insulana*; *E. vittella*; *Helicoverpa armigera*; *H. zea*; *Mamestra brassicae*; *Spodoptera* sp.; *Tiracola plagiata*; *Trichoplusia ni*; **Notodontidae**: *Clostera cupreata*; **Oecophoridae**: *Opisina arenosella*; **Pieridae**: *Eurema* sp.; **Pyralidae** (incl. **Crambidae**): *Chilo partellus*; *Corcyra cephalonica*; *Loxostege sticticalis*; **Sphingidae**: *Acherontia styx*; *Agrius convolvuli*; **Tortricidae**: *Cydia koenigana*; **Yponomeutidae**: *Plutella xylostella*.

Trichogramma adashkevitshi Sorokina

Identification: Sorokina (1984).
Distribution: Uzbekistan.
Hosts: **LEPIDOPTERA**: **Noctuidae**: Undetermined species.

Trichogramma agriae Nagaraja

Identification: Nagaraja (1973), Lin (1994).
Distribution: **Asia**: China; India; **Americas**: Trinidad and Tobago.
Hosts: **LEPIDOPTERA**: **Pyralidae**: *Corcyra cephalonica*; **Sphingidae**: *Agrius convolvuli*.

Trichogramma agrotidis Voegelé and Pintureau

Identification: Pintureau (2008).
Distribution: **Asia**: Russia; **Europe**: Bulgaria; France; Russia; Switzerland; **Americas**: Argentina (doubtful record).
Hosts: **LEPIDOPTERA**: **Lycaenidae**: *Cacyreus marshalli*; **Noctuidae**: *Agrotis* sp.; **Nymphalidae**: unspecified sp.; **Pyralidae**: *Anagasta* (=*Ephestia*) *kuehniella*; **Tortricidae**: *Eupoecilia ambiguella*; *Lobesia botrana*.

Trichogramma aldanense Sorokina

Identification: Sorokina (1989).
Distribution: **Asia**: Russia.
Hosts: **HYMENOPTERA**: **Tenthredinidae**: *Nematus* sp.

Trichogramma aomoriense Honda

Identification: Honda et al. (2006).
Distribution: **Asia**: Japan.
Hosts: **LEPIDOPTERA**: **Lycaenidae**: *Neozephyrus taxila*.

Trichogramma artonae Chen and Pang

Identification: Chen and Pang (1986).
Distribution: **Asia**: China.
Hosts: **LEPIDOPTERA**: **Zygaenidae**: *Artona funeralis*.

Trichogramma aurosum Sugonjaev and Sorokina

Identification: Pinto,(1999) Pintureau (2008), Sugonjaev and Sorokina (1976).
Distribution: **Europe**: Austria, Bulgaria, Denmark, France, Germany, Luxemburg, Moldova, Netherlands, Russia; **Asia**: Russia; **Americas**: Canada; U.S.A.
Hosts: **HYMENOPTERA**: **Cimbicidae**: *Cimbex femorata*; **Tenthredinidae**: *Ardis bipunctata*; *Caliroa cerasi*; *Nematus melanaspis*; *N. tibialis*; *Pristiphora geniculata*; *Profenusa thomsoni*; **LEPIDOPTERA**: **Noctuidae**: *Acronicta* sp.; *Trichoplusia ni*; **Notodontidae**: *Clostera curtula*; **Sphingidae**: undetermined species; **Tortricidae**: *Cydia inopinata*; *C. pomonella*.

Trichogramma bactrianum Sugonjaev and Sorokina

Synonymy: *Trichogramma bactriana*.
Identification: Sugonjaev and Sorokina (1976).
Distribution: *Asia*: Tadzhikistan.
Hosts: **LEPIDOPTERA**: **Noctuidae**: *Helicoverpa armigera*.

Trichogramma bilingense He and Pang

Identification: He and Pang (2000).
Distribution: *Asia*: China.

Hosts: **LEPIDOPTERA**: **Noctuidae**: *Spodoptera litura*; **Yponomeutidae**: *Plutella xylostella*.

Trichogramma bistrae Kostadinov

Synonymy: *Nuniella bistrae* Kostadinov.
Identification: Pintureau (2008).
Distribution: **Europe**: Bulgaria, France.
Hosts: None known.

Trichogramma bourarachae Pintureau and Babault

Identification: Pintureau (2008).
Distribution: **Europe**: Portugal; **Africa**: Morocco, Tunisia.
Hosts: **LEPIDOPTERA**: **Lymantriidae**: *Euproctis chrysorrhoea*; **Noctuidae**: *Helicoverpa armigera*; **Nymphalidae**: *Vanessa cardui*; **Pyralidae**: *Ectomyelois ceratoniae*; *Anagasta (=Ephestia) kuehniella*; *Palpita unionalis*; **Yponomeutidae**: *Prays oleae*; **NEUROPTERA**: **Chrysopidae**: *Chrysoperla carnea*.

Trichogramma brassicae Bezdenko

Synonymy: *Trichogramma maidis* Pintureau and Voegelé.
Identification: Pinto (1999); Pintureau (2008).
Distribution: **Asia**: China, Iran, Japan, Russia, Russia, Turkey; **Europe**: Austria, Belarus, Belgium, Bulgaria, France, Germany, Greece, Hungary, Italy, Moldova, Netherlands, Romania, Russia, Spain, Switzerland, Turkey, Ukraine, Yugoslavia (former, pre 1991). **Americas**: Argentina, Canada, U.S.A.; **Australasia**: Australia.
Hosts: **DIPTERA**: **Tachinidae**: *Lydella thompsoni*; **LEPIDOPTERA**: **Gelechiidae**: *Keiferia lycopersicella*; *Pectinophora gossypiella*; *Scrobipalpa ocellata*; *Sitotroga cerealella*; **Hesperiidae:** *Hesperia comma*; **Lycaenidae:** *Cyaniris semiargus*; *Plebejus idas*; *Polyommatus icarus*; **Lymantriidae:** *Laelia salicis*; **Noctuidae:** *Actebia fennica*; *Agrotis exclamationis*; *A. ipsilon*; *A. segetum*; *Archanara geminipunctata*; *Autographa gamma*; *Chrysodeixis chalcites*; *Helicoverpa armigera*; *H. zea*; *Mamestra brassicae*; *M. trifolii*; *Scotia ipsilon*; *Sesamia cretica*; *Sesamia nonagrioides*; *Trichoplusia ni*; **Notodontidae**: *Traumatocampa ispartaensis*; **Nymphalidae:** *Aphantopus hyperanthus*; *Argynnis adippe*; *A. niobe*; *Clossiana selene*; *C. titania*; *Coenonympha pamphilus*; *Erebia ligea*; *Hipparchia alcyone*; *Maniola jurtina*; *Melanargia galathea*; *Mellicta athalia*; *Vanessa atalanta*; **Oecophoridae:** *Maroga melanostigma*; *Papilionidae*: *Papilio demoleus*; *P. machaon*; **Pieridae**: *Pieris brassicae*; *P. napi*; *P. rapae*; **Pyralidae** (incl. **Crambidae**): *Anagasta kuehniella*; *Chilo phragmatellus*; *C. suppressalis*; *Ectomyelois ceratoniae*; *Ephestia cautella*; *Evergestis forficalis*; *Galleria mellonella*; *Ostrinia nubilale*; *Palpita unionalis*; *Plodia interpunctella*; *Spectrobates ceratoniae*; **Saturniidae:** *Samia cynthia*; Sphingidae: *Deilephila elpenor*; *Sphinx ligustri*; **Tortricidae:** *Cydia pomonella*; *Eupoecilia ambiguella*; *Lobesia botrana*; **Yponomeutidae:** *Plutella xylostella*; *Prays oleae*; **Zygaenidae:** *Zygaena filipendulae*.

Trichogramma brevifringiata Yousuf and Shafee

Identification: Yousuf and Shafee (1988).
Distribution: **Asia**: India.
Hosts: **LEPIDOPTERA**: **Crambidae**: *Chilo infuscatellus.*

Trichogramma buesi Voegelé

Synonymy: *Trichogramma brassicae* Voegelé.
Identification: Pintureau (2008).
Distribution: **Asia**: Turkmenistan, Uzbekistan; **Europe**: Bulgaria, France, Greece, Moldova; Romania, Switzerland; **Africa**: Egypt; **Americas**: Argentina, Canada.
Hosts: **COLEOPTERA**: **Cassidae**: *Cassida nebulosa*; *C. nobilis*; **LEPIDOPTERA**: **Noctuidae**: *Agrotis exclamationis; A. ipsilon; A. segetum; Autographa gamma; Busseola fusca; Chrysodeixis chalcites; Earias sp.; E. insulana; E. vittella; Eublemma amabilis; Helicoverpa armigera; Mamestra brassicae; M. configurata; M. trifolii; Spodoptera littoralis; Trichoplusia ni*; **Pieridae**: *Pieris rapae*; **Pyralidae** (incl. **Crambidae**): *Anagasta* (= *Ephestia*) *kuehniella*; **Yponomeutidae**: *Plutella xylostella.*

Trichogramma buluti Bulut and Kilinçer

Identification: Bulut and Kilinçer (1991).
Distribution: **Asia**: Turkey.
Hosts: **LEPIDOPTERA**: **Lasiocampidae**: *Malacosoma neustria.*

Trichogramma cacoeciae Marchal

Synonymy: *Trichogramma flavum* Marchal; *Trichogramma cacaeciae* Marchal; *Trichogramma telengai* Sorokina. N.b. UCD gives *T. telengai* as a synonym of *T. bezdencovii* – the latter species treated here as a junior synonym of *T. embryophagum*, following Lin (1994) and Pintureau (2008).
Hosts: **LEPIDOPTERA**: **Gelechiidae**: *Sitotroga cerealella*; **Noctuidae**: *Acronicta rumicis*; **Tortricidae**: *Cydia pomonella*; *Rhyacionia buoliana*; *Tortrix viridana.*
Identification: Lin (1994), Pinto (1999), Pintureau (2008).
Distribution: **Asia**: Armenia, China, Iran, Kazakhstan, Kirgizia, Russia, Syria, Turkey, Uzbekistan; **Europe**: Austria, Belarus, Bulgaria, Czech Republic Denmark, Estonia, France, Germany, Greece, Italy, Latvia, Lithuania, Moldova, Netherlands, Poland, Portugal, Russia, Slovakia, Switzerland, Ukraine, United Kingdom; Yugoslavia (former); **Africa**: Egypt, Morocco; Tunisia; **Americas**: Argentina, Chile, Cuba, Peru, U.S.A.
Hosts: **DIPTERA**: **Tachinidae**: *Strobilomyia anthracina*; **HEMIPTERA**: **Cimicidae**: *Cimex lectularius*; **HYMENOPTERA**: **Tenthredinidae**: *Caliroa cerasi*; **LEPIDOPTERA**: **Arctiidae**: *Arctia caja; Parasemia plantaginis; Hyphantria cunea*; **Gelechiidae**: *Sitotroga cerealella*; **Geometridae**: *Bupalus piniarius*; **Lasiocampidae**: *Dendrolimus pini; Malacosoma neustria*; **Lymantriidae**: *Euproctis phaeorrhoea; Leucoma salicis; Nygmia phaeorrhoea;*

Orgyia antiqua; Stilpnotia salicis; **Noctuidae**: *Acronicta aceris; A. rumicis; Agrochola circellaris; Agrotis exclamationis; A. segetum; Autographa gamma; Emmelia trabealis; Helicoverpa armigera; Mamestra brassicae; Noctua pronuba; Panolis flammea*; **Pieridae**: *Pieris brassicae*; **Pyralidae** (incl. **Crambidae**): *Anagasta* (= *Ephestia*) *kuehniella*; *Chilo sacchariphagus*; *Ectomyelois ceratoniae; Ephestia sp.; Ephestia elutella; Galleria mellonella; Loxostege sticticalis; Ostrinia nubilale; Plodia interpunctella; Spectrobates ceratoniae*; **Tortricidae**: *Adoxophyes orana; Archips crataeganus; A. murinana; A. rosanus; A. xylosteana; Argyrotaenia velutinana; Cacoecia rosana; Cacoecimorpha pronuba; Choristoneura murinana; Cydia funebrana; C. molesta; C. pomonella; C. strobilella; Enarmonia formosana; Epichoristodes acerbella; Eupoecilia ambiguella; Hedya nubiferana; Lobesia botrana; Pandemis heparana; P. ribeana; Rhyacionia buoliana; Spilonota ocellana; Tetramoera schistaceana; Tortrix viridana; Zeiraphera sp. Z. canadensis*; **Yponomeutidae**: *Acrolepiopsis assectella; Plutella xylostella; Prays oleae.*

Trichogramma carina Walker

Synonymy: *Calleptiles carina* (Walker).
Identification: Pintureau (2008). Regarded by Pinto (personal communication) as a *nomen dubium*, but listed as valid in the UCD.
Distribution: **Europe**: France.

Trichogramma cephalciae Hochmut and Martinek

Identification: Pintureau (2008).
Distribution: **Europe**: Czech Republic, Denmark, Italy, Moldova, Norway, Poland, Russia.
Hosts: **HYMENOPTERA**: **Pamphiliidae**: *Acantholyda erythrocephala; A. posticallis; Cephalcia abietis; C. arvensis; C. erythrogaster;* **LEPIDOPTERA**: **Pyralidae**: *Anagasta* (= *Ephestia*) *kuehniella.*

Trichogramma chilonis Ishii

Synonymy: *Trichogramma chelonis* Ishii; *Trichogramma chilones* Ishii; *Trichogramma confusum* Viggiani.
Identification: *T. chilonis* was initially misidentified as *T. australicum* Girault (Nagarkatti and Nagaraja 1971) following various previous authors' understanding of this species. The correct use of the name *T. chilonis* was established eight years later (Nagarkatti and Nagaraja 1979) after examination of the original type material. The species name has also been misspelled as *T. chinolis*, *T. chilones* and *T. chelonis*. *T. confusum* (Viggiani 1976) was correctly synonymised with *T. chilonis* (Nagarkatti and Nagaraja 1979), although the name was still widely incorrectly applied until relatively recently in many publications. *T. chilonis* exists as *Wolbachia*-induced thelytokous populations in parts of its distribution, as far as is known (Stouthamer et al. 1990a). For example, the introduced Hawaiian population exists as both thelytokous and arrhenotokous forms (Stouthamer et al. 1990b). Since taxonomy and identification relied until recently on the presence of males, it has been necessary

in some cases to "cure" the population of the *Wolbachia* in order to produce males (Stouthamer et al. 1990a) that can then be identified.

Trichogramma chilonis can be identified using the key to Indian *Trichogramma* by Nagaraja (1973, as *australicum*) in combination with the redescription of *T. chilonis* (Nagarkatti and Nagaraja 1979). Yousuf and Shafee's (1988) treatment of Indian *Trichogramma* unfortunately perpetuates Nagaraja's (1973) misidentification of *chilonis* as *australicum*, while also adding *T. chilonis* to the Indian fauna. Apparently these authors were not aware of Nagarkatti and Nagaraja's 1979 paper.

Distribution: **Asia**: Bangladesh, China, India, Indonesia, Japan, Malaysia, Nepal, Pakistan, Philippines, Taiwan, Thailand, Vietnam. **Europe**: Not present (one laboratory record from Germany). **Africa**: South **Africa** (introduced – establishment unconfirmed; Kfir 1994); **Americas**: Bahamas, Granada, Montserrat (introductions – establishment unconfirmed; Cock 1985). **Oceania**: Hawaii (introduced, established), Guam (introduced, established), Solomon Islands.

Trichogramma chilonis is an Asian species, its apparent area of endemism being south and southeast Asia. It has been widely moved around and released within its natural distribution area, and has also been introduced to South Africa, the Caribbean, Hawaii and Guam. It is established in the latter two islands, but its establishment in the other areas of its introduction require confirmation.

Hosts: **DIPTERA**: **Anthomyiidae**: *Atherigona soccata*; **Ephydridae**: *Hydrellia philippina*; **LEPIDOPTERA**: **Arctiidae**: *Amsacta moorei*; *Spilarctia obliqua*; *Spilosoma obliqua*; **Blastobasidae**: *Pseudohypatopa pulverea*; **Cossidae**: *Phragmataecia gummata*; **Danaidae**: *Anosia chrysippus*; **Eupterotidae**: *Apha aequalis*; **Gelechiidae**: *Pectinophora gossypiella*; *Phthorimaea operculella; Sitotroga cerealella*; **Geometridae**: *Boarmia variegata*; **Hesperiidae**: *Parnara guttata; Pelopidas mathias*; **Hyblaeidae**: *Hyblaea puera*; **Lasiocampidae**: *Dendrolimus punctatus*; *Gastropacha* sp.; **Lycaenidae**: *Deudorix epijarbas*; *Virachola isocrates; V. livia*; **Lymantriidae**: *Euproctis similis; Ivela auripes*; **Noctuidae**: *Achaea janata; Anomis flava; Asota ficus; Autographa nigrisigna; Busseola fusca; Chrysodeixis chalcites; Earias sp.; E. insulana; E. vittella; Eublemma amabilis; Helicoverpa armigera; H. assulta; Heliothis sp.; Heliothis virescens; Heliothis zea; Mamestra brassicae; Naranga aenescens; Othreis sp.; Plusia nigrisigna; Plusia orichalcea; Rivula atimeta; Sesamia inferens; Spodoptera sp.; Spodoptera exigua; S. litura; Tiracola plagiata; Trichoplusia ni*; **Notodontidae**: *Clostera cupreata*; **Nymphalidae**: *Hipolimnas anomala; H. anomala; H. bolina; Melanitis leda; Oecophoridae: Opisina arenosella; Papilionidae: Papilio demoleus; P. machaon; P. polytes; P. protenor; P. xuthus*; **Pieridae**: *Catopsilia pyranthe; Eurema sp.; Pieris brassicae; P. rapae*; **Pyralidae** (incl. **Crambidae**): *Anagasta* (= *Ephestia*) *kuehniella*; *Chilo sp.; C. auricilius; C. indicus; C. infuscatellus; C. partellus; C. sacchariphagus; C. simplex; C. suppressalis; C. venosatus; Chilotraea infuscatella; Cnaphalocrocis medinalis; Corcyra sp.; Corcyra cephalonica; Diaphania indica; Diatraea sp.; Diatraea saccharalis; Emmalocera depressella; Eoreuma loftini; Ephestia cautella; Hedylepta indicata; Hymenia recurvalis; Lamprosema indicata; Leucinodes* *orbonalis; Marasmia exigua; M. patnalis; Noorda albizonalis; Nymphula depunctalis; Ostrinia furnacalis; O. nubilale; Psara sp.; Raphimetopus ablutellus; Scirpophaga sp.; Scirpophaga excerptalis; S. incertulas; S. intacta; S. nivella*; **Saturniidae**: *Antheraea pernyi; A. yamamai; Dictyoploca japonica; Philosamia cynthia; P. ricini; Samia cynthia*; **Sphingidae**: *Acherontia styx; Agrius convolvuli; Cephonodes hylas; Clanis bilineata; Hippotion celerio; Theretra silhetensis*; **Tortricidae**: *Adoxophyes orana; Argyroploce schistaceana; Bactra sp.; Bactra truculenta; Cophoprora sp.; Coprophora sp.; Cydia pomonella; Eucosma schistaceana; Homona coffearia; Laspeyresia koenigana; Leguminivora glycinivorella; Pandemis heparana; Peronea crocepepla; Tetramoera schistaceana*; **Yponomeutidae**: *Plutella xylostella*; **NEUROPTERA**: **Chrysopidae**: *Chrysopa* sp.

Trichogramma chilotraeae Nagaraja and Nagarkatti

Identification: Nagaraja and Nagarkatti (1970).
Distribution: **Asia**: India, Indonesia, Malaysia, Philippines, Russia, Thailand; **Americas**: Antilles, Bahamas, Brazil, Colombia, Peru, Trinidad and Tobago, Venezuela.
Hosts: **LEPIDOPTERA**: **Cossidae**: *Phragmataecia gummata*; **Lycaenidae**: *Virachola isocrates*; **Noctuidae**: *Helicoverpa armigera; H. peltigera; Mamestra brassicae; Naranga aenescens; Rivula atimeta; Sesamia inferens*; **Pyralidae** (incl. **Crambidae**): *Anagasta* (= *Ephestia*) *kuehniella*; *Chilo* sp.; *C. partellus; C. sacchariphagus; C. suppressalis; C. tumidicostalis; Chilotraea infuscatella; Cnaphalocrocis medinalis; Corcyra* sp.; *C. cephalonica; Diatraea* sp.; *Loxostege sticticalis; Marasmia patnalis; Noorda albizonalis; Ostrinia furnacalis; Scirpophaga incertulas*; **Tortricidae**: *Tetramoera schistaceana*.

Trichogramma choui Chan and Chou

Identification: Chan and Chou (2000).
Distribution: **Asia**: Taiwan.
Hosts: **LEPIDOPTERA**: **Nymphalidae**: *Ariadne ariadne*.

Trichogramma chusniddini Sorokina and Atamirzaeva

Identification: Sorokina and Atamirzaeva (1993).
Distribution: **Asia**: Uzbekistan.
Hosts: None known.

Trichogramma closterae Pang and Chen

Identification: Lin (1994).
Distribution: **Asia**: China.
Hosts: **LEPIDOPTERA**: **Lasiocampidae**: *Gastropacha quercifolia*; **Limacodidae**: *Cnidocampa flavescens*; **Lymantriidae**: *Euproctis similis; Leucoma candida; Stilpnotia salicis*; **Noctuidae**: *Helicoverpa armigera*; **Notodontidae**:

Cerura erminea; Clostera anachoreta; Clostera anastomosis; Fentonia ocypete; Micromelalopha troglodyta; **Pyralidae**: *Corcyra cephalonica*; **Sphingidae**: *Clanis bilineata; Parum colligata; Smerinthus caecus; S. planus.*

Trichogramma cordubense Vargas and Cabello

Synonymy: *Trichogramma cordubensis* Vargas and Cabello.
Identification: Pintureau (2008).
Distribution: **Asia**: Iran; **Europe**: Portugal; Spain; **Africa**: Algeria, Egypt, Morocco.
Hosts: **LEPIDOPTERA**: **Lymantriidae**: *Euproctis chrysorrhea*; **Noctuidae**: *Autographa gamma; Chrysodeixis chalcites; Earias insulana; Helicoverpa armigera; Mamestra brassicae; Peridroma saucia*; **Papilionidae**: *Iphiclides podalirius*; **Pyralidae** (incl. **Crambidae**): *Anagasta* (= *Ephestia*) *kuehniella*; *Chilo suppressalis; Ectomyelotis ceratoniae; Ostrinia nubilale; Palpita unionalis; Udea ferrugalis*; **Sphingidae**: *Acherontia atropos*; **Yponomeutidae**: *Prays oleae.* **NEUROPTERA**: **Chrysopidae**: *Chrysopa carnea.*

Trichogramma cultellus Jose, Hirose and Honda

Identification: Jose, Hirose and Honda (2005).
Distribution: **Asia**: Japan.
Hosts: **LEPIDOPTERA**: **Pyralidae**: *Anagasta* (= *Ephestia*) *kuehniella.*

Trichogramma danubiense Birova and Kazimirova

Identification: Birova and Kazimirova (1997), Pintureau (2008).
Distribution: **Europe**: Slovakia; United Kingdom.
Hosts: **LEPIDOPTERA**: **Lasiocampidae**: *Macrothylacia rubi*; **Noctuidae**: *Mamestra brassicae*; **Pyralidae**: *Anagasta* (= *Ephestia*) *kuehniella*;.

Trichogramma daumalae Dugast and Voegelé

Synonymy: *Trichogramma domalae* Dugast and Voegelé.
Identification: Pintureau (2008).
Distribution: **Europe**: Bulgaria; France; United Kingdom.
Hosts: **LEPIDOPTERA**: **Gelechiidae**: *Sitotroga cerealella*; **Noctuidae**: *Mamestra brassicae*; **Pyralidae** (incl. **Crambidae**): *Anagasta* (= *Ephestia*) *kuehniella*; *Galleria mellonella; Ostrinia nubilale*; **Saturniidae**: *Samia cynthia*; **Tortricidae**: *Cydia pomonella; Lobesia botrana.*

Trichogramma dendrolimi Matsumura

Synonymy: *Trichogramma cacoeciae pallidum* Meyer; *Trichogramma cacoeciae pallida* Meyer; *Trichogramma dendrolimi* Matsumura; *Trichogramma dendrolimusi* Matsumura; *Trichogramma dendrolimi liliyingae* Voegelé and Pintureau; *Trichogramma pallida* Meyer.

Identification: In Europe, *T. dendrolimi* can be identified using the keys to European (Pintureau 2008) or Italian (Viggiani and Laudonia 1989) species, as well as notes by Sorokina (1991). In **Asia**, the key (in Chinese) and illustrated redescription by Lin (1994) can be used. Esterase patterns have also been analysed in this species using isozyme electrophoresis (Cao et al. 1988, Pintureau 1990, Pintureau and Keita 1989) as has RAPD-DNA analysis (Landry et al. 1993). More recently, species-specific primers for *T. dendrolimi* have been developed (Li and Shen 2001) using the ITS-2 region of ribosomal DNA, which should greatly facilitate the identification of *T. dendrolimi*. The chromosome number of *T. dendrolimi* is given by Liu et al. (1998).

Distribution: **Asia**: China, India, Iran, Japan, Kazakhstan, Korea (South), Pakistan, Russia, Taiwan, Turkey, Vietnam. **Europe**: Austria, Belarus, Bulgaria, France, Germany, Greece, Hungary, Italy, Latvia, Lithuania, Moldova, Netherlands, Poland, Romania, Russia, Turkey, Ukraine. *Trichogramma dendrolimi* is a Eurasian species, but like many *Trichogramma* species it has been widely moved around and released within its natural distribution area, and has also been introduced into Chile (Cerda and Gerding 1999, Gerding et al. 1996) though not, apparently, released and established. Unpublished records of its occurrence in, or introduction into, New Zealand and the Cook Islands require confirmation.

Hosts: **COLEOPTERA**: **Attelabidae**: *Deporaus betulae*. **HYMENOPTERA**: ***Pamphiliidae***: *Acantholyda posticalis*; **Tenthredinidae**: *Caliroa cerasi; Pristiphora pallipes*. **LEPIDOPTERA**: **Arctiidae**: *Hyphantria cunea; Spilarctia subcarnea*; **Attacidae**: *Dictyoploca japonica*; **Gelechiidae**: *Sitotroga cerealella*; **Geometridae**: *Buzura suppressaria; Phthonandria atrilineata*; **Hesperiidae**: *Parnara guttata*; **Hyblaeidae**: *Hyblaea puera*; **Hypsidae**: *Asota ficus*; **Lasiocampidae**: *Dendrolimus* sp.; *D. albolineatus; D. latipennis; D. pini; D.punctatus; D. sibiricus; D. spectabilis; D. superans; Gastropacha quercifolia; Lebeda nobilis; Malacosoma neustria*; **Limacodidae**; *Cnidocampa flavescens; Microleon longipalpis; Monema flavescens; Phrixolepia sericea; Scopelodes contractus*; **Lycaenidae**: *Lampides boeticus*; **Lymantriidae**: *Dasychira abietis Euproctis chrysorrhoea; E. pseudoconspersa; E. similis; Lymantria dispar; Orgyia antiqua;* **Noctuidae**: *Acronicta major; A. tridens; Aedia leucomelas; Agrotis exclamationis; A. ipsilon; A. segetum; A. ypsilon; Anomis flava; Arcte coerula; Asota ficus; Autographa gamma; Barathra brassicae; Cocytodes coerulea; Discestra trifolii; Helicoverpa armigera; H.assulta; Heliothis virescens; Leucania unipuncta; Mamestra brassicae; M. oleracea; M. trifolii; Mythimna separata; Panolis flammea; Phytometra aganata; Plusiodonta coelonota; Sesamia inferens; Spodoptera litura*; **Notodontidae**: *Cerura menciana; Clostera anachoreta; C. anastomosis; C. curtula; Fentonia ocypete; Lampronadata cristata; Micromelalopha troglodyta; Phalera assimilis; P. flavescens; Pheosia tremula*; **Nymphalidae**: *Inachis io*; **Oecophoridae**: *Opisina arenosella*; **Papilionidae**: *Papilio demoleus; P. protenor; P. xuthus*; **Pieridae**: *Pieris brassicae; P. rapae*; **Pyralidae** (incl. **Crambidae**): *Algedonia coclesalis; Anagasta kuehniella; Chilo partellus; C. sacchariphagus; C. suppressalis; Chilotraea polychrysa; Cnaphalocrocis medinalis; Corcyra sp.; C. cephalonica; Ephestia cautella; Galleria mellonella; Glyphodes pyloalis; Hymenia recurvalis;*

Loxostege sticticalis; Notarcha derogata; Ostrinia sp.; O. furnacalis; O. nubilale; Sylepte derogata; **Saturniidae**: *Actias artemis; A. selene; Antheraea pernyi; Caligula japonica; Dictyoploca japonica; Eriogyna pyretorum; Philosamia cynthia; Philosamia ricini*; **Sphingidae**: *Clanis bilineata*; **Tortricidae**: *Adoxophyes* sp.; *A. orana; A. privatana; Archips* sp.; *A. podana; A. rosanus; Cacoecia crataeganus; C. rosana; Cacoecimorpha pronubana; Capua reticulana; Carpocapsa pomonella; Choristoneura murinana; Cydia funebrana; Cydia pomonella; Eana argentana; Epichoristodes acerbella; Eupoecilia ambiguella; Grapholitha funebrana; Grapholitha molesta; Hedaya nubiferana; Homona coffearia; H. magnanima; Laspeyresia pomonella; L. splendana; L. strobilella; Leguminivora glycinivorella; Lobesia botrana; Pandemis cerasana; P. heparana; Rhyacionia buoliana; Spilonota lechriaspis; S. ocellana; Tetramoera schistaceana; Zeiraphera diniana*; **Yponomeutidae**: *Acrolepiopsis assectella; Plutella* sp.; *P. xylostella; Prays oleae.*

Trichogramma elegantum Sorokina

Identification: Pintureau (2008).
Distribution: **Asia**: Turkmenistan. **Europe**: Ukraine.
Hosts: **LEPIDOPTERA**: **Gelechiidae**: *Sitotroga cerealella*; **Noctuidae**: *Helicoverpa armigera.*

Trichogramma embryophagum (Hartig)

Synonymy: *Encyrtus embryophagus* Hartig; *Ichneumon* (*Encyrtus*) *embryophagum* (Hartig); *Ooencyrtus embryophagus* (Hartig); *Trichogramma bezdencovii* Bezdenko; *Trichogramma bezdenkovii* Bezdenko. *N.b.* UCD lists *T. bezdencovii* as a valid species, citing *T. telengai* as a junior synonym. *T. telengai* is treated here as a synonym of *T. cacoeciae* (Pintureau 2008).
Identification: Pintureau (2008).
Distribution: **Asia**: Armenia; China; Georgia; India; Iran; Israel; Kazakhstan; Kirgizia; Russia; Syria, Taiwan; Turkey; Turkmenistan; Vietnam; **Europe**: Albania; Austria; Belarus; Bulgaria; Czech Republic; France; Germany; Greece; Hungary; Italy; Latvia; Moldova; Morocco; Netherlands; Norway; Poland; Portugal; Romania; Russia; Slovakia; Spain; Ukraine; Yugoslavia (former, pre 1991); **Africa**: Algeria, Morocco. Recorded frequently in error from North America, particularly in older literature, and hence some of these records have been perpetuated on the UCD.
Hosts: **HYMENOPTERA**: **Diprionidae**: *Diprion pini*; **Pamphiliidae**: *Acantholyda erythrocephala*; *A. nemoralis; A. posticalis; Cephalcia* sp.; *C. abietis*; **Tenthredinidae**: *Nematus melanaspis*; **LEPIDOPTERA**: **Gelechiidae**: *Pectinophora gossypiella; Sitotroga cerealella*; **Geometridae**: *Bupalus piniarius; Ellopia prospiaria; Eupithecia* sp.; *Gonodontis bidentata; Semiothisa liturata; Thera firmata; T. obeliscata*; **Hyblaeidae**: *Hyblaea puera*; **Lasiocampidae**: *Dendrolimus pini; Gastropacha quercifolia; Malacosoma neustria; Selenephera lunigera*; **Lymantriidae**: *Dasychira pudibunda; Euproctis similis; Leucoma salicis; Orgyia antiqua*; **Noctuidae**: *Acronicta tridens; Agrotis segetum; Cirrhia togata; Helicoverpa zea; Lithophane ornitopus; Mamestra brassicae; Nycteola* ***Asia****tica;*

Panolis flammea; Sarrothripus **Asia***tica; Sesamia cretica; S. nonagrioides*; **Notodontidae**: *Clostera anachoreta; C. anastomosis; Thaumetopoea pityocampa; T. wilkinsoni*; **Oecophoridae**: *Opisina arenosella*: **Pieridae**: *Pieris brassicae; P. rapae*; **Pyralidae** (incl. **Crambidae**): *Anagasta* (= *Ephestia*) *kuehniella; Corcyra cephalonica; Ectomyelois ceratoniae; Ephestia elutella; Ostrinia furnacalis; O. nubilale*; **Saturniidae**: *Antheraea pernyi*: **Sphingidae**: *Macroglossum stellatarum*: **Tortricidae**: *Adoxophyes orana; A. reticulana; Ancylis achatana; Archips rosanus; Argyroploce nubiferana; Choristoneura murinana; Cydia pomonella; Eupoecilia ambiguella; Grapholitha funebrana; Grapholitha molesta; Grapholitha prunifoliae; Hedya nubiferana; Lobesia botrana; Pandemis heparana; Pandemis ribeana; Ptycholoma lecheana; Retinia resinella; Rhyacionia buoliana; Spilonota ocellana; Tortrix viridana; Zeiraphera diniana; Z. griseana.*

Trichogramma erosicornis Westwood

Synonymy: *Aprobosca erosicornis* (Westwood); *Trichogramma erosicorne* Westwood. Considered to be a *nomen dubium* by Pinto (personal communiction).
Distribution: **Asia**: Sri Lanka.
Hosts: **LEPIDOPTERA**: **Gelechiidae**: *Sitotroga cerealella*; **Pyralidae**: *Corcyra cephalonica; Ephestia cautella; E. kuehniella*; **Tortricidae**: *Homona coffearia.*

Trichogramma euproctidis (Girault)

Synonymy: *Pentarthron euproctidis* Girault; *Trichogramma turkestanica* Meyer; *Trichogramma meyeri* Sorokina; *Trichogramma voegelei* Pintureau.
Identification: Pintureau (2008), Rohi and Pintureau (2003).
Distribution: **Asia**: Armenia, China, Japan, Kazakhstan, Russia, Tadzhikistan, Turkmenistan, Turkey, Uzbekistan, Vietnam; **Europe**: Belarus, Bulgaria, Denmark, France, Greece, Italy, Moldova, Portugal, Russia, Ukraine; **Africa**: Egypt; **Americas**: Argentina, Chile, Cuba, Peru. The UCD additionally records *T. euproctidis* from U.S.A. Pinto (personal communication) points out that the species does not occur there, but that Girault's type series contains a mixture of European and American species. Other New World records should therefore also be treated as doubtful.
Hosts: **DIPTERA**: **Anthomyiidae**: *Erioischia brassicae*; **LEPIDOPTERA**: **Gelechiidae**: *Sitotroga cerealella*; **Lymantriidae**: *Euproctis chrysorrhoea; Nygmia phaeorrhoea; Orgyia antiqua*; **Noctuidae**: *Agrotis segetum; Amathes c-nigrum; Helicoverpa armigera; Mamestra brassicae; Sesamia nonagrioides; Syngrapha circumflexa*; **Nymphalidae**: *Vanessa cardui*; **Pieridae**: *Pieris brassicae*; **Pyralidae** (incl. **Crambidae**): *Anagasta* (= *Ephestia*) *kuehniella; Chilo agamenon; Corcyra cephalonica; Loxostege sticticalis; Margaronia quadristigmalis; Mescinia peruella; Ostrinia nubilale*; **Sphingidae**: *Agrius convolvuli*; **Tortricidae**: *Archips rosanus; Cydia pomonella; Epichoristodes acerbella; Grapholitha molesta; Laspeyresia pomonella; Lobesia botrana*; **Yponomeutidae**: *Plutella maculipennis; Prays oleae.* **NEUROPTERA**: **Chrysopidae**: *Chrysoperla carnea.*

Trichogramma evanescens Westwood

Synonymy: *Calleptiles latipennis* Haliday; *Calleptiles vitripennis* (Walker); *Pentarthron carpocapsae* Schreiner; *Pentarthron carpocapsai* Schreiner; *Pteroptrix evanescens* (Westwood); *Trichogramma barathrae* Skriptshinsky; *Trichogramma cacoeciae pini* Meyer; *Trichogramma carpocapsae* (Schreiner); *Trichogramma evanescens piniperda* Wolff; *Trichogramma pini* Meyer

Trichogramma piniperdae Wolff; *Trichogramma rhenana* Voegelé and Russo; *Trichogramma rhenanum* Voegelé and Russo; *Trichogramma vitripennis* Walker; *Trichogramma vitripenne* Walker.

Identification: The true identity of *T. evanescens* is still uncertain, but for practical reasons, most workers follow Pintureau's interpretation of this highly cited species (see Pintureau 2008).

Distribution: **Asia**: Armenia, Azerbaijan, China, Georgia, India, Iran, Israel, Kazakhstan, Oman, Pakistan, Philippines, Russia, Sri Lanka, Turkey, Turkmenistan, Uzbekistan, Vietnam; **Europe**: Austria, Belarus, Belgium, Bulgaria, Czech Republic, Denmark, France, Germany, Hungary, Italy, Lithuania, Macedonia, Madeira, Moldova, Netherlands, Poland, Portugal, Romania, Russia, Serbia, Slovakia, Spain, Sweden, Switzerland, Turkey, Ukraine, United Kingdom, Yugoslavia (former, pre 1991). **Africa**: Comores, Egypt, Libya, Madagascar, Mauritius, Morocco; **Americas**: Chile, Cuba. The UCD records *T. evanescens* additionally from Canada and U.S.A., while Pinto (personal communication) states that there is no evidence that *T. evanescens* occurs in North America. Early introductions of *T. evanescens* into North America were probably *T. brassicae*.

Hosts: **COLEOPTERA**: **Bruchidae**: *Bruchus obtectus*; **Chrysomelidae**: *Cassida deflorata; C. nebulosa; C. nobilis; C. vittata; Donacia simplex*; **Curculionidae**: *Rhynchaenus testaceus*; **Dermestidae**: *Dermestes maculatus*; **Rhynchitidae**: *Rhynchites auratus; R. betulae*; **Tenebrionidae**: *Tribolium castaneum*; **DIPTERA**: **Anthomyiidae**: *Atherigona soccata*; *Pegomya betae; P. hyoscyami*; **Stratiomyiidae**: *Oxycera* sp.; *Stratiomys* sp.; **Syrphidae**: *Melanostoma mellinum; Paragus quadrifasciatus; Syrphus* sp.; *S. balteatus; S. pyrastri; S. vitripennis*; **Tabanidae**: *Chrysops* sp.; *C. caecutiens; C. relictus*; *Tabanus* sp.; **HEMIPTERA**: **Cimicidae**: *Cimex lectularius*; **HYMENOPTERA**: **Pamphiliidae**: *Acantholyda erythrocephala; A. posticalis; A. stellata; Cephalcia abietis; C. arvensis; C. signata*; **Tenthredinidae**: *Caliroa cerasi; Croesus septentrionalis; Emphytus tener; Pteronidea ferruginea; P. ribesii*; **LEPIDOPTERA**: **Arctiidae**: *Arctia caja; Diacrisia obliqua; Eilema* sp.; *Hyphantria cunea; Spilosoma* sp.; **Bombycidae**: *Rondotia menciana*; **Danaidae**: *Danaus chrysippus*; **Gelechiidae**: *Pectinophora* sp.; *P. gossypiella; Phthorimaea operculella; Sitotroga cerealella*; **Geometridae**: *Boarmia grisescens; Bupalus piniarius; Cidaria bilineata; C. didymata; Crocallis elinguaria; Erannis defoliaria; Lambdina fiscellaria; Operophtera brumata*; **Glyphipterygidae**: *Anthophila atrilineata*; **Lasiocampidae**: *Cosmotriche potatoria; Dendrolimus pini; D. punctatus; D. segregatus; D. spectabilis; Malacosoma disstria; M. neustria*; **Leptidae**: *Atherix* sp.;

Lycaenidae: *Cacyreus marshalli; Thecla betulae; Virachola livia*; **Lymantriidae**: *Arctornis chrysorrhoea; Euproctis lunata; E. phaeorrhoea; Laelia salicis; Lymantria dispar; L. monacha; Nygmia phaeorrhoea; Orgyia antiqua; O. gonostigma; Stilpnotia salicis*; **Noctuidae**: *Acontia luctuosa; Acronicta aceris; A. major; A. rumicis; A. tridens; Agrotis exclamationis; A. ipsilon; A. segetum; Amathes c-nigrum; Antitype flavicincta; Apopestes spectrum; Autographa* sp.; *A. gamma; Barathra brassicae; Catocala elocata; Chrysodeixis chalcites; Cirrhia gilvago; Discestra trifolii; Earias* sp.; *E. cupreoviridis; E. fabia; E. insulana; Euxoa obelisca; E. segetum; Gonospileia glyphica; Helicoverpa armigera; H. assulta; H. virescens; H. zea; Mamestra sp.; M. brassicae; M. oleracea; M. trifolii; Naranga aenescens; Noctua pronuba; Oria musculosa; Panolis flammea; Parallelia algira; Phalaena typica; Phlogophora meticulosa; Phytometra gamma; Plusia gamma; Polia oleracea; P. pisi; P. suasa; Pyrrhia umbra; Rivula atimeta; Sarrothripus musculana; Scotia ipsilon; Sesamia cretica; S. nonagrioides; Spaelotis pronubana; Spodoptera sp.; S. littoralis; S. litura; Tholera popularis; Trachea triplicis; Trichoplusia ni; Triphaena pronuba*; **Notodontidae**: *Lampronadata cristata; Phalera bucephala; P. bucephaloides; Thaumetopoea pityocampa*; **Nymphalidae**: *Nymphalis polychloros*; **Oecophoridae**: *Depressaria nervosa; Endrosis lactella; Opisina arenosella*; **Papilionidae**: *Iphiclides podalirius; Papilio polytes*; **Pieridae**: *Aporia crataegi; Leptidea sinapis*; *Pieris* sp.; *P. brassicae; P. daplidice; P. napi; P. rapae*; **Pyralidae**: (incl. **Crambidae**): *Achroia grisella; Anagasta sp.; A. kuehniella; Cadra cautella; Chilo sp.; C. agamemnon; C. indicus; C. infuscatellus; C. partellus; C. sacchariphagus; C. simplex; C. suppressalis; Cnaphalocrocis medinalis; C. cephalonica; Crambus geniculeus; Diatraea sp.; Ectomyelois ceratoniae; Ephestia sp.; E. calidella; E. cautella; E. elutella; Etiella zinckenella; Evergestis forficalis; Galleria sp.; G. mellonella; Glyphodes pyloalis; Hymenia recurvalis; Loxostege sticticalis; Marasmia patnalis; Maruca vitrata; Ostrinia furnacalis; O. nubilale; Palpita unionalis; Phlyctaenia forficalis; Plodia interpunctella; Pyrausta machaeralis; Salebria semirubella; Spectrobates ceratoniae*; **Saturniidae**: *Attacus cynthia; A. ricini*; **Sphingidae**: *Celerio lineata*; *Manduca Sexta; Smerinthus populi; Sphinx pinastri*; **Tortricidae**: *Adoxophyes orana; A. reticulana; Archips crataeganus; A. pronubana; A. rosanus; Cacoecia rosanus; Cacoecimorpha pronubana; Carpocapsa pomonella; Choristoneura fumiferana; Clysia ambiguella; Cnephasia longana; C. pumicana; Cydia funebrana; C. pomonella; Epichoristodes acerbella; Epinotia pygmaeana; E. tedella; Eupoecilia ambiguella; Grapholitha delineana; Grapholitha funebrana; G. molesta; Gypsonoma aceriana; Homona coffearia; Laspeyresia microgrammana; L. molesta; L. nigricana; Lobesia botrana; Pandemis chondrillana; Pandemis heparana; Petrova resinella; Rhopobota naevana; Rhyacionia buoliana; Sparganothis pilleriana; Tetramoera schistaceana; Zeiraphera diniana*; **Yponomeutidae**: *Acrolepiopsis assectella; Argyresthia conjugella; Plutella sp.; P. maculipennis; P. xylostella; P. citri; Prays oleae*; **Zygaenidae**: *Theresimima ampelophaga; Zygaena* sp.; **MEGALOPTERA**: **Sialidae**: *Sialis lutaria*; **NEUROPTERA**: **Chrysopidae**: *Chrysopa carnea; C. ventralis; Chrysoperla* sp.; *Nothochrysa italica*.

Trichogramma flandersi Nagaraja and Nagarkatti

Identification: Nagaraja and Nagarkatti (1970).
Distribution: **Asia**: India.
Hosts: **LEPIDOPTERA**: **Crambidae**: *Chilo infuscatellus*; **Sphingidae**: *Agrius convolvuli.*

Trichogramma flavum Ashmead

Synonymy: *Trichogramma flavus* Ashmead (1880).
Identification: No reference found except for the original description. Considered to be a *nomen dubium* by Pinto (personal communiction).
Distribution: **Asia**: China (doubtful record); **Americas**: U.S.A.
Hosts: **HEMIPTERA**: **Coccidae**: *Coccus hesperidum* (almost certainly recorded in error).

Trichogramma forcipiforme Zhang and Wang

Synonymy: *Trichogramma forcipiformis* Zhang and Wang.
Identification: Lin (1994).
Distribution: **Asia**: China.
Hosts: **LEPIDOPTERA**: **Crambidae**: *Ostrinia furnacalis.*

Trichogramma fuzhouense Lin

Identification: Lin (1994).
Distribution: **Asia**: China
Hosts: None recorded.

Trichogramma gicai Pintureau and Stefanescu

Identification: Gibbs et al. (2004), Pintureau et al. (2000), Pintureau (2008).
Distribution: **Europe**: Spain, Madeira.
Hosts: **LEPIDOPTERA**: **Noctuidae**: *Mamestra brassicae*; **Nymphalidae**: *Pararge aegaeria*; *P. xiphia*; **Papilionidae**: *Iphiclides podalirius.*

Trichogramma hesperidis Nagaraja

Identification: Nagaraja (1973).
Distribution: **Asia**: India.
Hosts: **LEPIDOPTERA**: **Hesperiidae**: Undetermined species. **Pyralidae**: *Corcyra cephalonica.*

Trichogramma ingricum Sorokina

Identification: Pintureau (2008), Sorokina (1984).
Distribution: **Asia**: Russia.
Hosts: **LEPIDOPTERA**: **Gelechiidae**: *Sitotroga cerealella*; **Noctuidae**: *Acronicta rumicis.*

Trichogramma ivelae Pang and Chen

Synonymy: *Trichogramma ivalae* Pang and Chen.
Identification: Lin (1994).
Distribution: **Asia**: China; **Australasia**: Australia.
Hosts: **Limacodidae**: *Cnidocampa flavescens*: **Lymantriidae**: *Ivela ochropoda*: **Noctuidae**: *Helicoverpa punctigera; Mythimna salicis; M. separata*; **Notodontidae**: *Clostera anachoreta*; **Sphingidae**: *Callambulyx tatarinovii*; **Tortricidae**: *Cydia molesta; C. pomonella.*

Trichogramma japonicum Ashmead

Synonymy: *Neotrichogramma acutiventre* Girault. *Trichogramma ethiopicum* (Risbec), described from Cameroon, is another possible synonym.
Identification: *T. japonicum* can be identified using the key to Japanese *Trichogramma* by Honda *et al* (2006) and to Indian *Trichogramma* by Nagaraja (1973) and Yousuf and Shafee (1988). Lin (1994) provides a key to *Trichogramma* species known from China (in Chinese) including *T. japonicum.* Pinto (1999) redescribes the species from North America.
Distribution: **Asia**: Bangladesh, China, India, Indonesia, Japan, Korea, Malaysia, Myanmar, Philippines, Taiwan, Thailand, Vietnam; **Europe**: France (doubtful record – not recorded as European by Pintureau, 2008); **Australasia**: Australia; **Americas**: Antilles, Bahamas, Barbados, Brazil, Canada, Colombia, Grenada, Montserrat, Peru, Trinidad and Tobago, Venezuela; **Oceania**: Hawaii. Possibly also present in **Africa** as *Trichogramma ethiopicum* Risbec.
Hosts: **COLEOPTERA**: **Chrysomelidae**: *Dicladispa armigera*; **DIPTERA**: **Ephydridae**: *Notiphila* sp.; *N. similis; N. spinosa*; **Sciomyzidae**: *Sepedon sauteri*; *S. sphegea*; **LEPIDOPTERA**: **Arctiidae**: *Creatonotos gangis*; **Cossidae**: *Phragmataecia gummata*; **Gelechiidae**: *Brachmia modicella; Sitotroga cerealella*; **Geometridae**: *Boarmia variegata*; **Hesperiidae**: *Parnara guttata; P. mathias; P. pellucida*; **Noctuidae**: *Achaea janata; Anomis flava; Chrysodeixis eriosoma; Helicoverpa armigera; Leucania unipuncta; Naranga aenescens; Sesamia inferens; Trichoplusia ni*; **Notodontidae**: *Clostera cupreata*; **Oecophoridae**: *Opisina arenosella*; **Pieridae**: *Eurema* sp. **Pyralidae** (incl. **Crambidae**): *Achroia grisella; Ancylolomia chrysographella; Chilo sp.; C. auricilius; C. infuscatellus; C. partellus; C. polychrysus; C. sacchariphagus; C. simplex; C. suppressalis; Cnaphalocrocis medinalis; Corcyra sp.; C. cephalonica; Diatraea sp.; D. saccharalis; Ephestia cautella; Galleria mellonella; Leucinodes orbonalis; Marasmia exigua; Marasmia patnalis; Ostrinia nubilale; Pyralis farinalis; Raphimetopus ablutellus; Scirpophaga sp.; S. bipunctifer; S. excerptalis; S. incertulas; S. innotata; S. nivella*; **Saturniidae**: *Philosamia cynthia*; **Tortricidae**: *Laspeyresia koenigana*; *Tetramoera schistaceana*; **Yponomeutidae**: *Plutella xylostella.*

Trichogramma jaxarticum Sorokina

Identification: Sorokina (1984).

Distribution: **Asia**: Turkmenistan.
Hosts: **LEPIDOPTERA**: **Noctuidae**: *Helicoverpa armigera*.

Trichogramma jezoense Ishii

Synonymy: *Trichogramma jezoensis* Ishii.
Identification: Honda *et al* (2006).
Distribution: **Asia**: Japan.
Hosts: **LEPIDOPTERA**: **Pyralidae**: *Conogethes puntiferalis*; **Tortricidae**: *Grapholitha molesta*; *Leguminivora glycinivorella*.

Trichogramma kilinceri Bulut and Kilinçer

Identification: Bulut and Kilinçer (1991)
Distribution: **Asia**: Turkey.
Hosts: **LEPIDOPTERA**: **Tortricidae**: *Cydia pomonella*.

Trichogramma kurosuae Taylor, Yashiro, Hirose and Honda

Identification: Honda *et al* (2006).
Distribution: **Asia**: Japan.
Hosts: **LEPIDOPTERA**: **Lymantriidae**: *Ivela auripes*.

Trichogramma lacustre Sorokina

Identification: Pintureau (2008).
Distribution: **Asia**: Russia; **Europe**: Bulgaria; France; Russia; United Kingdom.
Hosts: **LEPIDOPTERA**: **Noctuidae**: *Mamestra brassicae*; **Tortricidae**: *Lobesia botrana*; **Yponomeutidae**: *Prays oleae*.

Trichogramma lenae Sorokina

Identification: Sorokina (1991).
Distribution: **Europe**: Russia.
Hosts: None recorded.

Trichogramma leptoparameron Dyurich

Identification: Pintureau (2008).
Distribution: **Europe**: Bulgaria, Moldova.
Hosts: **LEPIDOPTERA**: **Noctuidae**: *Lacanobia suasa; Mamestra brassicae*; **Pieridae**: *Pieris brassicae; P. rapae*; **Yponomeutidae**: *Plutella xylostella*.

Trichogramma leucaniae Pang and Chen

Identification: Lin (1994).
Distribution: **Asia**: China.

Hosts: **LEPIDOPTERA**: **Hesperiidae**: *Parnara guttata*; **Notodontidae**: *Clostera anachoreta*; **Noctuidae**: *Mamestra brassicae; Mythimna separata*; **Crambidae**: *Ostrinia furnacalis.*

Trichogramma lingulatum Pang and Chen

Identification: Pang and Chen (1974).
Distribution: **Asia**: China; Japan.
Hosts: **LEPIDOPTERA**: **Lasiocampidae**: *Dendrolimus sibiricus*; **Saturniidae**: *Samia cynthia.*

Trichogramma longxishanense Lin

Identification: Lin (1994).
Distribution: **Asia**: China.
Hosts: None recorded.

Trichogramma margianum Sorokina

Identification: Sorokina (1984)
Distribution: **Asia**: Turkmenistan.
Hosts: **LEPIDOPTERA**: **Noctuidae**: *Helicoverpa armigera.*

Trichogramma mirabile Dyurich

Identification: Dyurich (1987), Pintureau (2008).
Distribution: **Europe**: Moldova.
Hosts: **LEPIDOPTERA**: **Noctuidae**: *Autographa gamma; Helicoverpa armigera; Hydroecia micacia; Mamestra brassicae; M. oleracea.*

Trichogramma misiae Kostadinov

Identification: Kostadinov (1987), Pintureau (2008).
Distribution: **Europe**: Bulgaria.
Hosts: **LEPIDOPTERA**: **Noctuidae**: *Mamestra brassicae.*

Trichogramma nestoris Kostadinov

Synonymy: *Nuniella nestoris* Kostadinov.
Identification: Pintureau (2008).
Distribution: Bulgaria.
Hosts: None known.

Trichogramma neuropterae Chan and Chou

Identification: Chan *et al* (1996).
Distribution: **Asia**: Taiwan.
Hosts: **NEUROPTERA**: **Chrysopidae**: *Mallada basalis; M. boninensis.*

Trichogramma niveiscapus (Morley)

Synonymy: *Anagrus niveiscapus* Morley.
Identification: Pintureau (2008).
Distribution: **Europe**: England.
Hosts: **LEPIDOPTERA**: **Papilionidae**: *Papilio machaon.*

Trichogramma nubilale Ertle and Davis

Identification: Pinto (1999).
Distribution: **Asia**: China; **Americas**: U.S.A.
Hosts: **LEPIDOPTERA**: **Gelechiidae**: *Sitotroga cerealella*; **Noctuidae**: *Helicoverpa armigera; Heliothis virescens*; **Nymphalidae**: *Basilarchia* sp.; **Pyralidae** (incl. **Crambidae**): *Chilo infuscatellus; C. sacchariphagus; Cnaphalocrocis medinalis; Loxostege rantalis; Ostrinia furnacalis; O. nubilale*; **Sphingidae**: *Manduca sexta*; **Tortricidae**: *Argyroploce schistaceana.*

Trichogramma okinawae Honda

Identification: Honda *et al* (2006).
Distribution: **Asia**: Japan.
Hosts: **LEPIDOPTERA**: **Tortricidae**: *Tetramoera schistaceana.*

Trichogramma oleae Voegelé and Pointel

Identification: Voegelé and Pointel (1979), Pintureau (2008).
Distribution: **Europe**: France; Greece; Italy; Yugoslavia (former, pre 1991); **Americas**: Argentina; U.S.A. (though not recorded from U.S.A. by Pinto 1999).
Hosts: **LEPIDOPTERA**: **Pyralidae**: *Anagasta* (= *Ephestia*) *kuehniella*; *Ephestia calidella; E. cautella; Glyphodes unionalis; Palpita unionalis*; **Yponomeutidae**: *Plutella xylostella*; *Prays oleae.*

Trichogramma ostriniae Pang and Chen

Synonymy: *Trichogramma ostrinia* Pang and Chen.
Identification: Lin (1994), Pinto (1999).
Distribution: **Asia**: China; Japan; Korea (South); Russia, Taiwan; **Africa**: South Africa; **Americas**: U.S.A.
Hosts: **LEPIDOPTERA**: **Danaidae**: *Danaus plexippus*; **Gelechiidae**: *Sitotroga cerealella*; **Limacodidae**: *Cnidocampa flavescens*; **Noctuidae**: *Busseola fusca; Helicoverpa armigera; Heliothis virescens; Trichoplusia ni*; **Notodontidae**: *Lampronadata cristata*; **Papilionidae**: *Papilio* sp.; **Pyralidae** (incl. **Crambidae**): *A. kuehniella*; *Chilo partellus*; *C. sacchariphagus*; *Corcyra cephalonica; Ephestia cautella; Galleria mellonella; Ostrinia sp. O. furnacalis; O. nubilale*; **Saturniidae**: *Antheraea pernyi*; **Sphingidae**: *Macroglossum pyrrhostictum*; **Tortricidae**: *Adoxophyes fasciata; Argyroploce schistaceana; Leguminivora glycinivorella; Tetramoera schistaceana*; **Yponomeutidae**: *Plutella xylostella.*

Trichogramma pallidiventris Nagaraja

Identification: Nagaraja (1973).
Distribution: **Asia**: India.
Hosts: **LEPIDOPTERA**: **Pyralidae**: *Corcyra cephalonica; Scirpophaga incertulas.*

Trichogramma pangi Lin

Identification: Lin (1987, 1994).
Distribution: **Asia**: China.
Hosts: None recorded

Trichogramma papilionis Nagarkatti

Identification: Nagarkatti (1974).
Distribution: **Asia**: Japan; Malaysia; **Oceania**: Hawaii.
Hosts: **LEPIDOPTERA**: **Danaidae**: *Danaus plexippus*; **Lycaenidae**: *Neozephyrus taxila; Zizeeria maha*; **Noctuidae**: *Ctenoplusia albostriata; Mamestra brassicae; Plusia nigrisigna*; **Nymphalidae**: *Dichorragia nesimachus*; **Papilionidae**: *Papilio memnon; P. xuthus*; **Pieridae**: *Pieris rapae*; **Pyralidae** (incl. **Crambidae**): *Corcyra cephalonica; Ostrinia furnacalis*; **Sphingidae**: *Deilephila nerii; Macroglossum pyrrhostictum; Theretra silhetensis*. **Yponomeutidae**: *Plutella xylostella.*

Trichogramma parnarae Huo

Identification: Lin (1994).
Distribution: **Asia**: China.
Hosts: **LEPIDOPTERA**: **Hesperiidae**: *Parnara guttata.*

Trichogramma pelovi Kostadinov

Identification: Kostadinov (1986).
Distribution: **Europe**: Bulgaria.
Hosts: **LEPIDOPTERA**: **Tortricidae**: *Cydia pomonella.*

Trichogramma perkinsi Girault

Synonymy: *Trichogramma perkensi* Girault.
Identification: Pinto (1999).
Distribution: **Asia**: India; **Europe**: France; **Americas**: Argentina; Bahamas; Bolivia; Chile; Colombia; Cuba; Guatemala; Mexico; Peru; Venezuela; **Oceania**: Hawaii.
Hosts: **LEPIDOPTERA**: **Arctiidae**: *Diacrisia obliqua*; **Gelechiidae**: *Sitotroga cerealella*; **Lycaenidae**: *Deudorix epijarbas*; **Lymantriidae**: *Euproctis fraterna*; **Noctuidae**: *Achaea janata; Alabama sp.; Helicoverpa armigera; H. zea; Heliothis virescens; Spodoptera litura*; **Notodontidae**: *Clostera cupreata*; **Nymphalidae**: *Vanessa tammeamea*; **Pterophoridae**: *Exelastis atomosa*: **Pyralidae**: *Chilo*

partellus; Corcyra cephalonica; Diatraea sp.; D. saccharalis; Omiodes accepta; O. blackburni; O. swezeyi; **Tortricidae**: *Amorbia emigratella; Archips postvittana; Argyrotaenia sphaleropa; Bactra straminea; Platynota* sp.

Trichogramma piceum Dyurich

Identification: Pintureau (2008).
Distribution: **Europe**: Bulgaria, Italy, Moldova, Ukraine.
Hosts: **LEPIDOPTERA**: **Arctiidae**: *Hyphantria cunea*; **Noctuidae**: *Mamestra brassicae.*

Trichogramma piniperda Wolff

Synonymy: *Trichogramma pini* Meyer; *T. evanescens piniperda* Wolff; *Trichogramma piniperdae* Wolff.
Identification: Pintureau (2008).
Distribution: **Europe**: France, Germany, Russia.
Hosts: **LEPIDOPTERA**: **Geometridae**: *Bupalus piniarius*; **Lasiocampidae**: *Dendrolimus pini*; *D. segregatus*; **Noctuidae**: *Panolis flammea*; **Notodontidae**: *Thaumetopoea pityocampa.*

Trichogramma pintoi Voegelé

Identification: Lin (1994), Pinto (1999), Pintureau (2008).
Distribution: **Asia**: Armenia; China, India; Iran; Israel, Japan, Pakistan, Tadjikistan, Turkey; Uzbekistan; **Europe**: Belarus; Bulgaria; France; Greece, Moldova, Poland, Portugal; Romania, Russia; Spain; Ukraine; **Africa**: Tunisia; **Americas**: Argentina; Canada; Cuba; Peru; U.S.A.
Hosts: **COLEOPTERA**: **Cassidae**: *Cassida nebulosa*; **HYMENOPTERA**: **Pamphiliidae**: *Acantholyda posticalis*; **LEPIDOPTERA**: **Gelechiidae**: *Sitotroga cerealella*; **Lycaenidae**: *Icaricia acmon; Lycaeides melissa*; **Lymantriidae**: *Euproctis chrysorrhea; Leucoma salicis; Orgyia antiqua*; **Noctuidae**: *Actebia fennica; Agrotis segetum; Barathra brassicae; Helicoverpa armigera; H. zea; Mamestra brassicae; Plusia* sp.; **Nymphalidae**: *Vanessa* sp.; **Pieridae**: *Pieris brassicae*; **Pterophoridae**: *Platyptilia carduidactyla*; **Pyralidae** (incl. **Crambidae**): *Anagasta* (= *Ephestia*) *kuehniella*; *Corcyra cephalonica; Loxostege sticticalis; Ostrinia* sp.; *O. nubilale*; **Saturniidae**: *Antheraea pernyi*; *Philosamia cynthia*; **Tortricidae**: *Cydia nigricana; C. pomonella; Lobesia botrana*; **Yponomeutidae**: *Plutella xylostella; Prays oleae*; **NEUROPTERA**: **Chrysopidae**: *Chrysoperla carnea.*

Trichogramma plasseyense Nagaraja

Synonymy: *Trichogramma plasseyensis* Nagaraja.
Identification: Nagaraja (1973).
Distribution: **Asia**: India; Papua New Guinea

Hosts: **LEPIDOPTERA**: **Noctuidae**: *Helicoverpa armigera*; **Pyralidae** (incl. **Crambidae**): *Chilo infuscatellus; C. terenellus; Corcyra cephalonica; Ostrinia furnacalis.*

Trichogramma platneri Nagarkatti

Identification: Nagarkatti, 1975; Pinto (1999).
Distribution: **Asia**: Israel (introduction); **Americas**: Canada; U.S.A.; **Oceania**: Hawaii.
Hosts: **LEPIDOPTERA**: **Gelechiidae**: *Sitotroga cerealella*; **Geometridae**: *Boarmia selenaria; Lambdina fiscellaria; Sabulodes aegrotata*; **Lycaenidae**: *Atlides halesus*; **Lymantriidae**: *Orgyia sp.; O. antiqua; O. pseudotsugata*; **Noctuidae**: *Helicoverpa zea; Trichoplusia ni*; **Nymphalidae**: *Agraulis vanillae*; **Pyralidae** (incl. **Crambidae**): *Amyelois transitella; Anagasta* (= *Ephestia*) *kuehniella; Cryptoblabes gnidiella; Ephestia sp.; E. cautella; Ostrinia nubilale; Plodia interpunctella*; **Sphingidae**: *Manduca* sp.; *M. sexta*; **Tortricidae**: *Adoxophyes orana; Amorbia cuneana; Choristoneura rosaceana; Cydia pomonella; Pandemis heparana; P. limitata*; **Yponomeutidae**: *Plutella xylostella*; **NEUROPTERA**: **Chrysopidae**: *Chrysoperla carnea.*

Trichogramma poliae Nagaraja

Identification: Nagaraja (1973).
Distribution: **Asia**: India.
Hosts: **LEPIDOPTERA**: **Notodontidae**: *Clostera cupreata*; **Pyralidae** (incl. **Crambidae**): *Chilo infuscatellus; Corcyra cephalonica*; **Yponomeutidae**: *Plutella xylostella.*

Trichogramma polychrosis Chen and Pang

Identification: Lin (1994).
Distribution: **Asia**: China.
Hosts: **LEPIDOPTERA**: **Tortricidae**: *Polychrosis cunninghamiacola.*

Trichogramma principium Sugonjaev and Sorokina

Synonymy: *Trichogramma principia* Sugonjaev and Sorokina.
Identification: Pintureau (2008).
Distribution: **Asia**: Iran; Kazakhstan; Russia; Syria; Taiwan; Turkmenistan; Uzbekistan; **Europe**: Bulgaria; Czech Republic; France; Greece, Russia, Ukraine.
Hosts: **COLEOPTERA**: **Cassidae**: *Cassida* sp.; **LEPIDOPTERA**: **Gelechiidae**: *Sitotroga cerealella*; **Noctuidae**: *Acronicta rumicis; Agrotis segetum; Earias insulana; Helicoverpa armigera; Mamestra brassicae; M. oleracea; Spodoptera exigua*; **Pieridae**: *Pieris rapae*; **Pyralidae** (incl. **Crambidae**): *Anagasta* (= *Ephestia*) *kuehniella; Chilo suppressalis*; *Ephestia kuehniella*; *Loxostege sticticalis*; **Tortricidae**: *Cydia pomonella; Eupoecilia ambiguella; Lobesia botrana*; **Yponomeutidae**: *Plutella xylostella.*

Trichogramma psocopterae Chan and Chou

Identification: Chan and Chou (1996)
Distribution: **Asia**: Taiwan
Hosts: None recorded

Trichogramma raoi Nagaraja

Synonymy: *Trichogramma roi*
Identification: Lin (1994), Nagaraja (1973).
Distribution: **Asia**: China, India.
Hosts: **LEPIDOPTERA**: **Geometridae**: *Boarmia variegata*; **Noctuidae**: *Achaea janata; Naranga aenescens*; **Pyralidae**: *Corcyra cephalonica*; **Tortricidae**: *Laspeyresia koenigana*.

Trichogramma rossicum Sorokina

Identifcation: Pintureau (2008).
Distribution: **Europe**: Russia.
Hosts: **Noctuidae**: *Autographa gamma*.

Trichogramma savalense Sorokina

Identifcation: Pintureau (2008).
Distribution: **Asia**: Russia; Tadjikistan, Turkmenistan; Uzbekistan; **Europe**: Russia.
Hosts: **LEPIDOPTERA**: **Gelechiidae**: *Sitotroga cerealella*; **Noctuidae**: *Autographa gamma; Helicoverpa armigera*.

Trichogramma semblidis (Aurivillius)

Synonymy: *Oophthora semblidis* Aurivillius; *Trichogramma schuberti* Voegelé and Russo.
Identification: Lin (1994), Nagaraja (1973), Pinto (1999).
Distribution: **Asia**: China, India; Iran; Kazakhstan; Russia; Syria; **Europe**: Bulgaria; France; Germany; Hungary; Italy; Netherlands; Norway; Poland; Russia; Spain; Sweden; Switzerland; Ukraine; United Kingdom; **Americas**: Canada; U.S.A.
Hosts: **COLEOPTERA**: **Rhynchitidae**: *Rhynchites auratus*; **Scolytidae**: *Hylesinus crenatus*; *Leperesinus fraxini*; *L. orni*; **DIPTERA**: **Anthomyiidae**: *Pegomya hyoscyami*; **Sciomyzidae**: *Sepedon fuscipennis*; *S. sphegea*; *Tetanocera* sp.; **Tabanidae**: *Chrysops* sp. *C. aestuans*; *C. excitans*; *C. mitis*; *C. striatus*; *C. univittatus*; *Hybomitra* sp.; *Tabanus* sp.; *T. lasiophthalmus*; *T. macer*; **HYMENOPTERA**: **Pamphiliidae**: *Acantholyda erythrocephala*; *A. pinivora*; *A. posticalis*; **LEPIDOPTERA**: **Gelechiidae**:*Phthorimaea operculella*; *Sitotroga cerealella*; **Glyphipterygidae**: *Anthophila fabriciana*; **Noctuidae**: *Achaea janata*; *Agrotis segetum*; *Earias insulana*; *Helicoverpa armigera*; *Hypena proboscidalis*; *Mamestra brassicae*; *M. oleracea*; *Noctua pronuba*; *Trichoplusia ni*; *Xanthia togata*; **Notodontidae**: *Notodonta* sp.; *Thaumetopoea pityocampa*; **Nymphalidae**:

Aglais urticae; **Pieridae**: *Pieris rapae*; **Pyralidae** (incl. **Crambidae**): *Anagasta* (= *Ephestia*) *kuehniella*; *Chilo* sp.; *C. infuscatellus*; *Corcyra cephalonica*; *Ephestia* sp.; *Haritala ruralis*; *Loxostege sticticalis*; *Ostrinia nubilale*; **Tortricidae**: *Argyroploce lacunana; Cacoecimorpha pronubana; Cydia nigricana; C. pomonella; Epichoristodes acerbella; Eupoecilia ambiguella; Lobesia botrana; Notocelia uddmanniana*; **Yponomeutidae**: *Plutella xylostella*; **MEGALOPTERA**: **Sialidae**: *Chauliodes* sp; *C. rastricornis*; *Semblidis lutaria*; *Sialis* sp.; *S. californica*; *S. flavilatera*; *S. infumata*; *S. lutaria*; *S. rotunda*; **NEUROPTERA**: **Chrysopidae**: *Chrysopa* sp.

Trichogramma sericini Pang and Chen

Identification: Lin (1994), Pintureau (2008).
Distribution: **Asia**: China, Russia.
Hosts: **LEPIDOPTERA**: **Noctuidae**: *Heliocoverpa armigera*; **Papilionidae**: *Papilio polytes*; *Sericinus telamon*.

Trichogramma shaanxiensis Huo

Identification: Lin (1994).
Distribution: **Asia**: China.
Hosts: **LEPIDOPTERA**: **Hesperiidae**: *Parnara guttata*.

Trichogramma shchepetilnikovae Sorokina

Identification: Sorokina (1984).
Distribution: **Asia**: Tadzhikistan.
Hosts: **LEPIDOPTERA**: **Noctuidae**: *Helicoverpa armigera*.

Trichogramma sibiricum Sorokina

Synonymy: *Trichogramma sibericum* Sorokina.
Identification: Pinto (1999), Pintureau (2008) (n.b. both authors use the spelling *sibericum*).
Distribution: **Asia**: Russia; **Europe**: Modova; **Americas**: Canada, U.S.A.
Hosts: **HYMENOPTERA**: undetermined species; **LEPIDOPTERA**: **Noctuidae**: *Acronicta megacephala*; *Helicoverpa armigera*; *Trichoplusia ni*; **Notodontidae**: undetermined species; **Pyralidae**: *Anagasta* (= *Ephestia*) *kuehniella*; *Plodia interpunctella*; **Tortricidae**: *Choristoneura rosaceana*; *Cydia inopinata*; *Grapholitha inopinata*; *Rhopobota naevana*.

Trichogramma silvestre Sorokina

Identification: Pintureau (2008).
Distribution: **Asia**: Russia.
Hosts: **LEPIDOPTERA**: **Arctiidae**: *Hyphantria cunea*; **Gelechiidae**: *Sitotroga cerealella*; **Noctuidae**: *Acronicta rumicis*.

Trichogramma sogdianum Sorokina

Identification: Sorokina (1984).
Distribution: **Asia**: Uzbekistan.
Hosts: **LEPIDOPTERA**: **Noctuidae**: *Helicoverpa armigera*.

Trichogramma sorokinae Kostadinov

Identification: Pintureau (2008).
Distribution: **Asia**: Uzbekistan; **Europe**: Bulgaria.
Hosts: **LEPIDOPTERA**: **Pyralidae**: *Anagasta* (= *Ephestia*) *kuehniella*; **Tortricidae**: undeterminded species.

Trichogramma sugonjaevi Sorokina

Identification: Sorokina (1984).
Distribution: Uzbekistan.
Hosts: **LEPIDOPTERA**: **Noctuidae**: *Helicoverpa armigera*.

Trichogramma taiwanense Chan and Chou

Identification: Chan and Chou (2000).
Distribution: **Asia**: Taiwan.
Hosts: **LEPIDOPTERA**: **Noctuidae**: *Trichoplusia ni*.

Trichogramma talitzkii Dyurich

Identification: Dyurich (1987), Pintureau (2008).
Distribution: **Europe**: Moldova; United Kingdom.
Hosts: **NEUROPTERA**: **Chrysopidae**: *Chrysopa flava*.

Trichogramma tielingensis Zhang and Wang

Identification: Lin (1994).
Distribution: **Asia**: China.
Hosts: **LEPIDOPTERA**: **Crambidae**: *Ostrinia furnacalis*.

Trichogramma trjapitzini Sorokina

Identification: Sorokina (1984).
Distribution: **Asia**: Russia.
Hosts: **LEPIDOPTERA**: **Noctuidae**: *Mamestra brassicae*; **Pieridae**: *Euchloe* sp.; **Yponomeutidae**: *Plutella xylostella*.

Trichogramma tshumakovae Sorokina

Identification: Sorokina (1984).
Distribution: **Asia**: Iran; Kirgizia.

Hosts: **LEPIDOPTERA**: **Noctuidae**: *Mamestra brassicae*; **Crambidae**: *Chilo suppressalis*.

Trichogramma turkeiensis Bulut and Kilinçer

Identification: Bulut and Kilinçer (1991).
Distribution: **Asia**: Turkey.
Hosts: **LEPIDOPTERA**: **Gelechiidae**; *Sitotroga cerealella*; **Lymantriidae**: *Euproctis chrysorrhoea*; **Noctuidae**: *Agrotis segetum*; **Pyralidae**: *Anagasta* (= *Ephestia*) *kuehniella*; **Tortricidae**: *Cydia pomonella*.

Trichogramma umerus Jose, Hirose and Honda

Identification: Jose et al. (2005).
Distribution: **Asia**: Japan.
Hosts: **LEPIDOPTERA**: **Yponomeutidae**: *Plutella xylostella*.

Trichogramma urquijoi Cabello Garcia

Identification: Pintureau (2008).
Distribution: **Europe**: France, Spain.
Hosts: **LEPIDOPTERA**: **Noctuidae**: *Helicoverpa armigera*.

Trichogramma ussuricum Sorokina

Identification: Sorokina (1984).
Distribution: **Europe**: Russia.
Hosts: **LEPIDOPTERA**: **Noctuidae**: *Mamestra brassicae*.

Trichogramma yabui Honda and Taylor

Identification: Jose et al. (2005).
Distribution: **Asia**: Japan.
Hosts: **LEPIDOPTERA**: **Crambidae**: *Ostrinia furnacalis*; Saturniidae: *Samia cynthia*.

Trichogramma yawarae Hirai and Fursov

Identification: Hirai and Fursov (1998), Jose et al. (2005).
Distribution: **Asia**: Japan.
Hosts: **LEPIDOPTERA**: **Hesperiidae**: *Parnara guttata*; **Noctuidae**: *Asota ficus*; **Pyralidae** (incl. **Crambidae**): *Anagasta* (= *Ephestia*) *kuehniella*; *Chilo suppressalis*; *Marasmia exigua*.

Trichogramma zahiri Polaszek

Identification: Polaszek et al. (2002).
Distribution: **Asia**: Bangladesh.
Hosts: **COLEOPTERA**: **Chrysomelidae**: *Dicladispa armigera*.

Trichogramma zeirapherae Walter

Identification: Pintureau (2008).
Distribution: **Europe**: Germany.
Hosts: **LEPIDOPTERA**: **Pamphiliidae**: *Cephalcia* sp.; **Tortricidae**: *Zeiraphera diniana*.

Acknowledgements I am grateful to Roberto Zucchi for the opportunity to learn a little about *Trichogramma* species in order to prepare this chapter. John Pinto kindly corrected a large number of important errors in an earlier draft. Without access to John Noyes' Universal Chalcidoidea Database I would never have attempted this summary. The UCD is a resource that has empowered chalcid taxonomists globally.

References

Ashmead WH (1880) Orange insects: a treatise on the injurious and beneficial insects found in the orange trees in Florida :1–xv, 1–78, plates I–IV Jacksonville, Florida

Birova H, Kazimirova M (1997) *Trichogramma danubiense* sp. n. (Hymenoptera: Trichogrammmatidae), an egg parasitoid of *Macrothylacia rubi* (Lepidoptera: Lasiocampidae), with some data on its bionomics. Eur J Entomol 94:301–306

Bulut H, Kilinçer N (1991) Investigations on the species of *Trichogramma* spp. (Hym.: Trichogrammmatidae), egg parasites of important lepidoptera pests of fruit trees and their distribution in Ankara. Bitki Koruma Bülteni 29:19–46

Cabello T, Gallego JR, Vila E, Soler A, Del Pino M, Carnero A, Hernandez-Suarez E, Polaszek A (2009) Biological control of the south american tomato pinworm *Tuta absoluta* (Lep. Gelechiidae), with releases of *Triochogramma achaeae* (Hym.: Trichogrammatidae) in tomato greenhouses of spain. Integrated Control in Protected Crops, Mediterranean Climate. IOBC/WPRS Bull 49:225–230

Cao GL, Lu WQ, Long S (1988) Studies on comparison of esterase isozyme of different species of *Trichogramma*. Colloques de l'INRA 43:35–44

Cerda RC, Gerding PM (1999) Control biologico de *Rhyacionia buoliana* Den et Schiff (Lepidoptera: Tortricidae) con *Trichogramma* spp. Agro Ciencia 15:279–283

Chan ML, Chou LY (2000) The *Trichogramma* (Hymenoptera: Trichogrammatidae) of taiwan. Chin J Entomol 20:135–151

Chan ML, Chou LY, Chou KC, Chen CC (1996) Two new species of *Trichogramma* Westwood (Hymenoptera: Trichogrammmatidae) from Taiwan. Plant Protect Bull 38:143–148

Chan ML, Chou LY (2000) The *Trichogramma* (Hymenoptera: Trichogrammmatidae) of Taiwan. Chin J Entomol 20:135–151

Chen TL, Pang XF (1986) A new species of *Trichogramma* (Hymenoptera Trichogrammmatidae). Acta Entomol Sinica 29:89–90

Cock MJW (1985) (ed) A review of biological control of pests in the commonwealth caribbean and bermuda up to 1982. Tech Commun, Commonw Inst Biolo Control 9:1–218

Dyurich GF (1987) New species of the genus *Trichogramma* (Hymenoptera, Trichogrammmatidae) from Moldavia. Zoologicheskiy Zhurnal 66:780–784

Fursov VN (2000) Discovery of four species of *Trichogramma* (Hymenoptera, Trichogrammmatidae), new for the fauna of England. Vestnik Zoologii 34:07–113

Gerding M, Cisternas E, Céspedes C (1996) Use of *Trichogramma* in *Rhyacionia buoliana* control in chile. (Abstract 20–030). Proceedings XX International Congress of Entomology, Firenze, Italy, August 25–31, 620 1978

Gibbs M, Broad GR, Polaszek A (2004) *Trichogramma gicai* Pintureau and Stefanescu, 2000 (Hymenoptera: Trichogrammmatidae) reared as an egg parasitoid of the Madeiran endemic butterfly, *Parage xiphia* (Lepidoptera: Satryridae). Bocagiana 214:1–5

Girault AA (1912) On the identity of the most common species of the family Trichogrammmatidae (Hymenoptera). Bull Wisconsin Nat History Soc 9(4):135–165

He YR, Pang XF (2000) A new species of *Trichogramma* (Hymenoptera: Trichogrammmatidae). J South China Agric Univ 21:45–46

Hirai K, Fursov VN (1998) Description of *Trichogramma yawarae* Hirai et Fursov and redescription of *T. japonicum* Ashm. from Japan. J Ukrainian Entomolo Soc 4(3–4):36

Honda JY, Taylor L, Rodriguez J, Yashiro N, Hirose Y (2006) A taxonomic review of the Japanese species of *Trichogramma* (Hymenoptera: Trichogrammmatidae) with descriptions of three new species. Appl Entomol Zool 41:258–265

Jose J, Hirose Y, Honda JY (2005) Two new species of *Trichogramma* (Hymenoptera: Trichogrammmatidae) from the Ryukyu Islands, Japan. Proceedings of the Entomol Soc Washington 107:782–788

Kfir R (1994) Attempts at biological control of the stem borer *Chilo partellus* (Swinhoe) (Lepidoptera: Pyralidae) in South Africa. African Entomol 2:67–68

Kostadinov DN (1986) Two new species of genus *Trichogramma* (Hymenoptera: Trichogrammatidae) Acta Zool Bulg 31:71–74

Kostadinov DN (1987) Species of *Trichogramma* (Hymenoptera Trichogrammmatidae) in Bulgaria with description of a new species. Acta Zool Bulgarica 35:78–82

Kostadinov DN (1991) *Nuniella nestoris* – a new species of *Nuniella* Kastadinov (Hymenoptera, Lin, NQ (1987) Systematic studies of Trichogrammmatidae, I. On the species of *Trichogramma* Westwood from Fujian, south China. Wuyi Sci J 7:97–105

Landry BS, Dextraze L, Boivin G (1993) Random amplified polymorphic DNA markers for DNA fingerprinting and genetic variability assessment of minute parasitic wasp species (Hymenoptera: Mymaridae and Trichogrammatidae) used in biological control programs of phytophagous insects. Genome 36:580–587

Li ZX, Shen ZR (2001) rDNA-ITS2 sequencing and species-specific primer designing for *Trichogramma* spp. Chin J Biol Control 17:75–80

Lin NQ (1994) Systematic studies of Chinese Trichogrammmatidae. Contributions of the Biological Control Research Institute, Fujian Agricultural University. Special Publication No 4:362 pp. Chongqing Publishing House, Chongqing, China (Chinese with English summary)

Liu SS, Zhang GM, Zhang F (1998) Factors influencing parasitism of *Trichogramma dendrolimi* on eggs of the Asian corn borer, *Ostrinia furnacalis*. Biocontrol 43:273–287

Nagaraja H (1973) On some new species of Indian *Trichogramma* (Hymenoptera: Trichogrammmatidae). Orient Insects 7:275–290

Nagaraja H, Nagarkatti S (1970) Three new species of *Trichogramma* from India. Entomophaga 14:393–400

Nagarkatti S (1974) A new species of *Trichogramma* (Hymenoptera: Trichogrammatidae) parasitic on eggs of *Papillio* spp. In Japan. Orient Insects 8:391–393

Nagarkatti S (1975) Two new species of *Trichogramma* (Hym.: Trichogrammmatidae) from the USA. Entomophaga 20:245–248

Nagarkatti S, Nagaraja H (1971) Redescriptions of some known species of *Trichogramma* (Hym., Trichogrammmatidae), showing the importance of the male genitalia as a diagnostic character. Bull Entomol Res 61:13–31

Nagarkatti S, Nagaraja H (1977) Biosystematics of *Trichogramma* and *Trichogrammatoidea* species. Annu Rev Entomol 22:157–176

Nagarkatti S, Nagaraja H (1979) The status of *Trichogramma chilonis* Ishii (Hym.: Trichogrammmatidae). Oriental Insects 13:115–117

Noyes JS, Pinto JD, Stouthamer R (2000) *Trichogramma evanescens*: one of the best known species of parasitic Hymenoptera, or is it? Abstracts. Antonie van Leeuwenhoek Symposium. 7th European workshop on insect parasitoids, 1–6 October 2000, Teylers museum, Haarlem, The Netherlands. Wageningen University, The Netherlands

Pang HF, Chen TL (1974) *Trichogramma* of China. Acta Entomol Sinica 17:441–454.

Pinto JD (1999) Systematics of the North American species of *Trichogramma* Westwood (Hymenoptera: Trichogrammmatidae). Mem Entomol Soc Washington 22:1–287

Pinto JD, Platner GR, Oatman ER (1978) Clarification of the identity of several common species of North American *Trichogramma* (Hymenoptera: Trichogrammmatidae). Ann Entomol Soc North Am 71:169–180

Pinto JD, Oatman ER, Platner GR (1982) *Trichogramma australicum* Girault (Hymenoptera: Trichogrammmatidae): Redescription and lectotype deignation. Pan–Pacific Entomol 58:48–52

Pintureau B (1990) Polymorphisme, biogeographie et specificite parasitaire des trichogrammes Europeens. Bulletin de la Societe Entomologique de France 95:17–38

Pintureau B (2008) Les espèces européens des Trichogrammes. In Libro Veritas. 1–95

Pintureau B, Keita FB (1989) New esterases of the Trichogrammmatidae (Hymenoptera, Trichogrammmatidae). Biochem Systemat Ecol 17:603–608

Pintureau B, Stefanescu C, Kenis M (2000) Two new European species of *Trichogramma* (Hymenoptera: Trichogrammmatidae). Annales Société Entomologique de France 36:417–422

Polaszek A, Rabbi MF, Islam Z, Buckley YM (2002) *Trichogramma zahiri* (Hymenoptera: Trichogrammmatidae) an egg parasitoid of the rice hispa *Dicladispa armigera* (Coleoptera: Chrysomelidae) in Bangladesh. Bull Entomol Res 92:529–537

Rohi L, Pintureau B (2003) Are *Trichogramma bourarachae* and the *perkinsi* species–group really distinct from *Trichogramma buesi* and the *pintoi*–group respectively? J Appl Entomol 127: 265–268

Sorokina AP (1984) New species of the genus *Trichogramma* Westw. (Hymenoptera, Trichogrammmatidae) from the USSR. Entomologicheskoe Obozrenie 63:152–165

Sorokina AP (1989) A new species of the genus *Trichogramma* (Hymenoptera, Trichogrammmatidae) from East Siberia. Zool Z 68:151–152

Sorokina AP (1991) New data on knowledge of species of the genus *Trichogramma* Westw. (Hymenoptera, Trichogrammmatidae) of the USSR fauna with notes on Synonymy. Entomol Obozrenie 70:183–195

Sorokina AP (1993) Key to species of the genus *Trichogramma* Westw. (Hymenoptera, Trichogrammmatidae) of the World. Kolos, Moscow, 77 pp

Sorokina AP, Atamirzaeva TM (1993) A new species of the genus *Trichogramma* (Hymenoptera, Trichogrammmatidae) from central Asia. Zoologicheskiy Zhurnal 72(3):149–152 [1993: Entomol Rev Washington 72(7):136–138

Stouthamer R, Hu JG, Kan, FJPM van, Platner GR, Pinto JD, Hu JG (1998) The utility of internally transcribed spacer 2 DNA sequences of the nuclear ribosomal gene for distinguishing sibling species of *Trichogramma*. BioControl 43:421–440

Stouthamer R, Luck RF, Hamilton WD (1990a) Antibiotics cause parthenogenetic trichogramma (Hymenoptera; Trichogrammatidae) to revert to sex. Proc Natl Acad Sci USA 87:2424–2427

Stouthamer R, Pinto JD, Platner GR, Luck RF (1990b) Taxonomic status of thelytokous forms of trichogramma (Hymenoptera: Trichogrammatidae). Entomol Soc Am 83:475-481

Sugonjaev ES, Sorokina AP (1976) New species of the genus *Trichogramma* (Hymenoptera, Chalcidoidea) from middle Asia and Altai. Zool Z 55:777–779

Viggiani G (1976) Ricerche sugli hymenoptera chalcidoidea XLIX. Trichogramma confusum n. sp. per T. australicum Nagarkatti et Nagaraja (1968), nec Girault (1912), con note su trichogrammatoidea girault e descrizione di paratrichogramma heliothidis n. sp. . Bollettino del Laboratorio di Entomologia Agraria 'Filippo Silvestri', Portici. 23:182–187

Viggiani G, Laudonia S (1989) La specie italiane di *Trichogramma* Westwood (Hymenoptera: Trichogrammmatidae), con un commento sullo stato della tassonomia del genere. Bollettino del Laboratorio di Entomologia Agraria Filippo Silvestri 46:107–124

Voegelé J, Pointel JG (1979) *Trichogramma oleae* sp. n. a sibling species of *Trichogramma evanescens* Westwood (Hym. Trichogrammmatidae). Ann Soc Entomol France 15:643–648

Westwood JO (1833) Descriptions of several new British forms amongst the parasitic hymenopterous insects. Philos Mag 3:443–445

Yousuf M, Shafee SA (1988) Taxonomy of Indian Trichogrammmatidae (Hymenoptera: Chalcidoidea). Indian J Syst Entomol 4:55–200

Chapter 10
Mass Rearing of Egg Parasitoids for Biological Control Programs

José Roberto Postali Parra

Contents

10.1 Introduction

Among the egg parasitoids families of Hymenoptera, Mymaridae, Eulophidae, Aphelinidae, Evaniidae, Scelionidae, Encyrtidae and Trichogrammatidae, the former is by far the most worldwide produced in large scale. Species of Trichogrammatidae are used in more than 30 countries for the biocontrol of over 20 host-pests in crops such as corn, cotton, sugarcane, fruit trees, vegetables, forests, among others (Li 1994). Over 16 million ha receive parasitoid inundative releases (Hassan 1997, van Lenteren 2000), mostly represented by the egg parasitoid *Trichogramma*. It is very difficult to precisely estimate the correct size of the area in which egg parasitoids are released, as there is a continuous increase in the number

J.R.P. Parra (✉)
Departamento de Entomologia e Acarologia, ESALQ/USP, Av. Pádua Dias 11, 13418-900 Piracicaba, São Paulo, Brazil
e-mail: jrpparra@esalq.usp.br

F.L. Consoli et al. (eds.), *Egg Parasitoids in Agroecosystems with Emphasis on Trichogramma*, Progress in Biological Control 9, DOI 10.1007/978-1-4020-9110-0_10,

of countries in which biofactories are being established for the mass production and release of egg parasitoids, as those in Latin America (van Lenteren and Bueno 2003) (see Chapter 14). Yet, there is also the case of countries in which egg parasitoids are extensively used in Applied Biological Control, as socialist countries as Russia, China and Cuba, whose statistics are not always available. Therefore, estimates vary from author to author, but Smith (1996) estimated the use of *Trichogramma* spp. for biocontrol of insect pests on 32 million ha.

The reference to the species identity used is also very variable in the literature due to the issues discussed above, as well as taxonomic difficulties brought about by the minute size of these insects. Hassan (1988) reported 28 *Trichogramma* species being released on 28 crops, with *Trichogramma evanescens* Westwood, *T. pretiosum* Riley and *T. dendrolimi* Matsumura being the most commonly used, certainly due to their polyphagy and adaptability to a diversity of habitats (Li 1994) (see Chapters 7–9).

In this chapter, I will deal with the mass rearing of egg parasitoids emphasizing the rearing of *Trichogramma* spp. in Brazil, where the release area of parasitoids is increasing, especially for the control of the sugarcane borer with *T. galloi* Zucchi, and other species for several other crops (Parra and Zucchi 2004).

10.2 Definitions and Concepts

Mass rearing involves the production of millions of insects for control of insect pests as a support for Integrated Pest Management (IPM) programs. However, there are various definitions as to what mass rearing is (Parra 2008):

"Mass rearing is the economical production of millions of beneficial insects, on a production line, with the aim of rearing, with a minimum of man/hour and space, the maximum number of fertile females, in the shortest time possible and at a low cost" (Finney and Fisher 1964);

"The production of insects capable of reaching objectives with an acceptable cost/benefit relationship, and in numbers exceeding from 10 thousand to 1 million times the average productivity of the population of native females" (Chambers 1977);

"This is a systematic and automatic activity undertaken in integrated facilities, and with the aim of producing a relatively large amount of insects for distribution" (Leppla and Adams 1987).

These definitions are from the 1960s, 1970s and 1980s, when *Quality Control* was not taken into the rearing process. Nowadays, the correct definition of mass rearing should also add that the reared insect should have the same quality as those found in nature (van Lenteren 2003) (see also Chapter 12).

Therefore, mass rearing is really a synonymous of a factory, in which cost is of primary importance, especially as regarded to labor, which represents 70-80% of the total production costs. Hence, the personnel involved should be highly qualified in the area of "insect rearing techniques", an area which is still neglected nowadays. Thus, and especially in countries in which labor is more expensive, automation in the various stages of the process of natural enemy production becomes necessary.

Mass rearing initiates from smaller rearings, as *those used for research*, that requires only one or two persons, or from those of *intermediate size*, used in basic research on the target (an agricultural pest, for example) and on a natural enemy (parasitoid or predator). Therefore, it is necessary to rear two species, the pest and the natural enemy. However, natural enemy mass rearing became possible only after the development of artificial diets in the 60's of the last century, mainly for Lepidoptera, Coleoptera and Diptera (Singh 1977, Singh and Moore 1985, Cohen 2004, Schneider 2009).

On considering the Basic Concepts in Biological Control, in other words, *Introduction* (Classical Biological Control), *Conservation* (Natural Biological Control) and *Multiplication* (Applied Biological Control), mass rearing is more important in the latter. In Applied Biological Control (ABC), inundative release of natural enemies is undertaken, causing effects similar to the use of conventional insecticides, as there is a knockdown effect of the target host population. Inundative release of natural enemies is well accepted by farmers as it generates an effect similar to conventional pesticides, and may be used in the field and in greenhouse (seasonal release).

10.3 Background on the Use of *Trichogramma*

The first attempts on using *Trichogramma* in biocontrol were undertaken in 1900, with the introduction of two exotic species from Austria to control *Euproctis chrysorrhoea* L. (Lepidoptera: Lymantridae) in the USA (Luck and Forster 2003). At the same period, *T. minutum* Riley was collected naturally parasitizing eggs of *E. chrysorrhoea*, and started to be reared on the natural host and stored at low temperatures for release at the appropriate time, whenever host eggs would be available. Nevertheless, biocontrol of *E. chrysorrhoea* with *Trichogramma* was unsuccessful.

The most relevant advance for the practical application of *Trichogramma* was the report of Flanders (1927) on the possibility of rearing *Trichogramma* on a factitious host, in this case, the Angoumois grain moth *Sitotroga cerealella* (Oliver).

The first attempts on using *Trichogramma* in applied biological control programs were not successful, especially by the fact that the inter- and multi-disciplinary aspects of biocontrol were not considered. Because of that, many errors occurred, leading to the unsuccessful use of *Trichogramma*, such as: (i) the target pest egg density was not taken into consideration, (ii) there was no selection for the adaptability of strains or species of *Trichogramma* for the control of the target pest, (iii) there was no quality control of laboratory-reared insects, (iv) no attention was paid on the number of parasitoids released nor to the releasing devices, (v) the pest population dynamics, plant phenology and competition with resident natural enemies were ignored, and (vi) a wide range of broad spectrum pesticides were applied simultaneously with parasitoid release.

Therefore, it became clear the need for the re-assessment of all aspects involved in the mass rearing and utilization of *Trichogramma* for biocontrol. The first important aspect was related with the factitious host selection, with the

Mediterranean flour moth *Anagasta kuehniella* (Zeller) being found to be much more suitable host than *S. cerealella* as a rearing host for *Trichogramma*, even though the latter species was easier to rear. *T. pretiosum* parasitization capacity was reduced from 147.9 eggs/female, when reared on *A. kuehniella,* to 9.9 eggs/females, when reared on *S. cerealella.* Parasitoid longevity was similarly affected, being reduced from 19.9 to 4.5 days for adults reared on *A. kuehniella* and *S. cerealella,* respectively (Lewis et al. 1976). In China, the foremost user of *Trichogramma,* with an extensive knowledge acquired throughout the years, the rice moth *Corcyra cephalonica* (Stainton) eggs or unfertilized eggs (mature oocytes) from silkworms, such as *Antheraea pernyi* (Guérin-Méneville) and *Philosamia cynthia* Rebel, are preferred for rearing *Trichogramma* instead (Parra 1997, Luck and Forster 2003).

The correct identification of the parasitoid was also one of the major issues related to the initial unsuccessful attempts on the use of *Trichogramma.* Taxonomy of Trichogrammatidae remained very confused until 1970, when Nagarkatti and Nagaraja (1971) identified the male genitalia as a reliable morphological structure for species identification of these minute parasitoids. Therefore, becomes very difficult to ascertain for the species identification of *Trichogramma* and/or *Trichogrammatoidea* in most of the work published before Nagarkatti and Nagaraja (1971) (see also Chapter 7). A representative example of species misidentification was the report of *T. minutum* as an egg parasitoid of *D. saccharalis* in Brazil as shown by surveys in the 1980s, which showed that the predominant species on the sugarcane borer were *T. galloi* Zucchi or *T. distinctum* Zucchi, depending on the sampling region in Brazil (Zucchi and Monteiro 1997).

Hence, from then onwards there was the need for wider inter- and multidisciplinary programs, requiring different steps (Parra et al. 2002), such as (i) taxonomical and bio-ecological studies, (ii) an investigation on mass-rearing techniques, release techniques and selectivity of agrochemicals, as well as an (iii) evaluation on the cost/benefits, and (iv) the development of host-parasitoid models which would also attend differences in climatic conditions, plant architecture, competition from natural enemies, among others (see Chapter 15).

10.4 Ways of Obtaining Phytophages and Natural Enemies

10.4.1 Phytophages

Basically, there are three ways of obtaining insects: (i) collecting populations in the field; (ii) keeping populations on natural hosts (intact plants, leaves, roots, bulbs, stems, etc.) in the laboratory; and (iii) maintaining laboratory populations on artificial diets.

Collecting field populations is the oldest way and is well accepted by "conservative" researchers, as it pertains to wild populations. Nevertheless, the disadvantages on the use of such hosts is that there is no indication of periodicity, nor knowledge of origin, nutrition or age, all of which can be limiting factors in certain types of studies and applications. In certain cases, as in the study of plant resistance to insects, which takes time, such periodicity of occurrence may delay research programs still further.

Keeping populations on natural hosts demands hard work, but are required for insects such as Hemiptera and Thysanoptera. In this case, attention should be paid on plants that are easy to cultivate and manipulate, although they may not always be the natural host. Depending on the region, maintenance of the rearing host plants in greenhouses may require the temperature, relative humidity (RH) and photoperiod control to provide suitable conditions for growth year-round. Special care should also be taken in multiple rearing systems of small-sized species (thrips and whiteflies), as colony contamination becomes highly likely if they are not kept in separate rearing rooms and in rearing units protected with a fine mesh.

Keeping populations on artificial diets is advantageous for many reasons, but the reduction in labor is the most evident. However, diets are prepared in accordance with the insect feeding habits and mouth apparatus, and were successfully developed for Lepidoptera, Coleoptera and Diptera (Table 10.1). Diets should contain all usual required nutrients for insect growth and development (proteins, vitamins, mineral salts, carbohydrates, lipids and sterols), and may also require special nutrients, such as nucleic acids. Besides the need for adequate nutrient provision and balance, artificial diets must also have certain physical properties, and provide the necessary phagostimulation (both physical and chemical) for adequate insect development. Insects obligately associated to symbionts may require special artificial diet formulation and rearing may not be possible (Dadd 1977, Cohen 2004). A correctly formulated artificial diet contains the physical properties and chemicals necessary to stimulate and maintain nourishment, with balanced nutrients (both essential and non-essential) to yield optimum growth and development, and should be free from contaminating microorganisms.

Table 10.1 Insect Orders with the number of families and species reared on artificial diets (adapted from Singh 1985)

Order		Families	Species
Lepidoptera		34	556
Coleoptera		20	284
Diptera		29	279
Hemiptera	Heteroptera	8	22
	Homoptera	6	71
Orthoptera		2	24
Hymenoptera		13	67
Neuroptera		2	8
Siphonaptera		1	3
Isoptera		3	5
Mallophaga		1	3
Dictyoptera		2	5
Dermaptera		1	1
Phasmida		1	1

In the case of aphids the diet has to be liquid, whereas for chewing phytophages, diet must be semi-liquid, with a high water content, but with a consistency that offers resistance to the insect's mouth apparatus. On the other hand, for cockroaches or pests of stored grains, diet must be in a powder form or friable. The more water the diet contains, the greater is the problem with microorganism contamination.

An adequate artificial diet should attend to the following characteristics (Singh and Moore 1985):

- Offers high larval viability
- Allows for an immature developmental time similar to that obtained when insects are reared on their natural diet;
- Yield adults with high fertility capacity;
- Suitable for more than one species and, if possible, for more than one insect Order;
- Made of low cost components, easily acquired on the market;
- Overall viability should be superior to 75%;
- Allow for the maintenance of insect quality through successive generations.

10.4.2 Natural Enemies

There are three ways of rearing natural enemies: (i) on the natural host, (ii) on factitious or substitute hosts and (iii) on an artificial diet (in vitro).

Currently, the most common way to produce natural enemies worldwide is to rear them *on their natural host*, which requires the rearing of two species of insects, the host and the natural enemy. Thus, in order to rear the braconid *Cotesia flavipes* (Cameron), its natural host *D. saccharalis* must be also available. In this particular case, parasitoid rearing is facilitated as there are innumerous suitable artificial diets for the sugarcane borer (Parra and Mihsfeldt 1992), as for natural enemies of other Lepidoptera, Coleoptera and Diptera. In other instances, diets are unavailable and the host has to be reared on the host plant, such as the rearing of wheat aphids as hosts for their parasitoids, and the rearing of *Phyllocnistis citrella* Stainton on citrus seedlings for the production of *Ageniaspis citricola* Logvinovskaya (Chagas and Parra 2000).

Natural enemies can be reared *on factitious or alternative hosts* the same way as phytophages. There are many species of *Citrus* scales that can be easily reared on pumpkins or other Cucurbitaceae, which are then used as hosts for coccinelids and nitidulids. Loayza et al. (2003) reared the brown citrus scale *Selenaspidus articulatus* (Morgan) on *Citrullus silvestris*, a Cucurbitaceae imported from Peru that can be used for up to one year for rearing the brown scale. Other scale species, such as *Dysmicoccus cryptus* (Hempel), can be reared on potatoes sprouts.

In a similar way, there are cases in which the parasitoid is reared on a host which is not normally parasitized, but is suitable to allow parasitoid development, which is called factitious or alternative host. As an example, *Heliothis virescens* (Fabr.) larvae are used as hosts for rearing *Heliothis armigera* (Hübner) parasitoids. Similarly, eggs of *A. kuehniella*, *S. cerealella* and *C. cephalonica*, and mature oocytes of the silkworms *P. cynthia* and *A. pernyi*, are used for rearing *Trichogramma* (Parra 1997, Luck and Forster 2003). Factitious hosts can also be used in rearing egg-larval parasitoids, as *A. kuehniella* (Ozkan and Ozmen 2001) and *Plodia interpunctella* (Hübner) for rearing *Chelonus oculator* Panzer (Ozkan 2006). Eggs of stored-product moths are also used for rearing chrysopids, coccinelids, and earwigs, among others. Recently, *Diachasmimorpha longicaudata* (Ashmead), a parasitoid introduced into Brazil to control *Anastrepha*, was reared on the Mediterranean fruit fly, *Ceratilis capitata* (Wied.). *Lixophaga diatraeae* (Towns.), a tachinid parasitoid of *D. saccharalis,* is reared on *Galleria mellonella* (L.) in several parts of the world

In artificial media (in vitro). This would be the ideal way of rearing, as production costs would be reduced through the elimination of a series of steps (see Chapter 11).

Even though the qualitative nutritious needs of all insects are similar, independent of their respective systematic position and feeding habits (The Rule of Identity), few are the cases of the success of an artificial diet for either parasitoids or predators, independent of the host. In general, the several attempts at rearing the insect in vitro on a large scale have failed. According to Thompson (1986) and Thompson and Hagen (1999), this scarcity of results is due not only to the lack of basic knowledge, such as physiology, biology, nutrition, genetics and parasitoid metabolism, but also as to the nature of host-parasitoid interaction.

As both physiologically and biochemically, the parasitoid is well adapted to the host for its survival and development, it becomes very difficult to prepare an artificial medium. Even in the case of predators, whose development is independent of the host physiology, besides presenting lower specificity, as yet, a totally synthetic diet has not been obtained.

One of the few successful cases of rearing a parasitoid in vitro is that of *Trichogramma* by the Chinese in artificial eggs provided with a polyethylene chorion (Li et al. 1986). The artificial medium is composed of the pupal hemolymph of *A. pernyi* (or *P. cynthia*), the yolk of chicken eggs, milk and Neisenheimer salts (the latter as a stimulant for egg-laying). This medium can be used for rearing various species of *Trichogramma*, although there is the need for using plastic of different thicknesses, in accordance with the size of the ovipositor of the reared species. The parasitoids are produced (by machine) in plastic rings or cards. Nowadays, computer systems permit the mass-rearing of *Trichogramma* in vitro for release in large areas.

In the U.S.A. and other developed countries, in vitro rearing is being investigated very intensively, and there are cases of success (even though partial) in the Trichogrammatidae, Sarcophagidae, Tachinidae, Ichneumonidae, Braconidae, Chalcididae, Pteromalidae, Tetrastichidae, Encyrtidae and Scelionidae families, and in Chrysopidae and Coccinellidae predators, besides a few representatives of the Hemiptera (*Nabis* sp., *Geocoris* sp. and *Podisus).* Studies have proceeded well, and to date there are 73 parasitoids reared "in vitro" (partially or totally), these consisting of 16 Diptera and 57 Hymenoptera species, besides 44 predator species from the orders Coleoptera (30 species) and Neuroptera (14 species), as well as representatives from the Hemiptera-Heteroptera (8 species) (Cônsoli and Parra 2002).

In Latin America, the first report to appear was that of Parra and Cônsoli (1992) on the in vitro rearing of *T. pretiosum* by using an artificial diet based on the hemolymph of *Helicoverpa zea* (Boddie), egg-yolk, milk and an antibiotic. Cônsoli and Parra (1996) improved the rearing technique for *T. pretiosum*, and for the first time reared *T. galloi in vitro*, with results comparable to those obtained by the Chinese. *T. atopovirilia*, has recently been reared on an artificial diet in Brazil (data not published) (see Chapter 11). Comparisons have been done by Cônsoli and Parra (1996, 1997 a,b, 1999 a,b) and Cônsoli et al. (1999) of the hemolymph and holotissues of diverse insect species for the in vitro rearing of *T. galloi* and *T. pretiosum*.

Even though there has been an advance over latter years in work on in vitro rearing of parasitoids and predators, only the Chinese retain the production technology (Grenier 1994) necessary for the mass rearing of *Trichogramma* species (Li et al. 1986) (see Cônsoli and Parra 1999b).

Research in this area in the remaining countries, even nowadays, only serves as a support for studies on natural enemy-host interaction from their biochemical (enzymatic) nutritional, physiological, etc., aspects, thus being far distant from the mass- rearing of natural enemies on artificial diets.

Therefore, in order to rear parasitoids and predators with the knowledge available nowadays, it is necessary to rear two species of insects, in other words, the host and the natural enemy (Fig. 10.1). Nutrigenomics should lead to an advance in the area (Coudron et al. 2006), as well as cell-culture, as that undertaken by Heslin et al. (2005) for *T. pretiosum* starting from *T. australicum* Girault.

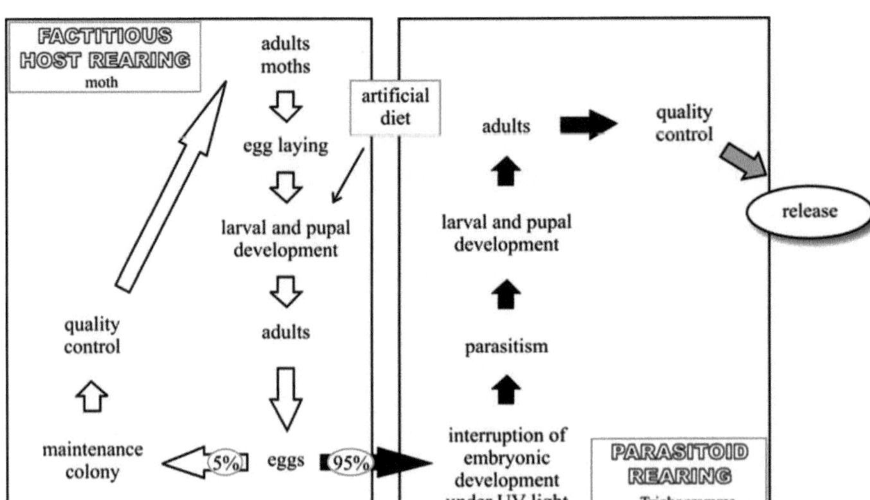

Fig. 10.1 Scheme for rearing *Trichogramma* on a factitious host (Parra et al. 2002)

10.5 Basic Studies on *Trichogramma*

In order to be able to use *Trichogramma* as a biocontrol agent, there is a series of steps to be followed (see Chapter 14).

Hence, after crop and natural enemy selection (in this case, the *Trichogramma* strain or species), basic studies from parasitoid taxonomy to quality control should be devised, including studies on the host (factitious in the present case) and on the parasitoid itself. Thus, the beginning is with taxonomy (classical and molecular) (Chapters 6–8), passing through the selection of strains, choice of the factitious host, thermal and hygrometric requirements, the best ratio parasitoid/host-egg for achieving the highest parasitism rate, adaptation to the factitious host, parasitization capacity, host selection behavior and quality control of the rearing process itself and of the product (natural enemy) to be released in the field (Leppla 2003).

Taxonomy. Recent studies (Pinto 2006) based on morphological and molecular research (Stouthamer et al. 1990a, b, 1993) refer to the existence of 210 described species of *Trichogramma.*

Strain or species selection is based on certain biological aspects, such as parasitism, life cycle development time (egg to adult), life cycle viability, sex ratio and longevity, whih are used for a cluster analysis or an analysis of the main components to define the most adequate strain/species (Cerutti and Bigler 1995, Pratissoli and Parra 2001).

In spite of the non-specificity of *Trichogramma*, the strain/species should preferably be collected on the target pest and in a region with climatic conditions similar to those of the target area, as the population growth capacity can be variable depending on the place of origin (Bleicher and Parra 1990) (Table 10.2). As mentioned by Kalyebi et al. (2005) for *Trichogramma* from Kenya, there are cases in which there is no correlation between altitude and climate. Furthermore, the presence of a taxonomist is indispensable, since incorrect identification can lead to a complete failure of a Biological Control program (see Chapter 7). A collection of strains from different geographical conditions and supervised by a taxonomist should be kept where the program is being developed (strain collection). Of the 210 *Trichogramma* species already described, 38 are found in South America (see Chapters 7 and 8).

Factitious host. The most adequate factitious host can vary depending on the species of *Trichogramma*. The predominant *Trichogramma* species in Brazil is *T. pretiosum*, which develops very well in *A. kuehniella*. On the other hand, *C. cephalonica* is the best host for *T. galloi* (Gomes and Parra 1998), a species which was released on 500,000 ha to control *D. saccharalis*. *Trichogramma*

Table 10.2 The effect of the collection site on the population growth capacity of *Trichogramma pretiosum* based on a fertility life table (Bleicher and Parra 1990)

Population	T	R_0	λ	rm
T. pretiosum (population from Icatu, CE)	15.5	102.13	1.3485	0.2990
T. pretiosum (population from Goiânia, GO)	14.2	44.38	1.3074	0.2680

atopovirilia can be reared on both *A. kuehniella* and *C. cephalonica*, but *T. bruni* and *Trichogrammatoidea annulata* De Santis prefer *C. cephalonica. S. cerealella* has always been the least suitable host for Brazilian species of *Trichogramma* (Parra et al. 1991, 1997, Dias et al. 2008).

Thermal requirements. Studies on thermal requirements are essential for forecasting production and for defining the association between pest cycle and natural enemy. These vary according to the species used as observed by Parra et al. (1991) and Parra et al. (2002) for Brazilian *Trichogramma* species and for the different factitious hosts that are used in Brazil.

Hygrometric requirements. Insect eggs are very sensitive to desiccation in humidities below 60%. *D. saccharalis* eggs can desiccate very fast under such low humidities, especially due to the large number of aeropyles, leading to an incomplete development of *T. galloi*, which could also affect field evaluation of this natural enemy efficiency (Parra et al. 1999).

The best ratio parasitoid/host egg experimentally defined in laboratory conditions and semi-field tests can vary from host to host, being 1.6 *T. galloi* females/*D. saccharalis* egg in sugarcane, 10.7 *T. pretiosum* females/*Helicoverpa zea* egg in corn, 5.3 *T. pretiosum* females/*Anticarsia gemmatalis* egg in soybean, and from 72 to 288 *T. atopovirilia*/*Gymnandrosoma aurantianum* Lima egg in citrus plants.

Part of the population should be reared on a natural host periodically in a rearing system to avoid preimaginal *adaptation to a factitious host* (Bigler 1994). Another aspect to be considered is that adaptation to a factitious host is not always easy under laboratory conditions, requiring successive rearing generations, leading to the selection of the most adapted individuals to the chosen rearing host (Leppla 2003, van Lenteren 2003).

Parasitization Capacity. Both the parasitism rhythm and parasitization capacity (variable from 70 to120 eggs/♀) should be assessed. The selection of strains/species that concentrate their parasitism activity during the first days of adult life is key to a successful control, especially in tropical regions where the parasitoid's lifespan is reduced due to more intense metabolic activity under conditions of higher temperatures.

Symbionts, especially *Wolbachia*, play an important role in *Trichogramma*, affecting the sex ratio through parthenogenesis induction (Stouthamer et al. 1990a, b, Werren 1997, Almeida 2004). As these bacteria can very often interfere in the physiology of the host and even in the parasitism capacity, it may be prudent to investigate their occurrence on selected strains. In the strain collection of the Department of Entomology and Acarology, 28% of the strains carry *Wolbachia*, 17% of which are telytokous and the reproductive mode in not affected in 11% of them (Salvador, Parra, Cônsoli unpublished data).

Quality control should be undertaken with insects reared in the laboratory, as conditions are different from those in nature in relation to temperature, humidity, nourishment, shelter, competition and pressure from other agents, human presence, the micro-environment, acceptance and search for mating, laying substrate, wind, dispersion, presence of chemicals and water, or food for adults (Singh and Moore 1985, van Lenteren 2003) (see Chapters 12 and 14).

In fact, two agents that can alter their characteristics throughout several laboratory generations are the factitious host and the parasitoid.

There are certain more important aspects also to be taken into account, especially those leading to changes in behavior, genetic deterioration and infection by pathogens, besides the fact that *Trichogramma* may contain bacteria and protozoa (Bjornson and Schütte 2003). A change in behavior is a consequence of the *quality of the host used* and of *storing* of the natural enemy when not in use, and which could lead to behavioral alterations. *Genetic deterioration* depends on genetic drift (founding effect), inbreeding and selection (Mackauer 1972, Bigler 1994, Leppla 2003). The *infection by pathogens* can be determinant in rearing, wherefore in many cases these can be viruses, bacteria and fungi as in *Cotesia*, or viruses, bacteria, protozoa and unidentified pathogens as in *Phytoseiulus* (Bjornson and Schütte 2003).

10.6 Mass Rearing

As earlier described, mass rearing involves millions of insects and is the result of initial rearing on a smaller scale (for research and intermediates, item Definitions and Concepts). The development of a rearing system for an egg parasitoid requires early studies on the host and parasitoid, including:

For the host:

- effects of temperature on host biology and determination of the host thermal requirements;
- ideal size of the rearing container for each development stage;
- techniques for host embryo development interruption;
- ideal combination of temperature for immature development and adult reproduction;
- time for exploiting collection boxes for adults and eggs;
- development of an adequate and economical diet;
- host storage techniques.

For the parasitoid:

- selection of *Trichogramma* strains with the correct identification, including that of symbionts;
- study of parasitoid biology at different temperatures and relative humidities in the most suitable factitious host;
- isolate rearing units for rearing different species;
- adequate parasitoid/host egg ratio;
- most suitable age of the host-egg to parasitism, in order to rear aggressive and competitive parasitoids;
- adaptation to the factitious host;
- parasitoid storage and quality control.

Following such basic studies, there is the need for an adjustment of scale, which can sometimes require a specific time, as it is very different to produce 2 or 3 g of moth's eggs when rearing for research, and then proceed to generating 2–4 kg of eggs for *Trichogramma* mass rearing (1 g of *A. kuehniella* eggs correspond to 36,000 eggs with an egg laying capacity of 445 eggs/♀ at 25°C) (Parra 1997).

The adjustment in the production scale should take into account early steps in the rearing process, such as the selection of nutritionally adequate and cheap artificial diets, up to the definition of the ideal number of eggs to be inoculated per tray to yield the highest number of heaviest adults (Fig. 10.2a), the correlation between the weight of the female and eggs laid (Fig. 10.2b), or further still, the correlation between the number of adults per rearing unit and the number of eggs laid (Fig. 10.2c). Generally, diets for moths used as factitious host for *Trichogramma* are of low cost and composed of whole wheat flour (97%) and yeast (3%), or wheat flour (40%) and corn flour (60%), for *A. kuehniella (*Daumal et al. 1975, 1985, Parra 1997), rice bran (94%), sugar (3%) and yeast (3%) for *C. cephalonica* (Bernardi et al. 2000), and whole wheat grains for *S. cerealella* (Flanders 1927). Further diets have been tested for *C. cephalonica* (Nathan et al. 2006) including millet, wheat, rice and sorghum.

One of the mass rearing system for *Trichogramma* spp. employed in Brazil (Fig. 10.3), using *A. kuehniella* as a factitious host, achieves a production of 5 kg of eggs per day, with a labor power of approximately 30 employees.

Once obtained and before being used for parasitism, embryonic development of the host eggs should be halted by UV exposure, as eclosed larvae from unparasitized eggs will canabilize both parasitized and unparasitized eggs, affecting the production efficiency. Eggs of *A. kuehniella* and *C. cephalonica* must be treated, and such treatment will not affect parasitoid acceptance and parasitization (Voegelé et al.1974).

Nowadays, *T. galloi* is released in 500,000 ha in Brazil, aiming at the control of *D. saccharalis* in sugarcane, besides the further release in more than 50,000 ha of *T. pretiosum* and *T. galloi* for controling *Tuta absoluta* (Meyrick) (Lep: Gelechiidae) in tomato plants and *D. saccharalis* in corn. In corn, *T. atopovirilia* is also released for *Spodoptera frugiperda* (J. E. Smith) (Lepidoptera: Noctuidae) control.

At present, there exists the mass rearing of natural enemies that are marketed as "biological insecticides". One of the most produced worldwide is *Trichogramma*, which is used to treat wide areas, especially in the socialist countries and in the New World (see Chapter 14).

Mass rearing involves the daily production of millions of insects and, in fact, resembles the production line of any other product. In the U.S.A., in the case of the control of *Cochliomyia hominivorax* (Coquerel) by means of the sterile insect technique, from 50 to 200 million sterile flies were produced and released weekly. At the end of the program, it was discovered that 49 sterile females had been released for each female in nature, a much higher ratio than that theoretically forecasted for this method, which is 1:9. More than three hundred workers were employed in this "factory". In such case, besides biological problems with rearing, others such as stock control, the purchase and storage of material, and installation and maintenance of equipments also appears. As the number of insects produced grows,

Fig. 10.2 Necessary specifications for scale adjustment in *Anagasta kuehniella* mass rearing. (**a**) Relationship eggs/tray vs. adult weight. (**b**) Relationship adult weight vs. eggs/female. (**c**) Total of adults/egg laying unit vs. eggs/female

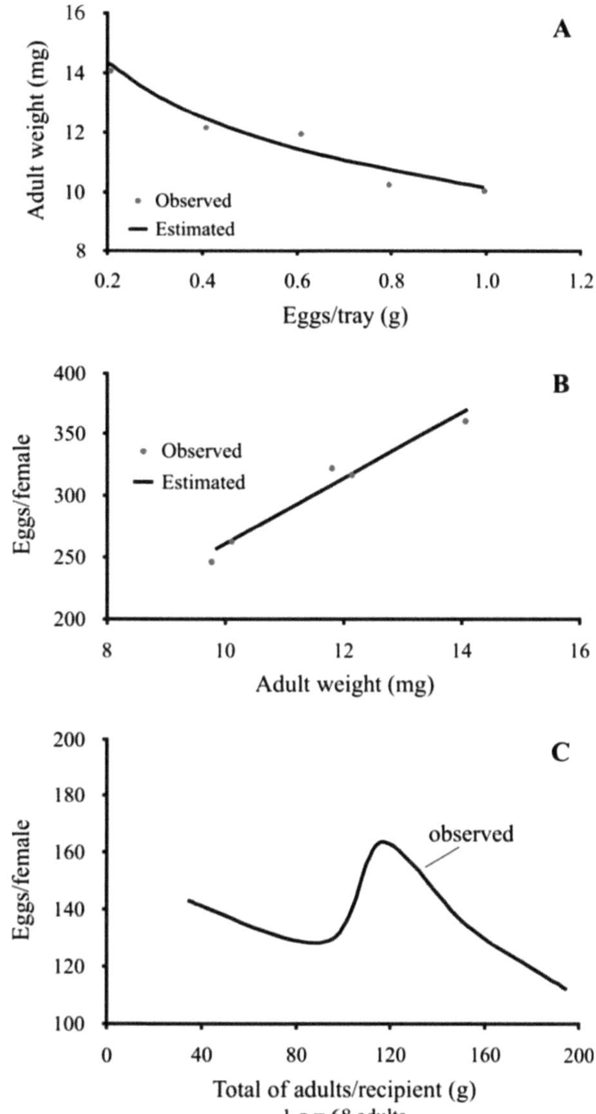

there is an increase in problems linked to facilities, costs, contaminating microorganisms and quality control of insects, which makes investments necessary on the automation (mechanization) of the rearing system. Automation should be incited when weekly production passes 3000–5000 adults. Details on the mass rearing of insects can be found in Smith (1966), Leppla and Ashley (1978), King and Leppla (1984), Singh and Moore (1985), Parrela et al. (1992), van Driesche and Bellows (1996), Ridgway and Inscoe (1998), Bellows and Fisher (1999), Etzel and Legner (1999), van Lenteren (2000), Cohen (2004) and Parra (2008).

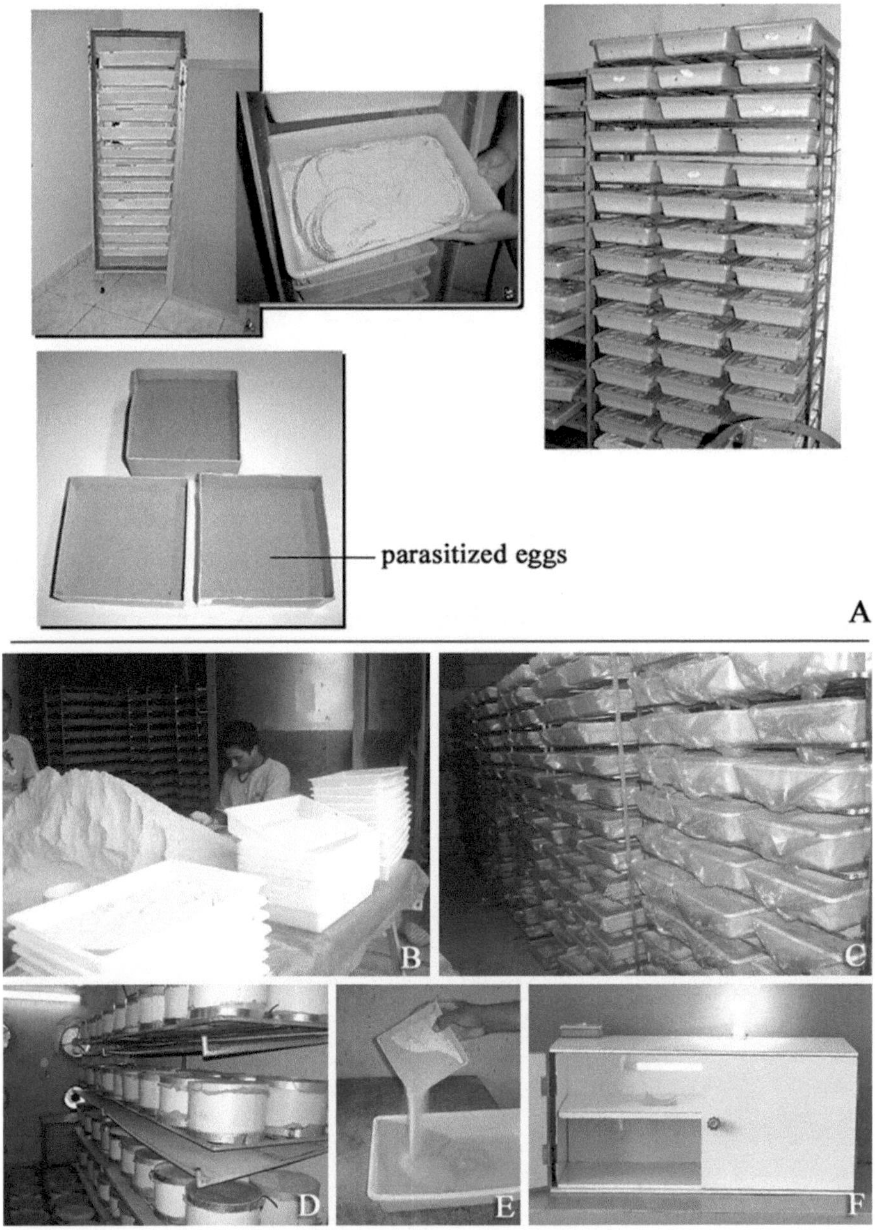

Fig. 10.3 An *in vivo* production line of *Trichogramma*. (**a**) Overall view. (**b**) Artificial diet for the factitious host. (**c**) Larval development. (**d**) Egg laying units. (**e**) Eggs of *Anagasta kuehniella*. (**f**) UV-treatment unit used for halting the host embryonic development

10.7 Constraints in the Mass Rearing of Natural Enemies

It is very difficult to separate the problems with the rearing of any insect whatsoever from those linked to the production of natural enemies. In the case of natural enemies, one is dealing with the rearing of two species, which is a highly complicating factor. Therefore, there is the need to know the biology, ecology, behavior, physiology, nutrition, etc., of both the host and the natural enemy, besides, evidently, their inter-relationships.

It is impossible to develop a mass rearing system without first having done research on a small scale rearing unit. The problems in any rearing activity grow as the number of insects produced increases. It is obvious that if the number of insects reared is to increase, the first problem will be related to the *facilities*. Whereas in a laboratory and in order to rear an insect on an artificial diet, three or four rooms are enough, while for rearing natural enemies, it will be necessary to have twice as much the area available for rearing the host. Details on facilities for insect rearing can be found in Leppla and Ashley (1978) and Cohen (2004).

The *cost* in the mass rearing system is another issue as 70–80% of the total expense refers to manpower. The larger the number of insects reared, the higher will be the expenses. Costs involved with manpower will be higher in first world countries, and may be prohibitive in rearing some species, making clear the necessity of automation of the rearing system. In Brazil, for example, it is common for rearing units of *C. flavipes* to have from 25 to 30 employees. However, enterprises are now beginning to appear in which more than 100 people work. In Europe, the main companies employ more than 70 people (van Lenteren 2003).

As most of the hosts used for rearing natural enemies are kept on artificial diets, *sanitary (contamination)* problems increase as the number of insects reared becomes larger. In this case, rooms for diet handling, larval and pupal development, host maintenance, parasitization and parasitoid development are to be carefully addressed in a mass rearing operation. Constant cleaning and the coating of walls, counters and floors with chemicals that allow for disinfection/purification is fundamental. Products containing quaternary compounds of ammonia, sodium hypochlorite, formaldehyde, among others, are commonly used. Sodium hypochlorite is the most used as it is cheap, have good stability and solubility, and is not very toxic to mammals. Besides these advantages, through being a strong oxidant, its use causes proteins to become innactive, besides eliminating viruses, bacteria, fungi, algae and protozoa. It is also interesting that floors receive daily treatment with this type of product. The entrance of personnel into the laboratory should also be rigorously controlled.

The sterilization of equipment and utilities through a physical source (autoclave, irradiation and dry heat) is fundamental for successful rearing. Nevertheless, the surface sterilization of eggs (sodium hypochlorite, formaldehyde, copper sulphate, glacial acetic acid, trichloroacetic acid, and of the artificial diet (methyl parahydroxybenzoate, butyl and propyl parahydroxybenzoate, sodium hypochlorite, propionic acid, sodium benzoate, benzoic acid, potassium sorbate, sorbic acid and antibiotics), constitute important steps to avoid contamination. Physical agents, such as irradiation or temperature, can also be important in sterilizing the medium. pH adjustment is an alternative method to avoid contamination, as low pHs favors the use of lower amounts of preservatives (Dunkel and Read 1991). In order to achieve efficiency in sanitary control, pupae should also be sterilized (sodium hypochlorite).

In any case and regarding artificial diets, certain aspects should be taken into consideration in the choice of the preservatives (Dunkel and Read 1991).

- dosage (test for each product);
- water content of the diet, pH and formulation of the product;
- stage of the exposed insect;
- effects on growth, in one and subsequent generations;
- effects on symbionts;
- insect Order.

Coleoptera and Diptera are more sensitive to preservatives, although there are families of Lepidoptera which are affected by sorbic acid (Dunkel and Read 1991). For further details see Greenberg (1970), Funke (1983), Sikorowski (1984), Soares (1992) and Sikorowski et al. (2001).

Quality of the insect produced is another important factor in mass rearing. To avoid problems with laboratory-reared insects, rearing should be initiated with a reasonable number of insects, from 200 to 500 individuals (Waage et al. 1985). In general, when the population is introduced into the laboratory, there will be a drop in genetic variability, due to genetic drift, selection and inbreeding during the first generations. It is only about the fifth to seventh generation that a recovery on the variability will occur due to mutation and recombination (Boller and Chambers 1977, Bartlett 1984, Leppla and Ashley 1978, Leppla and Fisher 1989, Leppla and Williams 1992, Bigler 1994).

Quality of lab-produced insects can be evaluated in several ways to check their fitness as compared to those in nature, and parameters to be evaluated would depend on the use the reared insect would have. Adaptability, mobility, sexual activity, breeding and colonization were referred as to be the quality components for quality assessment. Besides the evaluation of such parameters, it is recommended the periodical introduction of field-collected specimens into lab-maintained colonies.

There are also suggestions that rearing should be taken under fluctuating temperatures as a way to maintain quality, as adaptation to constant temperature, RH and photoperiod conditions could be avoided. Nevertheless, no differences were observed for *T. galloi* reared under constant and alternating temperatures (Cônsoli and Parra 1995a, b).

As previously mentioned, the parameters to measure fitness in a mass rearing procedure would be variable depending on the application seek for the reared insect. In cases of applied biological control, there should be taken into consideration characteristics of the *laboratory* (host preference and suitability), *semi-field* (searching capacity) and *field* (efficiency, adaptation, habitat and host location, host suitability and acceptance, and synchronization with the host) (Hassan 1997).

After selecting the species/strain of *Trichogramma*, the IOBC indicates the following tests during the production stage: determination of the percentage of emergence, sex ratio, duration of the cycle, morphological abnormalities (wings and abdomen), fertility (capacity to parasitize), longevity, locomotion (walking and flight) (see Bigler 1994 and Dutton and Bigler 1995), acceptance of the natural host and species/strain identification (Bigler 1994). There are specific guidelines available for *Trichogramma* and other natural enemies, which have baseline values for the measured parameters for measuring insect quality (van Lenteren, 2003). For *T. brassicae* Bezd., *T. cacoeciae* Marchal and *T. dendrolimi* using *A. kuehniella* as a factitious host, insects should score: parasitism $\geq$25 eggs/female in 48 h; emergence $\geq$ 80%; sex ratio $\geq$0.5; longevity $\geq$7 days (van Lenteren 1994) (see Chapter 12). However, there are reports indicating that insect quality can be measured with a smaller number of parameters and in an easier way. Prezotti et al. (2002) reported that longevity, parasitism and flight capacity were enough as indicators of the insect quality in rearing systems for these egg parasitoids. Furthermore, Prezotti et al. (2004) reported the number of founding females of *Trichogramma* can be very low, without any lost in quality during continuous rearing.

Storage is another problem in mass rearing, as entomologists find problems with the availability of insects throughout the year. One of the solutions found to overcome this difficulty would be the storage of insects in temperatures which can temporarily halt their development. Insect storage on such conditions would also facilitate insect exchange among laboratories of different regions, limiting the usual problems of hatching or emergence during shipping.

In continuous mass rearing of natural enemies, the host could be stored during the period of the year in which the natural enemy is not required. Trichogrammatids can be reared in eggs stored at −10°C for more than one year (Drooz 1981), but stored for up to nine months in liquid nitrogen (Huai-Yi 1988). On the other hand, eggs parasitized by *Trichogramma* can be kept for 20–22 days on temperatures between 8 and 12°C. The pupa is another stage of development that can be stored. Last-day pupae of *T. pretiosum* was stored from 4 to 10 days at 16.7°C, and up to 12 days if temperature was lowered to 15°C on the sixth day of exposure. Emergence after storage was 93% in a period of four hours at 26.7°C (Stinner et al. 1974).

One way of beginning the storage of one of the insect developmental stages (egg, pupa) is by means of its thermal requirements, as in theory, insect development would be halted at the lower threshold temperature without any harm, being the ideal temperature for insect storage. One other way would be the induction of diapause. In this case, favorable conditions would be made available at the sensitive stage when necessary.

Trichogramma evanescens Westwood pupae were stored from six months to a year, yielding over 90% emergence. Over shorter storage periods (three to six months) the emergence was 95%. The technique used is described as follows. Newly laid *A. kuehniella* eggs were offered to the parasitoid during four hours at 20°C, followed by incubation at 20°C for 24 h. Further incubation was undertaken for 24 h at 20°C with a photophase of 8 h. In order to induce diapause, eggs were subsequently kept for 40 d at 14°C with 8 h of photophase. Eggs were then stored at 3°C and

70% RH (see Boivin 1994). Diapause induction was also studied for *T. cordubensis* Vargas and Cabello (Garcia et al. 2002).

In Brazil, there is the need for more in depth studies on insect diapause (including parasitoids), as there are indications that *Trichogramma* species/populations, such as, *T. galloi* and *T. pretiosum* from Piracicaba, SP, do not present diapause (Rossi 1997). Until the mechanisms that allow for cessation of development in Neotropical species are known, one way to store these parasitoids is to keep newly-parasitized eggs at 18°C, and lower the temperature to 10°C on the 20th day. This procedure will allow the storage of parasitized eggs for a further 40 days, aiding on the scheduling of field releases (Parra 1997).

Although the general lack of information on diapause for Neotropical species, there are few examples in which mass reared insects can be stored in Brazil. Eggs of *Nezara viridula* (L.) for the production of *Trissolcus basalis* (Wollaston) are stored for periods up to 30 days at 5°C, or up to six months if stored in liquid nitrogen (Corrêa-Ferreira and Moscardi, 1993). Yet, adults of *T. basalis* can also be stored for 120 d at temperatures from 15 to 18°C (Foerster and Doetzer 2006).

Therefore, as the number of laboratory-reared insects increase, so do the problems, especially those related to *facilities, costs, automation, sanitation, insect quality* and *storage techniques*.

Even though there are peculiarities among parasitoids and predators, whether they be the natural enemies of eggs, larvae (nymphs), pupae or adults with extremely variable behaviors (Vinson 1997), including idio and koinobionts among parasitoids, and polyphages, stenophages, oligophages and monophages, among predators, it is necessary to know the biology and how to rear them in laboratory conditions. Therefore, there is no rule of thumb for rearing insects in the laboratory, as insects are diverse and have a diversity of habits. Nevertheless, microclimatic requirements, such as *temperature*, *relative humidity of the air (RH), light* and *aeration (ventilation)* should be taken into account in any rearing process. Besides these *abiotic* conditions, *biotic* factors, mainly *mating*, *oviposition* and *adult nutrition* also deserve attention. In certain regions, the *diapause* is a very important parameter, even though it is poorly studied in Brazil.

For each insect development stage for both host and natural enemy, there should be an optimum space for better development (Peters and Barbosa 1977). There must also be the best possible host/natural enemy ratio to avoid superparasitism. The most adequate age of the host for parasitism should be known (age of the egg for an egg parasitoid; the most adequate instar for a larval or nymphal parasitoid; most favorable age of the pupa for parasitism, or even the age of the adult to yield the highest possible parasitism).

In general, *Trichogramma* prefers newly-laid eggs for parasitism and can be maintained in laboratory culture by offering 10 newly-laid eggs/ *T. pretiosum* female and 4 eggs/*T. galloi* female (Parra 1997). Rearing should start with eggs from laboratories that observe rigorous quality control and by using adequate facilities and diets.

In all rearing processes, there are as much general problems, such as ants and mites, as specific. In the case of *Trichogramma* reared on *A. kuehniella* eggs, the ectoparasitoid *B. hebetor* and humidity are the main problems. The ectoparasitoid is attracted by the "frass" from *A. kuehniella* last instars (Parra et al. 1996). Either mass rearing must be done in installations protected by mesh-screens, or laboratories must be kept at low temperatures to avoid the entrance of *Bracon hebetor* (Parra 1997).

Trichogramma pretiosum lives 7.2 times longer with access to the host and a source of food (pure honey) (Bleicher and Parra 1991). Mac Dougall and Mills (1997) noted that *T. platneri* lived 9.5 days when fed on honeydew from *Dysaphis plantaginea* (Pass.) on apple, as compared to those fed only on water that lived only 1.2 days. Fuchsberg et al. (2007) also observed that *T. ostriniae* fed on honeydew parasitized more and lived longer in relation to those fed only water (see Jervis et al. 2005, Chapter 2 and Wäckers et al. 2008).

10.8 Forecasting Production

In certain countries, and for some time now, there are sophisticated follow-up and production programs for insects that make use of computerized systems (Singh and Clare 1992). Nevertheless, it is possible to draw up a scheme for production, simply taking as a base the insects' thermal requirements. These requirements are evaluated by the thermal constant (K) expressed in degrees/day, which has been used in forecasting plant growth for many years. This constant is based on the hypothesis that the duration of development by temperature is a constant, being the summation of the temperature computed starting from a lower thermal threshold for development (Tt – temperature threshold). As insects are *poikilothermic,* in other words, they comply with the temperature of their suroundings, the thermal constant can also be applied to insect development.

Thus,

$$K = D(T - Tt),$$

in which:

K = thermal constant (degree-days);
D = development time (days);
T = temperature of the environment (ºC);
Tt = lower temperature threshold for development.

Once the lower thermal threshold is defined, development can be undertaken according to Morris and Fulton (1970), and the insect's cycle can be estimated in a room where the temperature is registered or controlled. For example, in a rearing room kept at 25°C, it is possible to estimate the duration of *T. pretiosum*, which has a lower thermal threshold for development of 12.8°C (Tb) and a thermal constant of 133.1 degrees/day. In this case, the duration of development, when values are applied to the K formula, will be 10.4 days. The same can be applied for the factitious hosts (see Section 10.5).

10.9 Concluding Remarks

Mass rearing involves intrinsic problems (discussed in the present chapter), and as one is dealing with two species of insects to be reared, basic knowledge on both the host and parasitoid are imperative, including aspects regarding to biology, ecology, biochemistry, behavior, physiology and nutrition, and especially studies on thermal requirements for forecasting production. Other intrinsic problems are also involved, including, in this case, studies on selectivity (in crops where pesticides are applied), market and shipping logistics (especially in the case of large countries), competition with other natural enemies, climatic factors which affect parasitoid efficiency (Stiling 1993), time of release, releasing devices, dispersion capacity, and number of parasitoids per area, among others.

Biological control in the present-day philosophy of Integrated Pest Management and Integrated Fruit Production should always be regarded from a global point of view, involving both inter- and multidisciplinary actions. Within this context, mass rearing is becoming more important in Applied Biological Control programs.

As the number of insects produced in any rearing process increases, there is a rise in problems related to facilities, cost, microorganisms (contaminants) and insect quality control. Besides the need for automation (especially in countries where the use of manpower is more intensive), problems can appear with personnel due to allergic reactions arising from the manipulation of moths used for *Trichogramma* rearing.

Another problem arising from mass rearing is that the insect will be used in only one period of the year, making necessary the use of storage techniques for periods in which they are not required. Some storage techniques are available, as the use of diapause induction in *Trichogramma.* Storage in liquid nitrogen is also an alternative way for host egg storage. One of the great concerns that remain is with control of quality of the reared insect. Furthermore, the loss of genetic variability of mass-reared insects can lead to the loss of traits from the wild insect, making the laboratory-reared insect less competitive in nature. There are two types of processes that can result in genetic deterioration, alleatory and non-alleatory or adaptive. Among them, genetic drift (or founding effect), inbreeding and selection are the most relevant. Thus, accompanying production, the process itself and the final product stages throughout generations from the laboratory is key to the production of laboratory insects which will be competitive with those in nature. Standards are already available, as those established by the IOBC, which take into account these biological characteristics and molecular analyses, and which must be accompanied and legislated in order to avoid the appearance of ethical issues which could darken the image of Biological Control (van Lenteren 2003).

The ideal for mass rearing would be to eliminate the need of a host, rearing the parasitoid or predator on an artificial diet (in vitro). Even though the Chinese showed the world the advance in "in vitro" rearing in 1986, in the case of *T. dendrolimi* which was reared on an artificial diet, contrary to what was expected, there has been no further advance in the area, causing frustration in the scientific community.

Thus, there is the need for further studies as diets used nowadays are based on those developed in the 1960. The search for better diets should contemplate studies involving Nutrition Science, including nutritional requirements, such as analysis of the insect food matrix, but also taking into account pertinent technology and the equipment employed for large scale production. More refined bio-assays, using microscopic, nanotechnological, molecular and biochemical techniques, besides those of fermentation, could lead to an advance in the area. More sophisticated bio-assays could also improve the control of microorganisms in large-scale rearing processes, as well as elucidating the role of primary symbionts in insect nutrition. Running alongside, it is fundamental to give value to those working on insect rearing training at the undergraduate and graduate levels, as the use of natural enemies for pest control is a cultural process and requires qualified personnel for the acceptance of this extremely important option for the environment (Cohen 2004). This is fully possible, as the great majority of mass rearing is done by commercial enterprises nowadays, since government programs worldwide in this field lost ground over the last years.

Acknowledgments To Dr. Norman C. Leppla, Department of Entomology and Nematology, University of Florida, Gainesville, U.S.A., for the critical reading of the chapter.

References

Almeida RP (2004) *Trichogramma* and its relationship with *Wolbachia*. Identification of *Trichogramma* species, phylogeny, transfer and costs of *Wolbachia* symbionts. Ph D Thesis, Wageningen University, The Netherlands. 142p

Bartlet AC (1984) Genetic changes during insect domestication In: King EG, Leppla NC (eds) Advances and challenges in insect rearing. USDA, ARS, Washington, DC, pp 2–8

Bellows TS, Fisher TW (1999) Handbook of Biological Control. Academic, New York. 1046p

Bernardi EB, Haddad ML, Parra JRP (2000) Comparison of artificial diets for rearing *Corcyra cephalonica* (Stainton, 1865) (Lep., Pyralidae) for *Trichogramma* mass production. Revista Brasileira de Biologia 60:45–52

Bigler F (1994) Quality control in *Trichogramma* production. In Wajnberg E, Hassan SA (eds) Biological control with egg parasitoids. CAB International, Wallingford, CT

Bjornson S, Schütte C (2003) Pathogens of mass- produced natural enemies and pollinators, pp 133–165. In van Lenteren JC (ed), Quality control and production of biological control agents: Theory and testing procedures. CAB International, Wallingford, CT, pp 93–111

Bleicher E, Parra JRP (1990) Espécies de *Trichogramma* parasitóides de *Alabama argillacea*. II. Tabela de vida de fertilidade e parasitismo de três populações. Pesquisa Agropecuária Brasileira 25:207–214

Bleicher E, Parra JRP (1991) Efeito do hospedeiro de substituição e da alimentação na longevidade de *Trichogramma* spp. Pesquisa Agropecuária Brasileira 26:1845–1850

Boivin G (1994) Overwintering strategies of egg parasitoids, pp 219–244. In Wajnberg E, Hassan SA (eds) Biological control with egg parasitoids. CAB International, Wallingford, CT

Boller EF, Chambers DL (1977) Quality aspects of mass reared insects. In: Ridgway RL, Vinson SB (eds) Biological control by augmentation of natural enemies. Plenum Press, New York, pp 219–235

Cerutti F, Bigler F (1995) Quality assessment of *Trichogramma brassicae* in the laboratory. Entomol Exp Appl 75:19–26

Chagas MCM, Parra JRP (2000) *Phyllocnistis citrella* Stainton (Lepidoptera: Gracillariidae): técnica de criação e biologia em diferentes temperaturas. Anais da Sociedade Entomológica do Brasil 29:227–235

Chambers DL (1977) Quality control in mass rearing. Annu Rev Entomol 22: 289–308

Cohen AC (2004) Insect diets. Science and techonology. CRC Press, Cleveland, OH, 324p

Cônsoli FL, Parra JRP (1995a) Effects of constant and alternating temperatures on *Trichogramma galloi* Zucchi (Hym., Trichogrammatidae) biology. I – Developmental and thermal requirements. J Appl Entomol 119:415–418

Cônsoli FL, Parra JRP (1995b) Effects of constant and alternating temperatures on *Trichogramma galloi* Zucchi (Hym., Trichogrammatidae) biology. II – parasitism capacity and longevity. J Appl Entomol 119:667–670

Cônsoli FL, Parra JRP (1996) Comparison of hemolymph and holotissues of different species of insect as diet components for in vitro rearing of *Trichogramma galloi* Zucchi and *T. pretiosum* Riley. Biol Control 6 401–406

Cônsoli FL, Parra JRP (1997a) Development of an oligidic diet for in vitro rearing of *Trichogramma galloi* Zucchi and *Trichogramma pretiosum* Riley. Biol Control 8: 172–176

Cônsoli FL, Parra JRP (1997b) Produção "in vitro" de parasitóides: criação de *Trichogramma galloi* Zucchi e *T. pretiosum* Riley, no Brasil, I: Parra JRP, Zucchi RA (eds) *Trichogramma* e o Controle Biológico Aplicado. FEALQ, Piracicaba, Brasil, pp 259–302

Cônsoli FL, Parra JRP (1999a) Development of an artificial host egg for in vitro egg laying of *Trichogramma galloi* and *T. pretiosum* using plastic membranes. Entomol Exp Appl 91: 327–336

Cônsoli FL, Parra JRP (1999b) "In vitro" rearing of parasitoids: constraints and perspectives. Trends Entomol 2:19–32

Cônsoli FL, Parra JRP (2002) Criação in vitro de parasitóides e predadores. In Parra JRP, Botelho PSM, Corrêa-Ferreira BS, Bento JMS (eds) Controle biológico no Brasil: parasitóides e predadores. Editora Manole Ltda, Barueri, SP. 609p, pp 239–275

Cônsoli FL, Kitajima EW, Parra JRP (1999) Sensilla on the antenna and ovipositor of the parasitic wasps *Trichogramma galloi* Zucchi and *T. pretiosum* Riley (Hym., Trichogrammatidae). Microsc Res Tech 45:313–324

Corrêa-Ferreira BS, Moscardi F (1993) Técnicas de armazenamento de ovos do percevejo-verde visando à multiplicação do parasitóide *Trissolcus basalis* (Wollaston). Pesquisa Agropecuária Brasil 28:1247–1253

Coudron TA, Yocum GD, Brandt SL (2006) Nutrigenomics: a case study in the measurement of insect response to nutritional quality. Entomol Exp Appl 121:1–14

Dadd RH (1977) Qualitative requirements and utilization of nutrients: insects. In: Rechcigl Jr M (ed) Handbook series in nutrition and food, Section D, v.1. CRC Press, Cleveland, OH, pp 305–346

Daumal J, Voegelé J, Brun P (1975) Les *Trichogrammes*. II. Unité de production massive et quotidienne d'un hôte de substitution *Ephestia kuehniella* Zell. (Lepidoptera, Pyralidae). Ann Zool Ecol Anim 7:45–49

Daumal J, Marconi D, Chassain C (1985) Dispositif d'elevage miniaturise et automatise d' *Ephestia kuehniella* Zeller (Lepidoptera, Pyralidae). Bull Soc Linnean Lyon 54:7–12

Dias NS, Parra JRP, Costa Lima TC (2008) Seleção de hospedeiro alternativo para três espécies de tricogramatídeos tropicais. Pesquisa Agropecuária Brasil 43:1467–1473

Drooz AT (1981) Subfreezing eggs of *Lambdina pellucidaria* (Lepidoptera: Geometridae) alters status as factitious host for *Ooencyrtus ennomophagus* (Hymenoptera: Encyrtidae). Can Entomol 113:775–776

Dunkel FV, Read N.R. (1991) Review of the effects of sorbic acid on insect survival in rearing diets with reference to other antimicrobials. Am Entomol 37:172–173

Dutton A, Bigler F (1995) Flight activity assessment of the egg parasitoid *Trichogramma brassicae* (Hym.: Trichogrammatidae) in laboratory and field conditions. Entomophaga 40: 223–233

Etzel LK, Legner EF (1999) Culture and colonization. In Bellows TS, Fisher TW (eds) Handbook of biological control. Academic, San Diego. 1046p, pp 125–197

Finney GL, Fisher TW (1964) Culture of entomophagous insects and their host. In DeBach P, Sclinger EI (eds) Biological control of insect pests and weeds Chapman and Hall, London. 844p, pp 328–355

Flanders SE (1927) Biological control of the codling moth (*Carpocapsa pomonella).* J Econ Entomol 20:644

Foerster LA, Doetzer AK (2006) Cold storage of the egg parasitoid *Trissolcus basalis* (Wollaston) and *Telenomus podisi* Ashmead (Hymenoptera: Scelionidae). Biol Control 36:232–237

Fuchsberg JR, Yong TH, Losey JE, Carter ME, Hofmmann AA (2007) Evaluation of corn leaf aphid (*Rhopalosiphum maidis*; Homoptera: Aphididae) honeydew as a food for the egg parasitoid *Trichogramma ostriniae* (Hymenoptera: Trichogrammatidae). Biol Control 40:230–236

Funke BR (1983) Mold control for insect-rearing media. Bull Entomol Soc Am 29:41–44

Garcia P, Wajnberg E, Pizol J, Oliveira L (2002) Role of temperature in the induction of diapause in the egg parasitoid, *Trichogramma cordubensis.* J Insect Physiol 43:349–355

Gomes SM, Parra JRP (1998) The parasitization as a tool for factitious host selection for *Trichogramma galloi* Zucchi and *T. pretiosum* Riley. Mitt. Biol. Bundesaustalt. Land-Forstwirtsch. Berlin-Dahlem, H. 356, pp 13–23

Greenberg B (1970) Sterilizing procedures and agents, antibiotics and inhibitors in mass rearing of insects. Bull Entomol Soc Am 16:31–36

Grenier S (1994) Rearing of *Trichogramma* and other egg parasitoids on artificial diets. In: Wajnberg E, Hassan SA (eds) Biological control with egg parasitoids. CAB International, Wallingford, CT, pp 73–92

Hassan SA (1988) Guidelines of testing the side effect of pesticides on the egg parasite *Trichogramma cacoeciae.* IOBC/WPRS Bulletin 11:3–18

Hassan SA (1997) Seleção de espécies de *Trichogramma* para o uso em programas de controle biológico. In: Parra JRP, Zucchi RA (eds), *Trichogramma* e o Controle Biológico Aplicado. FEALQ, Piracicaba, Brasil, pp 183–206

Heslin LM, Kopittke RA, Merritt DJ (2005) The role of insect cell lines in an artificial diet for the parasitoid wasp, *Trichogramma pretiosum.* Biol Control 33:186–193

Huai-Yi MA (1988) Studies on long-term storage of hosts of propagating *Trichogramma.* Les Colloques de I'INRA 3:369–371

Jervis MA, Copland MJW, Harvey JA (2005) The life cycle. In: Jervis MA (ed) Insects as natural enemies: a practical perspective. Springer, Dordrecht, pp 73–165

Kalyebi A, Overholt WA, Achulthess F, Mueke JM, Hassan SA, Sithanantham S (2005) Functional response of six indigenous trichogrammatid egg parasitoids (Hymenoptera: Trichogrammatidae) in Kenya: influence of temperature and relative humidity. Biological Control 32:164–171

King EG, Leppla NC (1984) Advances and Challenges in Insect Rearing. USDA, ARS, 306p

Leppla NC (2003) Aspects of total quality control for the production of natural enemies. In: van Lenteren JC (ed) Quality control and production of biological control agents: Theory and testing procedures. CAB Publishing, Wallingford, CT, pp 19–24

Leppla NC, Adams F (1987) Insect mass-rearing technology, principles and applications. 20p.

Leppla NC, Ashley TR (1978) Facilities for insect research and production. USDA Technical Bulletin, vol 1576, 86p

Leppla NC, Fisher WR (1989) Total quality in insect mass production for insect pest management. J Appl Entomol 108:452–461

Leppla NC, Williams DW (1992) Mass rearing beneficial insects and the renaissance of biological control. Pesquisa Agropecuária Brasil 27:231–235

Lewis WJ, Nordlund DA, Gross Jr HR, Perkins WD, Voegelé J (1976) Production and performance of *Trichogramma* reared in eggs of *Heliothis zea* and other hosts. Environ Entomol 5: 449–452

Li LY (1994) Worldwide use of *Trichogramma* for biological control on different crops: a survey. In: Wajnberg E, Hassan SA (eds), Biological control with egg parasitoids. CAB International, Wallingford, CT, pp 37–53

Li LY, Wen-Hui Liu, Chao-Shian Chen, Shi-Tzou Han, Jia-Chi Shin, Hansun Du, Shu-Yi Feng (1986) In: vitro rearing *Trichogramma* spp.; *Anastatus* sp. in artificial "eggs" and the methods of mass production. In: International Symposium on *Trichogramma* and other eggs parasites, 2nd, Guangzhou, China (Abstracts)

Loayza RM, Parra JRP, Vendramim JD (2003) Biologia comparada de *Selenaspidus articulatus* (Morgan) (Hemiptera: Diaspididae) em cultivares de *Citrus sinensis* e em *Citrullus silvestris*. Neotrop Entomol 32:493–496

Luck RF, Forster LD (2003) Quality of augmentative biological control agents: a historical perspective and lessons learned from evaluating *Tricogramma*. In: van Lenteren JC (ed) Quality control and production of biological control agents: Theory and testing procedures. CAB International, Wallingford, CT, pp 231–246

Mackauer N (1972) Genetic aspects of insect production. Entomophaga 17:27–48

Mc Dougall SJ, Mills NJ (1997) The influence of hosts, temperature and food sources on the longevity of *Trichogramma platneri*. Entomol Exp Appl 83:195–203

Morris RF and Fulton WC (1970) Models for the development and survival of *Hypanthria cunea* in relation to temperature and humidity. Mem Entomol Soc Can 70:1–60

Nagarkatti S, Nagaraja H (1971) Redescriptions of some known species of *Trichogramma*, showing the importance of male genitalia as a diagnostic character. Bull Entomol Res 61:13–31

Nathan SS, Kalaivani K, Mankin RW, Murugan K (2006) Effects of millet, wheat, rice and sorghum diets on development of *Corcyra cephalonica* (Stainton) (Lepidoptera: Galleriidae) and its suitability as a host for *Trichogramma chilonis* Ishii (Hymenoptera: Trichogrammatidae). Environ Entomol 35:784–788

Ozkan C (2006) Laboratory rearing of the solitary egg-larval parasitoid, *Chelonus oculator* Panzer (Hymenoptera: Braconidae) on a newly recorded factitious host *Plodia interpunctella* (Hübner) (Lepidoptera: Pyralidae). J Pestic Sci 79:27–29

Ozkan C, Ozmen D (2001) A new record for Turkish fauna *Chelonus oculator* Panzer (Hymenoptera: Braconidae) and its two new hosts. Turkish J Entomol 25:263–265

Parra JRP (1997) Técnicas de criação de *Anagasta kuehniella*, hospedeiro alternativo para produção de *Trichogramma*. In: Parra JRP, Zucchi RA (eds) *Trichogramma* e o Controle Biológico Aplicado. FEALQ, Piracicaba, Brasil, pp 121–150

Parra JRP, Zucchi RA, Silveira Neto S, Haddad ML (1991) Biology and thermal requirements of *Trichogramma galloi* Zucchi and *T. distinctum* Zucchi, on two factitious hosts. Les Colloques de I'NRA 56:81–84

Parra JRP, Cônsoli FL (1992) In vitro rearing of *Trichogramma pretiosum* Riley, 1879. Ciência e Cultura 44:407–409

Parra JRP, Mihsfeldt LH (1992) Comparison of artificial diets for rearing the sugarcane borer. In: Anderson TE, Leppla NC (eds) Advances in insect rearing for research and pest management. Westview Press, Oxford, pp 195–209

Parra JRP, Vinson SB, Gomes SM, Cônsoli FL (1996) Flight response of *Habrobracon hebetor* (Say) (Hymenoptera: Braconidae) in a wind tunnel to volatiles associated with infestations of *Ephestia kuehniella* Zeller (Lepidoptera: Pyralidae). Biol Control 6:143–150

Parra JRP, Cônsoli FL, Hassan SA (1997) Effects of the factitious hosts, *Ephestia kuehniella* and *Sitotroga cerealella* on the quality of *Trichogramma pretiosum*. In: ANPP – Fourth international conference on pests in agriculture. Annales (Tome III), pp 735–740

Parra JRP (2008) Mass rearing of natural anemies. In: Capinera JL (ed) Encyclopedia of Entomology. 2 edn. Springer, 3, pp 2301–2305

Parra JRP, Milano P, Cônsoli FL, Zério NG, Haddad ML (1999) Efeito da nutrição de adultos e da umidade na fecundidade de *Diatraea saccharalis* (Fabr.) (Lepidoptera: Crambidae). Anais Sociedade Entomol Brasil 28:49–57

Parra JRP, Botelho PSM, Corrêa-Ferreira BS, Bento JMS (2002) Controle Biológico: Uma visão inter e multidisciplinar. In: Parra JRP, Botelho PSM, Corrêa-Ferreira BS, Bento JMS (eds)

Controle Biológico no Brasil – parasitóides e predadores Ed. Manole, Piracicaba. 609p, pp 125–142

Parra JRP, Zucchi RA (2004) *Trichogramma* in Brazil: feasibility of use after twenty years of research. Neotrop Entomol 33:271–284

Parrela MP, Heinz KM, Nunney L (1992) Biological control through augmentative releases of natural enemies: a strategy whose time has come. Am Entomol 38:172–178

Peters TM, Barbosa P (1977) Influence of population density on size, fecundity, and developmental rate of insects in culture. Annu Rev Entomol 22:431–434

Pinto JD (2006) A review of the New World genera of Trichogrammatidae (Hymenoptera). J Hymenoptera Res 15:38–163

Pratissoli D, Parra JRP (2001) Seleção de linhagens de *Trichogramma pretiosum* Riley (Hymenoptera, Trichogrammatidae) para o controle das traças *Tuta absoluta* (Meyrick) e *Phthorimaea operculella* (Zeller) (Lep., Gelechiidae). Neotrop Entomol 30:277–282

Prezotti L, Parra JRP, Vencovsky R, Dias CT, Cruz I, Chagas MCM (2002) Teste de vôo como critério de avaliação da qualidade de *Trichogramma pretiosum* Riley (Hymenoptera: Trichogrammatidae): adaptação de metodologia. Neotrop Entomol 31:411–418

Prezotti L, Parra JRP, Vencovsky R, Coelho ASG, Cruz I (2004) Effect of the size of the founder population on the quality of sexual populations of *Trichogramma pretiosum*, in laboratory. Biol Control 30:174–180

Ridgway RL, Inscoe MN (1998) Mass reared natural enemies for pest control: trends and challenges. In: Ridgway RL, Hoffmann MP, Incoe MN and Glenister CS (eds) Mass reared natural enemies: application, regulation and needs Thomas Say Publications in Entomology, Lanhan, 332p, pp 1–26

Rossi MM (1997) As interrupções de desenvolvimento de *Trichogramma*. In: Parra JRP, Zucchi RA (eds) *Trichogramma* e o Controle Biológico Aplicado. FEALQ, Piracicaba, Brasil, pp 151–172

Schneider JC (ed) (2009) Principles and procedures for rearing high quality insects. Mississipi State University, Mississipi. 352p

Sikorowski PP (1984) Microbial contamination in insectaries: occurrence, prevention and control. In: King EG, Leppla NC (eds) Advances and challenges in insect rearing. ARS/USDA, New Orleans. 306p, pp 143–153

Sikorowski PP, Lawrence AM, Inglis GD (2001) Effects of *Serratia marcescens* on rearing of the tobacco budworm (Lepidoptera: Noctuidae). Am Entomol 47: 51–61

Singh P (1977) Artificial diets for insects, mites and spiders. Plenun Press, Chicago. 594p

Singh P (1985) Multiple species rearing diets. In: Singh P, Moore RF (eds) Handbook on insect rearing, vol 1. Elsevier, Amsterdam, pp 19–44

Singh P, Clare GC (1992) Insect rearing management (IRM): an operating system for multiple-species rearing laboratories. In: Anderson TE, Leppla NC (eds) Advances in insect rearing for research and pest management. Westview Press, Boulder, pp 135–157

Singh P, Moore RF (1985) Handbook of insect rearing, vol 2. Elsevier, Amsterdam

Smith CN (1966) Insect colonization and mass production. Academic, San Diego, CA, 618p

Smith SM (1996) Biological control with *Trichogramma:* advances, successes and potential of their use. Annu Rev Entomol 41:375–406

Soares Jr GG (1992) Problems with entomopathogens in insect rearing. In: Anderson TE, Leppla NC (eds) Advances in insect rearing for research and pest management. Westview Press, Boulder, pp 289–322

Stiling P (1993) Why do natural enemies fail in classical biological control programs? Am Entomol 39:31–37

Stinner RE, Ridgway RL, Kinzer RE (1974) Storage, manipulation of emergence, and estimation of numbers of *Trichogramma pretiosum.* Environ Entomol 3:505–507

Stouthamer R, Pinto JD, Platner GR, Luck RF (1990a) Taxonomic status of thelytokous forms of *Trichogramma* (Hymenoptera: Trichogrammatidae). Ann Entomol Soc Am 83:475–481

Stouthamer R, Luck RF, Hamilton WD (1990b) Antibiotics cause parthenogenetic *Trichogramma* to revert to sex. Proc Nat Acad Sci USA 87:2424–2427

Stouthamer R, Breeuwer JAJ, Luck RF, Werren JH (1993) Molecular identification of microorganisms associated with parthenogenesis. Nature 361:66–68

Thompson SN (1986) Nutritional and in vitro culture of insect parasitoids. Ann Rev Entomol 31:197–219

Thompson SN, Hagen KS (1999) Nutrition of entomaphagous insects and other arthropods. In: Bellows TS, Fisher TW (eds) Handbook of Biological Control. Academic, New York. 1046p, pp 594–562

van Driesche RG, Bellows Jr TS (1996) Biological control. Chapman and Hall, London. 539p

van Lenteren JC (1994) Designing and implementing quality control of beneficial insects: towards more reliable biological pest control. Sting, Newsletter on Biological Control in Greenhouses Wageningen 14:3–24

van Lenteren JC (2000) Measures of success in biological control of arthropods by augmentation of natural enemies. In: Gurr G, Wratten S (eds) Biological control: measures of success. Kluwer, Dordrecht, 448p, pp 77–103

van Lenteren JC (2003) Commercial availability of biological control agents. In: van Lenteren JC (ed) Quality control and production of biological control agents: theory and testing procedures. London. CAB International, Wallingford, CT, 327p, pp 167–179

van Lenteren JC, Bueno VHP (2003) Augmentative biological control of arthropods in Latin America. BioControl 48:123–139

Vinson SB (1997) Comportamento da seleção hospedeira de parasitóides de ovos, com ênfase na família Trichogrammatidae. In: Parra JRP, Zucchi RA (eds) *Trichogramma* e o controle biológico aplicado. FEALQ, Piracicaba, Brasil, pp 67–119

Voegelé J, Daumal J, Brun P, Onillon J (1974) Action du traitement au froid et aux ultra-violets de I'oeuf d' *Ephestia kuehniella* (Pyralidae) sur le taux de multiplication de *Trichogramma evanescens* et *T. brasiliensis* (Hymenoptera: Trichogrammatidae) Entomophaga 19:341–348

Waage JK, Carl RP, Mills NJ, Greathead DJ (1985) Rearing entomophagous insects. In Singh P, Moore RF (eds) Handbook of insect rearing, vol 1. Elsevier, Amsterdam, pp 45–66

Wäckers FL, van Rijn PCJ, Heimpel GE (2008) Honeydew as a food source for natural enemies: making the best of a bad meal?. Biol Control 45:176–184

Werren JH (1997) Biology of *Wolbachia.* Annu Rev Entomol 42:587–609

Zucchi RA, Monteiro RC (1997) O gênero *Trichogramma* na América do Sul. In: Parra JRP, Zucchi RA (eds) *Trichogramma* e o controle biológico aplicado. FEALQ, Piracicaba, Brasil, pp 41–46

Chapter 11
In Vitro Rearing of Egg Parasitoids

Fernando L. Cônsoli and Simon Grenier

Contents

11.1 Introduction

Several fields of Entomology were benefited with the development of artificial diets for insect rearing late in the 1960s and the 1970s, and the variety of applications sought to artificial diets for insects led to the advance and development of rearing techniques to a number of insect orders, from herbivores to hematophages (Singh 1977, Anderson and Leppla 1992).

The use of natural enemies for biological control of insect pests was particularly benefited by the development of such rearing techniques, as the availability of suitable diets allowed the production of high quality insects in large scale, providing an adequate supply of hosts to sustain natural enemy production (Ridgway and Vinson 1976, Parra 2002). The requirements for production of large quantities

F.L. Cônsoli (✉)
Laboratório de Interações em Insetos, Departamento de Entomologia e Acarologia, ESALQ/USP, Av. Pádua Dias 11, 13418-900, Piracicaba, São Paulo, Brazil
e-mail: fconsoli@esalq.usp.br

F.L. Consoli et al. (eds.), *Egg Parasitoids in Agroecosystems with Emphasis on Trichogramma*, Progress in Biological Control 9, DOI 10.1007/978-1-4020-9110-0_11,

of high quality insects at a reasonable cost prompted the research into the development of automated systems, new packing materials and storage techniques to provide users the necessary number of natural enemies at a competitive cost to other available control measures (see Chapter 10).

Regardless the success achieved in the rearing of natural or factitious hosts for the mass production of natural enemies, the usual need to rear two different species when producing natural enemies based on in vivo rearing systems stimulated the investigation on their artificial rearing (see also Chapter 10). Nearly 127 species of entomophages have been completely or partially reared on artificial diets, including predators (53 species) and parasitoids (74 species). Most parasitoids reared on artificial diets were hymenopterans (61 species), with 30 of them being egg parasitoids. Among the egg parasitoids, the majority of the species tested on in vitro systems belongs to Trichogrammatidae (21 species), particularly of the genus *Trichogramma* (Fig. 11.1) (updated from Cônsoli and Parra 2002). Despite the large number of entomophagous insects reared on artificial diets, very few of them were successfully reared to allow their use in applied biological control (Li 1997b, Feng et al. 1999).

There are a number of extensive reviews on the in vitro rearing of parasitoids discussing several issues, from parasitoid nutrition to applications of in vitro rearing systems other than biological control (Mellini 1975, Thompson 1986, 1999, Grenier et al. 1994, Vinson 1994, Thompson and Hagen 1999, Cônsoli and Parra 1999a, 2002, Grenier 2009), including a few specifically focused on egg parasitoids (Grenier 1994, 1997, Cônsoli and Parra 1997). Artificial rearing systems were intensively investigated from 1970 to 2000, with a seven fold increase in the number of articles published in this period (including review articles and book chapters). However, there was a deep decrease since then, and the number of articles published in the last decade was reduced in more than a third (Fig. 11.2). The number of articles dealing with the development of in vitro rearing systems for egg parasitoids followed the same trend, decreasing from nearly 40 articles in 1991–2000 to about 10 in 2001–2010 (Fig. 11.2). While the reduction in the number of publications on parasitoids other than egg parasitoids could be understandable, as no successful case of in vitro rearing larval or pupal parasitoids for successive generations

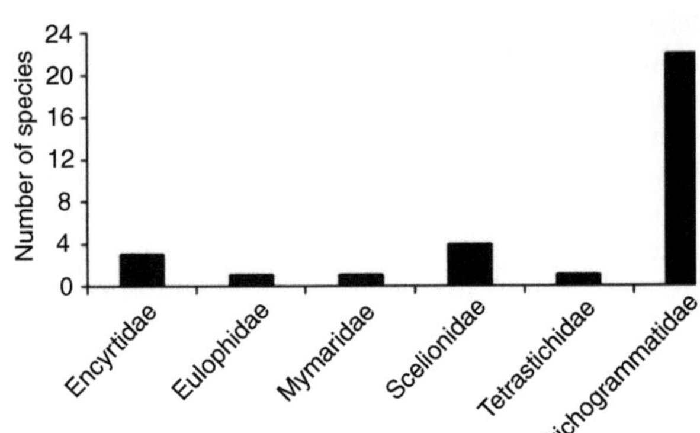

Fig. 11.1 Number of species in different families of hymenopteran egg parasitoids that were investigated for the development of artificial rearing systems

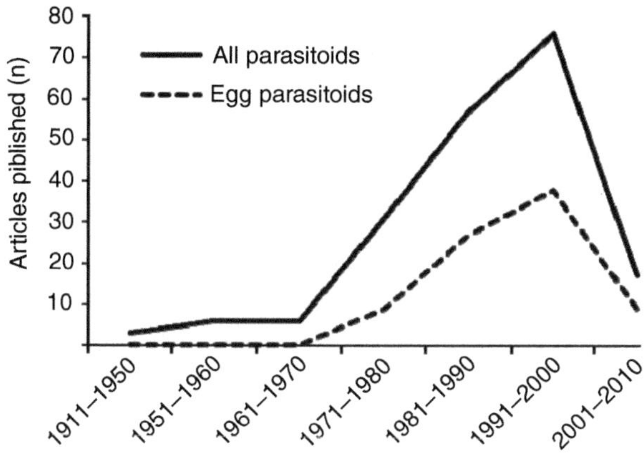

Fig. 11.2 Scientific and review articles and book chapters published from 1911 to 2010 on the development of artificial rearing systems for parasitoids

was ever achieved, the same would not be expected for egg parasitoids. Several egg parasitoids were successfully reared on artificial systems, yielding adults as fit as those reared in vivo. Besides, koinobiont parasitoids (larval parasitoids) are known to require a much more complex environment to sustain their development, as the chemical and physical composition of the host (internal milieu) changes during parasitoid growth and development, requiring an artificial system that could offer the different conditions (diet composition, diet consistency) necessaries to sustain the immature development (Hu and Vinson 1997a, b, 1998, Cônsoli and Vinson 2009).

In this chapter we will discuss the use of artificial diets for mass production of egg parasitoids by providing an overview of the main aspects of the work developed on rearing *Trichogramma* in vitro and focus our final discussion on issues that might be limiting the broad use of artificial diets in rearing systems for biological control purposes.

11.2 Artificial Rearing

11.2.1 Artificial Diets – Composition

The majority of the artificial diets utilized for rearing egg parasitoids included insect-derived molecules as one of the diet components, especially those in which parasitoids were successfully reared to adulthood (Grenier 1997, Grenier et al. 1994, Cônsoli and Parra 1997, 2002). These diets would usually include host egg derivatives, host hemolymph and/or pupal holotissues. Such components would either be used to stimulate egg laying by parasitoid females or to provide adequate nutrition and/or the necessary factors to stimulate parasitoid pupation (Nettles 1990). A variety of other components were mixed with such insect-derived molecules to ascertain for a suitable medium for egg laying and larval development, such as egg yolk, milk, yeast/casein hydrolysate, malt solution, bovine fetal serum, salt solutions, and antibiotics, among others (see Grenier 1994, 1997, Grenier et al. 1994, Cônsoli and Parra 1997, 2002 for review).

Insect-derived components could range from 20 to 70% of the diet composition, and most of them were derived from lepidopterans, but not necessarily natural hosts of *Trichogramma* species. The best results obtained for rearing *Trichogramma* in diets containing pupal holotissues were obtained by using pupae of species of the silk moths *Antheraea pernyi* (Guérin-Méneville) or *Phylosamia cynthia* Drury, or of the lepidopterous pests *Mamestra brassicae* (L.) and *Diatraea saccharalis* (F.) (Liu et al. 1979, Qin and Wu 1988, Dai et al. 1988, Grenier and Bonnot 1988, Cônsoli 1997, Gomes 2002) (Fig. 11.3). However, the quality of pupal holotissues as a component of artificial diets for *Trichogramma* was also dependent on the age at which the pupal tissues were collected. Older the pupae, lower the parasitoid development and survivorship (Cônsoli and Parra 2000). Although several *Trichogramma* species could be reared on diets containing hemolymph/holotissues of one of these insects, not all species or populations from a same species tested would properly parasitize artificial eggs filled with such diets, or complete their embryonic and/or larval development. Differences among species to exploit holotissues of a particular host species as a food source suggest the needs of *Trichogramma* may be quite diverse even in a genus of egg parasitoids recognized by its polyphagy (Cônsoli and Parra 1996a, Cônsoli 1997, Li et al. 1997a, Dias et al. unpublished).

The rate of larval and pupal successful development on these artificial diets was very variable depending on the substrate used as an artificial host, but being as high as 90% of larval and pupal viability. Despite the successful rate of development obtained, a number of adults also displayed malformations (abnormalities), especially an enlarged abdomen and/or unexpanded wings (see Cônsoli and Parra 1997, Grenier 1997, Li et al. 1997b for review).

Very few diets without any insect-derived components were developed that could successfully support the full development of egg parasitoids, basically of *Trichogramma* species (Wu et al. 1982, Xie et al. 1986a, Grenier et al. 1995). The first of such diets was composed of known solutions of amino acids, sugars, vitamins, salts, and organic acids, and bovine fetal serum, chicken embryo extract, wheat germ, yeast extract, hen egg yolk and Grace's culture medium. Although a pupation rate of 36–53% could be obtained when rearing *Trichogramma dendrolimi* Matsumura on such a diet, the emergence rate was very low (8–16%) (Wu et al. 1980, 1982, Qin and Wu 1988). Higher pupation (70%) and emergence (60%) rates could be later obtained by modifications in the proportions of the diet components or the addition of new molecules, such as casein hydrolysate (Grenier et al. 1995). It has also been shown that the concentration and source of protein added to the artificial diet may influence the utilization and assimilation of nutrients by *T.* pretiosum, with serum albumin resulting in better parasitoid growth than casein (Gomes 2002, Grenier et al. 2005).

In a tentative to have a better defined diet, several attempts were made in using insect cell culture media to replace/reduce the requirements for the growth factors associated with insect derivatives added to the composition of the artificial

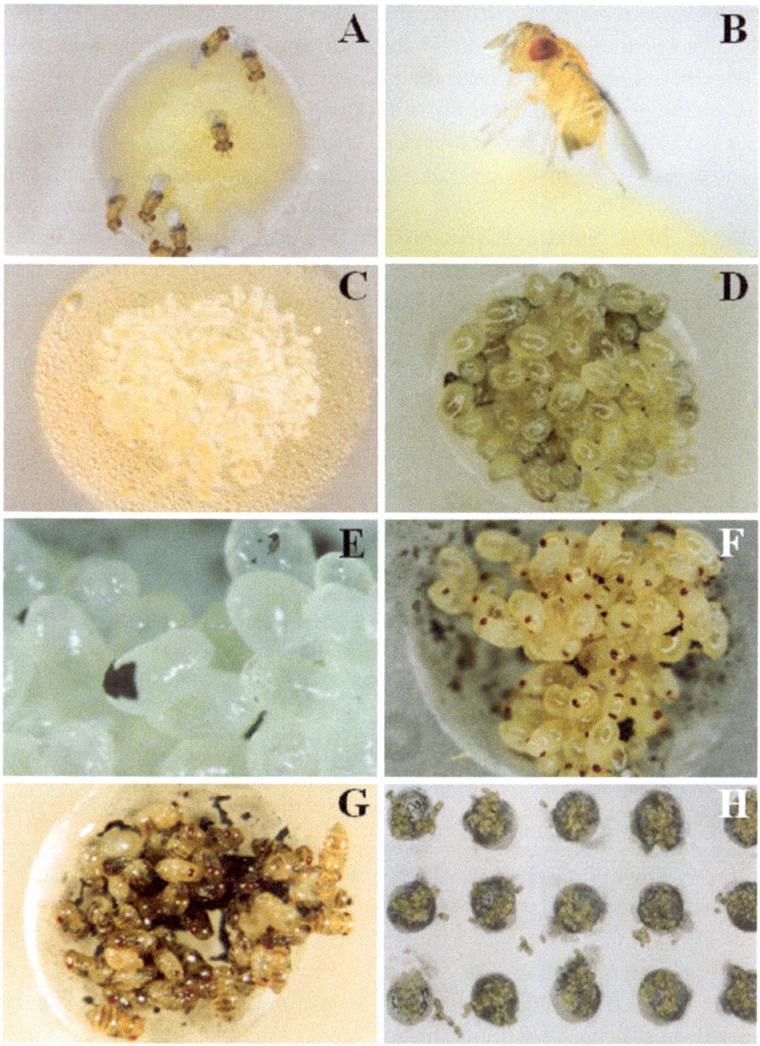

Fig. 11.3 Successful rearing of *Trichogramma galloi* on an artificial diet based on pupal holotissues of *Diatraea saccharalis*, using an artificial egg plastic card. (**a**) Parasitizing females; (**b**) *T. galloi* female drilling through the plastic membrane; (**c**) Early larvae developing into the artificial medium; (**d**) Late larvae after diet was completely consumed; (**e**) Detail of late larvae preparing for pupation – note the beginning of the deposition of the dark granules which are characteristics of eggs parasitized by *Trichogramma*; (**f**) Early pupae – note that all pupae have normal size abdomen; (**g**) Late pupae close to adult emergence; (**h**) General view of the artificial egg card close to wasps emergence

medium. The Grace's medium was used to supplement diets for *Trichogramma confusum* Viggiani and *Trichogramma dendrolimi* (Liu and Wu 1982) and *Trissolcus basalis* (Woll.) (Volkoff et al. 1992), BML-TC 10 for *Tetrastichus schoenobii* Ferrière (Ding et al. 1980), IPL-41 for rearing *Trichogramma pretiosum* Riley and

Anagrus breviphragma Soyka (Strand and Vinson 1985, Chiappini et al. 2004), Hink TN-HFK for *Telenomus heliothidis* Ashmead (Strand et al. 1988), and MGM-443 for *Ooencyrtus nezarae* Ishii (Takasu and Yagi 1992).

More recently, artificial diets containing insect cell lines were used as a substrate for parasitoid larval development as a replacement to hemolymph and pupal holotissues. The use of insect cell lines was also thought to improve parasitoid larval feeding, as *Trichogramma* larvae were demonstrated to ingest particulate food (Jarjees et al. 1998, Notarte and Merritt 2001). Further investigations on the use of cell lines for the development of artificial diets led to the development of a medium composed of hen egg yolk (31%), lepidopterous cell line (37.5%), 7% yeast extract (12.5%), 10% powdered cow milk (19%), and gentamycin (0.2%), yielding 60% larval survivorship, but less than 20% pupal survivorship (Heslin et al. 2005a, b).

11.2.2 Artificial Diets – Physicochemical Properties

Several other factors than nutrition have to be considered when designing artificial diets for endoparasitoids, as the nutritional medium is also the environment in which the embryo and larvae will interact with during their growth and development. The osmotic pressure, consistency and pH of the artificial diet may be as important as the nutrient composition of the diet. Artificial diets with unbalanced physicochemical properties may have deleterious effects on the parasitoid immature development, as they can affect parasitoid water balance, gas exchanges, nutrient acquisition and food digestion, for example (Grenier et al. 1986, Yazlovetsky 1992, Cônsoli and Parra 1999a). Although many diets were tested for egg parasitoids, few of them checked for the diet pH and osmotic pressure. The pH of most artificial diets used for *Trichogramma* ranged from 6.4 to 6.8, which is close to the values observed for insect hemolymph (Grenier 1997). Specific investigations of the ideal pH for rearing *T. confusum* and *T. pretiosum* reported the best values to be around 6.7–7.0. However, *T. pretiosum* successfully developed in diets with pH values ranging from 6.5 to 9.0 (Zhong and Zhang 1989, Nettles et al. 1985). The pH of the artificial medium can also affect the efficiency of the antimicrobials added to the diet, as the majority of the most common antimicrobials used in artificial diets is effective in acidic pHs (Cohen 2004c). Effective microbial control in artificial rearing systems of egg parasitoids can be a problem as several antifungals may be toxic to the developing parasitoid (Grenier and Liu 1990).

Most of the artificial diets utilized for rearing egg parasitoids were quite diluted, although studies on the larval feeding behavior of *Trichogramma* indicated the ingestion of solid food particles are common (Jarjees et al. 1998). Assays using artificial diets demonstrated that the use of solid particles in the diet improved parasitoid larval growth. These findings suggested that concentrated diets may be more suitable for parasitoid development than the diluted diets usually used, although the problems that may arise with the increase of the osmotic pressure (Wu et al. 2000). The osmotic pressure of the artificial diets utilized for rearing *Trichogramma* was mostly in the range of 300 – 350 mOsm/kg (Grenier and Bonnot 1988, Grenier et al.

1995, Heslin et al. 2005a, b). Osmotic pressures higher than 450 mOsm/kg were shown to be limiting for egg parasitoid development and hatching (Grenier et al. 1986), but *T. pretiosum* was successfully reared in an artificial diet with osmotic pressure higher than 460 mOsm/kg (Strand and Vinson 1985).

11.2.3 Ovipositional Stimulants

Except for few attempts to obtain on the one hand egg laying in superparasitized host eggs or in a first artificial medium, and on the other hand embryonic and larval development in another medium, most of the rearing systems tested used a single medium ensuring both egg laying and full immature development. One of the major obstacles of producing parasitoids in the latter conditions is the fact that the artificial culture medium has to provide the adequate stimuli to allow for female oviposition and be a suitable rearing medium for parasitoid embryonic and larval development. Diets containing insect-derived components were good elicitors for most *Trichogramma* species reared on artificial diets, with a few exceptions depending on the host species used or the parasitoid tested (Cônsoli 1997, Li et al. 1997a, Cônsoli and Parra 2000, Dias et al. unpublished).

Diets devoid of any insect-derived components also demonstrated a good capacity to elicit oviposition in females of *Trichogramma*, as these parasitoids were shown to respond to common cations, with potassium acting as an stimulant, while magnesium would act as a synergist (Nettles et al. 1982, 1983). KCl-$MgSO_4$ solutions were used as ovipositional stimulants to collect large amounts of eggs, which could later be transferred to ideal volumes of artificial diets (see discussion on *Artificial hosts*). Nevertheless, egg survivorship and further immature development were impaired by the difficulties in egg handling and sterilization (Xie et al. 1986a, b). Moreover, a too long stay of the eggs in the salt solutions was detrimental to their subsequent development (Xie personal communication). The efficiency of a KCl-$MgSO_4$ solution to elicit oviposition in *Trichogramma* was later shown to be negatively affected by the addition of glucose, casein hydrolysate and NH_4^+ (Nettles et al. 1985).

The amino acids leucine, isoleucine, phenylalanine and histidine were demonstrated to elicit oviposition in *T. dendrolimi* in artificial eggs, especially when presented in mixtures (Qin and Wu 1988). These amino acids are very common in the insect hemolymph (Wyatt 1961), and were found in egg macerates, hemolymph and pupal holotissues of several species of insects included as components in the artificial diet for *Trichogramma* (Cônsoli and Parra 1996a). These amino acids were reported in concentrations high enough to elicit oviposition in *Trichogramma* in several insect eggs, including non-host species (Barret and Schmidt 1991). *Trichogramma* seems to rely on very common compounds to elicit oviposition, facilitating the use of artificial diets also as substrate for egg laying.

11.2.4 Artificial Hosts

Artificial hosts for in vitro rearing of egg parasitoids imposes the same constraints as those utilized by larval endoparasitoids, as all egg parasitoids reared on artificial diets are endophages. Therefore, the artificial host has to serve as a container for the artificial diet at the same time that it provides the required cues to elicit parasitoid females to search, evaluate and accept it as a host. The host selection by egg parasitoid females can involve a number of chemical, physical and visual cues (see Chapters 2 and 4), which may be a limiting factor when developing artificial hosts and/or diets. The artificial host will also have to provide the required needs of the developing parasitoid immature to allow for gas exchanges and adult emergence. Besides fulfilling the required needs of parasitoid adult females for host selection and the immature needs for development, the artificial host must also work as a packing material, avoiding the artificial diet contamination and desiccation. Therefore, finding a suitable artificial host is as complex as developing a suitable artificial medium. There are so many behavioral, physiological and physical aspects involved on the interactions of the parasitoid female and parasitoid immature with the artificial host and artificial diet that could interfere with host acceptance, parasitoid egg laying activity, larval development, larval gas exchanges, pupation, and adult emergence, that addressing both issues, artificial host and artificial diet, concomitantly may interfere with the progress of the initial attempts in developing artificial rearing systems for parasitoids, including egg parasitoids.

Early studies on the development of artificial diets for *Trichogramma* focused on the collection of parasitoid eggs and their later transfer to the artificial diet. Several studies utilized superparasitized eggs of natural or artificial hosts as a source of parasitoid eggs to test artificial diets. Parasitoids were then reared on the artificial diet contained in depressed slides or on small drops of diets by using the hanging drop technique (Grenier and Bonnot 1988, Parra and Cônsoli 1992, Cônsoli and Parra 1996b). However, the use of host eggs as a substrate for oviposition always lead to the addition of a certain amount of the host egg yolk into the artificial diet, which could ultimately affect the appropriate evaluation of the diet suitability.

An option to avoid the addition of egg derivatives to the artificial diet being tested was the use of artificial wax eggs initially designed to contain the artificial diet to feed larvae of *Chrysopa* (Hagen and Tassan 1965). These wax eggs were tested for studying the parasitoid oviposition behavior and the potential of different solutions to elicit oviposition in *Trichogramma californicum* Nagaraja and Nagarkatti (Rajendram and Hagen 1974). A diversity of waxes and their mixtures were tested to adequate the hardness and thickness of the wax layer for parasitoid egg laying (Rajendram 1978a). They were later used to develop optimized oviposition stimulant solutions (Nettles et al. 1983, 1985) and to evaluate several nutritional factors that could affect the oviposition of *Trichogramma* into artificial diet-filled wax eggs (Rajendram 1978b). Therefore, wax eggs were used for collecting a large number of eggs which could later be isolated and transferred to the diet container for rearing large number of individuals, such as microplates or cell culture plates (Xie et al. 1986a). In these cases, both wax eggs and microplates are unsuitable substrates for parasitoid rearing. The first one does not allow for gas exchanges, while the second one cannot be used as a substrate for female oviposition.

The use of plastic eggs was also investigated. A technique to produce spherical artificial eggs that would be similar in size to lepidopterans eggs using paraxylilene was tested for *T. dendrolimi*. Although it was possible to produce these eggs with a size similar to those of lepidopteran hosts, it was very difficult to control the thickness of the layer of paraxylilene to cover the diet. Thick layers would limit gas exchange, while thin layers would allow diet desiccation. These eggs were suitable for oviposition and larval development, but no pupal development was obtained (Grenier and Bonnot 1988).

A number of plastic membranes were selected as substrates for oviposition. The first attempts utilized flat membranes of several polymers, some of them thicker than 40 μm (cellophane) (Li 1997a) or as thin as 1 μm (silicone-polycarbonate copolymer) (Morrison et al. 1983). However, the best substrates used as artificial host eggs were those using plastic membranes printed with dome shaped eggs which were maintained opened. Several studies were conducted on the size of the ovipositor of parasitoid females (Hsia and Wang 1979, Li et al. 1989, Parra et al. 1997, Grenier et al. 2001) and on the thickness of the chorion of natural and factitious hosts (Cônsoli et al. 1999), indicating the adequate thickness of the plastic membrane to be selected. In general, membranes ranged from 7–8 to 64 μm in thickness depending on the size of female ovipositor. Membrane stiffness and their surface characteristics after dome shaped eggs were printed could also affect membrane suitability as an artificial substrate for oviposition. Polyethylene and polypropylene membranes were the most successfully used in producing artificial egg cards for *Trichogramma* (Dai et al. 1988, Li et al. 1988, Grenier and Liu 1990, Cônsoli and Parra 2002), which yielded the best rates of parasitism, pupation, and emergence for many species of *Trichogramma* (Li et al. 1988, Grenier and Liu 1990, Cônsoli 1997).

Despite all the indications that *Trichogramma* females would use chemical cues in the host selection process (Jones et al. 1973, Zaborski et al. 1987, see also Chapter 4), high levels of parasitization on artificial eggs were obtained for *Trichogramma galloi* Zucchi and *T. pretiosum* without any particular chemical treatment (Cônsoli and Parra 2002) (Fig. 11.3a, b). However, the use of extracts of moth scales, insect hemolymph, a mixture of saturated hydrocarbons or Elmer's glue has been shown to improve female arrestment and acceptance of artificial eggs (Dai et al. 1988, Xie et al. 1997, Grenier et al. 1993). A concentrated solution of polyvinyl alcohol brushed over the artificial host egg surface could strongly enhance egg laying by *Trichogramma* females (Han et al. 1994, Dahlan and Gorth 1997).

The use of plastic membranes to shape artificial eggs also allowed for size control of the artificial host produced, and consequently, the volume of artificial diet/host egg. The control and standardization of the volume of the diet available to the immature parasitoid have direct implications on the successful rearing of egg parasitoids in vitro. By controlling the size of the artificial host (diet volume), it is possible to adjust the number of parasitoids eggs per volume unit of the artificial diet by using a proper ratio of female wasps to artificial egg during parasitization (Li 1997a, Cônsoli and Parra 1999b). Egg parasitoids, such as *Trichogramma*, will only initiate pupation if the surrounding diet is completely eliminated. Diet remains will induce larvae to overfeed, resulting in reduced pupation rates and increased adult malformations (enlarged abdomen, unextended wings). *Trichogramma* are very plastic semi-gregarious parasitoids, with a large diversity of host sizes. Species that can exploit a range of hosts will be more likely to develop on a wide range of conditions. Studies comparing the volume of the smallest and largest hosts of *T. galloi* and *T. pretiosum* and the clutch sizes allocated to them indicated the parasitoid density required to completely consume 1 μl of artificial diet should not be smaller than 9 or larger than 87 eggs (Cônsoli and Parra 1999b, Cônsoli et al. 1999). In the case of *T. dendrolimi*, the ideal number of parasitoid eggs to be allocated to the artificial egg was obtained by using a ratio of two parasitoid females/artificial egg, which produced 74.5 ± 30.8 eggs/artificial host, with 85.8 ± 3.2% larval eclosion and 45.3 ± 15.2% pupation (Li 1997a).

11.3 Quality Control of In Vitro Reared Parasitoids

Although the intense investigation on the development of in vitro rearing systems for parasitoids, very few of them were concerned with the evaluation of the quality of the insect produced on an artificial diet. The majority of those studies in which parasitoid quality was assessed, biological traits (parasitization capacity, longevity, adult size and weight, among others) of in vitro-reared insects were compared to those reared on factitious and/or natural hosts (see Grenier and De Clercq 2003 for a review). In general, in vitro-reared parasitoids displayed reduced pupal weight and adult size, reproductive capacity, and dispersion (Mellini and Campadelli 1996, Rojas et al. 1996, Carpenter and Greany 1998, Morales-Ramos et al. 1998). In the case of egg parasitoids, particularly *Trichogramma*, there are studies indicating females reared in artificial diets have a reduced fecundity as compared to those reared in vivo (Cônsoli and Parra 1996b), while others reported similar fecundity and longevity between females reared on the natural/factitious and in the artificial diet (Cônsoli 1997, Nordlund et al. 1997, Nurindah et al. 1997). Very few studies investigated the quality of insects produced in artificial diets by comparing the biochemical profile between adults reared in vivo and in vitro (Grenier et al. 1995).

The promising results reported by some of these studies in which in vitro-reared parasitoids were as fit as those reared in vivo should be cautiously analyzed when envisaging mass production for field application. For the majority of these studies, parasitoids were compared after one single generation on the artificial diet (Cônsoli 1997, Grenier et al. 1995, Nurindah et al. 1997), while few were able to evaluate parasitoid fitness after successive rearing in an artificial diet (Gao et al. 1982, Nordlund et al. 1997). In one of the most prolonged continuous rearing of a parasitoid species on an artificial diet, *T. dendrolimi* was reared on a diet composed of pupal holotissues of *Antheraea pernyi* (30%), egg yolk (14%), skimmed milk (26%) and distilled water (30%) for 41–50 generations, with high pupation and emergence rates (60–80%), while maintaining a reduced percentage of malformed adults (8%) (Dai et al. 1988).

Only a very limited number of studies were able to evaluate the parasitization capacity and efficiency of these egg parasitoids in the field, demonstrating in vitro rearing systems can be a feasible option for the mass production of egg parasitoids for applied biological control. More than 7 million adults of *T. confusum* were reared on an artificial diet and released in 6.7 ha experimental plots to control the sugarcane borers *Argyroploce schistaceana* (Snellen), *Chilo infuscatellus* Snellen and *Chilo sacchariphagus* (Bojer), with impressive parasitization rates (80–95%) (Liu et al. 1985). The field efficiency of in vitro-reared parasitoids was also assessed for *T. dendrolimi* and *T. confusum*, which were released against *Heliothis armigera* (Hübner) and *Dendrolimus punctata* Walker, and *C. sacchariphagus* and *C. infuscatellus*, respectively. The parasitization rates obtained by in vitro-reared wasps was always similar to those obtained for the in vivo-reared ones (eggs of *A. pernyi*). In the case of *T. dendrolimi* against *D. punctata*, the parasitization rate was twice as high for females produced on artificial diets than for those reared on a factitious host egg (Dai et al. 1988). The parasitization rates of corn borers by in vitro-reared *T. dendrolimi* and *T. chilonis* were similar (65 and 68%, respectively) to those by naturally reared parasitoids (Feng et al. 1999)

Two other egg parasitoids, *Anastatus japonicus* Ashmead and *Telenomus dendrolimusi dendrolimi* Chu, were reared by successively generations on artificial systems similar to those developed for *Trichogramma* by Chinese researchers. Their development on artificial diets was similar to that obtained on their respective natural/factitious hosts. In vitro-reared *A. japonicus* was further used in field release experiments to evaluate its potential in controlling eggs of the litchi stink bug *Tessaratoma papillosa* Drury at a rate of 125,000 adult wasps/ha, with a parasitization rate of 85%, a little over the 75% obtained in areas where in vivo-reared wasps were released (Li et al. 1997b).

11.4 What Are the Shortcomings in the Artificial Rearing of Egg Parasitoids?

One might be wondering why in vitro-reared egg parasitoids, or at least *Trichogramma*, are not being widely used in biological control programs if there are suitable artificial media, ovipositional stimulants, artificial eggs and a mechanized system in place to allow the production of good quality natural enemies in a large scale. The development and utilization of a machine by the Guangdong Entomological Institute and the Experimental Factory of Guangdong Academy of Sciences in 1985–1986 allowed a daily production of 6 million *Trichogramma* wasps or 300 thousand *A. japonicus* adults, with high percentages of parasitism, parasitoid pupation and adult emergence. An updated version of this machine was later designed by the Wuhan University team and had a capacity to produce 6,400 artificial egg cards, which would equal to 30 million *Trichogramma* wasps/day, with even higher rates of parasitization, parasitoid development and emergence (Li 1997b).

If egg parasitoids can be mass produced on artificial diets and efficiently used in biocontrol programs, why are they not been used? Is not the in vitro system profitable? Why the system developed for several *Trichogramma* species and *Anastatus* was not successfully extended to other egg parasitoids? *Trichogramma* and *Anastatus* are parasitoids of lepidopteran and hemipteran eggs, two major orders of pest insects. Both of these parasitoids were reported to be mass reared on similar diets containing silkworm moth pupal holotissues as the major component. Do *Trichogramma* and *Anastatus* have different requirements than other egg parasitoids? Well, it comes to a point that we ought to be asking if we really know enough about egg parasitoid biology and development.

We will first focus on the success development of egg parasitoids other than *Trichogramma* and *Anastatus*. There were several attempts to rear egg parasitoids in Encyrtidae (3) (Battisti et al. 1990, Masutti et al. 1993, Lee and Lee 1994, Takasu and Yagi 1992), Eulophidae (1) (Hu et al. 1998), Mymaridae (1) (Chiappini et al. 2004) and Scelionidae (4) (Strand et al. 1988, Li 1992, Volkoff et al. 1992, Shirazi 2006), but all of them with limited success, most of the time only up to the late instar or the pupae or from second instar larvae to pupae/adult. In those cases when adults were produced, a very low percentage of larval and pupal survivorship was obtained (Fig. 11.1). Once again, the only successful rearing was obtained by one of the Chinese research leading groups, who were able to rear *T. dendrolimusi dendrolimi* from egg laying to adulthood. In vitro-reared *T. dendrolimusi dendrolimi* were reported to be as fit as the in vivo-reared ones (Li 1982, Li 1997b).

The limited success obtained on the development of artificial diets and rearing system for egg parasitoids other than *Trichogramma* and the difficulties to have the immature parasitoid to develop through certain stages in the artificial medium, indicate egg parasitoids have quite distinct nutritional ecology and interactions with the developing host embryo (see Chapter 2 for further discussion). There are indications that egg parasitoids that release teratocytes upon larva eclosion improve their development on the artificial medium if these cells are added to the artificial medium (Strand et al. 1988). However, it also seems that these cells require themselves specific nutrients derived from the host egg (Cônsoli et al. 2001). If teratocytes are to be required for proper parasitoid development, the artificial medium should also be suitable for growth of these cells.

Yet, many parasitoid species prefer to attack their host in an early stage of the embryonic development, while others will do it later. Although the egg represents a defined-nutritional environment, the bioavailability of the nutrients may be affected as the yolk proteins available early in the embryonic development are further processed to give rise to tissues and structural proteins at a later stage of the embryonic development, affecting the nutritional value of the host (Cônsoli and Vinson 2009). The bioavailability of nutrients is one of the issues to affect the quality of artificial diets developed for insects (Cohen 2004a). Utilization of the host at different stages of development may correlate with the efficiency of ingestion and food utilization, as suggested by the improved feeding of *Trichogramma* on artificial diets containing food particles of a particular range (Wu et al. 2000).

Therefore, it is clear that the complexity of developing an artificial system for egg parasitoids is much higher, suggesting that the general concept that egg parasitoids behave as idiobionts should be revisited (see Chapter 2 for further discussion). The development of suitable artificial systems will require the understanding of how the egg parasitoid immature interacts with its environment (ooplasma) and how this environment changes its chemical and physical composition as the parasitoid development progresses. Particular differences may occur even within the same genus, as indicated by the differences in the rearing of several *Trichogramma* species, should it be on their acceptance of the artificial host, on the number of eggs laid/artificial host or on the successful development of the different immature stages in an artificial diet (Grenier and Bonnot 1988, Cônsoli and Parra 1996b, Cônsoli 1997, Li et al. 1988, 1997a, Xie et al. 1997, Dias et al. unpubl.). Such differences may even occur among different populations of a same species, as demonstrated for several strains of *T. pretiosum* (Cônsoli and Parra 1999c). However, the Chinese leading research group reported no specific differences in the nutritional requirements of 16 species of *Trichogramma* reared on artificial diets in China, regardless of their origin (China, France, Germany, Thailand, United States and the former USSR) (Li 1997b). Nevertheless, out of these 16 species, only *T. dendrolimi* and *T. confusum* were mass produced in vitro and field released for biological control purposes (Li 1997b). Life-history traits of *Trichogramma dendrolimi*, such as broad host range, gregarious development and plastic development to exploit a variety of host sizes certainly provided this species with biological traits that facilitated its development on artificial rearing systems.

But still, why artificially-reared *Trichogramma* have not been used worldwide for biocontrol programs? The in vitro- rearing system is certainly cheaper than the traditional rearing system (see Chapter 10), and there are estimates that the costs for producing *Trichogramma* on artificial diets are 50% lower than on their factitious/natural hosts (Dai et al. 1991). In vitro-rearing systems were also reported to reduce in 50% the costs of production of the larval parasitoid *Catolaccus grandis* (Burks) (King et al. 1996). These systems are also less expensive for the production of predators, and rearing systems based on encapsulated artificial diets allowed the production of *Geocoris punctipes* at a cost of U$0.63/1,000 adults (Cohen 1985) and the reduction of costs in the mass rearing system of *Chrysopa rufilabris*, from U$ 500/kg of host egg to U$6/kg of artificial diet (Cohen and Smith 1998).

The costs of *Trichogramma* production on artificial diets were calculated based on artificial diets that included pupal holotissues of silkworms as the major component, and on an automated system for production of artificial eggs (Dai et al. 1991). Although this system still requires the rearing of a second insect (silkworms) to serve as a substrate for diet preparation, silkworm pupae can be obtained at very low prices in countries that have a developed silk industry (as China has), as the insect pupae is considered a by-product. If the elimination of insect-derived components from the diet is a factor that could make the in vitro-rearing system unprofitable in countries were a silk industry is unavailable, and is limiting the worldwide production of these parasitoids on artificial diets, it is unknown and requires attention. The inexistence of automated systems for parasitoid production on artificial diets other than in China may certainly be a factor affecting the exploitation of these rearing systems, as automation is known to be one of the major factors to contribute with the reduction in cost of insect rearing systems (see Chapter 10, Cohen 2004b).

There may also have issues related to the scale up of the rearing systems developed in Australia, Brazil, France and United States, as the data reported in these countries on the in vitro rearing of *Trichogramma* were comparable to those reported in China (Nordlund et al. 1997, Cônsoli 1997, Grenier et al. 1998a, Heslin et al. 2005a,b). One of such issues can be the control of diet microbial contaminants, particularly mold, as several antimicrobials may affect parasitoid development (Grenier and Liu 1990). The truth is that regardless all the efforts conducted in laboratories in four different continents, the success obtained by Chinese researchers in mass producing *Trichogramma* was never met. The unsuccessful in reproduce and extend the results obtained in China for other egg parasitoids might be one of the reasons of the sharp decrease in investigating the development of such a rearing system for egg parasitoids (Fig. 11.2). But should we give up in pursuing an artificial system that could be used in the mass production of such natural enemies?

11.5 How to Address the Constraints of the Past and Solve the Difficulties that Lie Ahead?

In vitro rearing systems for egg parasitoids are a reality and the results obtained with these parasitoids were even better than those obtained with predators, which would be theoretically easier to rear on artificial diets due to their lower physiological dependence on the prey. Most of the problems with the development of artificial diets for egg parasitoids were related with the identification of chemical compounds that would be necessary to elicit artificial host acceptance and oviposition or be required for parasitoid proper development and/or pupation. The new technologies available nowadays for the fast and accurate characterization and identification of all sorts of molecules (CE-MS, LC-MS, LC-NMR, Maldi-TOF, MSn), and the development of bioinformatics applied to the study of metabolites (Shulaev 2006, Moco et al. 2007), are becoming increasingly accessible at several research units around the world. The characterization of the full metabolome of parasitoids, and the possibility of comparative studies on how different food sources may alter the insect metabolite profile and be correlated with the successful development of insects on a particular nutritional substrate will certainly enhance our knowledge on these biological systems. Molecular approaches, as the full genome sequencing of three *Nasomia* species (The *Nasomia* genome working group 2010), could also aid in understanding the nutritional requirements of parasitoids through the identification of missing routes in their metabolic pathways. The use of these technologies will certainly revolutionize the field of insect nutrition and give rise to a new era in the investigation of artificial diets for rearing natural enemies.

The application of high-throughput genomic tools for the study of nutrition (nutrigenomics), as already in place to understand how nutrition can affect and influence metabolic pathways and homeostatic control in humans, facilitating the development of dietary interventions to prevent diet-related diseases, can be one other way of stepping up the field of insect nutrition, as recently demonstrated for predatory insects. Nutrigenomics use microarray technology, providing valuable information on how nutrition may alter gene expression patterns. The identification and characterization of such genes could allow their utilization as molecular markers for an early identification of the insect response to different nutritional sources, aiding in the adaptation of the diet to suit the nutritional requirements of the insect to be reared (Coudron et al. 2006).

But as we have argued in previous sections of this chapter, the problems in developing artificial rearing systems for egg parasitoids are not constrained to the nutritional needs of these insects. The interactions among egg parasitoids and their hosts have to be seen into a new perspective, and deeply investigated. The increasing role shown to symbionts, such as *Wolbachia* on the development and reproductive biology of parasitoids should also be addressed (Grenier et al. 1998b, 2002, see also Chapter 6). There is also an urgent need for those working in the field of insect diet to include concepts of food science and food technology principles into their research programs. Cohen (2004d) pointed that future insect diet professionals should include in their curricula courses in food microbiology, food chemistry and physics, and food processing technology. He argued that the field of insect diet would be improved if processes such as technologies for fermentation, extrusion applications, modern flash sterilizer technologies, technologies for preservation using modified atmospheres, freeze-drying applications, and techniques for nutrient processing would be exploited when developing insect artificial systems. All these technologies and techniques of the food processing field would lead to the development of suitable packing materials (artificial hosts), that would provide the best texture, moisture and nutrient contents to the artificial diet.

In conclusion, there is future for the in vitro rearing of parasitoids and its application to parasitoid mass rearing systems to support biological control programs, but those involved must be aware that it is time to search for modern technologies and knowledge in other fields of science to succeed.

References

Anderson TE, Leppla NC (1992) Advances in insect rearing for research and pest management. Westview Press, Boulder, 517p

Barret M, Schmidt JM (1991) A comparison between the amino acid composition of an egg parasitoid wasp and some of its hosts. Entomol Exp Appl 59:29–41

Battisti A, Ianne P, Milani N, Zanata M (1990) Preliminary accounts on the rearing of *Ooencyrtus pityocampae* (Mercet) (Hym.; Encyrtidae). J Appl Entomol 110:121–127

Carpenter JE, Greany PD (1998) Comparative development and performance of artificially reared versus host-reared *Diapetimorpha introita* (Cresson) (Hymenoptera: Ichneumonidae) wasps. Biol Control 11:203–208

Chiappini E, Dindo ML, Negri I, Sighinolfi L (2004) In vitro rearing of *Anagrus breviphragma* (Hymenoptera: Mymaridae), an egg parasitoid of *Cicadella viridis* (Hemiptera: Cicadellidae), from second instar larva to adult on diets without insect components. Eur J Entomol 101: 419–422

Cohen AC (1985) Simple method for rearing the insect predator *Geocoris punctipes* (Heteroptera: Lygaeidae) on a meat diet. J Econ Entomol 78:1173–1175

Cohen AC (2004a) What makes a diet successful or unsuccessful? pp 47–74. In: Cohen AC (ed) Insect diets: science and technology. CRC, Boca Raton, FL, 324p

Cohen AC (2004b) Equipment used for processing insect diets: small-, medium-, and large-scale applications, pp 199–224. In: Cohen AC (ed) Insect diets: science and technology. CRC, Boca Raton, FL, 324p

Cohen AC (2004c) Microbes in the diet setting, pp 225–248. In: Cohen AC (ed) Insect diets: science and technology. CRC, Boca Raton, FL, 324p

Cohen AC (2004d) Future prospects for insect diets, pp 259–266. In: Cohen AC (ed) Insect diets: science and technology. CRC, Boca Raton, FL, 324p

Cohen AC, Smith LK (1998) A new concept in artificial diets for *Chrysoperla rufilabris*: the efficacy of solid diets. Biol Control 13:49–54

Cônsoli FL (1997) Criação in vitro de *Trichogramma galloi* Zucchi, 1988 e *T. pretiosum* Riley, 1879 (Hym.: Trichogrammatidae): desenvolvimento de um ovo artificial e aprimoramento de dietas artificiais. PhD Thesis, Piracicaba, ESALQ/USP, 153p

Cônsoli FL, Parra JRP (1996a) Comparison of hemolymph and holotissues of different species of insects as diet components for in vitro rearing of *Trichogramma galloi* Zucchi and *T. pretiosum* Riley. Biol Control 6:401–406

Cônsoli FL, Parra JRP (1996b) Biology of *Trichogramma galloi* Zucchi and *T. pretiosum* Riley reared "*in vivo*" and "in vitro" Ann Entomol Soc Am 89:828–834

Cônsoli FL, Parra JRP (1997) Produção in vitro de parasitóides: criação de *Trichogramma galloi* e *T. pretiosum* no Brasil, pp 259–302. In: Parra JRP, Zucchi RA (eds) *Trichogramma* e o controle biológico aplicado. FEALQ, Piracicaba, 324p

Cônsoli FL, Parra JRP (1999a) In vitro rearing of parasitoids: constraints and perspectives. Trends Entomol 2:19–32

Cônsoli FL, Parra JRP (1999b) Development of an artificial host egg for in vitro egg laying of *Trichogramma galloi* and *T. pretiosum* using plastic membranes. Entomol Exp Appl 91: 327–336

Cônsoli FL, Parra JRP (1999c) Egg laying and development of different strains of *Trichogramma pretiosum* Riley (Hymenoptera: Trichogrammatidae) in artificial eggs. Ann Soc Entomol Brasil 28:173–177

Cônsoli FL, Parra JRP (2000) Effect of the age of the pupal holotissue on the nutritional quality of artificial diets for *Trichogramma* spp. (Hymenoptera: Trichogrammatidae). Ann Soc Entomol Brasil 29:555–563

Cônsoli FL, Parra JRP (2002) Criação in vitro de parasitóides e predadores, pp 239–275. In: Parra JRP, Botelho PSM, Corrêa-Ferreira BS, Bento JMS (eds) Controle biológico no Brasil: parasitóides e predadores. Editora Manole Ltda, Barueri, SP, 609p

Cônsoli FL, Vinson SB (2009) Parasitóides (Hymenoptera), pp 837–873. In: Panizzi RA, Parra JRP (eds) Bioecologia e nutrição de insetos – Base para o manejo integrado de pragas. Embrapa Informação Tecnológica, Brasília, DF. 1164p

Cônsoli FL, Kitajima EW, Parra JRP (1999) Ultrastructure of the natural and factitious host eggs of *Trichogramma galloi* Zucchi and *T. pretiosum* Riley (Hymenoptera: Trichogrammatidae). Int J Insect Morphol Embryol 28:211–231

Cônsoli FL, Conti E, Dangott LJ, Vinson SB (2001) In vitro culture of the teratocytes of *Trissolcus basalis* (Hymenoptera: Scelionidae) and their requirements for host derived components. Biol Control 22:176–184

Coudron TA, Yocum GD, Brandt SL (2006) Nutrigenomics: a case study in the measurement of insect response to nutritional quality. Entomol Exp Appl 121:1–14

Dahlan AN, Gorth G (1997) Ovipositional synergists and artificial diets for *Trichogramma australicum* Girault (Hymenoptera: Trichogrammatidae). Aust J Entomol 36:383–390

Dai KJ, Zhang LW, Ma ZJ, Zhong LS, Zhan QX, Cao AH, Xu KJ, Li Q, Gao YG (1988) Research and utilization of artificial host egg for propagation of parasitoid *Trichogramma*. Les Colloques de l'INRA 43:311–318

Dai KJ, Ma ZJ, Zhang LW, Cao AH, Zhan QX, Xu KJ, Pan DS, Zhang JL (1991) Research on technology of industrial production of the artificial host egg of *Trichogramma*. Les Colloques de l'INRA 56:137–139

Dias NS, Parra JRP, Cônsoli FL (unpl.) Egg laying and development of neotropical trichogrammatid species (Hymenoptera: Trichogrammatidae) in artificial eggs. Entomol Exp Appl

Ding HC, Qui HG, Hwang CB (1980) In vitro rearing of an egg parasitoid, *Tetrastichus schoenobii* Ferrière. Contrib Shangai Inst Entomol 1:55–59

Feng JG, Tao X, Zhang AS, Yu Y, Zhang CW (1999) Study on using *Trichogramma* spp. on artificial host egg to control corn pests. Chin J Biol Control 15:97–99

Gao YG, Dai KJ, Shong LS (1982) Studies on the artificial host egg for *Trichogramma*. Les Colloques de l'INRA 9:181

Gomes SM (2002) Criação in vitro de *Trichogramma pretiosum* Riley, 1879 (Hymenoptera: Trichogrammatidae): aspectos nutricionais e bioquímicos. PhD Thesis, Piracicaba, ESALQ/USP, 100p

Grenier S (1994) Rearing of *Trichogramma* and other egg parasitoids on artificial diets. In: Wajnberg E, Hassan SA (eds) Biological control with egg parasitoids. CAB International, Wallingford, pp 73–92

Grenier S (1997) Desenvolvimento e produção in vitro de *Trichogramma*, pp 235–258. In: Parra JRP, Zucchi RA (eds), *Trichogramma* e o controle biológico aplicado. FEALQ, Piracicaba, 324p

Grenier S (2009) In vitro rearing of entomophagous insects – past and future trends: a minireview. Bull Insectol 62:1–6

Grenier S, Bonnot G (1988) Development of *Trichogramma dendrolimi* and *T. maidis* (Hymenoptera, Trichogrammatidae) in artificial media and artificial host eggs. Les Colloques de l'INRA 43:319–326

Grenier S, De Clercq P (2003) Comparison of artificially vs. naturally reared natural enemies and their potential for use in biological control, pp 115–133. In: van Lenteren JC (ed) Quality control and production of biological control agents: Theory and testing procedures. CABI Publishing, Wallingford, 327p

Grenier S, Liu WH (1990) Antifungals: mold control and safe levels in artificial media for *Trichogramma* (Hymenoptera. Trichogrammatidae). Entomophaga 35:283–291

Grenier S, Delobel B, Bonnot G (1986) Physiological considerations of importance to the success of in vitro culture: an overview. J Insect Physiol 32:403–408

Grenier S, Veith V, Renou M (1993) Some factors stimulating oviposition by the oophagous parasitoid *Trichogramma brassicae* Bezd. (Hym., Trichogrammatidae) in artificial host eggs. J Appl Entomol 115:66–76

Grenier S, Greany PD, Cohen AC (1994) Potential for mass release of insect parasitoids and predators through development of artificial culture techniques. In: Rosen D, Bennett FD, Capinera JL (eds) Pest management in the subtropics: Biological control – a Florida perspective. Intercept Ltd., Andover, Hampshire, 737p

Grenier S, Yang H, Guillaud J, Chapelle L (1995) Comparative development and biochemical analyses of *Trichogramma* (Trichogrammatidae: Hymenoptera) in artificial media with hemolymph or devoid on insect components. Comparat Biochem Physiol 111B:83–90

Grenier S, Han SC, Chapelle L, Liu WH, Guillaud J (1998a) In vitro development of *Trichogramma* spp. (Hymenoptera: Trichogrammatidae) in long-term stored, freeze-dried artificial media. Biocontrol Sci Technol 8:589–596

Grenier S, Pintureau B, Heddi A, Lassablière F, Jager C, Louis C, Khatchadourian C (1998b) Successful horizontal transfer of *Wolbachia* symbionts between *Trichogramma* wasps. Proc R Soc Lond 265B:1441–1445

Grenier S, Grille G, Basso C, Pintureau B (2001) Effects of the host species and the number of parasitoids per host on the size of some *Trichogramma* species (Hymenoptera : Trichogrammatidae). Biocontrol Sci Technology 11:21–26

Grenier S, Gomes SM, Pintureau B, Lassablière F, Bolland P (2002) Use of tetracycline in larval diet to study the effect of *Wolbachia* on host fecundity and clarify taxonomic status of *Trichogramma* species in cured bisexual lines. J Invertebr Pathol 80:13–21

Grenier S, Gomes SM, Febvay G, Bolland P, Parra JRP (2005) Artificial diet for rearing *Trichogramma* wasps (Hymenoptera: Trichogrammatidae) with emphasis on protein utilisation.

In: Proceedings of International Symposium on Biological Control of Arthropods, Davos Switzerland, Hoddle Compiler, pp 481–487
Hagen KS, Tassan RL (1965) A method of providing artificial diets to *Chrysopa* larvae. J Econ Entomol 58:999–1000
Han SC, Chen QX, Li LY (1994) A study of the oviposition synegists for in vitro rearing *Trichogramma* spp. Entomol Sin 1:333–338
Heslin LM, Kopittke RA, Merritt DJ (2005a) The role of insect cell lines in an artificial diet for the parasitoid wasp, *Trichogramma pretiosum*. Biol Control 33:186–193
Heslin LM, Kopittke RA, Merritt DJ (2005b) Refinement of a cell line based artificial diet for rearing the parasitoid wasp, *Trichogramma pretiosum*. Biol Control 33:278–285
Hsia PY, Wang MH (1979) The structure of oak silkworm egg shell and its relation to trichogrammatid parasitism. Acta Entomol Sinica 22:301–309
Hu JS, Vinson SB (1997a) In vitro development of *Campoletis sonorensis* (Hym.: Ichneumonidae), a larval endoparasitoid of *Heliothis virescens* (Lep.: Noctuidae) in an artificial medium with insect sources from egg to third larval instar. Entomophaga 42:405–415
Hu JS, Vinson SB (1997b) In vitro development of *Campoletis sonorensis* (Hym.: Ichneumonidae), a larval endoparasitoid of *Heliothis virescens* (Lep.: Noctuidae) in an artificial medium devoid of insect sources. Entomol Exp Appl 85:263–273
Hu JS, Vinson SB (1998) In vitro development of *Campoletis sonorensis* (Hym.: Ichneumonidae) in an artificial medium: importance of physical factors. J Insect Physiol 44:455–461
Hu JS, Gelman DB, Bell RA, Loeb MJ (1998) In vitro rearing of *Edovum puttleri*, an egg parasitoid of the Colorado potato beetle – development from egg through the pupal stage. BioControl 43:1–16
Jarjees E, Merritt DJ, Gordh G (1998) Anatomy of the mouthparts and digestive tract during feeding in larvae of the parasitoid wasp *Trichogramma australicum* Girault (Hymenoptera: Trichogrammatidae). Int J Insect Morphol Embryol 27:103–110
Jones RL, Lewis WJ, Beroza M, Bierl BA, Sparks AN (1973) Host-seeking stimulants (kairomones) for the egg parasite, *Trichogramma evanescens*. Environ Entomol 2:593–596
King EG, Coleman RJ, Morales-Ramos JA, Rojas GM, Summy KR, Wood L, Wendell L (1996) Biological control of the Boll weevil in cotton by mass propagation and augmentative releases of the wasp parasite, *Catolaccus grandis*. In: International Conference in Technology Transfer in Biological Control, Montpellier (IOBC WPRS Bulletin/Bull OILB, vol 19)
Lee HP, Lee KS (1994) Artificial rearing in vitro of *Ooencyrtus kuvanae* Howard (Hymenoptera: Encyrtidae): artificial media, oviposition and development. Korean J Entomol 24:311–316
Li LY (1982) *Trichogramma* sp. and their utilization in Peoples ́ Republic of China. Les Colloques de l'INRA 9:23–29
Li LY (1992) In vitro rearing of parasitoids of insects pests in China. Korean J Appl Entomol 31: 241–246
Li LY (1997a) Research and utilization of *Trichogramma* in China, pp 149–166. In: Li LY (ed) Parasitoids and predators (Insect) of Agricultural and forestry arthropod pests. Proceedings of the Guangdong Entomological Institute. Guangdong High Education Press, Guangzhou, China, 416p
Li LY (1997b) In vitro rearing of parasitoids of insect pests in China, pp 363–368. In: Li LY (ed) Parasitoids and predators (Insect) of agricultural and forestry arthropod pests. Proceedings of the Guangdong Entomological Institute. Guangdong High Education Press, Guangzhou, China, 416p
Li LY, Liu WH, Chen CS, Han SC, Shin JC, Du HS, Feng SY (1988) In vitro rearing of *Trichogramma* spp. and *Anastatus* sp. in artificial "eggs" and the methods of mass production. Les Colloques de l'INRA 43:339–352
Li LY, Chen QX, Liu WH (1989) Oviposition behavior of twelve species of *Trichogramma* and its influence on the efficiency of rearing them in vitro. Nat Enemies Insects 11:31–35
Li LY, Chen QX, Liu WH (1997a) Oviposition behavior of *Trichogramma* spp. and its influence on the efficiency of rearing them in vitro, pp 377–383. In: Li LY (ed) Parasitoids and predators (Insect) of agricultural and forestry arthropod pests. Proceedings

of the Guangdong Entomological Institute. Guangdong High Education Press, Guangzhou, China, 416p

Li LY, Xin JQ, Liu WH, Han SC, Chen QX, Zhang ML, Zhu DF, Xu X, Lin QF, Zhang SS (1997b) In vitro rearing of *Anastatus japonicus* Ashmead, its ovipositional behaviour, and effectiveness in controlling litchi stink bug, *Tessaratoma papillosa* (Drury), pp 358–369. In: Li LY (ed) Parasitoids and predators (Insect) of agricultural and forestry arthropod pests. Proceedings of the Guangdong Entomological Institute. Guangdong High Education Press, Guangzhou, China, 416p

Liu WH, Wu ZX (1982) Recent results in rearing *Trichogramma* in vitro with artificial media devoid of insectan additives. Acta Entomol Sin 25:160–163

Liu WH, Xie ZN, Xiao GF, Zhou YF, Ou Yang DH, Li LY (1979) Rearing of the *Trichogramma dendrolimi* in artificial diets. Acta Phytophyl Sin 6:17–25

Liu ZC, Sun YR, Wang ZY, Liu JF, Zhang LW, Dai KJ, Gao YG (1985) Field release of *Trichogramma confusum* reared on artificial host eggs against sugarcane borers. Nat Enemies Insects 3:1–5

Masutti L, Battisti A, Milani N, Zanata M, Zanazzo G (1993) In vitro rearing of *Ooencyrtus pityocampae* (Mercet) (Hym., Encyrtidae), an egg parasitoid of *Thaumetopoea pityocampa* (Lep., Thaumetopoeidae). Entomophaga 38:327–333

Mellini E (1975) Possibilità di allevamento di insetti entomofagi parassiti su diet artificiali. Boll dell'Istituto Entomol "Guido Grandi" della Università degli Studi Bologna 32:257–290

Mellini E, Campadelli G (1996) A first overall comparison between in vitro and in vivo production of the parasitoid *Exorista larvarum* (L.). Boll dell'Istituto Entomologia "Guido Grandi" della Università degli Studi Bologna 50:183–189

Moco S, Bino RJ, de Vos RCH, Vervoort J (2007) Metabolomics technologies and metabolite identification. Trends Anal Chem 26:855–866

Morales-Ramos JA, Rojas MG, Coleman RJ, King EG (1998) Potential use of in vitro-reared *Catolaccus grandis* (Hymenoptera: Pteromalidae) for biological control of the boll weevil (Coleoptera: Curculionidae). J Econ Entomol 91:101–109

Morrison RK, Nettles Jr WC, Ball D, Vinson SB (1983) Successful oviposition by *Trichogramma pretiosum* through a synthetic membrane. Southwestern Entomol 8:248–251

Nettles WC Jr (1990) In vitro rearing of parasitoids – role of host factors in nutrition. Arch Insect Biochem Physiol 13:167–175

Nettles WC Jr, Morrison RK, Xie ZN, Ball D, Shenkir CA, Vinson SB (1982) Synergistic action of potassium chloride and magnesium sulfate on parasitoid wasp oviposition. Science 218: 164–166

Nettles WC Jr, Morrison RK, Xie ZN, Ball D, Shenkir CA, Vinson SB (1983) Effect of cations, anions and salt concentrations on oviposition by *Trichogramma pretiosum* in wax eggs. Entomol Exp Appl 33:283–290

Nettles WC Jr, Morrison RK, Xie ZN, Ball D, Shenkir CA, Vinson SB (1985) Effect of artificial diet media, glucose, protein hydrolyzates and other factors on oviposition in wax eggs by *Trichogramma pretiosum*. Entomol Exp Appl 38:121–129

Nordlund DA, Wu ZX, Greenberg SM (1997) In vitro rearing of *Trichogramma minutum* Riley (Hymenoptera: Trichogrammatidae) for ten generations, with quality assessment comparisons of in vitro and in vivo reared adults. Biol Control 9:201–207

Notarte A, Merritt DJ (2001) Successful in vitro rearing of *Trichogramma australicum* (Hymenoptera: Trichogrammatidae) on artificial diets containing cultured insect cells. Bull Entomol Res 91:227–229

Nurindah, Gordh G, Cribb BW (1997) Oviposition behaviour and reproductive performance of *Trichogramma australicum* Girault (Hymenoptera: Trichogrammatidae) reared in artificial diet. Aust J Entomol 36:87–93

Parra JRP (2002) Criação missal de inimigos naturais, pp 143–164. In: Parra JRP, Botelho PSM, Corrêa-Ferreira BS, Bento JMS (eds) Controle biológico no Brasil: parasitóides e predadores. Editora Manole Ltda, Barueri, SP, 609p

Parra JRP, Cônsoli FL (1992) In vitro rearing of *Trichogramma pretiosum* Riley, 1879. Ciência Cultura 44:407–409

Parra JRP, Cônsoli FL, Hassan SA (1997) Effects of the factitious hosts, *Ephestia kuehniella* and *Sitotroga cerealella*, on the quality of *Trichogramma pretiosum*. In: Annales of the IV International Conference on Pests in Agriculture, vol 3. ANPP, Montpellier, France, pp 735–740

Qin J, Wu ZX (1988) Studies on cultivation of *Trichogramma* in vitro: ovipositional behavior and larval nutritional requirements of *T. dendrolimi*. Les Colloques de l'INRA 43:379–387

Rajendram GF (1978a) Oviposition behavior of *Trichogramma californicum* on artificial substrates. Ann Entomol Soc Am 71:92–94

Rajendram GF (1978b) Some factors affecting oviposition of *Trichogramma californicum* (Hymenoptera: Trichogrammatidae) in artificial media. Can Entomol 110: 345–352

Rajendram GF, Hagen KS (1974) *Trichogramma* oviposition into artificial substrates. Environ Entomol 3:261–267

Ridgway RL, Vinson SB (1976) Biological control by augmentation of natural enemies. Plenum Press, New York, 480p

Rojas MG, Morales-Ramos JA, King EG (1996) In vitro rearing of the boll weevil (Coleoptera: Curculionidae) ectoparasitoid *Catolaccus grandis* (Hymenoptera: Pteromalidae) on meridic diets. J Econ Entomol 89:1095–1104

Shirazi J (2006) Investigation on the in vitro rearing of *Trissolcus grandis* an egg parasitoid of *Eurygaster integriceps* by use of artificial diet. Pak J Biol Sci 9:2040–2047

Shulaev V (2006) Metabolomics technology and bioinformatics. Brief Bioinf 7:128–139

Singh P (1977) Artificial diets for insects, mites, and spiders. Plenum Press, New York, 594p

Strand MR, Vinson SB (1985) In vitro culture of *Trichogramma pretiosum* on an artificial medium. Entomol Exp Appl 39:203–209

Strand MR, Vinson SB, Nettles WC Jr, Xie ZN (1988) In vitro culture of the egg parasitoid *Telenomus heliothidis*: the role of teratocytes and medium consumption in development. Entomol Exp Appl 46:71–78

Takasu K, Yagi S (1992) In vitro rearing of the egg parasitoid *Ooencyrtus nezarae* Ishii (Hymenoptera: Encyrtidae). Appl Entomol Zool 27:171–173

The Nasonia Genome Working Group (2010) Functional and evolutionary insights from the genomes of three parasitoid *Nasonia* species. Science 327:343–348

Thompson SN (1986) Nutrition and in vitro culture of insect parasitoids. Annu Rev Entomol 31: 197–219

Thompson SN (1999) Nutrition and culture of entomophagous insects. Annu Rev Entomol 44: 561–592

Thompson SN, Hagen KS (1999) Nutrition of entomophagous insects and other arthropods, pp 594–652. In: Bellows TS, Fisher TW (eds) Handbook of biological control: principles and applications. Academic Press, New York, 1046p

Vinson SB (1994) Parasitoid in vitro rearing; success and challenges, pp 49–104. In: Ochieng-Odero JPR (ed) Techniques of insect rearing for the development of integrated pest and vector management strategies. ICIPE Science Press, Nairobi, Kenya, 787p

Volkoff AN, Vinson SB, Wu ZX, Nettles WC Jr (1992) In vitro rearing of *Trissolcus basalis* [Hym., Scelionidae) an egg parasitoid of *Nezara viridula* [Hem., Pentatomidae]. Entomophaga 37: 141–148

Wu ZX, Zhang ZP, Li TX, Liu DM (1980) Artificial media devoid of insect additives for rearing larvae of the endoparasitoid *Trichogramma*. Acta Entomol Sin 23:232

Wu ZX, Qin JD, Li TX, Chang ZP, Liu DM (1982) Culturing *Trichogramma dendrolimi* in vitro with artificial media devoid of insect materials. Acta Entomol Sin 25:128–133

Wu ZX, Cohen AC, Nordlund DA (2000) The feeding behavior of *Trichogramma brassicae*: new evidence for selective ingestion of solid food. Entomol Exp Applicata 96:1–8

Wyatt GR (1961) The biochemistry of insect haemolymph. Annu Rev Entomol 6:75–102

Xie ZN, Nettles Jr WC, Morrison RK, Irie K, Vinson SB (1986a) Three methods for the in vitro culture of *Trichogramma pretiosum* Riley. J Entomol Sci 21:133–138

Xie ZN, Nettles Jr WC, Morrison RK, Irie K, Vinson SB (1986b) Effect of oviposition stimulants and diets on the growth and development of *Trichogramma pretiosum* in vitro Entomol Exp Appl 42:119–124

Xie ZN, Nettles Jr WC, Saldaña G, Nordlund DA (1997) Elmers ́ school glue and Elmers ́ glue all: arrestants and probing/oviposition enhancers for *Trichogramma* spp. Entomol Exp Appl 82:115–118

Yazlovetsky IG (1992) Development of artificial diets for entomophagous insects by understanding their nutrition and digestion, pp 41–64. In: Anderson TE, Leppla WC (eds), Advances in insect rearing for research and pest management. Westview Press, Boulder, 517p

Zaborski E, Teal PEA, Laing JE (1987) Kairomone-mediated host finding by spruce budworm egg parasite, *Trichogramma minutum*. J Chem Ecol 13:113–122

Zhong LS, Zhang JL (1989) Influence of the diet pH on the development and efficacy of reproduction of *Trichogramma confusum* (Hym.: Trichogrammatidae) in artificial rearings. Chin J Biol Control 5:101–103

Chapter 12
Quality Control of Mass Reared Egg Parasitoids

Joop C. van Lenteren and Franz Bigler

Contents

12.1 Introduction

In this chapter we will present methods for determination of the quality of mass reared egg parasitoids. Before we describe quality control guidelines, we will first outline the development of quality control of natural enemies, summarize the basic considerations concerning management of laboratory populations, and discuss obstacles in mass rearing that may lead to deterioration of natural enemy quality. We consider a discussion of the above topics essential for

J.C. van Lenteren (✉)
Laboratory of Entomology, Wageningen University, PO Box 8031, 6700EH Wageningen, The Netherlands
e-mail: Joop.vanLenteren@wur.nl

F.L. Consoli et al. (eds.), *Egg Parasitoids in Agroecosystems with Emphasis on Trichogramma*, Progress in Biological Control 9, DOI 10.1007/978-1-4020-9110-0_12,

understanding practical quality control guidelines, because without a proper theoretical background, relevant guidelines cannot be developed. Where possible, we have illustrated background theory with examples from the egg parasitoid literature.

In the second part of this chapter we give an overview of research done on egg parasitoids for the development of quality control guidelines. This research is almost exclusively restricted to *Trichogramma* species. For *Trichogramma*, a basic guideline to test quality has been in place for several years. Current research is still being performed for the development of a flight test and simple laboratory tests that give a good indication of field performance. The quality control guideline for *Trichogramma* can be used as a model for development of guidelines for other egg parasitoids.

12.2 The IOBC Initiative on Quality Control

Although "modern" biological control of arthropod pests has been applied for more than 100 years (van Lenteren 2005), large-scale production of natural enemies only began in the 1970s (van Lenteren and Woets 1988, van Lenteren 1995, 2003). Initial mass rearing efforts involved the production of not more than several thousand individuals per week of two beneficial species. None of the early publications on commercial aspects of biological control mention the topic of quality control of natural enemies. Quality control in mass rearing of beneficial insects was first mentioned in the mid-1980s, and shortly after that the topic became popular (van Lenteren 1986a, Bigler 1989). The fifth workshop of the International Organization for Biological Control (IOBC) global working group "Quality Control of Mass-Reared Arthropods" in Wageningen, the Netherlands, formed the starting point for a heated discussion among producers of natural enemies and scientists on how to approach quality control in the commercial setting at that time (Bigler 1991).

A series of five IOBC workshops, partly financed by the European Union, followed and as a result of these meetings, quality control guidelines were written for about twenty species of natural enemies, which have been tested and adapted by commercial producers of biological control agents in Europe (van Lenteren 1998, van Lenteren and Tommasini 1999, van Lenteren et al. 2003). These guidelines cover features that are relatively easy to determine in the laboratory (e.g., emergence, sex ratio, life span, fecundity, adult size, and predation/ parasitism rate). Recently, the 20 old and 10 new guidelines have been published together with chapters providing the scientific background for the development of quality-control methods (van Lenteren 2003). Work is now focused on the development of flight tests and of a test relating these laboratory characteristics to field efficiency.

12.3 Trends in Commercial Mass Production of Natural Enemies

The appearance and disappearance of natural enemy producers have characterized commercialization of natural enemies over the past 40 years. Only a few producers active in the 1970s are still in business today. In addition to many small insectaries producing at the "cottage industry" level, four large facilities (i.e., having more than fifty persons employed) exist. At these four production sites, more than 5–10 million individuals per agent per week are produced (van Lenteren and Woets 1988, van Lenteren and Tommasini 1999, van Lenteren 2000, van Lenteren 2003). These facilities provide the full spectrum of natural enemies needed for an entire IPM program in a specific commodity (Albajes et al. 1999, van Lenteren 2000). As the sale of biological control agents is still an emerging market that is composed of small competing companies, product quality and prices are continuously affected by competitive pressure. While such pressure may be profitable in the short-term by lowering costs of natural enemies, it could lead in the long run to biological control failures. Natural enemies were properly evaluated before commercial use some thirty years ago, but recently some species of natural enemies have been sold without tests under practical cropping situations that determine if the natural enemies are effective against the target pest (e.g. van Lenteren and Manzaroli 1999). Lack of performance of natural enemies or inadequate guidance at the farm or greenhouse level has resulted in the sale and use of natural enemies of poor quality. These problems have in some cases resulted in failure of biological control.

Natural enemy producers are a rather diverse group. Rearing of natural enemies can be a full-time business, but can also be a part-time activity of farmers. Natural enemies may also be reared by companies in associated industries like seed companies or retailers of fertilizers. In some cases, production of natural enemies has been started by a research group with governmental support and later continued as a private endeavor. The number of individuals produced per natural enemy species and the number of biological control agents that are commercially available has increased dramatically over the past twenty-five years (Fig. 12.1).

Nowadays, about one hundred and fifty natural enemy species are on the market for biological control, and thirty of these are produced in commercial insectaries in very large quantities (van Lenteren 2003). These natural enemies are reared by some 50 companies, of which 26 are presently located in Europe. For an indication of prices of natural enemies in Europe and the United States, see van Lenteren et al. (1997) and Cranshaw et al. (1996), respectively.

Commercial natural enemy producers mainly rear predators and parasitoids. Only some companies produce nematodes, entomopathogenic fungi, bacteria, or viruses. Mass- rearing methods for parasitoids and predators are usually developed

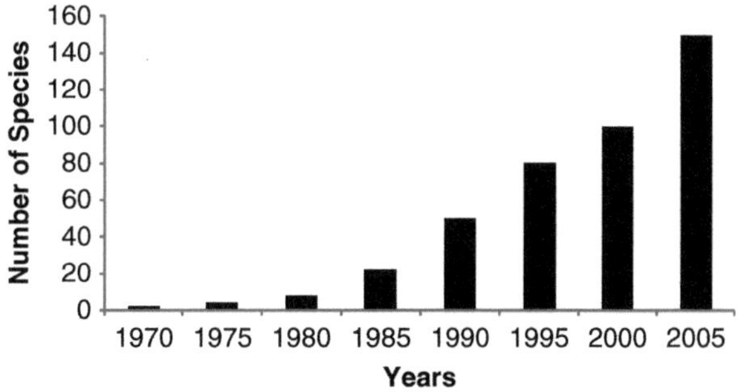

Fig. 12.1 Number of species of natural enemies commercially available for biological control in Europe

by a trial-and-error approach that may result in natural enemies of poor quality. The technology to rear natural enemies on artificial diets is not well developed yet and seems to be hampered not only by physiological problems, but also by ethological and ecological ones (requirements for associative learning of host-habitat and host finding cues; see chapters in van Lenteren 2003). Conflicts between attributes favored in mass rearing programs and those needed for field performance form another obstacle for cost-effective production of natural enemies. Artificial selection that occurs during mass rearing may lead to reduced performance. The suggested cures for this problem are often expensive and time consuming and are, therefore, very seldom applied.

Professional natural enemy producers may have research facilities, procedures for monitoring product quality, an international distribution network, promotional activities, and an advisory service. The market for high-quality, effective natural enemies will certainly increase with the growing demand for unsprayed food and a cleaner environment. The growing pesticide resistance problem will also move growers to adopt biological control methods.

Developments in the area of mass production, quality control, storage, shipment and release of natural enemies (Bueno 2000, van Lenteren 2000, 2003) have decreased production costs and led to better product quality, but much more can be done. Innovations in long-term storage (e.g., through induction of diapause), shipment, and release methods may lead to a further increase in natural enemy quality with a concurrent reduction in costs, thereby making biological control easier and more economically attractive to apply. But even if the natural enemies leave the insectary in good condition, shipment and handling by the producers, distributors, and growers may result in deterioration of the biological control agents before they are released.

Quality control programs that address not only natural enemy numbers, but also natural enemy quality (= field performance) are a necessity. Simple and reliable quality control programs for natural enemies are now emerging as a result of intensive cooperation between researchers and the biological control practitioners. These developments will result in an improvement of mass reared natural enemies and the biological control industry as a whole.

12.4 What Is Quality Control?

Quality control programs are applied to mass-reared organisms to maintain a good field performance of the population. The overall quality of an organism is defined as the ability to function as intended after release. The aim of quality control programs is to check whether the overall quality of a species is maintained, but that is too general a statement to be useful. Characteristics that affect overall quality have to be identified. These characteristics must be quantifiable and relevant for the performance of the parasitoid or predator. This is a straightforward statement, but very difficult to actually carry out (Bigler 1989).

Rather than discussing the development of quality control in strictly scientific terms (see for this approach chapters in van Lenteren 2003), this discussion will outline a more pragmatic approach. The aim of releases of mass-produced natural enemies is to control a pest. In this context the aim of quality control should be to determine whether a natural enemy is still in a condition to properly control the pest. Formulated in this way we do not need to consider terms like maximal or optimal quality, but rather acceptable quality. Some researchers believe the aim of quality control should be to keep the quality of the mass-reared population identical to that of the original field population. This is not only an illusion; it is an unnecessary and expensive goal to pursue. Another important consideration is that quality control programs are not applied for the sake of the scientist, but as a mere necessity. Leppla and Fisher (1989) formulated this dilemma as: "Information is expensive, so it is important to separate need to know from nice to know." Only if characteristics to be measured are very limited in number and directly linked to field performance, will natural enemy producing companies ever be able to apply quality control programs on a regular basis.

12.5 Basic Considerations for Quality Control

12.5.1 Genetic Changes in Laboratory Colonies

The problem of quality control of beneficial insects can be approached from either of two directions: (i) measure how well the biological control agent functions in its intended role. If it does not function well enough, trace the cause and improve the rearing method; or (ii) list what changes we can expect when a mass rearing system is started. Measure these and if the changes are undesirable, improve the rearing method.

The disadvantage of the first method is that changes may have occurred that cannot be corrected because the material has already changed so much that the original causes of the observed effects cannot be identified. The disadvantage of the second method is that too many measurements may be needed. The second approach has the advantage that potential problems are forecasted, and if seen, it may be possible to make changes in time to correct the problem. Bartlett (1984a), for example, approached the problem from the second viewpoint. He stated that many authors have suggested remedial measures for assumed genetic deterioration, but that causes for deterioration are not easily identified. Identification demands detailed genetic studies, and it is often difficult to define and measure detrimental genetic traits. He concluded, "I believe an unappreciated element of this problem is that the genetic changes taking place when an insect colony is started are natural ones that occur whenever any biological organism goes from one environment to another. These processes have been very well studied as evolutionary events and involve such concepts as colonization, selection, genetic drift, effective population numbers, migration, genetic revolutions and domestication theory."

In two other articles Bartlett (1984b, 1985) discussed what happens to genetic variability in the process of domestication, what factors might change variability, and which ones might be expected to have little or no effect. In the laboratory, domesticated insects are selected that have suitable genotypes to survive in this new environment, a process called *winnowing* by Spurway (1955) or, the less appropriate but widely used "forcing insects through a bottleneck." The changes that a field population may undergo when introduced into the laboratory are given in Table 12.1.

Variability in performance traits is usually abundantly present in natural populations and can remain large even in inbred populations (Prakash 1973, Yamazaki 1972, and chapters in van Lenteren 2003). But differences between field and laboratory environments will result in differences in variability. When natural enemy cultures are started, part of the "open population" from the field, where gene migration can occur and environmental diversity is large, is brought into the laboratory and becomes a "closed population." Thereafter, all future genetic changes act on the limited genetic variation present in the original founders (Bartlett 1984b, 1985). The size of the founder population will directly affect how much variation will be retained from the native gene pool. Although there is no agreement on the size of founder populations needed for starting a mass production, a minimum number of a thousand individuals is suggested (Bartlett 1985). Founder populations for commercial cultures of a number of natural enemies were, however, much smaller, sometimes fewer than twenty individuals. Fitness characteristics appropriate for the field environment will be different than those for the laboratory. These environments will place different values on the ability to diapause or to locate hosts/prey or mates. Such laboratory selection may produce a genetic revolution, and new balanced gene systems will be selected (Mayr 1970, Lopez-Fanjul and Hill 1973). One of the methods often suggested to correct for genetic revolutions is the regular introduction of wild individuals from the field. But if the rearing conditions remain the same in the laboratory, the introduced wild individuals will be subjected to the same process of genetic selection.

Table 12.1 Factors influencing changes in field populations after introduction into the laboratory

1. Laboratory populations are kept at constant environments with stable abiotic factors (light, temperature, wind, humidity) and constant biotic factors (food, no predation or parasitism). There is no selection to overcome unexpected stresses. The result is a change in the criteria that determine fitness and a modification of the whole genetic system (Lerner 1958)
2. There is no interspecific competition in laboratory populations, resulting in a possible change in genetic variability (Lerner 1958)
3. Laboratory conditions are made suitable for the average, sometimes even for the poorest genotype. No choice of environment is possible as all individuals are confined to the same environment. The result is a possible decrease in genetic variability (Lerner 1958)
4. Density-dependent behaviors (e.g., searching efficiency) may be affected in laboratory situations (Bartlett 1984b)
5. Mate-selection processes may be changed because unmated or previously mated females will have restricted means of escape (Bartlett 1984b)
6. Dispersal characteristics, specifically adult flight behavior and larval dispersal, may be severely restricted by laboratory conditions (Bush et al. 1976)

Furthermore, if a genetic differentiation has developed between laboratory and field populations, this may lead to genetic isolation (Oliver 1972). Also, positive correlation has been found between the incompatibility of such races and the differences between the environments (laboratory, field) where races occur, and for the length of time two populations have been isolated. Given these processes, introduction of native individuals to mass- rearing colonies is likely to be useless if incompatibility between field and laboratory populations is complete. If one wants to introduce wild genes, it should be done regularly and from the start of a laboratory rearing onwards. It should not be delayed until problems occur. Introducing field-collected insects into mass rearing also poses risks of introduction of parasitoids, predators or pathogens into the colony (Bartlett 1984b).

Another effect of laboratory colonization is inbreeding- mating of relatives and production of progeny that are more genetically homozygous than when random mating occurs in large populations. Genetically homozygous individuals often express harmful traits. The degree of inbreeding is directly related to the size of the founder population. Because artificial selection in the laboratory often results in an initial decrease in population size, the rate of inbreeding increases. The result is often a definite and rapid effect on the genetic composition of the laboratory population (Bartlett 1984b). Inbreeding can be prevented by various methods that maintain genetic variability (Joslyn 1984), including: (i) *Precolonization methods*: selection and pooling of founder insects from throughout the range of the species to provide a wide representation of the gene pool, resulting in a greater fitness of the laboratory material; *(ii) Postcolonization methods: (a) variable laboratory environments* (variation over time and space) – Although the concept of varying laboratory conditions is simple, putting it into practice is difficult. Consider for example the investments for rearing facilities with varying temperatures, humidities, and light regimes, or the creation of possibilities to choose from various diets or hosts, or the provision of space for dispersal, among others; *(b) gene infusion* – the regular rejuvenation of the gene pool with wild insects.

A fundamental question concerning inbreeding is: How large must the population size be to keep genetic variation sufficiently large? Joslyn (1984) said that to maintain sufficient heterogeneity, a colony should not decline below the number of founder insects. The larger the colony, the better. Very few data are available about effective population size; Joslyn mentioned a minimum number of five-hundred individuals. Prezotti et al. (2004) studied the effect of founder colony size on the quality of sexual populations of *Trichogramma pretiosum* Riley in the laboratory. Rearing units with one, five or ten founder couples were set up, each with five replicates. The populations were kept in a system of consanguineous crossings during 25 generations, monitoring at each generation the effective number of individuals and the coefficient of inbreeding. The populations were submitted to quality evaluations monitoring the number of parasitized eggs, emergence rate, sex ratio, longevity and percentage of deformed adults. The authors concluded that it is possible to start *T. pretiosum* cultures from a single couple without significant alterations in biological features during at least 25 generations. They also concluded that the initial population size is not so important, especially if the population size of the founder population increases significantly during the first generations. But they also stated

Table 12.2 Criteria to be considered before starting a mass rearing program

1.	The effective number of parents at the start of a mass- rearing is much lower than the number of founder individuals, so start with a large population
2.	Compensate for density-dependant phenomena
3.	Create a proper balance of competition, but avoid overcrowding
4.	Set environmental conditions for the best, not the worst or average genotype; use fluctuating abiotic conditions
5.	Maintain separate laboratory strains and cross them systematically to increase F_1 variability
6.	Measure frequencies of biochemical and morphological markers in founder populations and monitor changes
7.	Develop morphological and biochemical genetic markers for population studies
8.	Determine the standards that apply to the intended use of the insects, and then adapt rearing procedures to maximize those values in the domesticated strain

that the larger the number of founding individuals, the greater the possibility of obtaining high quality colonies, due to greater genetic diversity. This was shown by the larger rate of parasitism observed in populations initiated with 10 pairs.

The above discussion suggests several criteria that should be considered before a mass- rearing colony is started (Table 12.2) (Bartlett 1984b).

12.5.2 A Broader Approach to Quality Control

Chambers and Ashley (1984) and Leppla and Fisher (1989) put quality control in a much wider perspective. They presented some refreshing, and for most entomologists, new ideas. These authors approached quality control from the industrial side and considered three elements as essential: product control, process control, and production control. Product control rejects faulty products, and production control maintains consistency of production output. Process control tells how the manufacturing processes are performing. These elements of quality control are seldom applied to arthropod mass rearing programs.

Mass rearing, usually done by small private companies, is developed by trial-and-error. Knowledge on mass rearing techniques is often limited in such organizations and the time or money for extensive experimentation is lacking. If success is to be obtained, quality control of the end product is essential, but producers are generally more than happy if they can meet deadlines for providing certain numbers of natural enemies. Although most experts on quality control have adopted tools and procedures needed to regulate the processes of arthropod production so that product quality can be assured (Chambers and Ashley 1984), such tools and procedures are not yet widely used by the many small companies that compose 95% of all producers. The main reason most of the small companies do not develop and use such product, process, and production controls is that they lack the extra financial resources that are required. This limitation can be a serious constraint for new producers.

Chambers and Ashley (1984) stated that entomologists often concentrate too much on production control, while they are at best only partially controlling production processes and products. Quality control is frequently, but wrongly, seen as an alarm and inspection system that oversees and intimidates production personnel.

12.6 What Makes Quality Control so Difficult?

12.6.1 Obstacles in Mass- Rearing of Arthropods

Artificial selection forces in mass rearing may lead to problems related to performance of natural enemies if rearing conditions differ strongly from the situation in which natural enemies are to be released (Table 12.3). For example, if temperature in the mass rearing facility differs considerably from the field situation, synchronization problems can be expected. Also, rearing on non-target hosts or host plants can create problems with natural enemy quality or recognition by natural enemies of essential semiochemicals.

Any of the preceding obstacles (Table 12.3) may be encountered in mass production programs. One of the main obstacles to economic success seems to be the difficulty to produce qualitatively good natural enemies at a low price. But with a sharply decreasing number of available pesticides, increasing costs for chemical pesticides, and taxation of pesticides (as is presently taking place in several European countries), the price disadvantage of natural enemies may decrease.

Also, effective techniques to mass-produce natural enemies on artificial diets are often not available. Fewer than ten species of natural enemies can be produced on artificial diets, but generally their field performance is poorer than natural enemies reared on a host insect. Although mass production on artificial diets may lead to reduction of costs, the risks of changing natural enemy effectiveness should not be underrated.

Another obstacle for mass production is the lack of techniques to prevent selection pressures leading to genetic deterioration of the mass-produced organisms.

Table 12.3 Obstacles in mass rearing of natural enemies

1. Production of good quality natural enemies at low costs may be difficult (Beirne 1974)
2. Artificial diets are often not available for natural enemies (Beirne 1974)
3. Techniques that prevent selection pressures leading to genetic deterioration are usually lacking (Mackauer 1972, 1976)
4. Cannibalism by predators or superparasitism by parasitoids generally occurs
5. Rearing on unnatural hosts/prey or under unnatural conditions may cause behavioral changes in pre-imaginal and imaginal conditioning (Morrison and King 1977, Vet et al. 1990)
6. Reduced vigor can occur when natural enemies are reared on unnatural hosts (Morrison and King 1977)
7. Reduced vigor can also be the result when natural enemies are reared on hosts that are reared on an unnatural host diet (Morrison and King 1977)
8. Contamination by pathogens may occur (Bartlett 1984b)

Through such deteriorations, the natural enemy could loose its effectiveness (Boller 1972, Boller and Chambers 1977 and chapters in van Lenteren 2003).

Cannibalism among predators may make individual rearing or rearing at relatively high prey densities necessary and will lead to high costs. Superparasitism with parasitoids has the same effect.

Rearing of parasitoids and predators under "unnatural" conditions on "unnatural" hosts or prey, or on artificial media may change their reactions to natural hosts or host plant cues as a result of missing or improper pre-imaginal or imaginal conditioning. Rearing parasitoids on unnatural hosts may lead to reduced vigor as the result of an inadequate nutrient supply (quantity or quality) from the unnatural host. The same effect can occur when the host is reared on an unnatural diet, even if the host itself remains apparently unaffected.

Finally, the rearing colonies can become infected by pathogens. One of the problems often encountered in insect rearing is the occurrence of pathogens and microbial contaminants leading to high mortality, reduced fecundity, prolonged development, small adults, wide fluctuations in the quality of insects, or direct pathological effects. Goodwin (1984), Shapiro (1984), Sikorowski (1984), Singh and Moore (1985) and chapters in van Lenteren (2003) give information on the effects of microorganisms on insect cultures and the measures available to minimize or eliminate the pathogens or contaminants. Further, they discuss the recognition of diseases and microorganisms in insect rearing and the common sources of such microbial contaminants. The most common microbial contaminants encountered in insect rearing are fungi, followed by bacteria, viruses, protozoa, and nematodes. Field collected insects used to start a laboratory colony can be a major source of microbial contaminants. Another source is the various dietary ingredients. Disinfection of insects and dietary ingredients are recommended to prevent such contamination. Microbial contamination is usually rapidly recognized, but elimination of the pathogens from insect colonies is difficult (Bartlett 1984a).

Cases where inferior natural enemies that resulted from one of the factors mentioned above are well known among the biological control community, but are seldom published. Hassan and Wen (2001) report about the variability in quality of *Trichogramma* from commercial producers in Germany. The following text, concerning a failure in biological control with *Trichogramma* and the way it was solved by applying quality control, comes from Bigler (1994): "In Switzerland, *Trichogramma brassicae* has been mass produced since 1975 and applied commercially against the European corn borer, *Ostrinia nubilalis*, in maize since 1978. A significant loss in field efficacy was observed in 1980 (Fig. 12.2). By changing the mass-production system and the colony maintenance, it was possible to improve the performance of the strain and achieve the efficiency limit of at least 75% parasitisation in the field".

"A thorough analysis of the production system and the performance requirements of *T. brassicae* under the maize growing conditions in Switzerland led to the discovery of important traits which are crucial for a high efficacy. Since attributes like locomotory activity (Bigler et al. 1988), host acceptance (van Bergeijk et al. 1989), host suitability (van Bergeijk et al. 1989) and temperature tolerance were negatively

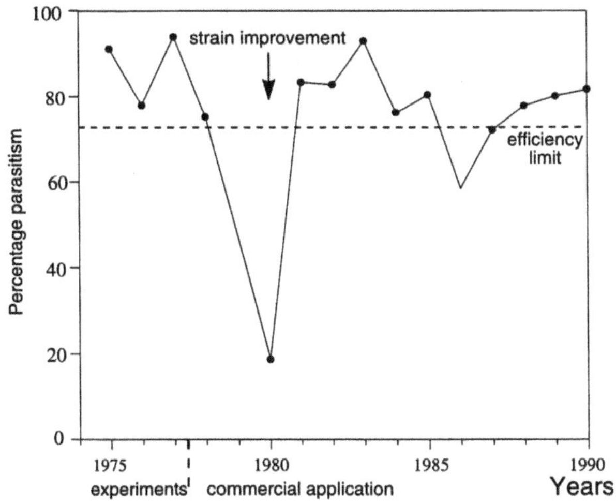

Fig. 12.2 Percentage parasitism of *Ostrinia nubilalis* eggs by *Trichogramma brassicae* in maize from 1975 to 1990 in Switzerland (after Bigler 1994)

affected by the former rearing system, we developed a new production unit. At the same time, risk evaluations of other deteriorations in the strain were performed and methods for measuring single traits and the field performance were developed.

In recent years, the production system of *T. brassicae* was changed from a short period production with a high daily output to a long period production with a low daily output. Improvements of the long-term storage of the parasitoids (diapause) have prolonged the mass production period from 2 to 9 months per year.

The total production system consists of three subunits, namely the European corn borer rearing unit as natural host of the wasps, the *Anagasta* (*Ephestia*) *kuehniella* (Zeller) unit as factitious (mass rearing) host and the *T. brassicae* unit. Each of these subunits has its own quality control system. Quality control procedures in the host units concentrate on: the rearing diet quality (origin, storage, grind, mixture), egg hatch, larval and pupal weight per unit rearing diet, egg production, egg sterilisation efficiency, temperature and humidity regulation, ventilation control to prevent health hazards by insect scales and sanitation procedures to avoid insect diseases.

Eggs of the European corn borer are used to produce the sting stock of *T. brassicae*. This population is permanently reared under semi-field conditions, i.e. in a field insectary in summer and in a greenhouse with fluctuating temperatures in winter; corn borer egg masses are attached to corn plants and the adults of the parental generation emerge 2–4 m away from the plants so that the egg masses must be reached by flight. A portion of the sting stock is regularly used to develop the strain on eggs of *A. kuehniella*. From our experience we know that the performance decreases with an increasing number of generations reared on the factitious host (van Bergeijk et al. 1989). Therefore, production is recommended for no more than six to seven generations before releasing the parasitoids for biological control purposes.

Since the sting stock is reared under near-natural conditions, deterioration is not expected. However, the "quality control" parameters and, in addition, the field efficacy (percentage parasitism) is assessed once a year. Since our previous experiments indicated that a change of quality attributes does not occur (or is not measurable) within the first generations on *A. kuehniella* eggs, the parameters are evaluated only once a year. The sixth generation (F6) is normally sold to the farmers, and a few rapid tests (parasitism, emergence, and sex ratio) are made on each batch when the parasitized eggs are shipped immediately. The final user (the farmer) is usually unable to do any performance tests. Therefore, government institutes, with financial support of the *Trichogramma* producers, accomplish the tests."

Bigler (1994) concludes that "quality control in *Trichogramma* mass- rearing is one of the measures used to avoid failures in biological control with these parasitoids. The extremely artificial rearing conditions, compared to the habitat where they are released, call for the establishment of sophisticated quality control concepts. [. . ..]. The importance of single performance attributes has to be established and related to field performance. The methods must be quick, simple and reliable. A single trait will never predict the overall performance accurately and therefore, the best combination of a set of laboratory methods must be developed. Whereas performance of the parasitoids in the field is the best indication of a good rearing system, low field efficacy does not tell us the causes. Regular performance control, carried out in the laboratory, will either indicate deterioration of performance and initiate corrections, or make us confident to produce wasps that are within the quality specifications."

The above summarized work done by the group of Bigler (1994) has initiated the development of the *Trichogramma* guideline as presented later in this chapter.

Deterioration of field performance does not apparently always have to be based on the development of a lower acceptance of the field host in the case of *Trichogramma*. Recently, Kölliker-Ott et al. (2003) showed that field collected *Trichogramma brassicae* reared for 27 generations on the laboratory host *A. kuehniella* accepted the target host better than when reared for 24 generations on the target host.

12.6.2 Behavioral Variation in Natural Enemies

The variation and changes in behavior of natural enemies caused by rearing conditions need a more detailed discussion as exciting new information has emerged in recent years. The lead question is: Can erratic behavior of natural enemies be prevented or cured? Recently, several papers have appeared on how to interpret and deal with variability in natural enemy behavior (Lewis et al. 1990, Vet et al. 1990, Vet and Dicke 1992, and chapters in van Lenteren 2003). Most ecologists are aware that variability in natural enemy behavior occurs frequently. It is important to know how natural enemies function in agroecosystems. Such understanding may help in designing systems where natural enemies play an even more important role in inundative and seasonal inoculative releases made in the field or greenhouses. Basic natural enemy behaviors for host-habitat and host location show great variability that often leads to inconsistent results in biological control. Most studies aimed at understanding such variability have focused on extrinsic factors as causes for any inconsistencies seen in foraging behavior. Typically, however, foraging behavior remained irregular even when using precisely the same set of external stimuli. These irregularities are caused by intraspecific individual variation in behavior. In order to understand erratic behavior and to be able to manipulate such variation, biological control researchers need to know the origins and width of variation.

Two types of adaptive variation have been recognized in the foraging behavior of natural enemies (Lewis et al. 1990): genetically fixed differences and phenotypic plasticity.

12.6.2.1 Genetically Fixed Differences

Differences exist among individuals (so called fixed-behaviors or innate responses). For example, natural enemy strains have different capabilities for searching in different habitats or have different host acceptance patterns. Such variation is now used in commercial selection of natural enemies. Genetically different strains of the same natural enemy species may react in very different ways to the same set of chemical stimuli from hosts and their host plants. To choose the best strain of a natural enemy for a particular task, it is important to have knowledge of these inherited preferences for particular environments and to match such inherited preferences with the stimuli present in the environment where natural enemy strains will be released. For a population of natural enemies to provide consistent biological control, the strain must have a proper blend of genetic traits appropriate to the target environment, and these traits must occur with sufficient uniformity in the population. This statement has been broadly recognized as true, but has been used only at a very gross level in applied programs (e.g., climate, habitat, and host matching).

12.6.2.2 Phenotypic Plasticity

Unfixed, learned or plastic behaviors arise as a result of experience accumulated by natural enemies as individuals. These learned behaviors allow the natural enemies to forage more effectively in any one of a variety of circumstances that might be encountered. Preferences develop for habitats in which suitable hosts have been encountered. The response of a foraging natural enemy can be quite plastic and can be modified within the bounds of its genetic potential and its experience as an individual. Modifications can be initiated during pre-imaginal ("larval") stages and at emergence.

The response of a *naive* adult (one which has not yet encountered a host) will already be affected by its rearing conditions. Such alterations have seldom been quantified, but changes in preference have been observed as a result of different rearing hosts or host diets. Particularly for inundative and seasonal inoculative programs of biological control, quantification of this variability is essential. An individual can often change its inherited response range, so it can develop an increased response for particular foraging environments as a result of experience with stimuli of these environments. Absence of reinforcement (i.e., absence of contact with host-related stimuli) will result in waning of the level of that response and a reversion to the naive preference. Natural enemies are plastic in their behavior, but operate within genetically defined boundaries.

Only recently have we begun to appreciate the extent to which natural enemies can learn. Many parasitoid species are able to acquire by experience an increased preference for and ability to forage in a particular environmental situation (Vet et al. 1990, Vet and Dicke 1992). There is some evidence for learning by immature stages and abundant evidence for learning by adults of various natural enemies. Learning is mostly by a matter of association of two and more stimuli, such as the odors of a host-plant and a host encounter, resulting later in orientation towards the odors of that specific host-plant species. Foraging behavior can continuously be modified according to the foraging circumstances encountered (Vet and Dicke 1992).

12.6.2.3 Physiological State of the Natural Enemy

Foraging behavior can also be strongly influenced by the physiological condition of the natural enemy. Natural enemies face varying situations in meeting their food, mating, reproductive and safety requirements. Presence of strong chemical, visual or auditory cues related to the presence of enemies of the natural enemy, and temporary egg depletion can all reduce or disrupt the response to cues used to find hosts. For example, hunger may result in increased foraging for food and decreased attention to hosts. In that case, the reaction to food and host cues will be different than when the natural enemy is well fed.

The sources of intrinsic variation in foraging behavior (genetic, phenotypic and those related to the physiological state) are not mutually exclusive, but overlap extensively, even within a single individual. The eventual foraging success of a natural enemy is determined by how well the natural enemy's intrinsic net condition is matched with the foraging environment in which it operates (Lewis et al. 1990, and chapters in van Lenteren 2003).

12.6.3 How to Manage Variability in Behavior of Natural Enemies

In order to be "efficient" as biological control agents, natural enemies must be able to (i) effectively locate and attack a host and (ii) stay in a host-infested area until most hosts are attacked. "Efficient" is used here in the anthropocentric sense (i.e., an agent able to provide pest control), that does not necessarily mean efficiency in a natural selection sense.

Management of natural enemy variation is particularly important when species are mass produced in the laboratory, especially if rearing is done in factitious hosts (species different from the target pest). Such laboratory rearing removes natural enemies from the context of natural selection and expose them to artificial selection for traits that are useless in the greenhouse or field (van Lenteren 1986a). In addition to the genetic component, associative learning may lead to many more changes in behavioral reactions. Bigler et al. (1997) have summarized several of the above mentioned aspects and their implications for quality control on *Trichogramma*.

12.6.3.1 Managing Genetic Qualities

Successful predation or parasitism of a target host in a confined situation does not guarantee that released individuals will attack that host under field conditions (see chapters in van Lenteren 2003). When selecting among strains of natural enemies, we need to ensure that the traits of the natural enemies are appropriately matched with the targeted use situations in the field. Natural enemy populations chosen for mass rearing should perform well on the target crop and under the specific field climate conditions.

12.6.3.2 Managing Phenotypic Qualities

Without care, insectary environments lead to agents with weak or distorted responses. If we understand the mechanism of natural enemy learning and the stimuli that affect it we can, in theory, provide the appropriate level of experience before releasing a natural enemy to correct defects from mass rearing. Prerelease exposure to important stimuli can help improve the responses of natural enemies through appropriate associative learning, leading to reduction in the escape response and an increase in natural enemy arrestment in target areas. An example of this approach is the exposure of *Trichogramma* wasps to cues associated with the intended target pest after they have been mass reared on an alternative laboratory host species.

12.6.3.3 Managing Physical and Physiological Qualities

Natural enemies should be released in a physiological state in which they are most responsive to herbivore or plant stimuli and will not be hindered by deprivations that might interfere with host searching. Thus, adult parasitoids should be well fed (honey or sugar source available in mass rearing), have had opportunities to mate, and have had time for the pre- oviposition period before releases are made.

12.6.4 Laboratory Rearing and Field Performance of Natural Enemies

In view of all these obstacles, it is clearly best to rear natural enemies under conditions that are as natural as possible, a conclusion that is supported by a number of researchers with experience in mass production (e.g. Bigler et al. 1997, van Lenteren 2003). Another important conclusion based on the new information about rearing is that the host habitat and the host should provide the same cues in mass rearing as in the field. If this is not possible, the natural enemies should be exposed to these cues

Table 12.4 Conflicting requirements concerning performance of natural enemies in a mass rearing colony and under field conditions

Features valued in mass rearing	Features important for field performance
Polyphagy (makes rearing on unnatural host/prey easier)	Monophagy or oligophagy (more specific agents often have a greater pest reduction capacity)
High parasitism or predation rates at high pest densities	High parasitism or predation rates at low pest densities
No strong migration as a result of direct or indirect interference	Strong migration as a result of direct or indirect interference
Dispersal behavior unnecessary and unwanted, ability to disperse minimal	Dispersal behavior essential
Associative learning not appreciated	Associative learning appreciated

after rearing, but before being released in the field. The problems that remain, even when rearing is done as naturally as possible, are related to obstacles 3, 4, 5 and 8 (Table 12.3). Anyone starting a mass-rearing facility should be prepared to not only overcome these obstacles, but should also recognize the conflicting requirements placed on natural enemies in mass production and during field performance (Table 12.4).

12.7 Quality Control Guidelines for Egg Parasitoids

Natural enemies are often mass-produced under greenhouse or laboratory conditions that are not necessarily similar to those found in commercial crops, e.g. at much higher pest densities. Because of this difference, most of the points listed in Table 12.4 are applicable and must be considered in quality control programs.

The development of quality control programs for natural enemy production has been rather pragmatic. Guidelines have been developed for 30 species of natural enemies, and descriptions of the various quality control tests included in these guidelines can be found in van Lenteren (1996, 1998, 2003) and van Lenteren and Tommasini (1999). The standard elements of quality control are given in Table 12.5. All current guidelines specify the conditions under which the tests have to be performed and, generally, information on the following characteristics has to be obtained: quantity of organisms in the package, sex ratio, fecundity, longevity, adult size, rate of predation or parasitism, and the package expiration date. Flight tests are not yet included in most tests, though they are available for *Trichogramma* and will be summarized below.

In Table 12.6 an example of a quality control test for *Trichogramma* is presented. Although there are other egg parasitoids used in commercial control than *Trichogramma* spp. only, quality control guidelines have been currently developed only for three species of *Trichogramma* (van Lenteren et al. 2003). The main reason

Table 12.5 General quality control criteria for rearing arthropod parasitoids or predators in mass culture

Quantity: predators: number of live predators in container
parasitoids: if delivered as adults: number of live parasitoids
if immatures: number of emerging adults in a certain period
Sex ratio: minimum percentage females; male biased ratio may indicate poor rearing conditions
Fecundity: number of offspring produced during a certain period;
for parasitoids fecundity is an indication of the potential maximum host kill rate
Longevity: minimum longevity in days
Predation: number of prey eaten during a certain period
Adult size: hind tibia length, sometimes pupal size (size is often a good indication for longevity, fecundity, and predation capacity if natural enemy is not negatively affected when manipulated during harvest, packaging, shipment, and release)
Flight: short-range flight (natural enemy can still fly)
Flight: long-range flight + predation/ parasitization capacity (can fly and perform)
Field performance: capacity to locate and consume prey or parasitize hosts in crop under field conditions

- Fecundity, longevity, and predation capacity tests can often be combined
- Expiration date for each shipment is given on packaging material
- All numbers, ratios, and sizes are mentioned on the container or packaging material
- Quality control is done under standardized test conditions of temperature, relative humidity, and light regime that are specified for each test

Table 12.6 Quality control guideline for *Trichogramma brassicae*. Similar guidelines for *T. cacoeciae Marchal* and *T. dendrolimi Matsumara* can be found in van Lenteren *et al.* (2003).

Trichogramma brassicae Bezd. (=*T. maidis*) (Hymenoptera: Trichogrammatidae)

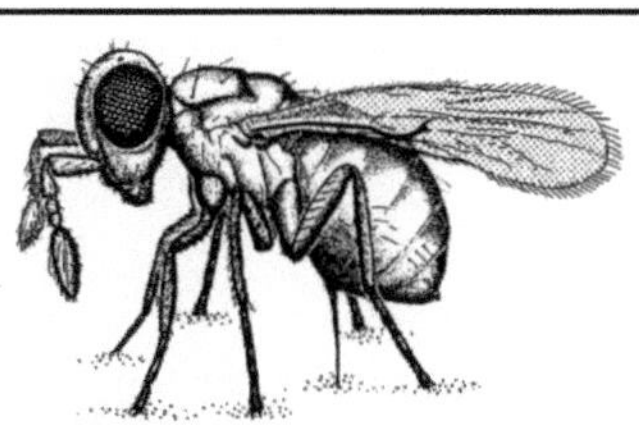

Test conditions	Temperature: 23±2°C RH: 75±10% Light regime: 16L:8D Rearing hosts: *Anagasta (=Ephestia) kuehniella* *Sitotroga cerealella*
Species identification:	The species is specified on the label and verified by the producer[a]
Quality control criteria	
Sex-ratio	≥ 50% females; 100 adults assessed on 10 release units each or 5 x 100 adults of bulk material; at least weekly or batch-wise test if batches were exposed to special treatments (e.g. storage)
Number of females[b]	As indicated on label; determined as for sex-ratio
Fecundity and longevity	≥40 offspring/7 days/female; 80% of females should live at least 7 days; monthly or batch-wise test; n=30.
Natural host parasitism	≥10 parasitised hosts/4 h/female

Table 12.6 (continued)

Trichogramma brassicae Bezd. (=*T. maidis*) (Hymenoptera: Trichogrammatidae)	
Description of testing methods	
Fecundity and longevity	30 females (age 24 h) are confined individually in glass tubes; at least 200 factitious host eggs (< 24 h) are glued with water on a small cardboard strip; a small droplet of honey and a droplet of water are added directly to the wall of the vial. Eggs of *E. kuehniella* (< 24 h old) are UV irradiated and provided at day 1 and removed after day 7; fresh eggs of *S. cerealella* are provided at day 1, 3 and 5. The number of living adults is recorded after day 7. Egg-cards are incubated and the number of black eggs is counted not earlier than at day 10. Minimum fecundity after day 7 is 40 offspring/female; mortality after day 7 is < 20%; at least monthly test or batch-wise if batches were exposed to special treatments (e.g. storage procedures, long-range shipments).
Natural host parasitism	30 females (age 24 h) are confined individually in tubes; two fresh egg-masses of at least 20 eggs/egg-mass of *Ostrinia nubilalis* (< 24 h-old) are added for 4 h; honey and water are provided as described above; after separation of the egg-masses from the females they are incubated for 3 days; the number of black eggs is counted; the mean number of black eggs is ≥10 per female. The host cluster acceptance rate (= females parasitising at least one host egg) should be ≥80%. This measure is important because parasitism drops drastically if a high proportion of females does not accept their hosts. This is especially true at low host densities and when hosts occur in batches. Often, parasitoids find only one egg mass during their lifetime and a high percentage acceptance is therefore crucial. This test is an indirect measure of the acceptance and suitability of the natural host egg. The test should be performed 2–4 times/year depending on the rearing system (number of generations reared on the factitious hosts).
Comments	[a]Molecular techniques are available at Laboratory of Entomology, Wageningen University, The Netherlands. Test necessary once a year, sample size min. 30 individuals [b]The emergence period and pattern depend on the mixture of developmental stages released together and must be specified on the label.
Initial design: F Bigler	

for this is that *Trichogramma* are applied on a vast area worldwide (van Lenteren 2000), though other species of egg parasitoids are, for example, used against the soybean bug in soybean in Latin America (van Lenteren and Bueno 2003). To our information, quality control guidelines have not yet been developed for these other species. Elements of the guideline presented in Table 12.6 have been used to evaluate the quality of batches of *Trichogramma* by producers, but data are seldom, if ever, published. The IOBC working group on Quality Control of Natural Enemies proposes that if a batch does not meet the minimum standards, it should be destroyed, though in exceptional cases, adaptations can be made to correct for lower quality (e.g. by releasing higher numbers of natural enemies or at a higher frequency).

A short-range flight test based on the one used for *Encarsia formosa* (van Lenteren et al. 2003) has been developed and tested for *Trichogramma* by Dutton and Bigler (1995) (Fig. 12.3). They tested four strains of *T. brassicae* in this cylinder: strain 1 was reared for 2 and strain 2 for over 35 generations on *A. kuehniella*, a diapause strain (strain 3) reared for 6 generations on *A. kuehniella* and a commercial strain (strain 4) was also reared on *A. kuehniella*, but rearing conditions and number of generations on the host were unknown. Clear differences in flight activity (= number of adults reaching the glass plate at the top of the cylinder) were observed. Strains 1 and 4 showed the highest flight activity in the cylinder, and flight in a field cage of 1.5 × 1.5 meter was also better than that of strains 2 and 3. These promising results indicate that a simple laboratory flight test might assist in testing the potential field efficiency of *Trichogramma*.

Though shortly addressed in the previous paragraph as well as in Section 12.6 of this chapter [based on information from Bigler (1994)], field performance of a natural enemy is rarely scientifically tested in the field. Honda et al. (1999) and Silva et al. (2000) described and used an interesting test that was initially developed by

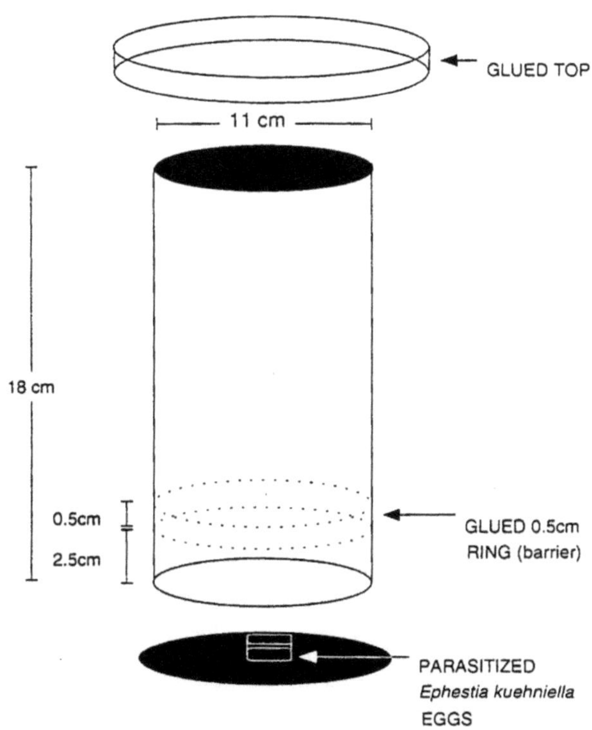

Fig. 12.3 Setup of short-range flight test for *Trichogramma* spp. The cylinder (11 cm diameter and 18 cm high) is covered with a Petri dish sprayed with glue at the inner side. Inner sides of bottom and cylinder are covered with a black plastic sheet to attract the adults to the light source at top. A 0.5 cm glue ring is sprayed on the plastic sheet of the cylinder 2.5 cm from the lower end of the unit as a barrier for walking adults. Adults emerge from parasitized *Anagasta* (*Ephestia*) *kuehniella* eggs placed in a small container on the bottom in the centre of the cylinder [see Dutton and Bigler (1995) for details]

Greenberg (1991) to evaluate searching and dispersal ability of *Trichogramma* in a "maze" in the laboratory. Ideally it would appear that the most useful and practical bioassays would be ones which are inexpensive, quick and repeatable. The objective of the studies by Honda et al. (1999) and Silva et al. (2000) was to develop such a bioassay and determine if it is possible to: *i*) distinguish differences in dispersal between *Trichogramma* populations in the laboratory and *ii*) determine if a relationship exists between our bioassay results and performance in the greenhouse. They chose to evaluate dispersal as a parameter because of its importance for field and greenhouse efficacy. Earlier, Bigler et al. (1988) worked on this issue, and found that a correlation existed between locomotion rate (= travel speed) and parasitization (= efficiency) in the field of different *Trichogramma maidis* Pintureau and Voegelé strains. Although they developed a scientifically attractive approach, the methodology was rather time consuming to apply and was considered too complicated for mass rearing companies.

Honda et al. (1999) and Silva et al. (2000) developed an inexpensive chamber based on the "maze" designed by Greenberg (1991) and tested it first as an evaluative tool to monitor *Trichogramma cordubensis* Vargas and Cabello dispersal in the laboratory (Fig. 12.4). The chamber consisted of a continuous winding channel which was cut into an aluminium block. Wasps were released at one end of the channel and allowed to walk in the channel for 21 h and to parasitize *Mamestra brassicae* eggs placed 3.4 m from the point of wasp introduction. Comparisons between two *T. cordubensis* populations demonstrated that one population (population A) dispersed more in the chamber and located host eggs more successfully than the other population (population B). Subsequent greenhouse releases in the set-up described

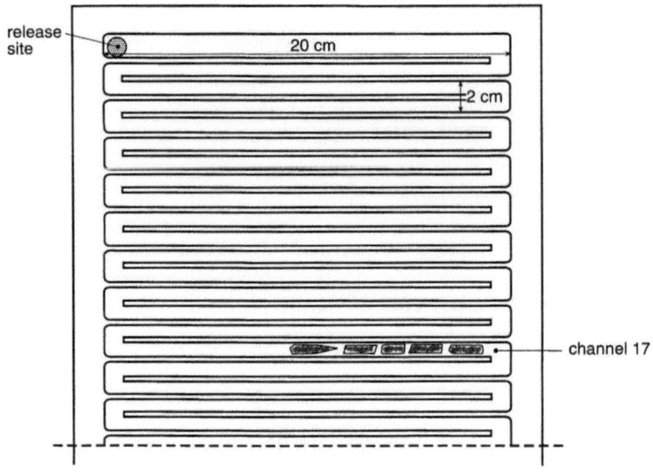

Fig. 12.4 Section of aluminium block (top view) showing continuous winding channel cut into metal block. Wasps were allowed to walk in the channel for 21 h and to parasitize *Mamestra brassicae* egg masses present in sub-channel 17, ca. 3.4 m from the entrance. Total length 8 m (sub-channel dimensions 1 cm × 1 cm × 20 cm)

in Fig. 12.5 confirmed that the population A dispersed more readily and had significantly higher parasitism rates on sentinel *A. kuehniella* eggs on tomato plants than the population B.

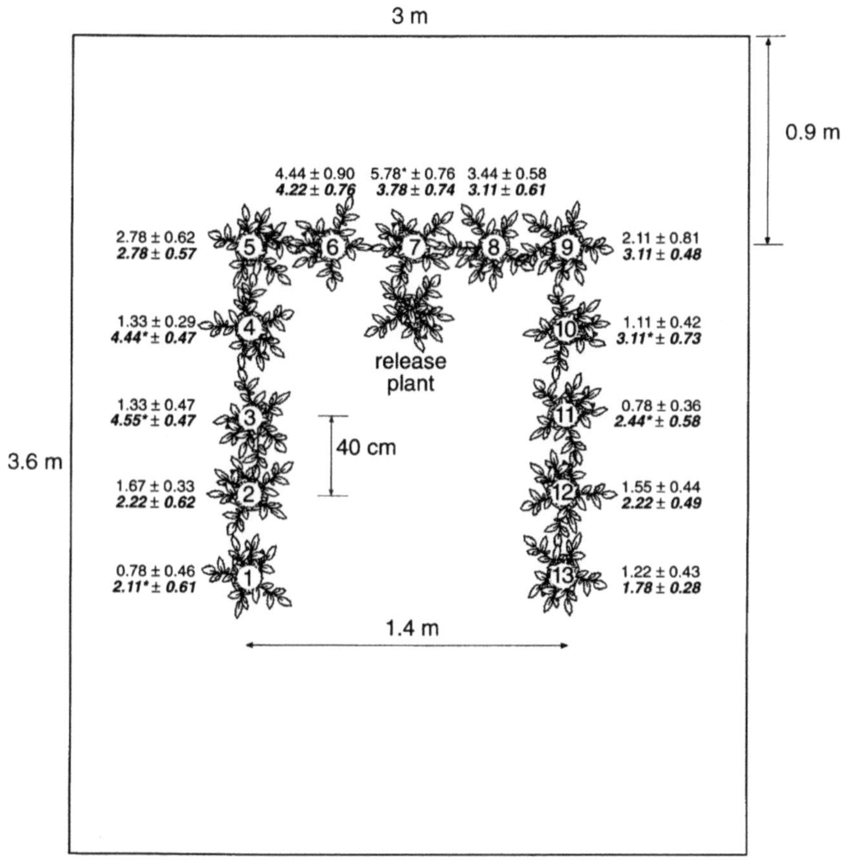

Fig. 12.5 Position of tomato plants and parasitism patterns of two *Trichogramma cordubensis* populations (A and B) on 13 tomato plants in nine paired greenhouse releases. Numbers represent mean numbers of sentinel egg masses parasitized (±SEM) per each tomato plant (italics=population A, normal= population B). * indicates significant differences between means for each plant (t-test, $p < 0.05$). Approximately 700–800 wasps of each population were released per replicate

The chamber used in the laboratory bioassay to evaluate *Trichogramma* dispersal possesses a number of advantages. First, the experiments themselves are easy to run as there is no need for continuous observations. Data may be collected daily and scored within 30 minutes per replicate. Second, the apparatus is inexpensive and has few parts; chamber construction costs less than 100 Euro and multiple chambers can be made and run simultaneously. Finally, based on the present results, the authors conclude that this bioassay may serve as a quick first stage screening process to evaluate *Trichogramma* populations in the laboratory. Poorly performing species or populations can be quickly and easily distinguished in this chamber from superior populations, which may then be more rigorously tested under field or greenhouse conditions.

An interesting approach for a field performance test was described by van Schelt and Ravensberg (1990). Their goal was to compare diapause-stored or freshly reared *T. maidis* capacity to control *O. nubilalis*. In the laboratory, percentage emergence, sex ratio and fecundity were determined for diapaused and freshly-reared parasitoids. Vials with parasitoids of the same samples as the laboratory material were put at a central release point in a corn crop. From the release point, cards with sentinel *O. nubilalis* eggs were hung on corn plants in 8 directions, with an interval of 1 meter and up to 10 meters away from the release point. Percentage parasitism was determined on these cards. The laboratory results showed no differences in emergence and fecundity between the diapaused and fresh parasitoids, but the sex ratio (% females) of the diapaused parasitoids was lower than that of fresh ones. The field tests showed that diapaused and fresh parasitoids dispersed in all directions, but that percentage parasitism by fresh parasitoids was higher than that of diapaused parasitoids (van Schelt and Ravensberg 1990).

12.8 Concluding Remarks

Quality control procedures for the most important commercially used species of natural enemies, including *Trichogramma*, have been developed and tested, and are used today by producers of biological control agents (van Lenteren 2003). The quality control criteria now employed relate to product control procedures, not to production or process control, and are based on laboratory measurements that are easy to carry out. They were designed to be as uniform as possible so they can be used in a standardized manner by many producers, distributors, pest management advisory personnel, and farmers. These tests should preferably be carried out by the producer after all handling procedures and just before shipment. It is expected that the user (farmer or grower) only needs to perform a few quality tests, e.g., checking the percent emergence or number of live adults in the package. Some tests are to be carried out frequently by the producer, i.e., on a daily, weekly or per-batch basis. Others will be done less frequently, i.e., on an annual or seasonal basis, or when rearing procedures are changed. These criteria have to be complemented with flight tests and field performance tests. Such tests are needed to show the relevance of the laboratory measurements. Laboratory tests are only adequate when a good correlation has been established between the laboratory measurements, flight tests and field performance.

Quality control programs should be designed to obtain acceptable quality, not necessarily the best possible quality. The number of necessary tests will be smallest if the natural enemies are reared under the same conditions as those under which they also have to function in the field in terms of same climate, host and host plant. The more artificial the rearing conditions become and the more the natural enemies are "handled" before use (removed from the plant or host, counted, put in containers, glued to a substrate, manipulated to induce diapause, shipped, released, etc.), the larger the number of tests that will have to be performed.

Companies just beginning the production of a natural enemy are often rather ignorant about the obstacles and complications entailed in mass rearing programs. New producers are even more ignorant about the need to develop and apply quality control testing criteria to their products. A special point of concern is the lack of knowledge about the sources of variability of natural enemy behavior and methods to prevent genetic deterioration of natural enemies. If the biological control industry is to survive and flourish, the production of reliable natural enemies that meet basic quality standards is essential. This is particularly important for egg parasitoids like *Trichogramma*, as these egg parasitoids are the most mass-reared and released natural enemies worldwide (van Lenteren and Bueno 2003).

Acknowledgements European producers of natural enemies are thanked for cooperation in development of quality control guidelines and for providing unpublished data on quality control. Development of quality control guidelines was financially supported by the Commission of the European Communities, Directorate General for Agriculture, DG VI, Concerted Action "Designing and Implementing Quality Control of Beneficial Insects: Towards more reliable biological pest control."

References

Albajes R, Gullino ML, van Lenteren JC, Elad Y (1999) Integrated pest and disease management in greenhouse crops. Kluwer, Dordrecht, The Netherlands

Bartlett AC (1984a) Establishment and maintenance of insect colonies through genetic control. In: King EG, Leppla NC (eds) Advances and challenges in insect rearing, vol 1. U.S. Department of Agriculture, Agricultural Research Service, Southern Region, New Orleans, p 1

Bartlett AC (1984b) Genetic changes during insect-domestication. In: King EG, Leppla NC (eds) Advances and challenges in insect rearing, vol 1. U.S. Department of Agriculture, Agricultural Research Service, Southern Region, New Orleans, pp 2–8

Bartlett AC (1985) Guidelines for genetic diversity in laboratory colony establishment and maintenance. In: Sing P, Moore RF (eds) Handbook of Insect Rearing, vol 1. Elsevier, Amsterdam, pp 7–17

Beirne BP (1974) Status of biological control procedures that involve parasites and predators. In: Maxwell FG, Harris FA (eds) Proceedings of the Summer Institute on Biological Control of Plant Insects and Diseases. University Press of Mississippi, Jackson, Mississipi, pp 69–76

Bigler F (1989) Quality assessment and control in entomophagous insects used for biological control. J Appl Entomol 108:390–400

Bigler F (1991) Fifth Workshop of the IOBC Global Working Group, Quality control of mass reared arthropods. IOBC, Wageningen, The Netherlands

Bigler F (1994) Quality control in *Trichogramma* production. In: Wajnberg E, Hassan SA (eds) Biological control with egg parasitoids. CABI International, Wallingford, pp 93–111

Bigler F, Bieri M, Seidel K (1988) Variation in locomotion between laboratory strains of *Trichogramma maidis* and its impact on parasitism of eggs of *Ostrinia nubilalis* in the field. Entomol Exp Appl 49:283–290

Bigler F, Suverkropp BP, Cerutti F (1997) Host searching by *Trichogramma* and its implications for quality control and release techniques. In: Andow DA, Ragsdale DW, Nyvall RW (eds) Ecological interactions and biological control. Westview Press, London, pp 240–253

Boller EF (1972) Behavioral aspects of mass-rearing of insects. Entomophaga 17:9–25

Boller EF, Chambers DL (1977) Quality aspects of mass-reared insects. In: Ridgway RL, Vinson SB (eds), Biological control by augmentation of natural enemies. Plenum, New York, pp 219–236

Bueno VHP (2000) Controle Biológico de Pragas: Produção Massal e Controle de Qualidade. Editora UFLA, Lavras

Bush GL, Neck RW, Kitto GB (1976) Screwworm eradication: Inadvertent selection for noncompetitive ecotypes during mass rearing. Science 193:491–493

Chambers DL, Ashley TR (1984) Putting the control in quality control in insect rearing. In: King EG, Leppla NC, Advances and Challenges in Insect Rearing. U.S. Department of Agriculture, Agricultural Research Service, Southern Region, New Orleans, LA, pp 256–260

Cranshaw W, Sclar DC, Cooper D (1996) A review of 1994 pricing and marketing by suppliers of organisms for biological control of arthropods in the United States. Biol Control 6:291–296

Dutton A, Bigler F (1995) Flight activity assessment of the egg parasitoid *Trichogramma brassicae* (Hym.: Trichogrammatidae) in laboratory and field conditions. Entomophaga 40:223–233

Goodwin RH (1984) Recognition and diagnosis of diseases in insectaries and the effects of disease agents on insect biology. In: King EG, Leppla NC (eds) Advances and Challenges in Insect Rearing. U.S. Department of Agriculture, Agricultural Research Service, Southern Region, New Orleans, LA, pp 96–129

Greenberg SM (1991) Evaluation techniques for *Trichogramma* quality. In: Bigler F (ed) Quality Control of Mass Reared Arthropods. Proceedings 5th Workshop IOBC Global Working Group "Quality Control of Mass Reared Arthropods". IOBC, Wageningen, The Netherlands, pp 138–145

Hassan SA, Wen QZ (2001) Variability in quality of *Trichogramma brassicae* (Hymenoptera: Trichogrammatidae) from commercial suppliers in Germany. Biol Control 22: 115–121

Honda JY, Silva IMMS, Vereijssen J, Stouthamer R (1999) Laboratory bioassay and greenhouse evaluation of *Trichogramma cordubensis* strains from Portugal. BioControl 44:1–11

Joslyn DJ (1984) Maintenance of genetic variability in reared insects. In: King EG, Leppla NC (eds), Advances and Challenges in Insect Rearing. U.S. Department of Agriculture, Agricultural Research Service, Southern Region, New Orleans, LA, pp 20–29

Kölliker-Ott UM, Bigler F, Hoffmann A (2003) Does mass rearing of field collected *Trichogramma brassicae* wasps influence acceptance of European corn borer eggs? Entomol Exp Appl 109:197–203

Leppla NC, Fisher WR (1989) Total quality control in insect mass production for insect pest management. J Appl Entomol 108:452–461

Lerner I (1958) Genetic basis of selection. Wiley, New York

Lewis WJ, Vet LEM, Tumlinson JH, van Lenteren JC, Papaj DR (1990) Variations in parasitoid foraging behavior: essential element of a sound biological control theory. Environ Entomol 19:1183–1193

Lopez-Fanjul C, Hill WG (1973) Genetic differences between populations of *Drosophila melanogaster* for quantitative trait. II. Wild and laboratory populations. Genet Res 22: 60–78

Mackauer M (1972) Genetic aspects of insect control. Entomophaga 17:27–48

Mackauer M (1976) Genetic problems in the production of biological control agents. Ann Rev Entomol 21:369–385

Mayr E (1970) Populations, species, and evolution. Harvard University Press, Cambridge, MA

Morrison RK, King EG (1977) Mass production of natural enemies. In: King EG, Leppla NC (eds), Advances and Challenges in Insect Rearing. U.S. Department of Agriculture, Agricultural Research Service, Southern Region, New Orleans, LA, pp 183–217

Oliver CG (1972) Genetic and phenotypic differentiation and geographic distance in four species of Lepidoptera. Evolution 26:221–241

Prakash S (1973) Patterns of gene variation in central and marginal populations of *Drosophila robusta*. Genetics 75:347–369

Prezotti L, Parra JRP, Vencovsky R, Coelho ASG, Cruz I (2004) Effect of the size of the founder population on the quality of sexual populations of *Trichgramma pretiosum*, in laboratory. Biol Control 30:174–180

Shapiro M (1984) Microorganisms as contaminants and pathogens in insect rearing. In: King EG, Leppla NC (eds) Advances and challenges in insect rearing. U.S. Department of Agriculture, Agricultural Research Service, Southern Region, New Orleans, LA, pp 130–142

Sikorowski PP (1984) Microbial contamination in insectaries. In: King EG, Leppla NC (eds) Advances and challenges in insect rearing. U.S. Department of Agriculture, Agricultural Research Service, Southern Region, New Orleans, LA, pp 143–153

Silva IMMS, van Meer MMM, Roskam MM, Hoogenboom A, Gort G, Stouthamer R (2000) Biological control potential of *Wolbachia*-infected versus uninfected wasps: laboratory and greenhouse evaluation of *Trichogramma cordubensis* and *T. deion* strains. Biocontrol Sci Technol 10:223–238

Singh P, Moore RF (1985) Handbook of insect rearing, vol 1 and 2. Elsevier, Amsterdam

Spurway H (1955) The causes of domestication: An attempt to integrate some ideas of Konrad Lorenz with evolution theory. J Genet 53:325–362

van Bergeijk KE, Bigler F, Kaashoek NK, Pak GA (1989) Changes in host acceptance and host suitability as an effect of rearing *Trichogramma maidis* on a factitious host. Entomol Exp Appl 52:229–238

van Lenteren JC (1986a) Evaluation, mass production, quality control, and release of entomophagous insects. In: Franz JM (ed) Biological plant and health protection. Fischer, Stuttgart, Germany, pp 31–56

van Lenteren JC (1986b) Parasitoids in the greenhouse: Successes with seasonal inoculative release systems. In: Waage JK, Greathead DJ (eds) Insect parasitoids. Academic Press, London, pp 341–374

van Lenteren JC (1995) Integrated pest management in protected crops. In: Dent DR (ed) Integrated pest management: principles and systems development. Chapman and Hall, London, pp 311–343

van Lenteren JC (1996) Designing and implementing quality control of beneficial insects: Towards more reliable biological pest control. In: Proceedings for Quality Control Meeting. IOBC/EU, Antibes, France

van Lenteren JC (1998) Quality control guidelines. Sting, Newsletter on Biological Control in Greenhouses 18:1–32

van Lenteren JC (2000) Measures of success in biological control of arthropods by augmentation of natural enemies. In: Gurr G, Wratten S (eds) Measures of success in biological control. Kluwer, Dordrecht, pp 77–103

van Lenteren JC (2003) Quality control and production of biological control agents: theory and testing procedures. CABI Publishing, Wallingford

van Lenteren JC (2005) Early entomology and the discovery of insect parasitoids. Biol Control 32:2–7

van Lenteren JC, Bueno VHP (2003) Augmentative biological control of arthropods in Latin America. BioControl 48:123–139

van Lenteren JC, Manzaroli G (1999) Evaluation and use of predators and parasitoids for biological control of pests in greenhouses. In: Albajes R, Gullino ML, van Lenteren JC, Elad Y (eds), Integrated pest and disease management in greenhouse crops. Kluwer, Dordrecht, the Netherlands, pp 183–201

van Lenteren JC, Tommasini MG (1999) Mass production, storage, shipment and quality control of natural enemies. In: Albajes R, Gullino ML, van Lenteren JC, Elad Y (eds), Integrated pest and disease management in greenhouse crops. Kluwer, Dordrecht, the Netherlands, pp 276–294

van Lenteren JC, Woets J (1988) Biological and integrated pest control in greenhouses. Annu Rev Entomol 33:239–69

van Lenteren JC, Roskam MM, Timmer R (1997) Commercial mass production and pricing of organisms for biological control of pests in Europe. Biol Control 10:143–149
van Lenteren JC, Hale A, Klapwijk JN, van Schelt J, Steinberg S (2003) Guidelines for quality control of commercially produced natural enemies. In: van Lenteren JC (ed) Control and production of biological control agents: theory and testing procedures. CABI Publishing, Wallingford, UK, pp 265–303
van Schelt J, Ravensberg WJ (1990) Some aspects on the storage and application of *Trichogramma maidis* in corn. Colloques l'INRA 56:239–242
Vet LEM, Dicke M (1992) Ecology of infochemical use by natural enemies in a tritrophic context. Annu Rev Entomol 37:141–172
Vet LEM, Lewis WJ, Papaj DR, van Lenteren JC (1990) A variable-response model for parasitoid foraging behavior. J Insect Behav 3:471–490
Yamazaki T (1972) Detection of single gene effect by inbreeding. Nature 240:53–54

Chapter 13
Biological Control with Egg Parasitoids other than *Trichogramma* – the Citrus and Grape Cases

Jorge E. Peña, Josep A. Jacas, Serguei Triapitsyn, Bryan J. Ulmer, and R.E. Duncan

Contents

13.1 Introduction

Egg parasitoids of economic significance are found in the families Trichogrammatidae, Mymaridae, Scelionidae and Platygastridae. Research on egg parasitoids other than *Trichogramma* has increased at a slower level during the last 15 years. For instance, the last International Organization of Biological Control review on egg parasitoids summarizing references from 1997 to 2003 listed 77 references on Trichogrammatidae, 50 on Mymaridae, 72 on Scelionidae and none on Platygastridae (Table 13.1). Within the Trichogrammatidae, 94% of these

J.E. Peña (✉)
Department of Entomology and Nematology, Tropical Research and Education Center, University of Florida, Homestead, FL 33031, USA
e-mail: jepe@ifas.ufl.edu

F.L. Consoli et al. (eds.), *Egg Parasitoids in Agroecosystems with Emphasis on Trichogramma*, Progress in Biological Control 9, DOI 10.1007/978-1-4020-9110-0_13,

Table 13.1 References listing hymenopteran families and genera within the Trichogrammatidae between 1997 through June, 2004 (after Herz et al. 2003)

Family	Genus	2003	2002	2001	1997–2000	Total
Mymaridae		3	18	21	8	50
Scelionidae		1	13	11	8	33
Trichogrammatidae	*Trichogramma*	2	38	25	7	72
	Brachyufens	0	0	1	0	1
	Uscana	0	0	2	2	4
	Haeckeliania	0	0	0	0	0

Table 13.2 References listing several genera within the Trichogrammatidae between 1960 and 2004, using Coleoptera, Cicadellidae and Thripidae eggs as hosts (after Agricola 2003)

Genera	Coleoptera	Cicadellidae	Thripidae	Miridae	Lepidoptera
Trichogramma	3	0	0	0	185
Brachyufens	1(Curculionidae)	0	0	0	0
Chaetogramma	1	0	0	0	0
Oligosita	0	16	0	1	0
Megaphragma	0	0	3	0	0
Paracentrobia	0	8	0	0	0
Ufens	0	2	0	0	0
Zagella	0	2	0	0	0
Haeckeliania	0	0	0	0	0

Source: Agricola, http://md2.csa.com/htbin/ids65/procskel/cgi

references are on the genus *Trichogramma*, 1% on the genus *Brachyufens* and 5% on the genus *Uscana* and none on *Haeckeliania* (Herz et al. 2003). Moreover, from the late 60s through mid 2004, Agricola (2003) cites a total of 219 publications on the Trichogrammatidae; of these, 86% are references on the genus *Trichogramma*, 0.4% on *Brachyufens*, 0.4% on *Chaetogramma* Doutt, 8% on *Oligosita*, 3.6% on *Megaphragma*, 0.8% on *Ufens*, 0.8% on *Zagella* and none on *Haeckeliania* (Table 13.2). The major objective of this chapter is to discuss biological control of root weevils affecting primarily citrus and biological control of cicadellids affecting grape with egg parasitoids other than *Trichogramma*.

13.2 Biological Control of Citrus Root Weevils

Several species of weevils (Curculionidae) affect citrus in the Americas (Schroeder and Beavers 1977). These are the West Indian sugarcane rootstalk borer weevil, *Diaprepes abbreviatus* (L.), the fuller rose beetle *Pantomorus cervinus* (Boheman), the citrus root weevil, *Pachneus litus* (Germar), the northern citrus root weevil *Pachneus opalus* (Olivier), the little leaf notcher *Artipus floridanus* Horn and species of *Lachnopus, Exopthalmus* and *Compsus* (Woodruff 1964, 1968, 1985, Coto et al. 1995, Otero et al. 1995, Knapp et al. 1996). In Asia and Africa, *Hypomeces squamasus* Fabricus occurs in Southeast Asia and citrus snout weevil *Sciobius marshalls* Schoeman in South Africa. In Australia, *Eutinophea bicristata, Orthrorhinus cylindrirostris, Noemerimnetes sobrinus, Maleuterpes spinipes, Otiorhynchus cribicollis* are weevil species attacking citrus (Smith et al. 1997) (Table 13.3). In the Neotropics the different curculionid genera (*Diaprepes, Pachnaeus, Artipus, Lachnopus, Exopthalmus, Compsus*) adopted citrus as one of their major hosts in those areas where it developed as a monoculture. However, they are reported to affect a larger number of plant species (Simpson et al. 1996). Species of weevils within these genera have a similar life cycle. The above ground stages are the free living adults that feed on leaves, often leaving a characteristic pattern of notches around the edges, except for Fuller rose weevil, which deposits its eggs beneath the calyx of the fruit (Knapp et al. 1999). Female root weevils lay their eggs on clusters on leaves and, by secreting a sticky substance, cement leaves together with a gelatinous secretion leaving the egg mass concealed and protected.

Among these weevils, *D. abbreviatus* is perhaps the most destructive species. *Diaprepes abbreviatus* was introduced into Florida in 1964 (Woodruff 1968) and has become a serious pest of citrus throughout much of central and southern Florida, and it has also now spread to the citrus growing regions of Texas and California (Grafton-Cardwell et al. 2004). A single female of *Diaprepes* may lay as many as 5,000 eggs during her life of three to four months. In seven or eight days the eggs hatch and larvae leave the cluster falling to the soil. After they enter the soil, they may stay there for 8–12 months feeding on plant roots. After pupating in the soil, adults emerge and start the cycle over again, eating leaves and laying eggs. The costs of fighting a *Diaprepes* infestation in a citrus grove are considerable. Currently, it involves a fall application of nematodes for larval control, use of spring and fall foliar sprays to kill adults (McCoy and Simpson 1994) and use of fungicides in the spring, summer and fall to control root rot diseases (*Phytophthora* spp.) encouraged by the lesions caused by the larval feeding. Adding up these costs could come up to $369 an acre. Factoring a theoretical loss of 10% of the citrus trees and a 25% decrease in harvests, the estimated a loss reaches $1,270/acre/year.

13.2.1 Egg Parasitoids

A lack of native egg parasitoids for this weevil in Florida (Hall et al. 2001), and past failures of classical biological control (Beavers et al. 1980), triggered renewed efforts to introduce, evaluate, and release candidate egg parasitoids from the Caribbean Region into Florida (Peña et al. 1998, Peña and Amalin 2000, Hall et al. 2002, Jacas et al. 2005, Castillo et al. 2006). The first introduction of parasitic wasps, e.g., *Quadrastichus haitiensis* (Gahan) (Armstrong 1987) (formerly in *Tetrastichus*) collected from Caribbean countries into continental US was not considered effective (Beavers et al. 1980). However, Etienne et al. (1990) and Etienne and Delvare (1991) reported that the egg parasitoids appeared to provide control of the genus *Diaprepes* and recommended their introduction into other Caribbean Islands and into the US. During 1997, a biological control component of a Florida state-wide

Table 13.3 Pest status of weevil species (Coleoptera: Curculionidae) occurring in different citrus growing areas of the world

Weevil species	Northern America	Central and South America	Southeast Asia	Australia	Southern Africa	Mediterranean Basin
Artipus floridanus Horn	Minor?					
Asynonychus cervinus (Boheman)	Major	Minor		Minor	Minor	Minor
Brachycerus citriperda Marshall		Minor			Minor	
Compsus spp.	Minor?	?				
Diaprepes abbreviatus (L)	Major	?				
Exophthalmus spp.	Minor?	?				
Eutinophaea bicristata Lea				Minor		
Hypomeces squamasus F			Minor			
Lachnopus spp.	Minor?	?				
Lixus algirus L						Minor
Maleuterpes dentipes Heller			Minor			
Maleuterpes spinipes Blackburn				Minor		
Myllocerus dentifer F			Minor			
Neomerimnetes sobrinus Lea				Minor		
Ortorrhinus cylindrirostris (F)				Minor		
Otiorrhynchus aurifer Bohemann						Minor
Otiorrhynchus cribricollis Gyllenhal				Minor		Minor
Pachnaeus litus (Germar)	Minor?	?				
Pachnaeus opalus (Olivier)	Minor?	?				
Protostrophus avidus Marshall					Minor	
Sciobius granosus Fahrer					Minor	
Sciobius marshalli Schoeman					Minor	

program was initiated to develop and implement strategies to manage the *Diaprepes* root weevil. The classical biological control effort was collaborative involving state, federal, private institutions and international cooperators working on foreign exploration, quarantine, mass production, release, and recovery efforts (Peña et al. 2004). Several species within Eulophidae, Trichogrammatidae, Platygastridae and Scelionidae were collected in the Caribbean and introduced into Florida for testing (Beavers and Selhime 1975, Peña et al. 2001, Ulmer et al. 2006a). Few of these were released for control of *Diaprepes* weevil eggs. Here we discuss the characteristics, biology and rearing of these species, results of releases and pitfalls of this program.

13.2.1.1 Eulophidae

The Eulophidae include 297 genera placed in 4 subfamilies as follows: Euderinae, Eulophinae, Entedoninae and Tetrastichinae. The Eulophinae are regularly idiobiont ectoparasitoids of larvae or pupae, whereas the Entedoninae are principally solitary or gregarious, primary or secondary endoparasitoids of concealed larvae, or less commonly eggs or pupae. The Tetratischinae are usually primary endoparasitoids of eggs, larvae or pupae of Diptera, Hymenoptera, Lepidoptera and Coleoptera. Some species of the Euderinae are primary ectoparasitoids of concealed larvae of Lepidoptera or Coleoptera (Noyes 2008). The following Eulophidae (Tetrastischinae) were introduced in Florida: *Quadrastichus haitiensis* (Gahan) and *Aprostocetus vaquitarum* Wolcott.

Quadrastichus haitiensis (Gahan)

Efforts to establish *Diaprepes* egg parasitoids began during the 1970's with releases of the endoparasitoid *Q. haitiensis* (Sutton et al. 1972, Beavers et al. 1980). *Quadrastichus haitiensis* failed to establish (Beavers and Selhime 1975, Hall et al. 2001) at that time, but efforts to establish the parasitoid were reinitiated in 1997 (Peña et al. 2004). This species was originally described from specimens reared from eggs of *Exopthalmus quadrivitattus* Schauff (Coleoptera: Curculionidae) from Port-au-Prince, Haiti (Schauff 1987). It is a primary egg parasitoid that has also been reared from *D. abbreviatus* and *Pachnaeus litus* (Germar) (Coleoptera: Curculionidae) (Armstrong 1987, Schauff 1987). According to van Whervin (1968), it is the most common parasitoid of citrus weevil eggs in Jamaica and at times may parasitize up to 100% of the weevil egg masses. It was also the most abundant egg parasitoid of citrus weevils found in Guadeloupe (Ulmer et al. 2008). This parasitoid has also been collected from the Dominican Republic, Puerto Rico, Andros Island, St. Lucia and Cuba (Schauff 1987, Ulmer et al. 2006c). During 1998, the parasitoid was re-introduced in Florida from Puerto Rico by D. G. Hall and Nguyen (Peña et al. 2004). Subsequently, releases of *Q. haitiensis* were made beginning in 2000 into citrus and ornamental fields in Florida. The parasitoid is now established in southern Florida (Miami-Dade and Broward Counties) in open-field ornamental plant nurseries and citrus groves, but it has failed to establish elsewhere (Peña et al. 2001, 2004).

Rearing

Adults of *Diaprepes* root weevils can be maintained either in 30 x 30 x 30 cm Plexiglas cages with water and foliage of *Conocarpus erectus* L. (Myrtales: Combretaceae) as a food source and an oviposition substrate (Castillo et al. 2006) or fed with a bouquet of tender citrus leaves and green bean pods (Nguyen et al. 2003). Cages can be kept at 26°C with 14:10 L:D or 12:12 L:D (Nguyen et al. 2003, Castillo et al. 2006). The oviposition substrate consists of either wax paper strips (3 × 10 cm) stapled together or one piece of wax paper as described by Étienne et al. (1990) or of one piece of parafilm stapled together as described by Nguyen et al. (2003). Strips with eggs of the *Diaprepes* root weevil are removed daily from the cage and either hung inside a similar cage containing a colony of *Q. haitiensis*, water and smears of honey (Castillo et al. 2006), or cut into small pieces containing eggs, which are introduced in 50-dram snap capped vials containing *Q. haitiensis* with smears of honey (Nguyen et al. 2003). Eggs are removed from the ovipositing cage or chamber after approximately 3 days, the strips are opened and the parasitized eggs exposed to facilitate parasitoid emergence.

Biology

Eggs of *Q. haitiensis* are translucent white, oblong, 0.28 ± 0.01 mm long, typically hymenopteriform. First instar larvae are fusiform translucent white, advanced larval instars are opaque light yellowish, larval size during last instars ranges from 1.40 to 1.47 mm. Newly formed prepupae are similar in shape to the larvae, but darker. Average prepupal length is 1.42 ± 0.01 mm. Pupae are exarate and yellowish translucent, with red eyes when recently formed. As they mature, the thorax and one third of the lower abdomen turn dark brown-greenish (Castillo et al. 2006).

Only one parasitoid emerges from a single host egg. Overall adult length for females is 1.32 ± 0.09 mm and 1.21 ± 0.07 mm for males (Castillo et al. 2006). In both sexes the thorax is black and the abdomen is yellowish anteriorly, becoming darker to the posterior end which is gradually pointed. The adult was described in detail by Schauff (1987). The sex ratio (female:male) during the first 3 days of emergence is male biased 0.12–0.40, compared to the female biased sex ratio during the last 3 days of emergence: 0.60–0.63 (Castillo et al. 2006).

Duration of development from egg to adult decreases from 40.0 to 13.6 days (mean) as temperature increases from 20° to 33°C, respectively. No development occurs from 5 to 15°C. Fecundity is highest at 25° and 30°C (70–73 eggs per female), but is reduced at 33°C (21.5 eggs per female). *Quadrastichus haitiensis* accepts 0–7 days old host eggs for oviposition, but is most prolific when parasitizing 1–4 day old eggs. Very few adult *Q. haitiensis* emerge from host eggs that are 5–7 days old. (Castillo et al. 2006).

Aprostocetus vaquitarum Wolcott

In the Caribbean Region, one of the most important natural enemies of *D. abbreviatus* is *Aprostocetus vaquitarum*. The tetrastichinae *A. vaquitarum* (formerly *A. gala*) was previously known as *Tetrastichus gala* (Walker) and misidentified as *T. marylandensis* Girault (Schauff 1987). Specimens of *A. vaquitarum* were collected in the Dominican Republic on *Diaprepes* spp. eggs in citrus during 2000 (Peña and McCoy, pers. obs.) and introduced into Florida. Subsequent to importation and screening under quarantine conditions, the University of Florida (UF) commenced mass-rearing as well as release and evaluation.

Rearing

Conocarpus erectus L. (Myrtales: Combretaceae) leaves containing *D. abbreviatus* eggs collected from stock colonies as described above are placed inside a 30 x 30 x 30 cage in a room held at 26.5 ± 1°C, 12:12 L:D, and approximately 78% RH. Adults of *A. vaquitarum* are introduced into the cage and provided honey and water. Parasitized eggs are removed from the cage 4–5 days later and placed in emergence containers for approximately 14 days (Ulmer et al. 2006b, Jacas et al. 2005).

Biology

Females of *A. vaquitarum* deposit their eggs in close contact with those of its host by introducing their ovipositor through the sealed leaves that protect the egg masses of *D. abbreviatus*. On eclosion, larvae of *A. vaquitarum* feed externally on several eggs of *D. abbreviatus* to complete their preimaginal development. Then, after completing 4 larval instars they pupate within the sealed leaves and emerge 10–14 days later as adults by chewing an emergence hole through the leaf (Ulmer et al. 2006b). Thus, *A. vaquitarum* is an ectoparasitoid behaving as a predator, a common feature observed among other tetrastichinae (Noyes 2008).

Temperature has a dramatic affect on *A. vaquitarum* development. The egg stage lasts approximately 7 times longer at 15°C than at 25 or 30°C and the larval stage takes significantly less time at each increasing temperature from 15 to 30°C (Ulmer et al. 2006b). *Aprostocetus vaquitarum* development from egg to adult takes approximately 16 days at an optimal temperature of 30°C. Temperatures of 35°C do not affect egg eclosion of *A. vaquitarum* eggs, but appear to reduce survival of the first instar (Ulmer et al. 2006b). No *A. vaquitarum* development past the egg stage occurs at 5 or 40°C. Different photoperiods do not significantly affect developmental time from egg to adult. Oviposition is greatest at 30°C, a large number of eggs are also laid at 25 and 35°C, but oviposition is greatly reduced when temperatures drop to 20°C and very few eggs are laid below 20 or above 35°C (Ulmer et al. 2006b). The developmental response of *A. vaquitarum* to various temperatures is comparable, though less pronounced, to that of *Q. haitiensis* (Hymenoptera: Eulophidae) (Castillo et al. 2006). A similar relationship, peaking at 30°C, has also been observed between temperature and development for several other Eulophid parasitoids (Rahim et al. 1991, Kfir et al. 1993, Acosta and O'Neil 1999, Urbaneja et al. 2002, Bazzocchi et al. 2003, Urbaneja et al. 2003). At a constant temperature of 15°C, *A. vaquitarum* does not complete development past the pupal stage. Similarily, *Q. haitiensis* was shown not to develop past the prepupal stage at 15°C (Castillo et al. 2006). Given the Caribbean origin of *D. abbreviatus* and its egg parasitoids, it is not surprising that sustained temperatures of 15°C or below are lethal (Ulmer et al. 2006b). At constant temperatures of 35°C and above, *A. vaquitarum* does not survive; similarly, host *D. abbreviatus* eggs do not survive at these temperatures. Lapointe (2001) also found that developmental rate of *D. abbreviatus* eggs increases with temperature up to 30°C but larvae do not emerge at temperatures of 32°C or higher (Lapointe 2001). *Aprostocetus vaquitarum* is an important primary parasitoid of *D. abbreviatus* in its native range (Jacas et al. 2005) and it is likely that the comparable relationship between temperature and developmental rates for parasitoid and host are a result of close evolutionary ties.

Aprostocetus vaquitarum adult females live approximately 15 days laying an average of 53 eggs during that time, though they may lay over 120 (Jacas et al. 2005). Parasitoid fecundity is significantly affected by the age of the host egg mass. Host eggs 0–3 days old are more readily accepted by *A. vaquitarum* than those aged 4–6 days.

Aprostocetus vaquitarum as an Egg Mortality Factor

Release and recovery efforts were aimed primarily at establishing *A. vaquitarum* using open field releases. Adult wasps obtained from a laboratory culture were released from 2000 through 2003 in 7 Florida counties that included areas and commodities severely affected by the weevil (Simpson et al. 1996, Mannion et al. 2003). The total number of adult wasps released in Florida was approx. 700,000 and releases in ornamental sites in southern Florida Miami-Dade County, from April 2001 to September 2003 totaled 230,270 adults. In Southeast Florida 78–91% mortality of *Diaprepes* eggs has been attributed to *A. vaquitarum* as it has established. Percent mortality before the release of *A. vaquitarum* was approximately 7.5% (2.20 $\pm$ 0.68 dead eggs/clutch; $n = 487$ egg masses, 1999; J.E. Peña and R. Duncan, unpublished results). In areas where *A. vaquitarum* is dispersing, percent mortality is higher than in the Caribbean islands where the parasitoid was originally collected. For instance, during 2002 and 2003 percent parasitism by *A. vaquitarum* in the Dominican Republic and Dominica was 69% and 43%, respectively (J.E. Peña, unpublished results). One of the reasons for this could be a reduced impact of hyperparasitoids (e.g., *Horismenus bennetti* Schauff) in Florida that are commonly found in parasitized *Diaprepes* egg masses in the Caribbean islands (J.E. Peña unpublished results). Because *A. vaquitarum* was considered established in Southern Florida, wasp releases were suspended in that area after September 2003 (Peña et al. 2004).

13.2.1.2 Platygastridae

Members of Platygastridae are regularly endoparasitoids of Coleoptera, Diptera and Homoptera (Clausen 1940, Notton 1998). Parasitoids can be gregarious or solitary, with a monoembrionic or polyembrionic development. Several species of *Fidiobia* have been collected as parasitoids of Coleoptera eggs. *Fidiobia* species are parasitic of curculionid and chrysomelid eggs (Notton 1998), but very little is known on their biology. For instance, from 1997 through 2003, no references on *Fidiobia* were listed by the International Organization for Biological Control (IOBC) in their publication Egg Parasitoid News, where most of the information on egg parasitoids is publicized (Herz et al. 2003) (Table 13.2). In the genus *Fidiobia,* host records are known for 6 of the 13 known species. Of these, five species are only known to parasitize eggs of weevils belonging to the genus *Diaprepes*, *Entypotrachelum*, *Hypera* (alfalfa weevil), or *Naupactus* (Curculionidae) and one species is known to parasitize eggs of *Fidia viticida*, a leaf beetle (Chrysomelidae) (Crawford 1916, Szabo 1958, Nixon 1969, Ellis 1973, Loiacono 1982, Buhl 1998, 1999, 2002, Evans and Peña 2005). The genus *Fidiobia* is also reported parasitizing weevil eggs found in citrus in Brazil (Guedes et al. 2001) and in grape in Chile (Gonzalez 1983). Readers are referred to Masner and Huggert (1989) for a key to the genera of Platygastridae which includes diagnosis, discussion and illustrations for each platygastrid genus, and to Schauff (1987) for the key to the parasites of citrus weevils.

Fidiobia citri (Nixon) is reported as a common parasitoid of the fuller rose weevil *Asynonychus cervinus* (Boheman) by Smith et al. (1997), though in Australia, this parasitoid has not been found attacking other citrus weevil eggs (i.e., *Eutinophea bicristata, Orthrorhinus cylindrirostris, Noemerimnetes sobrinus, Maleuterpes spinipes, Otiorhynchus cribicollis*) (Smith et al. 1997). *Fidiobia citri* parasitized up to 50% of each egg mass of the fuller rose weevil in California (Anonymous 2004) and this parasitoiod was also collected in southern Florida during 1999 from eggs of the blue green weevil, *Pachnaeus* spp., (Duncan, Pers. Comm.). Laboratory trials to determine if *F. citri* from Florida also parasitizes *D. abbreviatus* failed (Duncan, Pers. Comm.). However, *F. citri* (referred here as Jamaican Strain) was collected from curculionid eggs in Jamaica during 2000 by J. E. Peña and subsequently introduced into the UF-TREC quarantine facility. *Diaprepes abbreviatus* eggs and *Pachnaeus* eggs were exposed to *F. citri* (Jamaican Strain). This Jamaican Strain was reared successfully on *Pachnaeus* eggs, producing an average of 11.6 parasitoids per egg mass, compared to 0.34 adult parasitoids obtained from *D. abbreviatus* eggs. Thus, it does not appear that *F. citri* can use *D. abbreviatus* effectively as a host (Duncan and Peña, unpublished data). Moreover, trials exposing eggs of *Diaprepes* in citrus groves across the state of Florida to determine possible native egg parasitoids provided no parasitism (Hall et al. 2001). No parasitism from *F. citri* or any other 'resident' parasitoid species was obtained from 1997 through 1999 when native plants infested with *D. abbreviatus* eggs were exposed in natural habitats of south Florida to parasitoids (Peña and Duncan, unpublished data). The genus *Fidiobia* has been released against other curculionid pests (e.g., *Exopthalmus vittatus, Panthomorus cervinus, Asynonychus godmani*) (Coulson et al. 2000), but results of these releases are unknown (Evans, Pers. Comm.).

Fidiobia dominica (Evans and Peña)

In 2003, a new species of *Fidiobia* (Platygastridae), *F. dominica* (Evans and Peña) was collected in Dominica from eggs of *Diaprepes doublierii,* and transported to the quarantine facility in Homestead, Florida for testing and subsequent release in Florida. This species behaves as a solitary idiobiont endoparasitoid. It has a stalked egg and two larval instars (Jacas et al. 2007). The first instar larva is cyclopoid and 7-segmented whereas the second one is hymenopteriform and 11-segmented. Mandibulae are conspicuous in both instars. On completion of the larval development, the host egg turns amber transparent, making parasitized eggs easily recognizable. The pupa is exarate. *Fidiobia dominica* is a protandrous species and once emerged, males help females to emerge. *F. dominica* successfully develops between 9.6 and 30.0°C, but can endures higher temperatures provided that they do not coincide with critical stages such as pupation. As expected for closely evolved species, such as host-parasitoid assemblages, thermal limits roughly fit within those of the host *D. abbreviatus* (Jacas et al. 2008).

Biology

In a series of laboratory assays, *F. dominica* sex ratio was male biased at 15°C whereas it was female-biased at higher temperatures (Duncan et al. 2007, Jacas et al. 2008). Because primary sex ratio (the sex ratio at oviposition) was the same for all temperatures, the observed adult sex ratio indicates a greater mortality experienced by females at 15°C while immature. These results are indicative that close to limit zones, *F. dominica* has an extra constraint for its successful establishment than other egg parasitoids of *D. abbreviatus* studied so far (Castillo et al. 2006, Ulmer et al. 2006b) do not have.

Based on the thermal constants estimated, and provided that host eggs are available during the whole season, *F. dominica* would be able to complete 18 generations annually (with a maximum of almost two per month in July and August, and more than one even during the coldest months) in south Florida. However, host availability during critical periods should not be forgotten. Lapointe et al. (in press) have recently demonstrated that air temperature ≤12°C is lethal to eggs of *D. abbreviatus*, both those already oviposited and those present in the ovaries prior to oviposition. Such a limitation for the reproduction of *D. abbreviatus* can severely affect the ability of egg parasitoids to establish and successfully control *D. abbreviatus* in areas where winter temperatures fluctuate around the aforementioned threshold.

13.2.1.3 Trichogrammatidae

Hackeliania

Among the 80 genera within the Trichogrammatidae, most of the information is in the genus *Trichogramma* spp., and very little is known about other genera, such as *Hackeliania* (Table 13.2). From the late 1960s through mid 2004, three references cite *Trichogramma* using Coleopteran as hosts, while 185 references cite Lepidoptera as hosts instead. Conversely, some genera within the Trichogrammatidae appear to be more specific (Jarjees and Merritt 2002, 2003).

For instance, *Brachyufens* is only listed from Curculionidae, *Oligosita* from families within the orders Homoptera and Hemiptera and *Megaphragma* from Thripidae. For more information, see Doutt and Viggiani (1968).

Information on *Haeckeliania* is recorded by Pinto (1997) which followed Viggiani's (1992) description of the only species described from the New World (*H. minuta* Viggiani). Original descriptions of the genus and various other species in Australia and SE Asia are reported in Noyes (2008).

Pinto (2005) cited Noyes (2001, but see also Noyes 2008) stating that eight species of *Haeckeliania* have been described to date. This included four from Australia, three from Asia and only a single species, *H. minuta*, from the New World. According to Pinto (2005), this modest number fails to adequately portray the diversity of the genus. Pinto (2005) stated that the examination of Trichogrammatidae from throughout the world suggests that *Haeckeliania* is one of the largest genera in the family. It is likely than no more than 5–10% of the fauna is described. *Haeckeliania* is considered particularly diverse and abundant in Australasia and the New World Tropics. It also occurs in Africa and Asia. The genus has yet to be recorded from Europe. Pinto (2005) is only aware of two to three undescribed species in the United States.

Until recently, no references were found on host interactions for *Haeckeliania* (Table 13.2). Therefore, the literature shows that some genera within the Trichogrammatidae use only species within specific orders or families as hosts, while some, such as *Trichogramma*, are highly polyphagous (Agricola 2003).

The lack of references on *Haeckeliania* is not surprising. For instance, during 1999, the senior author collected 4 different species of *Haeckeliana* from eggs of *Compsus viridilineatus* Jekel (Coleoptera: Curculionidae), a species close to *Diaprepes* sp., in Colombia. These specimens were introduced into the UF-TREC quarantine facility and exposed to 338 eggs of *Diaprepes*; none of the parasitoids accepted *Diaprepes* as a host. This demonstrated that the *Haeckeliania* spp. from Colombia shows specificity using curculionid species as hosts (Peña and Duncan, unpublished). An unidentified species of *Haeckeliania* was collected from weevil eggs in Florida by B.J. Ulmer in 2005. In the laboratory, these specimens were offered and parasitized eggs of the weevils *Artipus floridanus, Diaprepes abbreviatus* and *Pachnaeus littus,* but they did not develop on *D. abbreviatus* eggs (Ulmer et al., unpubl) (Table 13.4).

Haeckeliania sperata Pinto was collected from *Diaprepes* eggs in the island of Dominica. This new species has been successfully reared through several generations on *Diaprepes* eggs (Peña et al. 2004), and tested against eggs of two species of Lepidoptera, one species of coccinellid and eggs of another species of curculionid. The parasitoid did not accept either Lepidoptera, or other Coleoptera eggs as hosts. Even though these tests were limited to 2 genera within Lepidoptera (eggs deposited on the leaf surface, eggs not inserted, deposited or concealed inside of plant host tissue, as is the case of *Diaprepes* eggs), the results of non-acceptance of non-concealed eggs were encouraging (Peña et al. 2004).

Table 13.4 Egg mortality (%)[a] each parasitoid inflicted on each weevil and parasitoid successful development to the adult stage (Y = yes; N = no)

	Parasitoid						
Weevil	*A.vaquitarum*	*B. osborni*	*F. dominica*	*H. sperata*	*Q. haitiensis*	*B. fennahi*	Native Trichogrammatidae
A. floridanus	23 (Y*)	95 (Y)	22 (Y*)	57 (?)	51 (Y)	n/a	99 (Y)
D. abbreviatus	77 (Y)	93 (N)	29 (Y)	47 (Y)	69 (Y)	81 (Y)	20 (N)
L. floridanus	0 (N)	n/a	91 (Y)	63 (?)	79 (Y)	0 (N)	n/a
P. litus	72 (Y)	100 (Y)	14 (Y)	68 (?)	85 (Y)	89 (Y)	48 (Y)

*The parasitoid was able to develop to an adult, but it was very uncommon

[a]Percentage egg mortality is the mean percentage of eggs that were killed by parasitism in each egg mass (n = average 20)

Biology

Most species of *Haeckeliania* are uniformly dark brown in color, compact and gibbous in shape (Pinto 2005). *H. sperata* represents a less common phenotype characterized by a lighter body color, and a more slender apically attenuate and elongated body (Pinto 2005). *Haeckeliania sperata* searches for weevil egg masses on the upperside of leaves where eggs are laid. Upon detection of the egg mass the parasitoid inserts its ovipositor through the leaf until it reaches the weevil egg. The endoparasitoid egg development may take just a few hours, larval stage lasts approx. 14 days and pupal stage lasts 2 days. Adults will live for 2–4 days after emergence (Duncan and Peña unpubl.). Under quarantine conditions, an average of 36 *H. sperata* emerge from each parasitized egg mass, the sex ratio is 1:2 (M:F). Under quarantine conditions, *H. sperata* causes approximately 50% mortality per egg mass.

Rearing

Ten centimeter green buttonwood shoot tips with less than 1-day-old *D. abbreviatus* eggs (collected as described above) are grouped in bouquets and placed in a 500-ml plastic container with water and exposed in plexiglass cages (as described above) to about 750 adults of *H. sperata* for 6 hours. Shoot tips are subsequently removed, washed with water and further checked under the microscope to remove all adults. They are then placed in emergence containers or experimental units until adult emergence.

Host Plant Effect on Successful Parasitism

The reproductive success of *Haeckeliania sperata* appears to be affected by the host plant that *D. abbreviatus* selects for oviposition (Carrillo et al. 2008). Six host plants with varying degrees of pubescence were used to determine successful parasitism and the effect of leaf trichomes on the searching behavior of *H. sperata*. No-choice tests revealed that *H. sperata* was able to parasitize *Diaprepes* eggs laid on the six host plants; however, the plants that bare a high trichome density in their leaves had a lower percent of parasitism than the plants with smoother leaves ($p<0.01$). Experiments removing trichomes from a host plant revealed that the presence of some leaf trichomes had a negative effect on the overall searching efficiency of *H. sperata*. These results are similar to those obtained by Amalin et al. (2005) working with another trichogrammatid parasitoid of *D. abbreviatus*, *Ceratogramma etiennei* Delvare. In a no-choice experiment with four of the same host plants that we used in our experiments, *C. etiennei* showed a high parasitism of *D. abbreviatus* eggs on lime, pygmy palm and green buttonwood, and a low parasitism on the pubescent silver buttonwood. Interestingly, Mannion et al. (2003) found that *D. abbreviatus*, when given a choice, prefers to oviposit on silver buttonwood, where it has better larval survivorship. These results suggest that a low parasitism by *H. sperata* could be expected if *D. abbreviatus* chooses silver buttonwood or other pubescent plants as a host for oviposition. The results obtained on silver buttonwood and the other pubescent plants (i.e. elephant grass and loquat) suggested that plants that have a high density of trichomes in their leaves are not good candidates for releasing *H. sperata*.

Ceratogramma etiennei Delvare

Ceratogramma etiennei was described by from Guadeloupe (Delvare 1988) and it is a highly specific parasitoid known to occur in Basse Terre and in Grande Terre which are respectively humid and dry part of Guadeloupe. *Ceratogramma etiennei* was introduced into Martinica (Etienne et al. 1990) and the Dominican Republic (Etienne et al. 1992) for biological control of *Diaprepes* spp.

Biology

Adults live between 2 and 11 days, females lay single eggs in each *Diaprepes abbreviatus* egg. Development from egg to adult lasts 14–20 days, sex ratio is 1: 0.66 (female: male). Adults of *C. etiennei* consume the egg chorion and leaf tissue and copulate in few hours after emergence. The genus *Ceratogramma* is known from Central and South America, and from the West Indies. According to Delvare (1988), the genus was described by De Santis in 1957. Six species are recognized. Antennal and genitalcharacters distinguish *Ceratogramma* from other trichogrammatid genera (Pinto and Viggiani 1991). In Guadeloupe, *C. etiennei* is reported to parasitize *Diaprepes abbreviatus*, *D. famelicus* (Olivier) and *D. marginatus* (Fabricius).

Rearing

Ceratogramma etiennei was maintained in the laboratory at 26.5± 1°C, 12:12 L:D, and approx 78% RH, on eggs of the root weevil laid on strips of wax paper following Etienne et al. (1990).

Ceratogramma etiennei was introduced from Guadeloupe into Florida, USA in 1997. Specificity tests on species of Lepidoptera, i.e., *Eumaeus atala florida* Roeber (a threatened species), *Papilio cresphontes* Cramer, *Battus polyadmus lucayus* (Rostchild and Jordan) (Papilionidae), *Phoebis philea* (L.) (Pieridae), *Danaus plexippus* (L.) (Danaiidae), *Heliconius charitonius tuckeri* Constock and Brown, *Agraulis vanilla nigrior* Michener (Heliconiidae) and *Oxyops vitiosa* (Coleoptera: Curculionidae), which has been released in Florida against the noxious weed *Melaleuca leucadendron*, resulted in lack of egg parasitism of any of the Lepidopteran species or eggs of *O. vitiosa. Ceratogramma etiennei* was released in south Florida during 1998 in citrus, ornamental fields and natural habitats infested with the root weevil. Recoveries of *C. etiennei* occurred in 1999 in citrus and ornamentals, one year after the first release, but no more recoveries were obtained between 2000 and 2002 at the same locations (Peña et al. 2004, 2006).

Brachyufens osborni (Dozier)

Brachyufens osborni (Hymenoptera: Trichogrammatidae: Trichogrammatinae) is a weevil egg idiobiont endoparasitoid restricted to the Nearctic region (Pitkin 2003). It was first described in 1932 from weevil eggs collected in Puerto Rico (Dozier 1932), and its first record in continental USA dates from 1959, when it was recovered from the indigenous root weevil *Pachnaeus litus* (Germar) (Coleoptera: Curculionidae: Entiminae) (Baranowski 1960). This endoparasitoid is common in Florida and is presumed to occur on weevil egg masses of *Diaprepes abbreviatus*, *P. litus* and *P. opalus* (Olivier) (Schauff 1987). Parasitism on *P. litus* can be as high as 81% (Baranowski 1960).

Biology

The immature stages of *Brachyufens osborni* reared in eggs of *Pachnaeus litus* had a developmental time of 14.7 days. Egg hatching occurred within 15 h from oviposition, and mandible measurements indicated the existence of two larval instars. First instar larva was mymariform whereas the second instar was sacciform. Pupation ocurred 5–6 days after eclosion and this stage lasted for 8 days. On completion of the preimaginal development, the meconium was explelled and emergence took place by biting a hole in the chorion of the host egg. Both sexes emerged simultaneously and sex ratio was female biased (3:1) (Jacas et al. 2009). *B. osborni* successfully reproduced on both *P. litus* and *A. floridanus* in non-choice laboratory studies, though females did oviposit in *D. abbreviatus* eggs, no successful adult emergence occurred (Ulmer, unpublished).

Factors Affecting Diaprepes Egg Parasitoid Establishment in Florida

The areas where both *Q. haitiensis* and *A. vaquitarum* have established in southeast Florida have higher mean temperatures and less volatile temperatures throughout the year, comparable to their native Caribbean range, than do the areas in central Florida where these parasitoids have been released but shown no signs of establishment. During the winter months, the mean temperature of various *Diaprepes* infested areas across central Florida is below 20°C, and the mean minimum temperature drops to below 10°C (Anonymous 2005). Temperatures below 20°C are less than ideal for both *Q. haitiensis* and *A. vaquitarum* oviposition or development and sustained temperatures of 15°C and below are lethal. It is likely that the relatively cool winters and the more extreme temperatures experienced throughout the year in central Florida are a constraint to the establishment of these parasitoids across some regions of *Diaprepes* infestation. However, efforts to locate populations of these parasitoids adapted to lower temperatures may provide an opportunity to expand their range in Florida (Ulmer et al. 2006a, b). Similar constraints may face other egg parasitoid candidates; however, a potential advantage of both *F. dominica* and *H. sperata* when compared to the eulophids *Q. haitiensis* (Castillo et al. 2006) and *A. vaquitarum* (Jacas et al. 2005, Ulmer et al. 2006b), is their higher tolerance to both minimum and maximum temperatures. If *F. dominica* and *H. sperata* are eventually recovered from orchards located at a higher latitude, this would reveal that their thermal plasticity is higher than that of the other parasitoids introduced up till now.

Several other factors including host availability and the seasonal changes in relative humidity may also be unfavorable for the establishment of Caribbean egg parasitoids in all *Diaprepes* infested areas in Florida and California. The abundance of *D. abbreviatus* adults is not as stable throughout the year in central Florida as it is in the extreme southeast (Ulmer et al. 2006b). Even if adult parasitoids could endure temperatures below the estimated LDT, the absence of host eggs for prolonged periods of time during winter months in central Florida counties where *D. abbreviatus* occurs could preclude the establishment of these parasitoids. Lapointe et al. (2007) found winter periods of up to 141 days when no egg masses were observed in an orchard located in Fort Pierce (27°26.47'10"N lat, 80°19.33'04"W long, 5 m high). Under these circumstances, only the presence of alternative host eggs could allow the introduced parasitoids to reproduce during such periods. There are 11 genera of weevils associated with citrus in Florida and the West Indies (Woodruff 1985), of which at least five species occur in Florida. However, whether these weevils or other closely related species could be alternative hosts for the imported parasitoids remains unknown and deserves further research. Preliminary work (Ulmer, unpublished) shows that several of the egg parasitoids discussed can in fact reproduce on at least one other weevil species found in Florida aside from *D. abbreviatus* (Table 13.4). Studies on alternative hosts could help increase the chances of completing a biological control program against *D. abbreviatus*. Additionally, the selection of races or biotypes of these parasitoids from higher elevation areas of the Caribbean Islands could yield specimens with increased cold-tolerance, which could be useful not only in limit zones of Florida, but also in Texas and California, if needed.

Given the relatively short life history of these parasitoids, extended periods without the presence of host eggs could be catastrophic. Host-free periods and low temperatures during the winter months have hindered the establishment of *Edovum puttleri* Grissell in the northeastern United States (Obrycki et al. 1985) and *Tetrastichus gallerucae* (Hymenoptera: Eulophidae), an egg parasitoid of the elm leaf beetle (Coleoptera: Chrysomelidae), in northern California (Dreistadt and Dahlsten 1991). Neither species is known to diapause; similarly, neither *Q. haitiensis* or *A. vaquitarum* have been shown to enter diapause under various light and temperature regimes in >5 years of insectary production and they may not be able to overcome extended winter periods of cool temperature (Ulmer et al. 2006b).

Another possible limitation for the oviposition of *Diaprepes* egg parasitoids in Florida is the narrow window for parasitism offered by its host. Successful parasitism by some egg parasitoids occurs only if newly laid eggs are attacked (Strand 1986). For *A. vaquitarum* this is almost limited to *D. abbreviatus* eggs less than 2 days old and for *Q. haitiensis* to those less than 4 days. Oviposition on older egg masses always meant an immature mortality over 50%. The size of the pupae obtained from these egg masses was not different from that obtained on younger eggs. However, occurrence of runts in egg parasitoids feeding on hosts where resources for parasitoid development are just plentiful enough to prevent death have already been reported (Salt 1941, Jackson 1958). Therefore, further detrimental effects of older eggs on *A. vaquitarum* performance can not be completely excluded, and deserve further research. Based on the results presented, hypothetical immature survival rates used for the estimation of demographic parameters ranged from 100 to 60%. Between these values, generation time almost did not change (2.2% increase), but both net fecundity and the intrinsic rate of increase dropped by 40.0 and 15.5%, respectively. However, even the lowest values lie within the range common among Eulophidae (Urbaneja et al. 2001, 2002). These differences should not hamper the biological control of *D. abbreviatus* by *A. vaquitarum* or *Q. haitiensis*, but rather the unavailability of suitable egg masses for parasitism. Only continuous generations of *D. abbreviatus* can guarantee the supply of freshly deposited eggs for *A. vaquitarum* to parasitize, but if discrete generations exist, as it seems to happen in areas of Florida north of Miami-Dade County with winter temperatures below 15°C, its beneficial role could be seriously impaired. Further research dealing with temperature-development studies of *A. vaquitarum* and *Q. haitiensis* on *D. abbreviatus* are needed to clarify this situation.

13.2.1.4 Effect of Pesticides on Parasitism

Applications of organophosphate, carbamate and pyrethroid pesticides might have a negative impact on the natural control of *D. abbreviatus* by the introduced parasitoids (Amalin et al. 2004). Carrillo et al. (2009) reported that organophospate, carbamate and pyrethroid pesticides showed a rapid and strong toxic effect on *H. sperata*. Ulmer et al. (2006a) reported similar results with organophospate and carbamate insecticides tested at label rates on *A. vaquitarum* adults. However, *H. sperata* was affected similarly by carbamate, organophosphate and pyrethroid pesticides, whereas pyrethroid insecticides were less toxic to *A. vaquitarum* than the organophosphate or carbamate pesticides.

13.3 Biological Control of Leafhoppers (Cicadellidae) on Grapes in California, USA, Using Fairyfly Egg Parasitoids (Mymaridae)

The glassy-winged sharpshooter, *Homalodisca vitripennis* (Germar) (hereafter GWSS) [formerly known as *H. coagulata* (Say)], is a well-known vector of the plant diseases caused by the phytopathogenic bacterium *Xylella fastidiosa* (Blua et al. 1999), which causes Pierce's disease on grapes. Glassy-winged sharpshooter is a self-introduced pest in California from southeastern USA (Blua et al. 1999). The establishment of *H. coagulata* in California in the 1990s, later in Hawaii (USA) and French Polynesia, and even more recently in Easter Island (Chile) (Pilkington et al. 2005) prompted interest in proconiine sharpshooter investigations, including studies of their egg parasitoids in North America (Triapitsyn and Phillips 1996, 2000,

Triapitsyn et al. 1998, Phillips et al. 2001, Triapitsyn et al. 2002ab), mainly for classical biological control purposes (Morgan et al. 2000, Jones 2001, Triapitsyn and Hoddle 2001, 2002, Morgan et al. 2002, Pilkington et al. 2004, 2005). Most of the reported mymarid egg parasitoids of *H. vitripennis* are members of *Gonatocerus* Nees (Turner and Pollard 1959, Triapitsyn et al. 1998, Triapitsyn and Phillips 2000, Triapitsyn et al. 2002a, Triapitsyn 2006), particularly its *ater* species group (Triapitsyn 2002ab). All the North American *Gonatocerus* species that parasitize eggs of Proconiini are solitary parasitoids, with an exception of *G. fasciatus* Girault, which is a gregarious parasitoid (Triapitsyn et al. 2003).

Triapitsyn (2003) reviewed the trichogrammatid egg parasitoids of proconiine sharpshooters in southeastern USA; none of those was selected for introduction into California as biological control agents because of difficulties with their rearing.

The initial surveys of the egg parasitoids of GWSS in California (Triapitsyn and Phillips 1996, Triapitsyn et al. 1998, Morgan et al. 2000, Phillips et al. 2001) revealed the presence of a small complex of Mymaridae (several species of *Gonatocerus* dominated by *G. ashmeadi* Girault) and Trichogrammatidae (two species of *Ufens* Girault). The latter were later described as *Ufens ceratus* Owen and *Ufens principalis* Owen (Al-Wahaibi et al. 2005). Yet, egg parasitization rate of the spring brood of the GWSS in California remained quite low, although during the summer months it was usually very high, sometimes reaching almost 100% (Triapitsyn et al. 1998). Therefore, a biological control program was implemented, which initially focused on classical biological control (Triapitsyn et al. 1998, Morgan et al. 2000). Later, it was complemented by a neoclassical biological control program (Jones 2001, Jones et al. 2005b, Logarzo et al. 2005).

For the classical biological control program, egg parasitoids of *H. coagulata* and other Proconiini were obtained in the United States through survey activities conducted in northern Florida and Louisiana in 1997 (Triapitsyn et al. 1998), southeastern Texas during 1999–2001 (Triapitsyn and Phillips 2000), Louisiana, northern Florida, southern Georgia, and southeastern Texas in 2000 (Morgan et al. 2000, Triapitsyn 2006), throughout Florida and in Texas in 2001 (Triapitsyn and Hoddle 2001, Triapitsyn et al. 2002a, b), Louisiana and Mississippi in 2002 (Triapitsyn and Hoddle 2002, Triapitsyn et al. 2003), Georgia, Illinois, Kentucky, North Carolina, South Carolina and Tennessee (Hoddle and Triapitsyn 2003, 2004b). Surveys of egg parasitoids of Proconiini in the Nearctic part of northeastern Mexico (Coahuila, Nuevo León, Tamaulipas, San Luis Potosí, Veracruz) were conducted during 1999–2005 (Morgan et al. 2000, Triapitsyn and Phillips 2000, Triapitsyn and Hoddle 2001, 2002, Triapitsyn et al. 2002b, Hoddle and Triapitsyn 2004a, Pilkington et al. 2004, 2005).

For the neoclassical biological control program, egg parasitoids of Proconiini were obtained mostly in Argentina (Jones 2001, Logarzo et al. 2005), and also in Peru (Logarzo et al. 2003) and Chile (Logarzo et al. 2006), and more recently in northwestern Mexico (mainly in Sinaloa and Sonora) (Triapitsyn and Bernal 2009). Currently, three species of *Gonatocerus* (two from Argentina and one from Mexico) as well as an undetermined species of Trichogrammatidae (*Pseudoligosita* sp. from Mexico) are under quarantine evaluation at UCR.

The following species of mymarid egg parasitoids were released in California by the California Department of Food and Agriculture (CDFA 2008) after their colonies were established at the University of California, Riverside (hereafter UCR) quarantine facility on GWSS eggs under appropriate federal and state permits, evaluated, and the released permits were obtained (Triapitsyn 2006, CDFA 2008). The current status of establishment and possible impact of these species in mostly unclear at this stage, although preliminary data indicate the establishment of *G. morrilli* (Howard) at least (CDFA 2008). It is important to mention that because the GWSS is able to oviposit on a wide range of host plants, the biological control efforts have to be more community-based, rather than concentrate on the releases of the natural enemies only in the vineyards. In fact, the sources of GWSS infestations of the vineyards are often nearby citrus orchards and urban areas, which harbor large populations of the GWSS.

13.3.1 Anagrus epos Girault

A colony of this species was established by in UCR quarantine from the adults that emerged from the parasitized egg masses of *Cuerna fenestella* Hamilton, collected in Minnesota by Roman Rakitov at the end of May – beginning of June 2004 (Triapitsyn and Rakitov 2005, Triapitsyn 2006). Mated females were exposed to fresh eggs of *H. vitripennis* (laid in leaves of *Euonymus japonica*), on which they successfully reproduced, producing up to 14 individuals per each host egg in the progeny. This gregarious trait was considered useful for a biological control agent (even though its developmental time from egg to adult, which sometimes exceeds one month, is considerably longer than that of *Gonatocerus* spp.), and *A. epos* was then released in California against the GWSS (Pilkington et al. 2005, Triapitsyn 2006, CDFA 2008). Host specificity of *A. epos* was studied (under laboratory conditions) by Krugner et al. (2008).

13.3.2 Gonatocerus ashmeadi Girault

Populations of this species from southeastern USA and northeastern Mexico were released in California (CDFA 2008) even though this species was already present in California when GWSS arrived there in the 1990s. It is by far the most abundant natural enemy of GWSS in California, particularly during summer and fall months. It is quite common in southeastern USA and probably not native to California; according to molecular studies, the likely origin of the California population is Texas (Vickerman et al. 2004, de León and Jones 2005). This species likely established itself in California long before the arrival of GWSS on eggs of the congeneric sharpshooter native to California, the smoke tree sharpshooter *H. liturata* Ball; it simply switched back to its original host after GWSS invaded the state (Triapitsyn 2006). *Gonatocerus ashmeadi* successfully suppressed the GWSS infestation in Hawaii, where it is apparently self-introduced (together with its host); it also introduced itself, apparently together with GWSS, into Easter Island, Chile (Triapitsyn 2006). More recently *G. ashmeadi* was with great success released in French Polynesia against GWSS (Grandgirard et al. 2008).

13.3.3 Gonatocerus morrilli (Howard)

Populations of this species from Texas and Tamaulipas (Mexico) were introduced into California (Morgan et al. 2002, Triapitsyn et al. 2002b, Pilkington et al. 2005) although a later report (de León and Morgan 2005) indicated that it was *G. walkerjonesi* Triapitsyn, a species native to California, that actually was released in California due to contamination of the cultures of the insectary-reared *G. morrilli* (following its release from quarantine) with this similarly looking native species. Initially, *G. walkerjonesi* was misidentified taxonomically as *G. morrilli* (Phillips et al. 2001), but the very useful molecular data by de León et al. (2004, 2006) and also Hoddle and Stouthamer (2005) helped differentiate these two cryptic species, allowing for the taxonomic differentiation and description of *G. walkerjonesi* (Triapitsyn 2006). Releases of the true *G. morrilli* (from southern Texas) were initiated in 2005, and the species appears to have established in California (CDFA 2008).

13.3.4 Gonatocerus fasciatus Girault

This is the only known gregarious species among the North American *Gonatocerus* egg parasitoids of Proconiini (Triapitsyn et al. 2003) able to produce up to 7 adult wasps per each GWSS egg. Its exit holes can be easily recognized by their number (two, rarely three) and position (at the opposite ends of the host egg if only two holes are present) per each host egg. Some other aspects of the biology of *G. fasciatus* were studied by Irvin and Hoddle (2005a, b). The recent discovery of *G. fasciatus* in northern California suggests that this species is native there (Triapitsyn 2006), but it has never been collected elsewhere in California prior to its introduction. The species was first introduced into California from Louisiana (Triapitsyn et al. 2003) and later released against *H. vitripennis* (Pilkington et al. 2005, CDFA 2008).

13.3.5 Gonatocerus triguttatus Girault

This mainly Neotropical species also occurs in USA (Florida and Texas) and northeastern Mexico (Triapitsyn et al. 2002b, Triapitsyn 2006). In southern Texas, it appears to be the most common and effective natural enemy of GWSS (Triapitsyn and Phillips 2000). It was introduced and established in California against *H. vitripennis* (initially from Tamaulipas, Mexico, and then from Texas)

(Morgan et al. 2000, Morgan et al. 2002, Triapitsyn et al. 2002b, Pilkington et al. 2005). Some aspects of the biology of *G. triguttatus* were studied by Irvin and Hoddle (2004, 2005a, b).

13.3.6 Host Associations and Biology of *Gonatocerus* Species, Egg Parasitoids of GWSS

All above-mentioned *Gonatocerus* species are naturally narrowly oligophagous, i.e., they are known from several species belonging to the proconiine leafhopper genera *Cuerna*, *Homalodisca*l, *Oncometopia*, *Paraulacizes*, and *Pseudometopia* (Triapitsyn 2006). However, they would readily parasitize other species within the same host genus, such as, for instance, *Homalodisca liturata*, a species native to California. Moreover, if given a chance, they would also parasitize some factitious host species from other, non-native genera of Proconiini: *G. triguttatus* from Peru, originally reared from a *Pseudometopia* sp., was easily reared on GWSS in quarantine (Logarzo et al. 2005), while several species of *Gonatocerus* from Argentina, originally reared from *Tapajosa rubromarginata* (Signoret), were reared on GWSS eggs for many generations in the quarantine laboratories in California and Texas (Jones et al. 2005a, b).

Biology of several *Gonatocerus* species has been studied, as follows: *G. fasciatus* (the only gregarious species) by Triapitsyn et al. (2003); *Gonatocerus ashmeadi*, *Gonatocerus triguttatus*, and *Gonatocerus fasciatus* by Irvin and Hoddle (2004, 2005a, b); and *G. tuberculifemur* (Ogloblin) by Virla et al. (2005). Depending on sex and temperature, development of these species from egg to adult usually takes 11–14 days, with males developing a little faster than females. Females of *G. tuberculifemur* can parasitize eggs of all ages (from 4 to 190 h old), although wasps did not emerge from eggs older than 96 hours (Virla et al. 2005). In high density laboratory colonies, a female of *G. tuberculifemur* usually guards the newly parasitized egg mass of the GWSS from attempts by other females to superparasitize it (S. V. Triapitsyn, unpublished).

13.3.7 Biological Control of the Variegated Grape Leafhopper in California

The variegated grape leafhopper, *Erasmoneura variabilis* (Beamer) (formerly known as *Erythroneura variabilis* Beamer), is a pest of grapes in California as well as in Baja California and Sonora, Mexico (Kido et al. 1984). For many years, before *E. variabilis* became a problem in California's San Joaquin Valley, the mymarid *Anagrus epos* Girault had been regarded as the only species of egg parasitoid responsible for a relatively good natural control of the native leafhopper species, the Western grape leafhopper *Erythroneura elegantula* Osborn, in central California vineyards. However, because the native egg parasitoids were unable to provide adequate control of *E. variabilis*, several forms of the genus *Anagrus* Haliday, known then as "biotypes of *A. epos*", were collected in Arizona, Colorado, New Mexico, USA, as well as in Baja California and Sonora, Mexico, released, and "became established" in selected vineyards in the San Joaquin Valley in an attempt to enhance control of *E. variabilis* (González 1988, González et al. 1988, Pickett et al. 1987, 1989), although their establishment was never properly documented by voucher specimens, etc. Earlier, Pickett et al. (1987) made an assumption that some of the "biotypes" of *A. epos* may in fact be different species. Indeed, that was the case, and specimens reared by D. González from eggs of *E. variabilis* on grape in Coachella Valley (California, USA) and Mexicali (Baja California, Mexico) were described by Trjapitzin and Chiappini (1994) as a new species, *A. erythroneurae* Trjapitzin and Chiappini. The identities of the other forms were later determined by Triapitsyn (1998), who identified the two native species in California parasitizing eggs of both *E. variabilis* and *E. elegantula* as *A. erythroneurae* and *A. daanei* Triapitsyn. Thus, it was partially *A. erythroneurae* that was introduced from New Mexico, USA, and Baja California, Mexico, into California by González et al. (1988); because it was impossible to distinguish the imported wasps from the native parasitoids belonging to the same species, it is also impossible now to prove if any of them actually got established. In Arizona, eggs of *E. variabilis* are parasitized by *A. tretiakovae* Triapitsyn, and this species has never been collected in California. The species of *Anagrus* collected by González et al. (1988) in Colorado, USA, and Sonora, Mexico, were initially misidentified by Triapitsyn (1998) as *A. epos*, but recently proven to belong to different, undescribed forms that were recognized using molecular methods by Morse and Stouthamer (2007). None of these was ever collected in California. Thus, it would be probably safe to assume that the classical biological control program against *E. variabilis* in California using *Anagrus* spp. (González et al. 1988) was not successful.

References

Acosta NM, O'Neil RJ (1999) Life history characteristics of three populations of *Edovum puttleri* Grissell (Hymenoptera: Eulophidae) at three temperatures. Biol Control 16:81–87

Agricola (2003) Downloaded as http://md2.csa.com/htbin/ids65/procskel/cgi. Accessed 6 July 2003

Al-Wahaibi AK, Owen AKand, Morse JG (2005) Description and behavioural biology of two *Ufens* species (Hymenoptera: Trichogrammatidae), egg parasitoids of *Homalodisca* species (Hemiptera: Cicadellidae) in southern California. Bull Entomol Res 95:275–288

Amalin DM, Stansly Pand, Peña JE (2004) Effect of Micromite® on the egg parasitoids *Ceratogramma etiennei* (Hymenoptera: Trichogrammatidae) and *Quadrasticus haitiensis* (Hymenoptera: Eulophidae). Florida Entomol 87:222–224

Amalin DM, Peña JEand, Duncan RE (2005) Effects of host age, female parasitoid age, and host plant on parasitism of *Ceratogramma etiennei* (Hymenoptera: Trichogrammatidae). Florida Entomol 88:77–82

Anonymous (2004) http://www,ars-grin.gov/cgi-bin/nirgp/probl./taxon.P73657. Accessed 8 April 2004

Anonymous (2005) Southeast Regional Climate Center. Historical Climate Summaries for Florida (www.dnr.state.sc.us/climate/sercc/climateinfo/historical/historical_fl.html) Accessed 13 Oct 2005

Armstrong A (1987) Parasitism of *Tetrastichus haitiensis* Gahan on egg masses of *Diaprepes abbreviatus* in Puerto Rico. J Agric Univ Puerto Rico 71:407–409

Baranowski RM (1960). Notes on a parasite of the citrus root weevil Pachnaeus litus (Germ.). Florida Entomol 43:197

Bazzocchi GG, Lanzoni A, Burgio G, Fiacconi MR (2003) Effects of temperature and host on the pre-imaginal development of the parasitoid *Diglyphus isaea* (Hymenoptera: Eulophidae). Biol Control 26:74–82

Beavers JB, Selhime AG (1975) Futher attempts to establish the weevil egg parasite, *Tetrastichus haitiensis* in Florida. Florida Entomol 58:29–31

Beavers JB, Lovestrand SA, Selhime AG (1980). Establishment of the exotic parasite *Tetrastichus haitiensis* (Hym: Eulophidae) and recovery of a new *Trichogramma* (Hym: Trichogrammatidae) from root weevil egg masses in Florida. Entomophaga 25:91–94

Blua MJ, Phillips PA, Redak RA (1999) A new sharpshooter threatens both crops and ornamentals. Calif Agric 53:22–25

Buhl PN (1998) On some new or little known NW European species of Platygastridae (Hymenoptera, Proctotrupoidea). Fragmenta Entomol 30:295–334

Buhl PN (1999) A synopsis of the Platygastridae of Fennoscandia and Denmark (Hymenoptera, Platygastroidea). Entomofauna Z Entomol 20:17–52

Buhl PN (2002) Contributions to the platygastrid fauna of Panama. Entomofauna Z Entomol 23:309–332

Carrillo D, Peña JE, Capinera JL (2008) Effect of host plants on successful parasitism of *Haeckeliania sperata* (Hymenoptera: Trichogrammatidae) on *Diaprepes abbreviatus* (Coleoptera: Curculionidae) eggs. Environ Entomol 37:1565–1572

Carrillo D, Peña JE, Rogers M (2009) Relative susceptibility of *Haeckeliania sperata* Pinto (Hymenoptera: Trichogrammatidae) to pesticides used in citrus and ornamental systems in Florida. J Econ Entomol 102:905–912

Castillo J, Jacas JA, Peña JE, Ulmer BJ, Hall DG (2006) Effect of Temperature on Life History of *Quadrastichus haitiensis* (Hymenoptera: Eulophidae), an Endoparasitoid of *Diaprepes abbreviatus* (Coleoptera: Curculionidae). Biol Control 36:189–196

CDFA (2008) Pierce's Disease Control Program. Biological control http://www.cdfa.ca.gov/pdcp/Biological_Control.html. Accessed 9 July 2008

Clausen CP (1940) Entomophagous insects. McGraw Hill, New York

Coto D, Saunders JL, Vargas C, King AC (1995) Plagas invertebradas de cultivos tropicales en America Cental Centro Agronomico Tropical de Investigacion y Ensenanza, Manual Tecnico 12, Costa Rica, 66p

Coulson JR, Vail PV, Dix ME, Norlund DA, Kauffman W (2000) 110 years of biological control research and development in the United States Department of Agriculture 1883–1983 USDA, ARS, 645p

Crawford JC (1916) Some new American Hymenoptera. Insecutor Insectiae Menstruus 4:135–144

de León JH, Jones WA (2005) Genetic differentiation among geographic populations of *Gonatocerus ashmeadi*, the predominant egg parasitoids of the glassy-winged sharpshooter, *Homalodisca coagulata*. J Insect Sci 5:1–9

de León JH, Morgan DJW (2005) Small scale post-release evaluation of *Gonatocerus morrilli* program in California against the glassy-winged sharpshooter: utility of developed molecular diagnostic tools. In: Tariq MA, Blincoe P, Mochel M, Oswalt S, Esser T (eds) Proceedings of the 2005 Pierce's Disease Research Symposium; and December 5–7, 2005, San Diego Marriott Hoteland, Marina, San Diego, California, pp 306–309. Organized by California Department of Food and Agriculture, Copeland Printing, Sacramento, California, 399p. (Available online at http://www.cdfa.ca.gov/phpps/pdcp/ResearchSymposium/gw2005symp.htm)

de León JH, Jones WA, Morgan DJW (2004) Molecular distinction between populations of *Gonatocerus morrilli*, egg parasitoids of the glassy-winged sharpshooter from Texas and California: Do cryptic species exist? J Insect Sci 4:1–7 de León JH, Jones WA, Sétamou M, Morgan DJW (2006) Genetic and hybridization evidence confirms that a geographic population of *Gonatocerus morrilli* (Hymenoptera: Mymaridae) from California is a new species: egg parasitoids of the glassy-winged sharpshooter *Homalodisca coagulata* (Homoptera: Cicadellidae). Biol Control 38: 282–293

Delvare G (1988) *Ceratogramma etiennei* n. sp., parasite a la Guadeloupe de *Diaprepes abbreviatus* L. (Hymenoptera: Trichogrammatidae; Coleoptera: Curculionidae) Rev Fr Entomol 10:1–4

Doutt RL, Viggiani G (1968) The classification of the Trichogrammatidae (Hymenoptera: Chalcidoidea). Proc Calif Acad Sci 35:477–586

Dozier HL (1932) Descriptions of new Trichogrammatid (Hymenoptera) egg parasites from the West Indies. Proc Entomol Soc Wash 34:29–37

Dreistad SE, Dahlsten DL (1991) Establishment and overwintering of *Tetrastichus gallerucae* (Hymenoptera: Eulophidae), an egg parasitoid of the elm leaf beetle (Coleoptera: Chrysomelidae) in northern California. Environ Entomol 20:1711–1719

Duncan RE, Ulmer B, Peña JE, Lapointe S (2007) Reproductive biology of *Fidiobia dominica* (Hymenoptera: Platygastridae), an egg parasitoid of *Diaprepes abbreviatus* (Coleoptera: Curculionidae). Environ Entomol 36:376–382

Ellis CR (1973) Parasitism of *Hypera postica* eggs at Guelph, Ontario, by *Patasson luna* and *Fidiobia rugosifrons*. J Econ Entomol 66:1059–1061

Etienne J, Delvare G (1991) Les parasites de *Diaprepes abbreviatus* (Coleoptera: Curculionidae) aux Antilles Françaises. Bull Soc Entomol France 36:295–299

Etienne J, Mauleon H, Pintureau B (1990) Biologie et dynamique de *Ceratogramma etiennei* (Hymenoptera: Trichogrammatidae) parasite de *Diaprepes abbreviatus* (Coleoptera:Curculionidae) en Guadeloupe. Colloques 58:458–468

Etienne J, Reyes M, Castillo M, Diaz F, Abud-Antun A (1992) Posibilidad de lucha biologica Contra *Diaprepes abbreviatus* (L.) (Coleoptera: Curculionidae) en República Dominicana. Caribbean Food Crops Soc Congr 28:104–112

Evans GA, Peña JE (2005) A new *Fidiobia* species (Hymenoptera: Platygastridae) reared from eggs of *Diaprepes doubleirii* (Coleoptera: Curculionidae) from Dominica. Florida Entomol 88:61–66

González D (1988) Biotypes in biological control – examples with populations of *Aphidius ervi, Trichogramma pretiosum* and *Anagrus epos* (Parasitic Hymenoptera). In: Gupta V (ed) Advances in parasitic Hymenoptera research. EJ Brill, Leiden, New York, pp 475–482

González D, Cervenka V, Moratorio M, Pickett C, Wilson LT (1988) Biological control of variegated leafhopper in grapes. Calif Agric 42:23–25

Gonzalez RH (1983) Manejo de plagas de la vid. Ciencias agricolas 13, Universidad de Chile, 115 p

Grafton-Cardwell EE, Godfrey KE, Peña JE, McCoy CW, Luck RF (2004) *Diaprepes* root weevil. University of California, Div. Agric. Nat. Res. Publication 8131. Oakland, CA. (http://anrcatalog.ucdavis.edu.)

Grandgirard J, Hoddle MS, Petit JN, Roderick GK, Davies N (2008) Engineering an invasion: classical biological control of the glassy-winged sharpshooter, *Homalodisca vitripennis*, by the egg parasitoid *Gonatocerus ashmeadi* in Tahiti and Moorea, French Polynesia. Biol Invasions 10:135–148

Guedes JVC, Parra J, Loiacono M (2001) Parasitismo natural de posturas de curculionideos da fraiz dos citros por *Fidiobia* spp (Hym: Platygastrioidea). In VII Simposio de Control Biologico 2001. Poços de Caldas, MG, Livro Resumos, 2001, p 340

Hall D, Peña J, Franqui R, Nguyen R, Stansly P, Mccoy C, Lapointe S, Adair R, Bullock R (2001) Status of biological control by egg parasitoids of *Diaprepes abbreviatus* (Coleoptera:Curculionidae) in citrus in Florida and Puerto Rico. BioControl 46:61–70

Hall DG, Eger J, Peña JE, Duncan R, O'Brien C, Evans G, McCoy C (2002) Exploration in Belize for parasitoids attacking eggs of citrus weevils and an evaluation of *Pediobious irregularis* and *Horismenus bennetti* (Hymenoptera: Eulophidae) as potential biological control agents of *Diaprepes abbreviatus* and *Pachnaeus litus* (Coleoptera: Curculionidae). Florida Entomol 85:663–666.

Herz A, Zimmerman O, Hassan S (2003). Egg parasitoid News. International Organization for Biological Control 15, 56p

Hoddle MS, Stouthamer R (2005) Is the glassy-winged sharpshooter parasitoid *Gonatocerus morrilli* one species or a complex of closely related sibling species? In: Tariq MA, Blincoe P, Mochel M, Oswalt S, Esser T (eds) Proceedings of the 2005 Pierce's Disease Research Symposium, andDecember 5–7, 2005, San Diego Marriott Hoteland, Marina, San Diego, California, pp 338–340. Organized by California Department of Food and Agriculture, Copeland Printing, Sacramento, California, 399p. (Available online at http://www.cdfa.ca.gov/phpps/pdcp/ResearchSymposium/gw2005symp.htm)

Hoddle MS, Triapitsyn SV (2003) Searching for and collecting egg parasitoids of the Glassy-winged Sharpshooter in the central and eastern USA. In: Tariq MA, Oswalt S, Blincoe P, Spencer R, Houser L, Ba A, Esser T (eds) Proceedings of the Pierce's disease research symposium, and December 8–11, 2003, Coronado Island Marriott Resort, Coronado, California, pp 261–262. Organized by California Department of Food and Agriculture (compiled), Copeland Printing, Sacramento, California, 323p. (Available online at http://www.cdfa.ca.gov/phpps/pdcp/ResearchSymposium/gw2003symp.htm)

Hoddle MS, Triapitsyn SV (2004a) Searching for and collecting egg parasitoids of the Glassy-winged Sharpshooter and other *Homalodisca* species in southeastern and southwestern Mexico. In: Tariq MA, Oswalt S, Blincoe P, Ba A, Lorick T, Esser T (eds) Proceedings of the 2004 Pierce's Disease Research Symposium, andDecember 7–10, 2004, Coronado Island Marriott Resort, Coronado, California, pp 339–341. Organized by California Department of Food and Agriculture, Copeland Printing, Sacramento, California, 391p. (Available online at http://www.cdfa.ca.gov/phpps/pdcp/ResearchSymposium/gw2004symp.htm)

Hoddle MS, Triapitsyn SV (2004b). Searching for and collecting egg parasitoids of Glassy-winged Sharpshooter in the central and eastern USA. In: Tariq MA, Oswalt S, Blincoe P, Ba A, Lorick T, Esser T (eds) Proceedings of the 2004 Pierce's Disease Research Symposium, andDecember 7–10, 2004, Coronado Island Marriott Resort, Coronado, California, pp 342–344. Organized by California Department of Food and Agriculture, Copeland Printing, Sacramento, California, 391p. (Available online at http://www.cdfa.ca.gov/phpps/pdcp/ResearchSymposium/gw2004symp.htm)

Irvin NA, Hoddle MS (2004) Oviposition preference of *Homalodisca coagulata* for two *Citrus limon* cultivars and influence of host plant on parasitism by *Gonatocerus ashmeadi* and *G. triguttatus* (Hymenoptera: Mymaridae). Florida Entomol 87:504–510

Irvin NA, Hoddle MS (2005a) Determination of *Homalodisca coagulata* (Hemiptera: Cicadellidae) egg ages suitable for oviposition by *Gonatocerus ashmeadi, Gonatocerus triguttatus*, and *Gonatocerus fasciatus* (Hymenoptera: Mymaridae). Biol Control 32: 391–400

Irvin NA, Hoddle MS (2005b) The competitive ability of three mymarid egg parasitoids (*Gonatocerus* spp.) for glassy-winged sharpshooter (*Homalodisca coagulata*) eggs. Biol Control 34:204–214

Jacas JA, Peña JE, Duncan RE (2005) Successful oviposition and reproductive biology of *Aprostocetus vaquitarum* (Hymenoptera: Eulophidae): a predator of *Diaprepes abbreviatus* (Coleoptera: Curculionidae). Biol Control 33:352–359

Jacas JA, Peña JE, Duncan RE (2007) Morphology and devekopment of the immature stages of *Fidiobia dominica* Evansand, Peña (Hymenoptera: Platygastridae: Sceliotracheminae). Ann Entomol Soc Am 100:413–417

Jacas JA, Peña JE, Duncan RE, Ulmer BJ (2008) Thermal requirements of *Fidiobia dominica* (Hymenoptera: Platygastridae) and *Haeckeliania sperata* (Hymenoptera: Trichogrammatidae), two exotic egg parasitoids of *Diaprepes abbreviatus* (Coleoptera:Curculionidae). BioControl. doi:10.1007/s10526-007-9082-4

Jacas JA, Peña JE, Duncan RE (2009) Morphology and development of the immature stages of *Brachyufens osborni* (Dozier) (Hymehoptera: Trichogrammatidae: Trichogrammatinae), a parasitoid of different broad-nosed weevil eggs (Coleoptera:Curculionidae: Entiminae). Ann Entomol Soc Am 102:112–118

Jackson DJ (1958) Observation of the biology of *Caraphractus cinctus* Walker (Hymenoptera: Mymaridae), a parasite of the eggs of Dytiscidae. I. Methods of rearing and numbers bred on different host eggs. Trans R Entomol Soc Lond 110:533–566

Jarjees EA, Merritt DJ (2002) Development of *Trichogramma australicum* Girault (Hymenoptera: Trichogrammatidae) in *Helicoverpa* (Lepidoptera: Noctuidae) host eggs. Aust J Entomol 41:310–315

Jarjees EA, Merritt DJ (2003) Structure of the gut contents of *Trichogramma australicum* Girault (Hymenoptera: Trichogrammatidae) larvae fixed in situ in eggs of its host *Helicoverpa armigera* (Hübner) (Lepidoptera: Noctuidae). Aust J Entomol 42:203–209

Jones WA (2001) Classical biological control of the glassy-winged sharpshooter. In: Proceedings of the Pierce's Disease Research Symposium, December 5–7, 2001, Coronado Island Marriott Resort, San Diego, California, pp 50–51. California Department of Food and Agriculture, Copeland Printing, Sacramento, California, 141p. (Available online at http://www.cdfa.ca.gov/phpps/pdcp/gwSymposium.htm)

Jones WA, Logarzo GA, Triapitsyn SV, Casas M, Virla EG, Purcell AH (2005a) Biology and host range of 2 South American egg parasitoids (Hymenoptera: Mymaridae), possible biocontrol agents for Glassy-Winged Sharpshooter, *Homalodisca coagulata* (Say) (Hemiptera: Cicadellidae; Proconiini). In: Submitted Papers T1–T21, 12th International Auchenorrhyncha Congress and 6th International Workshop on Leafhoppers and Planthoppers of Economic Significance. University of California, Berkeley, 8–12 August 2005, Submitted Papers T7–T8. (Available online at http://nature.berkeley.edu/hoppercongress/)

Jones WA, Logarzo GA, Virla EG and, Luft E (2005b) Environmental risk assessment of egg parasitoids from South America: Nontarget field and laboratory host range in Argentina and the United States. In: Tariq MA, Blincoe P, Mochel M, Oswalt S, Esser T (eds) Proceedings of the 2005 Pierce's Disease Research Symposium, andDecember 5–7, 2005, San Diego Marriott Hoteland, Marina, San Diego, California. Organized by California Department of Food and Agriculture, Copeland Printing, Sacramento, California, 399p. (Available online at http://www.cdfa.ca.gov/phpps/pdcp/ResearchSymposium/gw2005symp.htm)

Kfir R, Gouws J, Moore SD (1993) Biology of *Tetrastichus howardi* (Olliff) (Hymenoptera: Eulophidae): a facultative hyperparasitoid of stem borers. Biocontrol Sci Technol 3: 149–159

Kido H, Flaherty DL, Bosch DF, Valero KA (1984) The variegated grape leafhopper in the San Joaquin Valley. Calif Agric 38:31–32

Knapp J, Noling JW, Timmer LW, Tucker DP (1996) Florida Citrus IPM. In: Rosen D, Bennett Fand, Capinera J (eds) Integrated management: a Florida perspective. Intercept, Andover, pp 317–348

Knapp J, Bergh J, Oswalt T, Salyani M, Parsons L, Tucker D, Wheaton T, Bullock R, Childers C, Hall D, McCoy C, Rogers J, Stansly P, Browning H, Graham J, Timmer L, Pena J, Futch S, Duncan L, Nolling J, Roberts P, McMillan R, Sonoda R, Brown G, Davies F, Ismail M, Stover E (1999) Florida Citrus Pest Management Guide. University of Florida, Gainesville, FL Institute of Food and Agricultural Sciences, Florida Cooperative Extension Service, SP-43

Krugner R, Johnson MW, Groves RL, Morse JG (2008) Host specificity of *Anagrus epos*: a potential biological control agent of *Homalodisca coagulata*. BioControl 53:439–449

Lapointe SL (2001) Effect of temperature on egg development of *Diaprepes abbreviatus* (Coleoptera:Curculionidae). Florida Entomol 84:298–299

Lapointe SL, Borchert DM, Hall DG (2007) Effect of low temperatures on mortality and oviposition in conjunction with climate mapping to predict spread of the root weevil *Diaprepes abbreviatus* and introduced natural enemies. Environ Entomol 36:73–82

Logarzo G, Triapitsyn SV, Jones WA (2003) New host records for two species of *Gonatocerus* (Hymenoptera: Mymaridae), egg parasitoids of proconiine sharpshooters (Hemiptera: Clypeorrhyncha: Cicadellidae), in Peru. Florida Entomol 86:486–487

Logarzo GA, de León JH, Triapitsyn SV, González RH, Virla EG (2006) First report of a proconiine sharpshooter, *Anacuerna centrolinea* (Hemiptera: Cicadellidae), in Chile, with notes on its biology, host plants, and egg parasitoids. Ann Entomol Soc Am 99: 879–883

Logarzo GA, Virla E, Jones WA (2005) Egg parasitoids from Argentina, potential candidates for the biological control of glassy-winged sharpshooter in the United States. In: Hoddle MS (ed) Second International Symposium on Biological Control of Arthropods, Davos, Switzerland, September 12–16, 2005. USDA Forest Service Publication FHTET–2005–08, vol 3, pp 115–116. (Available online at http://www.bugwood.org/arthropod2005/vol3/index.html)

Loiacono MS (1982) Un nuevo platygastrido (Hymenoptera-Platygastridae) criado de huevos de Naupactus xanthographus Germ. (Coleoptera-Curculionidae). Rev Soc Entomol Argentina 41:85–88

Mannion C, Hunsberger A, Peña J, Osborne L (2003) Oviposition of *Diaprepes abbreviatus* on ornamental plants in south Florida. Florida Entomol 86:165–173

Masner L, Huggert L (1989) World review and keys to genera of the subfamily nostemmatinae with reassignment of the taxa to the Platygastrinae and Sceliotrachelinae (Hymenoptera: Platygastridae). Mem Entomol Soc Canada 147:1–214

McCoy C, Simpson S (1994) Past and current IPM strategies to combat the spread of *Diaprepes abbreviatus* in Florida citrus. Proc Caribb Food Soc 30:247–255

Morgan DJW, Triapitsyn SV, Redak RA, Bezark LG, Hoddle MS (2000) Biological control of the glassy-winged sharpshooter: current status and future potential, pp 167–171. In: Hoddle MS (ed) Proceedings of the California Conference on Biological Control II, 205p

Morgan DJW, Simmons GS, Higgins LM, Shea K (2002) Glassy-winged sharpshooter biological control in California: building framework for active adaptive management, pp 140–143. In: Hoddle MS (ed) Proceedings of the 3rd California Conference on Biological Control, 162p

Morse JG, Stouthamer R (2007) The *Anagrus epos* complex: a likely source of effective classical biological agents for glassy-winged sharpshooter control. In: Esser T (ed) Proceedings of the 2007 Pierce's Disease Research Symposium. San Diego, California, pp 94–97. Organized by California Department of Food and Agriculture. PIP Printing and Document Services, Sacramento, California. (Available online at http://www.cdfa.ca.gov/pdcp/Research_Symposium_Index.html)

Nguyen R, Hall DG, Peña JE, Amalin D, McCoy CW, Lapointe S, Adair R, Stansly P (2003) Rearing methods for *Quadrastichus haitiensis* (Gahan) (Hymenoptera: Eulophidae) for biological control of *Diaprepes abbreviatus* (L.) (Coleoptera: Curculionidae) Florida Department of Agricultural and Consumer Services. (http://www.doacs.state.fl.us/pi/methods/diaprepesposter.html)

Nixon GEJ (1969) Two new species of *Platystasius* Nixon with a note on the generic relationship between *Platystasius* and *Fidiobia* (Hymenoptera:Platygasteridae). Proc Entomol Soc Wash 7:445–449

Notton D (1998) Platygastroidea. In: Taxonomy and Biology of parasiticHymenoptera, 18–25 April 1998, Department of Entomology, The Natural History Museum, London and the Department of Biology, Imperial College, University of London

Noyes JS (2008) Universal Chalcidoidea database. http://www.nhmac.uk/entomology/chalcidoids/index.html. Accessed 23 Apr 2008

Obrycki JJ, Tauber MJ, Gollands B (1985). *Edovum puttleri* (Hymenoptera: Eulophidae), an exotic egg parasitoid of the Colorado potato beetle (Coleoptera: Chrysomelidae): responses to temperature zone conditions and resistant potato plants. Environ Entomol 14: 48–54

Otero O, Montes M, Arteaga E, Rodriguez N, Gonzalez C, Cabrera R, Broche R, Castellanos A, Fernandez O (1995) Manual de orientaciones para el manejo fitosanitario de las principales plagas y enfermedades de los citricos. Departamento de Proteccion de Plantas, Instituto de Investigaciones de Citricos, La Habana, Cuba, 21p

Peña JE, Amalin DM (2000) Biological control of *Diaprepes abbreviatus* by parasitoids, pp 66–76. In: Futch SH (ed) *Diaprepes* Short Course. Cooperative Extension Service Florida Agricultural Experiment Station. Citrus Researchand, Education Center, Lake Alfred, FL

Peña JE, Etienne J, Duncan R, Pinto J (1998) Introduction of *Ceratogramma etiennei* (Hymenoptera: Trichogrammatidae) for biological control of *Diaprepes abbreviatus* in Florida, USA, pp 145–148. In: Hassan S (ed) Egg parasitoids, 5th International Symposium. IIBC, Berlin, 356p

Peña JE, Hall D, Nguyen R, Duncan RE, Amalin D, Stansly P, McCoy CW, Adair R, Lapointe SE, Browning H, Knapp J (2001) Efforts toward establishment of biological control agents of *Diaprepes* root weevil. University of Florida, Cooperative Extension Service, IFAS, EDIS, ENY-643. (http://edis.ifas.ufl.edu/BODY_IN122)

Peña JE, Hall DG, Nguyen R, McCoy CW, Amalin D, Stansly P, Adair R, Lapointe S, Duncan R, Hoyte A (2004) Recovery of parasitoids (Hymenoptera: Eulophidae and Trichogrammatidae) released for biological control of *Diaprepes abbreviatus* (Coleoptera: Curculionidae) in Florida. Proc Int Citrus Congr 3:879–884

Peña JE, Ulmer BJ, Jacas JA, Duncan RE, McCoy CW (2006) Biological control of Neotropical citrus root weevils. Memorias of the IV Congreso Internacional de Control Biologico, 31 May to 2 June 2006, Palmira, Colombia

Phillips P, Triapitsyn S, Hoddle MS (2001) Survey for egg parasitoids of glassy-winged sharpshooter in California. In: Proceedings of the Pierce's Disease Research Symposium, December 5–7, 2001, Coronado Island Marriott Resort, San Diego, California, p 95. California Department of Food and Agriculture, Copeland Printing, Sacramento, California, 141p. (Available online at http://www.cdfa.ca.gov/phpps/pdcp/gwSymposium.htm)

Pickett CH, Wilson LT, González D, Flaherty DL (1987) Biological control of variegated leafhopper. Calif Agric 41:14–16

Pickett CH, Wilson LT, Flaherty DL, González D (1989) Measuring the host preference of parasites: an aid in evaluating biotypes of *Anagrus epos* [Hym: Mymaridae]. Entomophaga 34:551–558

Pilkington LJ, Irvin N, Boyd EA, Hoddle MS, Triapitsyn S, Carey BG, Jones WA, Morgan DJW (2004) Biological control of glassy-winged shrpshooter in California. In: Hoddle MS (ed) Proceedings of the California Conference on Biological Control IV, pp 133–136, UC Berkeley, California, 152p

Pilkington LJ, Irvin NA, Boyd EA, Hoddle MS, Triapitsyn SV, Carey BG, Jones WA, Morgan DJW (2005) Introduced parasitic wasps could control glassy-winged sharpshooter. Calif Agric 59:223–228

Pinto JD (1997) Trichogrammatidae. In: Gibson GAP, Huber JTand, Woolley JB (eds) Annotated keys to the Genera of nearctic Chalcidoidea (Hymenoptera), NRC Research Press, Ottawa, Canada, pp 726–752

Pinto J (2005) Descriptions of additional New World Trichogrammatidae (Hymenoptera): the genus *Nicokavespa* and a new species of *Haeckeliania*. Proc Entomol Soc Wash 107:627–641

Pinto DJ, Viggiani G (1991) A taxonomic study of the genus *Ceratogramma* (Hymenoptera: Trichogrammatidae). Proc Entomol Soc Wash 93:719–732

Pitkin BR (2003) Trichogrammatidae. Universal Chalcidoidea database. http://www.nhm.ac.uk/research-curation/projects/chalcidoids/. Accessed 24 Aug 2007

Rahim A, Hashmi AA and, Khan NA (1991) Effects of relative humidity on longevity and development of *Ooencyrtus papilionis* Ashmead (Hymenoptera: Eulophidae), a parasite of the sugarcane pest, *Pyrilla perpusilla* Walker (Homoptera: Cicadellidae). Environ Entomol 2:774–775

Salt G (1941) The effects of hosts upon their insect parasites. Biol Rev 16:239–264

Schauff ME (1987) Taxonomy and identification of the egg parasites (Hymenoptera: Platygastridae, Trichogrammatidae, Mymaridae, and Eulophidae) of citrus weevils (Coleoptera: Curculionidae). Proc Entomol Soc Wash 89:31–42

Schroeder WJ, Beavers JB (1977) Citrus root weevils in Florida: Identification, biology and control. Proc Int Soc Citriculture 2:498–500

Simpson S, Nigg H, Coile N, Adair R (1996). *Diaprepes abbreviatus* (Coleoptera: Curculionidae): Host plant associations. Environ Entomol 25:333–349

Smith D, Beattie GAC, Broadley R (1997) Citrus pests and their enemies: integrated pest management in Australia. Queensland Department of Primary Industries Series QI97030, 272p

Strand MR (1986) The physiological interactions of parasitoids with their hosts and their influence on reproductive strategies. In: Waage JKand, Gratehead D (eds) Insect parasitoids. Academic, London, pp 97–136

Sutton RA, Selhime G, McCloud W (1972) Colonization and release of *Tetrastichus haitienisis* as a biological control agent for citrus root weevils. J Econ Entomol 65:184–185

Szabo JB (1958) Erganzende Beobachtungen uber die holarktische Gattung *Fidiobia* Ashmead 1894. Folia Entomol Hungarica 11:457–464

Triapitsyn SV (1998) *Anagrus* (Hymenoptera: Mymaridae) egg parasitoids of *Erythroneura* spp. and other leafhoppers (Homoptera: Cicadellidae) in North American vineyards and orchards: a taxonomic review. Trans Am Entomol Soc 124:77–112

Triapitsyn SV (2002a) Species-level taxonomy of Mymaridae (Hymenoptera): current status and implications for biological control of leafhoppers of economic importance. In: Gand M, Thuróczy C (eds) Parasitic wasps: evolution, systematics, biodiversity and biological control. Agroinform Kiadó és Nyomda Kft., Budapest, pp 89–94

Triapitsyn SV (2002b) Taxonomy and host associations of *Gonatocerus* spp. (Mymaridae) – egg parasitoids of proconiine leafhoppers. Egg Parasitoid News 14:10

Triapitsyn SV (2003) Taxonomic notes on the genera and species of Trichogrammatidae (Hymenoptera) – egg parasitoids of the proconiine sharpshooters (Hemiptera: Clypeorrhyncha: Cicadellidae: Proconiini) in southeastern USA. Trans Am Entomol Soc 129:245–265

Triapitsyn SV (2006) A key to the Mymaridae (Hymenoptera) egg parasitoids of proconiine sharpshooters (Hemiptera: Cicadellidae) in the Nearctic region, with description of two new species of *Gonatocerus*. Zootaxa 1203:1–38

Triapitsyn SV, Bernal JS (2009) Egg parasitoids of Proconiini (Hemiptera: Cicadellidae) in northwestern Mexico, with description of a new species of *Gonatocerus* (Hymenoptera: Mymaridae). J Insect Sci 5:1–9

Triapitsyn SV, Hoddle MS (2001) Search for and collect egg parasitoids of glassy-winged sharpshooter in southeastern USA and northeastern Mexico. In: Proceedings of the Pierce's Disease Research Symposium, December 5–7, 2001, Coronado Island Marriott Resort, San Diego, California, pp 133–134. California Department of Food and Agriculture, Copeland Printing, Sacramento, California, 141p. Available online at http://www.cdfa.ca.gov/phpps/pdcp/gwSymposium.htm

Triapitsyn SV, Hoddle MS (2002) Search for and collect egg parasitoids of glassy-winged sharpshooter in southeastern USA and northeastern Mexico. In: Proceedings of the Pierce's Disease Research Symposium, December 15–18, 2002, Coronado Island Marriott Resort, San Diego, California, pp 94–95. California Department of Food and Agriculture, Digital Logistix, Sacramento, California, 177p. (Available online at http://www.cdfa.ca.gov/phpps/pdcp/gw2002symp.htm)

Triapitsyn SV, Phillips PA (1996) Egg parasitoid of glassy-winged sharpshooter. Citrograph 81:10

Triapitsyn SV, Phillips PA (2000) First host record of *Gonatocerus triguttatus* (Hymenoptera: Mymaridae) from eggs of *Homalodisca coagulata* (Homoptera: Cicadellidae), with notes on the distribution of the host. Florida Entomol 83:200–203

Triapitsyn SV, Rakitov RA (2005) Egg parasitoids (Hymenoptera: Mymaridae and Trichogrammatidae) of *Cuerna* sharpshooters (Hemiptera: Cicadellidae) in the USA,

Posters P11–12. In: Poster Abstracts P1–P43, 12th International Auchenorrhyncha Congress and 6th International Workshop on Leafhoppers and Planthoppers of Economic Significance, University of California, Berkeley, 8–12 August 2005. (Available online at http://nature.berkeley.edu/hoppercongress/)

Triapitsyn SV, Mizell III RF, Bossart JL and, Carlton CE (1998) Egg parasitoids of *Homalodisca coagulata* (Homoptera: Cicadellidae). Florida Entomol 81:241–243

Triapitsyn SV, Hoddle MS, Morgan DJW (2002a) A new distribution and host record for *Gonatocerus triguttatus* in Florida, with notes on *Acmopolynema sema* (Hymenoptera: Mymaridae). Florida Entomol 85:654–655

Triapitsyn SV, Bezark LG, Morgan DJW (2002b) Redescription of *Gonatocerus atriclavus* Girault (Hymenoptera: Mymaridae), with notes on other egg parasitoids of sharpshooters (Homoptera: Cicadellidae: Proconiini) in northeastern Mexico. Pan-Pac Entomol 78:34–42

Triapitsyn SV, Morgan DJW, Hoddle MS, Berezovskiy VV (2003) Observations on the biology of *Gonatocerus fasciatus* Girault (Hymenoptera: Mymaridae), egg parasitoid of *Homalodisca coagulata* (Say) and *Oncometopia orbona* (Fabricius) (Hemiptera: Clypeorrhyncha: Cicadellidae). Pan-Pac Entomol 79:75–76

Trjapitzin SV, Chiappini E (1994) A new *Anagrus* (Hymenoptera: Mymaridae), an egg parasitoid of *Erythroneura* spp. (Homoptera: Cicadellidae). Entomol News 105:137–140

Turner WF, Pollard HN (1959) Life histories and behavior of five insect vectors of phony peach disease. Tech Bull, United States Dept Agric 1188:1–28

Ulmer BJ, Lapointe S, Peña JE, Duncan R (2006a) Toxicity of pesticides used in citrus to *Aprostocetus vaquitarum* (Hymenoptera: Eulophidae), an egg parasitoid of *Diaprepes abbreviatus*. Florida Entomol 89:10–19

Ulmer B, Jacas J, Peña JE, Duncan RE, Castillo J (2006b) Effect of temperature on life history of *Aprostocetus vaquitarum* (Hymenoptera:Eulophidae), an egg parasitoidof *Diaprepes abbreviatus* (Coleoptera:Curculionidae). Biol Control 39:19–25

Ulmer B, Peña JE, Lapointe S, Mathurin G (2006c) The occurrence of parasitoids attacking citrus weevils eggs on Saint Lucia. Florida Entomol 89:407–409

Ulmer B, Duncan RE, Pavis C, Pena JE (2008) Parasitoids attacking citrus weevil eggs in Guadeloupe. Florida Entomol 91:311–314

Urbaneja A, Llácer E, Garrido A, Jacas JA (2001) Effect of temperature on the life history of *Cirrospilus* sp. near *lyncus* (Hymenoptera: Eulophidae), a parasitoid of *Phyllocnistis citrella* (Lepidoptera: Gracillariidae). Biol Control 21:293–299

Urbaneja A, Hinarejos R, Llácer EAH, Garrido A, Jacas J (2002) Effect of temperature on life history of *Cirrospilus vittatus* (Hymenoptera: Eulophidae), an ectoparasitoid of *Phyllocnistis citrella* (Lepidoptera: Gracillariidae). J Econ Entomol 95:250–255

Urbaneja A, Morales C, DeMendoza AH, Garrido A, Jacas J (2003) Effect of temperature on development and survival of *Citrostichus phyllocnistoides* (Hymenoptera: Eulophidae), a parasitoid of *Phyllocnistis citrella* (Lepidoptera: Gracillariidae). Biocontrol Sci Technol 13:127–130

van Whervin LW (1968) The citrus weevils of Jamaica and some of their parasites. University West Indies, St. Augustine, Trinidad, Tech Bull 1:1–23

Vickerman DB, Hoddle MS, Triapitsyn S, Stouthamer R (2004) Species identity of geographically distinct populations of the glassy-winged sharpshooter parasitoid *Gonatocerus ashmeadi*: morphology, DNA sequences, and reproductive compatibility. Biol Control 31: 338–345

Viggiani G (1992) New species of Trichogrammatidae (Hymenoptera: Chalcidoidea) from South America. Redia 75:253–265.

Virla EG, Logarzo GA, Jones WA, Triapitsyn S (2005) Biology of *Gonatocerus tuberculifemur* (Hymenoptera: Mymaridae), an egg parasitoid of the sharpshooter, *Tapajosa rubromarginata* (Hemiptera: Cicadellidae). Florida Entomol 88:67–71

Woodruff RE (1964) A Peurto Rican weevil new to the United States (Coleoptera: Curculionidae). Florida Department of Agriculture, Division of Plant Industry, Entomol Circular 30: 1–2

Woodruff RE (1968) The present status of a West Indian weevil (*Diaprepes abbreviatus*) in Florida (Coleoptera: Curculionidae). Florida Department of Agriculture and Consumer Services, Plant Industry Entomol. 77, Gainesville, Fl

Woodruff RE (1985) Citrus weevils in Florida and the West Indies: Preliminary report on systematics, biology, and distribution (Coleoptera: Curculionidae). Florida Entomol 68:370–379

Chapter 14
Egg Parasitoids Commercialization in the New World

José Roberto Postali Parra

Contents

14.1 Introduction

Following the Green Revolution in the second half of the 20th Century, biological control has resurfaced with renewed force in recent years, especially by the adoption of integrated pest management (IPM) programs. These programs were implemented as a consequence of the indiscriminate use of agrochemicals, which led to a number of problems, such as insect and mite resistance to insecticides and acaricides, as well as environmental contamination (Parra et al. 2002, Guillon 2008).

Within this new context in which measures are adopted in order to keep pests below their economic-injury level, taking into account economic, ecological, and social criteria, biological control, either classic or applied, assumes an important role. However, for biological control to be fully utilized, it is essential for the user to have insects available whenever they are required, in order to use them in inoculative (classical biological control), inundative (applied biological control), or seasonal

J.R.P. Parra (✉)
Departamento de Entomologia e Acarologia, ESALQ/USP, Av. Pádua Dias 11, 13418-900 Piracicaba, São Paulo, Brazil
e-mail: jrpparra@esalq.usp.br

F.L. Consoli et al. (eds.), *Egg Parasitoids in Agroecosystems with Emphasis on Trichogramma*, Progress in Biological Control 9, DOI 10.1007/978-1-4020-9110-0_14,

inoculative releases (biological control in protected crops) (van Lenteren 2000) to control the target pest.

Therefore, after universities or research institutions develop technological packages for the biological control of a given pest, that package should be transferred to private companies or to cooperatives (or their equivalent), which will then multiply and make those natural enemies (parasitoids or predators) available. Nevertheless, for such organizations to emerge, a cultural process must be established; education on the advantages of biological control is, unfortunately, not yet completely mature in many New World countries. Above all, it is essential that sales of natural enemies are not viewed exclusively from a commercial perspective, but are supported by technical knowledge in order to enjoy good credibility with the producer.

Economic and social conditions, as well as those pertaining to technological knowledge create differences from country to country and from region to region with regard to the natural-enemy market. In Europe, several biological control agents have been used since the 1970s, especially in greenhouses (van Lenteren et al. 1997). In the USA, five species (or groups) of parasitoids, six species of predators, five species of predatory mites, and two genera of entomopathogenic nematodes were the primary biocontrol agents sold in the 1990s (Cranshaw et al. 1996). However, according to Warner and Getz (2008), no new insect- rearing facilities have since appeared in the USA; the market has remained stable during this period, and many such businesses have failed due to economic problems.

On the other hand, there has been great excitement in the New World about the expansion of areas where parasitoids, predators, and pathogens are released, especially with respect to releases under field conditions. This is mainly due to the prospects for these countries to become exporters of foods and fruit crops; in this case, importer requirements with regard to the absence of chemical residues are higher and will require the use of other alternatives to agrochemicals, with emphasis on biological input. In addition, organic-farming areas have increased in Latin America, and currently represent 20.8% of the agricultural land as compared to 6.7% in the USA (Guillon 2008).

Therefore, the existing differences previously pointed out converge towards the same point, and constitute challenges to increase the use of biological control worldwide. These challenges will be emphasized in this chapter, which deals with the commercial trade in parasitoid species that attack the eggs of crop pests in the New World.

14.2 Current Status of the Worldwide Commercialization of Natural Enemies

14.2.1 USA and Europe

The beginning of the natural-enemy pest-control business in many countries coincided with the end of the "dark period" of pest control, when only agrochemicals were employed. This period lasted from 1939, when DDT was first synthesized, and continued with the uncontrolled use of agrochemicals until the time when society became concerned about environmental issues. This concern arose due to biological imbalances caused by the calendar-based applications (Kogan 1998). This beginning also coincided with the development of artificial diets for insects, especially Lepidoptera, Coleoptera, and Diptera (Smith 1966, Singh 1977, Singh and Moore 1985, Cohen 2004, Schnider 2009).

Since the 1960s, ladybugs (*Hippodamia convergens* Guérin-Méneville) and praying mantids have been available for sale in the USA to be released in gardens and nurseries (De Bach and Hagen 1964). However, it was not until the 1970s that this industry experienced significant development, greatly increasing in the 1980s, although Ridgway and Vinson (1977) already listed 95 insect-selling companies in the USA and Canada in the previous decade. In 1992, 24 species of predatory mites, 8 species of mealybug parasitoids, 15 species of lepidopteran parasitoids, 12 species of fly predators and parasitoids, 4 species of entomopathogenic nematodes, and 43 species of other beneficial organisms were available for sale (Hunter 1994). Cranshaw et al. (1996) listed the main species available for sale in the USA with their unit prices, noting that for large quantities, the prices could be dramatically reduced. Hunter (1997) reported that 130 species were available in the market in North America (USA, Canada, and Mexico), produced by 142 companies. Tauber et al. (2000) reported that *Chrysoperla* spp. was the top-selling predator in the USA, representing 44% of the total numbers sold and utilized. Warner and Getz (2008) reported that currently there are 22 insect- rearing facilities in the USA, producing 38 natural-enemy species. The number of such facilities is decreasing in North America, as discussed by the authors. A large concentration of companies is in California (50%), while 37% are in other states and 13% are in Canada. The natural-enemy market in the USA represents an 8-million dollar revenue. It consists of small companies. Only 5 of the 22 facilities produce more than three species of natural enemies, and all of them combined represent fewer than 200 jobs.

In Europe, according to van Lenteren and Woets (1988), the natural-enemy trade began in 1968 with the use of *Phytoseiulus persimilis* Athias-Henriot to control tetranychids. European countries frequently use biological control in greenhouses; natural enemies were released on about 14,000 ha, as compared to 200 ha covered in 1970 (van Lenteren 2000). In 1997, according to van Lenteren et al. (2003), approximately 26 companies produced over 80 species of natural enemies, about 29 of these in high demand. van Lenteren (2003) reported that about 125 species are produced and sold worldwide, which indicates significant growth, apparently differing from the situation in North America. Today, more than 170 natural enemy species are on the market for biological control in the world (Cock et al. 2010), and Europe is the most important market (75% of the market value) (Bolckmans 2008).

Surveying the number of companies that sell natural enemies, although their names are often available on the Internet, is not a reliable procedure because these are small companies, often without continuity, that emerge only to disappear shortly afterwards. Some very solid companies do exist, such as Rincón-Vitova (USA), Koppert (Netherlands), Biotop (France), Biobest (Belgium), Bioplanet (Italy), and Biobee (Israel), among others. In these lists, a predominance of American and

Canadian companies is common, together with European ones, but companies also exist in Asia, India, Australia, New Zealand, and Latin America.

Most natural enemies sold are parasitoids and predators, even though many companies also produce microbial agents such as nematodes, entomopathogenic fungi, bacteria, and viruses. Bacteria are often produced by larger companies (pesticide industries) because of the difficulty of quality control.

With regard to parasitoids that attack the eggs of plant pests, the focus of this chapter, *Trichogramma evanescens* Westwood, *T. brassicae* Bezdenko, *T. cacoeciae* Marchal, and *T. dendrolimi* Matsumura are sold in Europe to control lepidopterans in greenhouses, *Ostrinia nubilalis* Hübner in corn, and other lepidopterans in orchards (van Lenteren 2003). In the USA, *T. pretiosum* Riley and *T. platneri* Nagarkatti are sold to control several lepidopterans (the first 2 species), while *T. minutum* is used in orchards and *T. platneri* is used against avocado pests (Cranshaw et al. 1996).

14.2.2 New World

van Lenteren and Bueno (2003) reviewed the use of augmentative biological control in Latin America. The following egg parasitoids are reared and sold in the New World:

Trichogrammatidae: *Trichogramma galloi* Zucchi, *T. atopovirilia* Oatman and Platner, *T. pretiosum* Riley, *T. fuentesi* Torre, *T. nerudai* Pintureau and Gerding, *T. exiguum* Pinto and Platner, *T. pintoi* Voegelé, and *T. platneri* Nagarkatti;

Scelionidae: *Telenomus remus* Nixon and *Trissolcus basalis* Wollaston;

Encyrtidae: *Ageniaspis citricola* Logvinovskaya.

14.2.2.1 Brazil

Parra et al. (2002) described the most important studies on parasitoids and predators in Brazil done during the past 30 years. Important applied biological control programs have been developed in this period. Three examples can be mentioned. The first is the release of the exotic parasitoid *Cotesia flavipes* (Cam.) to control *Diatraea saccharalis* (Fabr.). *Cotesia flavipes* is released annually in areas equivalent to 1,700,000 ha (Botelho and Macedo 2002). The second case relates to the application of *Metarhizium anisopliae* on 1,000,000 ha to control the spittlebugs *Mahanarva fimbriolata* and *M. posticata* (Hemiptera: Cercopidae) in sugarcane (Alves and Lopes 2008). The third case is the use of *Baculovirus anticarsia* Hubner in soybeans, which during the 1980s and 1990s was applied to 2 million ha (Moscardi 1999).

Several studies have been conducted on *Trichogramma* spp. in recent years in Brazil (Parra and Zucchi 1997). Their use has increased, especially for the control of *Diatraea saccharalis* in sugarcane, in areas where egg predation is low or in areas where *C. flavipes* has not provided good control, mainly because of climatic adaptation problems (in newly planted areas). Releases are carried out on 300,000 ha of sugarcane with the species *T. galloi*, produced on the factitious host

Anagasta kuehniella (Zeller) [or *Corcyra cephalonica* (Stainton)]. In 2009, this area is expected to increase to 500,000 ha. The same species of *Trichogramma* is beginning to be used to control *D. saccharalis* in corn. *T. atopovirilia* is already employed to control *Spodoptera frugiperda* (J. E. Smith) in corn, and *T. pretiosum* to control *Tuta absoluta* (Meyrick) in tomatoes. These uses amount to more than 50,000 ha, with over 200,000 ha "treated" with *Trichogramma* spp. in 2007–2008.

Trissolcus basalis is used on approximately 20,000 ha of organic soybean or in micro-basin areas to control soybean stink bugs (Heteroptera: Pentatomidae) (Corrêa-Ferreira 2002); in this case, they are produced by cooperatives or producer associations, not by private companies. There are rearing problems related to increased soybean stink bug populations, particularly *Nezara viridula* (L.): laboratory populations of these heteropterans degenerate after a few generations, requiring new infusions from wild populations to maintain their genetic variability.

In Brazil, two companies presently produce egg parasitoids; only one of them produces *T. galloi*, together with *T. pretiosum* and *T. atopovirilia*. Many laboratories that produce *C. flavipes* are part of sugar and ethanol mills (Fig. 14.1).

14.2.2.2 Colombia

The companies that produce natural enemies are located in the Valle del Cauca. *Trichogramma* was released in the past on 200,000 ha in cotton, soybean, cassava (manioc), tomato, sorghum, and sugarcane areas. With the invasion of the boll

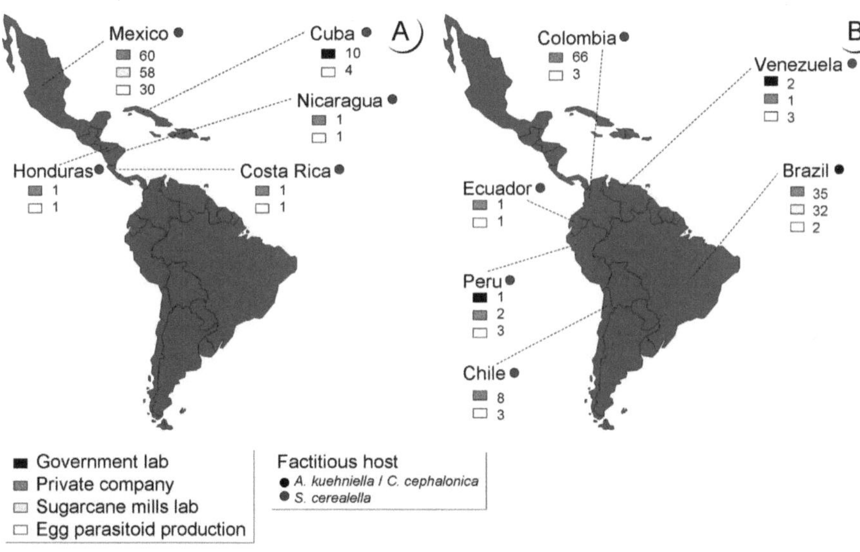

Fig. 14.1 Number of laboratories that sell natural enemies of crop pests in the New World. (**a**) Chile, Peru, Ecuador, Colombia, Venezuela, and Brazil; (**b**) Honduras, Mexico, Cuba, Nicaragua, and Costa Rica

weevil, *Anthonomus grandis* Boh., the area where releases were made in cotton decreased from 30,000 to 5,000 ha in recent years.

An association of *Trichogramma* and larval parasitoids such as *Lydella minense* (Townsend), and *Paratheresia claripalpis* Wulp. [*sic*] is employed to control *D. saccharalis* on 130,000 ha (van Lenteren and Bueno 2003). Colombians have many parasitoids that control the coffee berry borer, *Hypothenemus hampei* (Ferr.), as well as flowers and ornamentals (Garcia 1996), in addition to larval parasitoids for the control of *Tuta absoluta* in tomatoes.

14.2.2.3 Mexico

In 2007, Mexico had 60 natural-enemy rearing laboratories, and reared 37 species, of which 70% were entomophagous and 30% entomopathogens. Fifty per cent of the Mexican companies produce *Trichogramma,* a parasitoid used for over 40 years in that country [including the species *T. exiguum, T. pintoi, T. pretiosum*, and *T. platneri* against *D. saccharalis, Plutella xylostella* L., lepidopterans in general, and species of *Diatraea* and *Cydia pomonella* (L.), respectively]. A single laboratory produces *Telenomus remus* to control *S. frugiperda.*

14.2.2.4 Cuba

Trichogramma species are released on approximately 685,000 ha to control lepidopterans in pastures, cassava, sugarcane, and vegetables.

14.2.2.5 Ecuador

Trichogramma (local species) is used to control sugarcane pests.

14.2.2.6 Venezuela

Telenomus remus is used in association with *T. pretiosum* for the control of *Spodoptera frugiperda* on approximately 1,600 ha. *Cotesia flavipes* is also produced to control *D. saccharalis* on 40,000 ha.

14.2.2.7 Peru

Peru has a long biological-control tradition (Altieri and Nicholls 1999). At one time it had 82 government-sponsored laboratories for the mass production of parasitoids and predators, and 27 laboratories for the production of pathogens (van Lenteren and Bueno 2003). The country has long produced *Trichogramma* spp. and *T. remus* to control pests in asparagus, sugarcane, rice, and corn.

Costa Rica and Guatemala use *Trichogramma* spp. to control sugarcane and cotton pests; in Guatemala, *Trichogramma* was once used on 14,000 ha of cotton. Other countries such as Uruguay, Argentina, Bolivia, Honduras, Nicaragua, and Panama have prospects for the use of egg parasitoids, although their use is still limited. A general overview of laboratories in the New World is presented in Fig. 14.1.

14.3 Reasons for the Limited Use of Biological Control (and Commercial Biological Products)

In spite of the progress in biological control worldwide, this alternative is little used compared with chemicals. There are several reasons for this, because a number of problems must be overcome before the product can reach the end user. Even in the case of conservation (natural biological control) of biological agents, studies are still needed on selectivity, maintenance and refuges, and detection of feeding sites for parasitoids and predators. In the case of classical (introduction) or applied (multiplication) biological control, insects must be reared: in the former case, on a smaller scale (inoculative releases); and in the second, on a larger scale (mass rearing) targeted to inundative releases (Parra 2002).

Whenever an insect is reared in the laboratory, some precautions must be taken, because several biotic and abiotic factors may interfere with rearing, changing the traits of the laboratory-reared insects in relation to wild ones (Table 14.1) (Singh

Table 14.1 Different factors that affect laboratory insects and wild (natural) insects (Singh and Moore 1985)

Factor	Laboratory	Nature
Temperature	Stable or periodical	Fluctuating
Light	Artificial, constant	Natural
Humidity	Stable or associated with temperature control	Fluctuating
Food	Provided artificially	Natural, freely sought
Shelter	Not provided or inadequate	Actively sought when needed
Predator-parasitoid	No pressure	Constant or fluctuating pressure
Competition	Absent	May be present and intense
Human presence	Intense and continuous exposure	Sporadic exposure
Micro-environment	Medium	Optimum is actively sought
Search for a mate	Easy to find	May require time, at the expense of energy
Mate acceptance	Promiscuous (forced mating)	Complex in terms of behavior
Egg-laying (site)	Artificial, restricted	Natural, free-choice
Dispersal	Restricted	Dependent on density or induced by the environment
Wind	Generally absent or with low velocity	Variable, but generally present
Chemicals	Concentrated pheromones, sometimes at small concentrations	Diffuse pheromones, some types of plants at high concentrations
Free water	One source (generally a sugar at a single concentration)	Various sources (variable concentrations and types)

and Moore 1985). Two species of insects must be reared: the natural host, i.e., the target pest (or, in a few cases such as *Trichogramma*, a factitious host), and the natural enemy. Although progress has been made in the area of in vitro production (rearing the natural enemy on an artificial diet) (Cônsoli and Parra 2002; see also Chapter 11), current knowledge is still insufficient for the large-scale production of insects or as a routine laboratory activity.

The process involves steps performed in the laboratory, semi-field, and the field. Particularly in the field, different ecological factors interact and regulate natural-enemy populations (exogenous and endogenous factors) (Table 14.2).

In Brazil, in many cases such as in *Trichogramma* (Parra and Zucchi 2004), there is a wealth of laboratory results and a scarcity of semi-field and field investigations for many crops. The reasons for the limited use of biological control can be listed as follows:

Tradition. Without question, a tradition exists in controlling pests with chemical products. Since the synthesis of DDT in 1939, there was a feeling that controlling pests with chemical products would be the final solution, which proved false over time. Brazil, differently from other countries of North America that had a strong biological-control influence from American universities such as Berkeley and Riverside, began entomological research with a focus on agrochemicals.

Specificity. Specificity of biological products. In a crop that is subject to several pests, the use of various biological-control agents may lead to higher costs, since the grower may choose to use a broad-spectrum insecticide and kill all the pests at once. On the other hand, in a crop where three pests occur, three biological agents (parasitoids, for example) should be used.

Credibility. Because the culture of chemical control is deeply rooted in Brazil, there is little information and a complete lack of knowledge on biological control. Therefore, if biological control is to be used, the grower must be educated about what is taking place by means of demonstration fields in which he or she can learn how parasitism occurs, what are the characteristics of the natural enemy, etc. Only then will credibility be developed, since growers expect pests to "drop dead", which is how the efficiency of a control method is evaluated in chemical control.

Technological knowledge. Without doubt, this is a major hindrance to the widespread application of biological control. There are large gaps in knowledge with regard to what biological control is all about and how it should be applied. Besides the need to develop a biological-control program over the long term until it finally reaches the farmer, such a program requires detailed knowledge about the pest's bioecology on the part of the user, since egg parasitoids must be released when

Table 14.2 Exogenous and endogenous factors that interact in the regulation of natural-enemy populations (Hajek 2004)

Exogenous	Endogenous
Natural enemies	Sex and age
Food supply	Physiology
Climate	Behavior
Shelter	Genetics

the eggs of the target insect are present in the field. The same is true for parasitoids that attack nymphs, caterpillars, pupae, or adults.

Therefore, according to Parra et al. (2002), the following steps occur in an applied biological-control program involving *Trichogramma*, an egg parasitoid. These steps are also valid for parasitoids that attack insects at other stages of development. They are:

- collection, identification, and maintenance of *Trichogramma* spp. strains;
- selection of a factitious host for the mass rearing of the parasitoid;
- biological and behavioral aspects of *Trichogramma* spp.;
- egg dynamics of the target pest;
- parasitoid releases: number of released parasitoids and release points; season and form of release;
- selectivity of agrochemicals;
- efficiency evaluation;
- parasitoid/pest simulation model.

Availability (commercialization) of the biological input. Differently from chemical products that are formulated and packaged, biological products must be released alive. Therefore, the number of companies that sell natural enemies of plant pests, such as those in countries of Europe, the USA, Australia, Canada, etc., should be increased. In Brazil, the first companies of this kind are beginning to appear (Parra 2004). However, differently from countries in Europe, where the trade is strong, in Brazil the distances are very large, and, without planning, biological products may arrive at the application site in an unsuitable condition for use.

Quality. For biological control to enjoy credibility, the biological product should be a "quality" product, that is, it must be "competitive" in relation to the insects in nature (van Lenteren 2003). There have been cases in Brazil where biological control was no longer used because of poor product quality, sometimes produced by small companies that could not provide adequate quality control. This situation was common in the early 1970s with the production of the fungus *M. anisopliae*. In order to preserve quality, the quality control should be performed by public institutions (universities and/or research institutions). Laws in this regard are still incipient in Brazil, although scientific societies and environmental institutions are engaged in creating a law dealing specifically with biological input.

Technology transfer. This issue exists worldwide, and congresses are held about the best way to transfer this technology. However, many studies are disseminated only among researchers, and never reach the farmer. A sequence of steps should be followed in the case of applied biological control (Fig. 14.2).

Selectivity. Biological control can operate side by side with chemical control in crops with a large number of pests. However, this case, requires the application of *selective products* that will not harm the natural enemies released or those that exist naturally. There is a lack of lists of selective products for the various crops. The application of non-selective products or products with persistent residual action

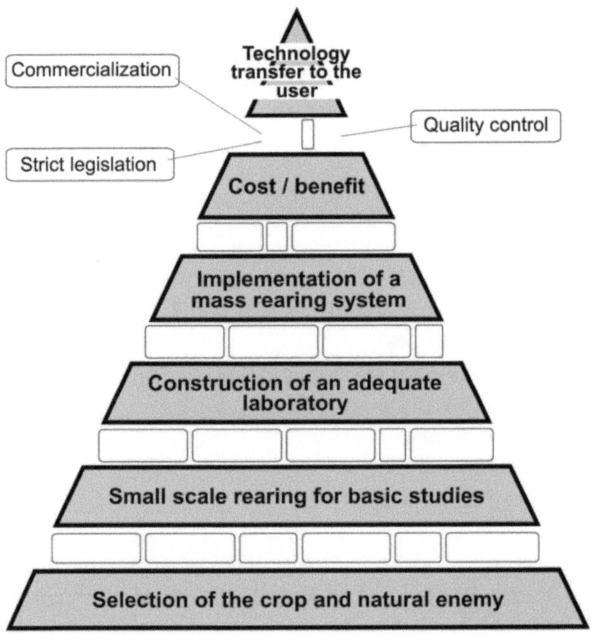

Fig. 14.2 Sequence of steps for the success of an applied biological control program (Parra et al. 2002)

after the natural enemies are released can limit the success of biological control programs.

Timeliness. Users are accustomed to seeing the "shock" effect of many chemical products, such as pyrethroids. In this case, regardless of the pest's population level, the products are effective. With biological agents, because of their slower action (even with applied biological control), the product should be released or used when pest populations are still at low levels.

Predation. If the release system is inadequate, that is, when the parasitoids or predators are exposed, unsuccessful releases can certainly occur. In tropical countries such as Brazil, this problem is aggravated, especially because of the action of ants, although other predators such as earwigs, chrysopids, ladybugs, etc. can also be important. Release methods involving capsules, which prevent the action of predators, or even the release of adults instead of pupae or other stages of development, can minimize predation.

Release techniques. Among the various steps of a biological control program, *the form of release*, the *number of individuals released per hectare* (parasitoids and predators), as well as the *interval between releases*, in addition to the *number of release points*, are essential to achieve a control level comparable to conventional chemical methods. The negative results obtained with *Trichogramma* in the USA in the period from 1950 to 1970 resulted from the lack of consideration of those factors.

Cost/benefit. Agriculture is an investment, and as such, investors expect returns. Therefore, if no *immediate* advantage or even an *equivalence* with the *cost of application and the investment return* is achieved, the farmer will hardly choose a biological alternative, even when he or she is very sensitive to environmental issues.

Selection of lines (strains). Some parasitoids, like *Trichogramma*, are nonspecific and attack eggs of a large number of lepidopteran species. When *Trichogramma* was initially employed, species of this parasitoid would be collected from a particular (sometimes cold) region and released in areas with a different climate (sometimes hot). Today it is known that specificities do exist with respect to microclimates; within the same species, strains with different climatic adaptations must be separated. These strains should be maintained in collections to be multiplied and then released into the same areas where they were collected. Preference should also be given to *Trichogramma* collected in eggs of the target pest species (Parra and Zucchi 1997, Parra and Zucchi 2004).

Ecological factors. Among ecological factors, climate (rainfall, relative humidity, temperature, wind, air pressure, etc.) constitutes the greatest obstacle in biological control programs (34.5%), followed by competition with the native fauna (20.3%) and lack of food (releases at unsuitable seasons) (16.9%) (Stiling 1993). Dispersal of the natural enemy into other areas and lack of refugia for the natural enemies could also be important causes of failure in the use of biological-control agents (Collier and van Steenwyk 2004).

In addition to all the reasons mentioned above, especially in the New World:

- there is a lack of basic studies dealing with the biology, physiology, nutrition, host/natural enemy relations, symbionts, and environmental impact analyses for parasitoids. In general, rearings are started, sometimes at a large scale (mass rearings), without sufficient knowledge about the pest and the natural enemy;
- programs lack continuity;
- projects are poorly planned and often represent individual efforts (without interdisciplinary or multidisciplinary characteristics);
- no national policy has been established with respect to defining priorities or investments in the industry.

14.4 Current Needs

Although some cases of commercial success in trading biological-control agents do exist, some points must be taken into consideration to foster this sector, which is essential in integrated pest-management programs, especially in developing countries.

Without question, in order to increase the use of biological control, it is essential to consolidate a biological-control culture. Such a consolidation will only occur if this control measure becomes *credible*.

Some of these requirements are:

- development of basic studies about the host (either natural or facultative) and the natural enemies, including risk analysis and environmental-impact studies;
- an increased number of laboratories working on insect- rearing techniques;
- a closer relationship between private companies and universities and research institutions;
- relate the country's economy to a definition of priorities; rear several natural-enemy species to withstand difficult economic periods;
- sell natural enemies in situations where they will be able to compete with chemical products;
- quality control of the natural enemies produced. Such control should be accomplished in cooperation with universities and research institutions to provide credibility. In Europe, this function is performed by the IOBC and by biological-control companies; in the USA by the American Society for Testing and Materials (ASTM); and in Colombia by governmental research institutions. In Brazil, the foundation of the Brazilian Association for Biological Control (*Associação Brasileira de Controle Biológico*) was an initiative of private companies. Standards have been defined by the IOBC for several species of natural enemies (van Lenteren et al. 2003, Leppla 2003, Prezotti et al. 2004).
- training of growers in using the biological input, by means of demonstration fields;
- production scale adjustment from the research laboratory to the company. Sometimes long periods are required to go from a laboratory production of 20 g of *Anagasta kuehniella* eggs (720,000 eggs) in research, to a daily yield of 2–3 kg (72,000,000–108,000,000 eggs) in the company;
- storage techniques for periods during which the produced insect is not used [diapause (Pizzol 1978), liquid nitrogen (Ma Huai-Yi 1988 and Corrêa-Ferreira and Oliveira 1998), low temperatures (Drooz 1981), etc.];
- government support for starting small companies;
- improve the quality of qualified personnel working on rearing techniques, including studies on temperature requirements (Insect Rearing Management) to make production forecasts, preventing the appearance of incompetent firms;
- strict laws to maintain the quality of the biological input produced;
- logistical studies on distribution and transport, especially in large countries such as Brazil.

14.5 Concluding Remarks

The commercialization of natural enemies of crop pests is a reality in our modern world, and growers are increasingly in search of *high-quality* natural enemies to replace conventional agrochemicals. On the other hand, biological-control organisms are often sold by small companies that cannot compare with larger companies that produce insecticides, fungicides, or herbicides. Such small companies are often unaware about insect rearing techniques, or are not associated with universities or research institutions to keep abreast of new developments in this area. In order to rear natural enemies, two species of insects must always be reared, that is, the natural host (or an factitious host in a few cases, such as *Trichogramma*), and the natural enemy, since in vitro rearing did not result in significant progress as expected during the 1980s and the 1990s (see Chapter 11).

In Brazil, in the past 40 years, a reasonable critical mass of researchers has been trained in the area of biological control: of the 2,000 researchers who obtained post-graduate degrees, about 25% (including MSc and PhD degrees) specialized in this area.

Business management systems are often inadequate, which causes many companies to go out of business. The majority of companies only rears a few species, and is very vulnerable, depending on the occurrence of the pest and on the economic characteristics of a particular period.

This is still a small, emerging market compared with chemicals, with an annual revenue of only 350 million dollars, with expectations to double or triple in the beginning of this century (van Lenteren 2003). This market is also quite variable from country to country, as a function of economic and social characteristics. In Brazil and the New World, in general, labor is cheaper: one release of *T. galloi* to control *D. saccharalis* in sugarcane costs around 14 dollars, including release of the parasitoid in the field, a value that is much lower than costs in Europe and the USA. On the other hand, this market is unquestionably a place for emergent labor. In Brazil, it is common for companies that produce *Cotesia flavipes* to have up to 100 employees, a number equivalent to those of large companies in Europe (van Lenteren 2003).

The use of egg parasitoids is higher in the New World than in developed countries, especially for *Trichogramma* spp. In Brazil, the results are quite interesting, as well as in Colombia, Mexico, and other countries of Latin America. Good results have been obtained with *Trissolcus basalis* in soybeans and *Telenomus remus* in corn (against *S. frugiperda*). *Ageniaspis citricola* Logvinovskaya, an imported parasitoid, has provided excellent results in the control of *Phyllocnistis citrella* Stainton, the citrus leafminer (Parra et al. 2004). In light of these successes, the prospects for increases in release area are substantial, especially if we consider that the sugarcane area in Brazil for ethanol production is expected to double in a short period of time (from 7 to 14 million ha). Therefore, the prospects for the use of *T. galloi* are significant, exceeding the 500,000 ha expected for 2009–2010.

Consequently, both new and existing companies must meet several requirements to achieve credibility, and, above all, evaluate the characteristics of the market. In addition to emphasizing ethical aspects (Hoy et al. 1991), companies must have a thorough knowledge of the subject, including not only the technology for the production of the natural enemy but also the market for selling the input for the crop in which the parasitoid or predator will be used. Evaluating the quality of the insects produced should be particularly important. As in other countries, monitoring all the steps involved, including production, processing, and the final product are equally important activities (Penn et al. 1998, Leppla 2003, van Lenteren 2003).

Therefore, in order to start the commercialization of natural enemies in the right manner, avoiding the pitfalls incurred by other countries, joint action is essential among representatives of academia, government, and industry in monitoring the various steps, since aspects such as quarantine, risk evaluation, quality control, and definition of priorities (selection of natural enemies) should be carried out after a mutual discussion. Regulations on monitoring of the quality of insects produced by universities or research Institutions are indispensable (Cook et al. 1998). Even in the USA, a country with a well-established tradition in this industry, deviations frequently occur in the sex ratio of insects reared by private companies (Heimpel and Lundgren 2000). It is common in studies involving *Trichogramma* that species different from those intended are released, because of their minute size and taxonomic difficulties.

Consequently and above all, it is necessary to think through an appropriate manner of transferring technology (one of the major problems in this sector, given the low level of farmer education in developing countries) that would be compatible with the level of the receptor (this could be done by means of demonstration fields). Efficiency should be also monitored, including the number of release points, number of released parasitoids, form of release, etc. Only by adopting these precautions in the implementation of companies that specialize in selling natural enemies will this alternative become better accepted and incorporated into integrated pest-management programs for other crops. This is in addition to the points cited in the present text, since augmentative biological control is a reality in the modern world (van Lenteren 2006), despite some criticism, not always coherent, of the success of this control measure (Collier and van Steenwyk 2004). The biodiversity of tropical regions is enormous and should be exploited, although some questions have not been completely clarified, such as those related to the behavior of natural enemies as areas cultivated with genetically modified plants increase (see Chapter 17).

Acknowledgments To Dr. Marcos Gerding (Chile), Dr. F R Ferrer Wurst (Venezuela), Dr. César Basso (Uruguay), Dr. Katherine G. Pérez (Colombia), Dr. Hugo C A Bernal (Mexico), Dr. Rosmarina Marin L (Peru), Dr. L C Almeida (CTC – Brazil), and Dr. E. Botto (Argentina) for their valuable information on the use of egg parasitoids in biological control in their respective countries.

References

Altieri MA, Nicholls CI (1999) Classical biological control in Latin America. In: Bellows TS, Fisher T W (eds) Handbook of biological control. Academic, San Diego, CA, 1454p, pp 975–991

Alves SB, Lopes RB (2008) Controle Microbiano de Pragas na América Latina. Avanços e desafios. FEALQ, Piracicaba, Brasil. 414p

Bolckmans KJF (2008) De insectenfrabriek. In: Osse J, Schoonhoven L, Dicke M, Buiter R (eds) Natuur als bondgenoot: biologische bestrijding van ziekten en plagen. Bio-Wetenschappen en Maatschappij, Den Haag, 80p, pp 51–52

Botelho PSM, Macedo N (2002) *Cotesia flavipes* para o controle de *Diatraea saccharalis*. In Parra JRP, Botelho PSM, Corrêa-Ferreira BS, Bento JMS (eds) Controle biológico no Brasil: parasitóides e predadores. Editora Manole, São Paulo, 609p, pp 409–425

Cock MJW, van Lenteren, JC, Brodeur J, Barratt BIP, Bigler F, Bolckmans K, Cônsoli FL, Hass F, Mason PG, Parra JRP (2010). Do new access and benefit sharing procedures under the convention on biological diversity threaten the future of biological control? BioControl 55:199–218

Cohen AC (2004). Insect diets. Science and technology. CRC Press, Boca Raton. 324p

Collier T, van Steenwyk R (2004) A critical evaluation of augmentative biological control. Biological Control 31:245–256

Cônsoli FL, Parra JRP (2002) Criação *in vitro* de parasitóides e predadores. In: Parra JRP, Botelho PSM, Corrêa-Ferreira BS, Bento JMS (eds) Controle biológico no Brasil: parasitóides e predadores. Editora Manole, São Paulo. 609p, pp 239–275

Cook FR, Inscoe MN, Ridgway RL (1998) Marketed natural enemies: regulatory issues. In: Ridgway RL, Hoffmann MP, Inscoe MN, Glenister CS (eds) Mass-reared natural enemies: application, regulation and needs. Tomas Say Publications in Entomology, Lanham. 332p, pp 231–241

Corrêa-Ferreira BS (2002) *Trissolcus basalis* para o controle de percevejos da soja. In: Parra JRP, Botelho PSM, Corrêa-Ferreira BS, Bento JMS (eds), Controle biológico no Brasil: parasitóides e predadores. Ed. Manole, São Paulo. 609p, pp 449–476

Corrêa-Ferreira BS, Oliveira MCN (1998) Viability of *Nezara viridula* (L.) eggs for parasitism by *Trissolcus basalis* (Woll.) under different storage techniques in liquid nitrogen. Anais da Sociedade Entomológica do Brasil 27:101–107

Cranshaw W, Casey Sclar D, Cooper D (1996) A review of 1994 pricing and marketing by suppliers of organisms for biological control of arthropods in the United States. Biol Control 6:291–296

De Bach P, Hagen KS (1964) Manipulation of entomophagous species. In: De Bach P, Schlinger EI (eds) Biological control of insect pests and weeds. Chapman and Hall, London. 844p, pp 429–458

Drooz AT (1981) Subfreezing eggs of *Lambdina pellucidaria* (Lepidoptera: Geometridae) alters status as factitious host for *Ooencyrtus ennomophagus* (Hymenoptera: Encyrtidae). Can Entomol 113:775–776

Garcia F (1996) El Control Biológico Aplicado en Colombia. In: Zapater MC (ed) El Control Biológico en América Latina. IOBC, Buenos Aires, pp 31–33

Guillon M (2008) Current world situation on acceptance and marketing of biological control agents (BCAS). mg.pres.ibma@club-internet.fr: www.ibma.ch. Accessed 11 Jan 2010

Hajek A (2004) Natural enemies: an introduction to biological control. Cambridge University Press, Cambridge, 378p

Heimpel GE, Lundgren JG (2000) Sex ratios of commercially reared biological control agents. Biol Control 19:77–93

Hoy MA, Nowierski RM, Johnson MW, Flexner JL (1991) Issues and ethics in commercial releases of arthropod natural enemies. Am Entomol 37:74–75

Hunter CD (1994) Suppliers of beneficial organisms in North America. Sacramento, California Environmental Protection Agency, 30p

Hunter CD (1997) Suppliers of beneficial organisms in North America. California Environmental Protection Agency, Sacramento. (Available at http://www.cdpr.ca.gov/docs/pestmgt/ipminov/bensuppl.htm) (date of last access: 03/03/2010 – list last uptaded in 1997)

Kogan M (1998) Integrated pest management: historical perspectives and contemporary developments. Annu Rev Entomol 43:243–270

Leppla NC (2003) Aspects of total quality control for the production of natural enemies. In van Lenteren (ed), Quality control and production of biological control agents: Theory and testing procedures. CABI Publishing, Cambridge, 327p, pp 19–24

Ma Huai-Yi (1988) Studies on long-term storage of hosts for propagating *Trichogramma*. Les Colloques de I'NRA 43:369–371

Moscardi F (1999) Assessment of the application of baculoviruses for control of Lepidoptera. Annu Rev Entomol 44:257–289

Parra JRP (2002) Criação massal de inimigos naturais. In: Parra JRP, Botelho PSM, Corrêa-Ferreira BS, Bento JMS (eds) Controle Biológico no Brasil: parasitóides e predadores. Editora Manole, São Paulo. 609p, pp 143–164

Parra JRP (2004) Controlando pragas com inimigos naturais. Ciência Hoje 210:18–23
Parra JRP, Zucchi RA (2004) *Trichogramma* in Brazil: feasibility of use after twenty years of research. Neotrop Entomol 33:271–284
Parra JRP, Bento JMS, Chagas MCM, Yamamoto PT (2004) O controle biológico da larva-minadora-dos-citros. Visão Agrícola 2:64–67
Parra JRP, Zucchi RA (1997) *Trichogramma* e o controle biológico aplicado. FEALQ, Piracicaba, Brasil. 324p
Parra JRP, Zucchi RA (2008) Utilização de *Trichogramma* no Brasil: situação atual e desafios. In: Venzon M, Paula TJ Jr, Pallini A (eds.) Avanços no controle alternativo de pragas e doenças. EPAMIG-UFV, Viçosa, 283p, pp 1–29
Parra JRP, Botelho PSM, Corrêa-Ferreira BS, Bento JMS (2002) Controle Biológico no Brasil: parasitóides e predadores. Editora Manole, São Paulo. 609p
Penn SL, Ridgway RL, Scriven GT, Inscoe MN (1998) Quality assurance by the commercial producer of arthropod natural enemies. In: Ridgway RL, Hoffman MP, Inscoe MN, Glenister CS (eds) Mass-reared natural enemies: application, regulation, and needs. Thomas Say Publications in Entomology, Lanham, 332p, pp 202–230
Pizzol J (1978) La diapause chez *Trichogramma evanescens* Westwood (Hymenoptera: Trichogrammatidae), ecotype Moldave, parasite oophage de la pyrale du mais. Institut de Montpellier (France) 75p. (Diplôme présenté pour obtenir le grade d'Elève diplomé)
Prezotti L, Parra JRP, Vencovsky R, Coelho ASG, Cruz I (2004) Effect of the size of the founder population on the quality of sexual populations of *Trichogramma pretiosum*, in laboratory. Biol Control 30:174–180
Ridgway RL and Vinson SB (1977) Biological control by augmentation of natural enemies. Plenum Press, New York, 480p
Schneider JC (ed) (2009) Principles and procedures for rearing high quality insects. Mississipi State University, Mississipi, 352p
Singh P (1977) Artificial diets for insects, mites and spiders. Plenum Press, New York, 594p
Singh P, Moore RF (1985) Handbook of insect rearing, vol. 2. Elsevier, Amsterdam, 514p
Smith CN (1966) Insect colonization and mass production. Academic, New York, 618p
Stiling P (1993) Why do natural enemies fail in classical biological control programs? Am Entomol 39:31–37
Tauber MJ, Tauber CA, Daane KM, Hagen KS (2000) Commercialization of predators: recent lessons from green lacewings (Neuroptera: Chysopidae: *Chrysoperla*). Am Entomol 64:26–38
van Lenteren JC (2000) Critérios de seleção de inimigos naturais a serem usados em programas de controle biológico. In: Bueno VHP (ed) Controle biológico de pragas: produção massal e controle de qualidade. Editora UFLA, Lavras, 196p, pp 1–19
van Lenteren JC (2003) Quality control and production of biological control agents: theory and testing procedures. CABI Publishing, Cambridge, 327p
van Lenteren JC (2006) How not to evaluate augmentative biological control. Biol Control 39: 115–118
van Lenteren JC, Bueno VHP (2003) Augmentative biological control of arthropods in Latin America. BioControl 48:123–139
van Lenteren JC, Woetz J (1988) Biological and integrated control in greenhouses. Annu Rev Entomol 33:239–269
van Lenteren JC, Roskam MC, Timmer R (1997) Commercial mass production and pricing of organisms for biological control of pests in Europe. Biol Control 10:143–149
van Lenteren JC, Hale A, Klapwijk JN, van Schelt J, Steinberg S (2003) Guidelines for quality control of commercially produced natural enemies. In: van Lenteren JC (ed) Quality control and production of biological control agents: theory and testing procedures. CABI Publishing, Cambridge, 327p, pp 265–303
Warner KD, Getz C (2008) A socio-economic analysis of the North American commercial natural enemy industry and implications for augmentative biological control. Biol Control 45:1–10

Chapter 15
Egg Parasitoids in Biological Control and Integrated Pest Management

Nick Mills

Contents

15.1 Introduction

Egg parasitoids, by definition, are parasitoids that both attack and complete their development within a host egg (Mills 1994a). They may be either solitary or gregarious, but in all cases prevent the host egg from hatching and use only a single host individual to complete their development. This distinguishes true egg parasitoids from other guilds of parasitic Hymenoptera such as egg-pre-pupal parasitoids that attack the host egg but delay development to kill the host just before pupation (Mills 1994a), and egg predators that consume multiple eggs within an ovisac or egg pod (Quicke 1997).

N. Mills (✉)
Department of Environmental Science, Policy and Management, University of California, Berkeley, CA 94720-3114, USA
e-mail: nmills@berkeley.edu

F.L. Consoli et al. (eds.), *Egg Parasitoids in Agroecosystems with Emphasis on Trichogramma*, Progress in Biological Control 9, DOI 10.1007/978-1-4020-9110-0_15,

True egg parasitoids have been used in the biological control of agricultural pests for over a century, with most of this activity focusing on inundative releases of *Trichogramma* species in both annual and perennial crops. Historically, the most important development in the use of egg parasitoid in pest management was the realization that *Trichogramma* species could be mass reared in large numbers using the eggs of stored product moths as alternative hosts. This recognition stemmed largely from Flanders' (1930) initial studies on mass production of *Trichogramma minutum* Riley on eggs of the Angoumois grain moth *Sitotroga cerealella* (Oliver). The very short generation times of *Trichogramma* species, and the relative ease with which they can be mass reared on hosts that feed on stored food products, without the need for living plants, are perhaps the two most important characteristics that have subsequently lead to the commercial production of egg parasitoids as biological control agents in many different parts of the world (Hassan 1993, Smith 1996, Parra and Zucchi 2004).

Another key attribute of egg parasitoids that has captured attention in their development as biological control agents is the fact that they not only attack host eggs, but also kill the host in the egg stage. The notion of killing a pest before it can cause damage to the crop is intuitively appealing and has also been an important contributing factor in the development of egg parasitoids in biological control and integrated pest management (Hassan 1993). While this is of particular concern in the context of inundative biological control, the aim of which is to achieve more immediate and localized pest suppression, it is far less relevant in the context of classical biological control which aims to provide sustained long-term suppression of an invasive pest.

While much of the emphasis on egg parasitoids as biological control agents has been on inundative biological control, there are also a number of examples of the very successful use of egg parasitoids as classical biological control introductions. In addition, a number of studies have addressed ways in which the crop environment can be manipulated to enhance the activity of egg parasitoids through conservation biological control. In this chapter we will focus on the use of egg parasitoids in classical and augmentative biological control, and their integration with other control option in integrated pest management.

15.2 Egg Parasitoids in Classical Biological Control

Egg parasitoids that have been used in the classical biological control of agricultural insect pests belong to one of six families of the parasitic Hymenoptera, three of which are exclusively egg parasitoids (Mymaridae, Scelionidae and Trichogrammatidae), and four that include species that are egg parasitoids (Elasmidae, Encyrtidae, Eulophidae and Platygastridae). The BIOCAT database includes records of parasitoid introductions on a global scale from the earliest introductions through to about 1990 (Greathead and Greathead 1992). This database includes records for 91 egg parasitoid species that have been introduced for the control of 77 different pest species. There are many instances of multiple introductions of the same parasitoid species against the same pest species in different regions of the world, and so in analyzing this record, we remove the bias of repeat introductions by focusing on the single best outcome for each parasitoid species-pest species combination (Mills 1994a). Two important measures of success are the rate of establishment, the percentage of unique egg parasitoid – pest combinations that at least resulted in the establishment of the parasitoid, and the rate of impact, the percentage of the established egg-parasitoid – pest combinations that resulted in partial, substantial or complete control of the pest.

Of 157 unique combinations of egg parasitoid and pest species recorded as introductions in the BIOCAT database, only 33 have an unknown outcome, leaving 124 that have a known outcome that can be analyzed. Overall, for all parasitoid families and all orders of pest insects, the rate of establishment of egg parasitoids is 46% and the rate of impact from those that did establish is 28%, which combine to give an overall rate of success of 13%. For comparison, the complete record of parasitoid introductions for all parasitoid guilds (*sensu* Mills 1994b) that attack the full range of life stages of the target pests provides a rate of establishment of 38% and a rate of impact of 44%, generating a 17% overall rate of success. Thus egg parasitoids tend to establish a little more readily than parasitoids as a whole, but their potential to suppress the abundance of a pest appears to be rather more limited than for other parasitoid guilds.

Taking a closer look at the rates of establishment and impact of the different families of egg parasitoids and the different orders of pests that have been targeted for egg parasitoid introductions, we see that there is considerable variation among groups (Fig. 15.1). Parasitoid families (Fig. 15.1a) show significant variation in their rates of establishment (Fisher's exact test, $\chi^2 = 23.7$, df $= 5$, $P < 0.001$), but not in the rate of impact of those that were successfully established (Fisher's exact test, $\chi^2 = 1.15$, df $= 3$, $P = 0.76$, with Eulophidae and Eupelmidae excluded due to the small sample sizes and total absence of impact). Trichogrammatidae have been introduced more frequently than any other family of egg parasitoids and perhaps surprisingly, they have the lowest rate of establishment despite being polyphagous in most cases. While eupelmid egg parasitoids appear to have the greatest rate of establishment, the sample size is too small to have confidence in this outcome, leaving the Mymaridae as the family with greatest establishment rate. Finally, encyrtid, eulophid and scelionid egg parasitoids showed a very similar intermediate rate of establishment.

In the context of pests as targets for introductions of egg parasitoids (Fig. 15.1b), there is significant variation between the different orders of pests in the rate of establishment of egg parasitoids (Fisher's exact test, $\chi^2 = 10.24$, df $= 4$, $P = 0.03$), but again not in the their rate of impact (Fisher's exact test, $\chi^2 = 0.59$, df $= 4$, $P = 0.98$). While Orthoptera have been the subject of very few egg parasitoid introductions the rate of establishment appears to be equally as low as for the Lepidoptera which have been the subject of the greatest number of introductions. The rate of establishment has been greatest for the Homoptera corresponding closely with the rate of establishment of mymarid egg parasitoids. That of egg parasitoids introduced against Heteroptera and Coleoptera are then intermediate.

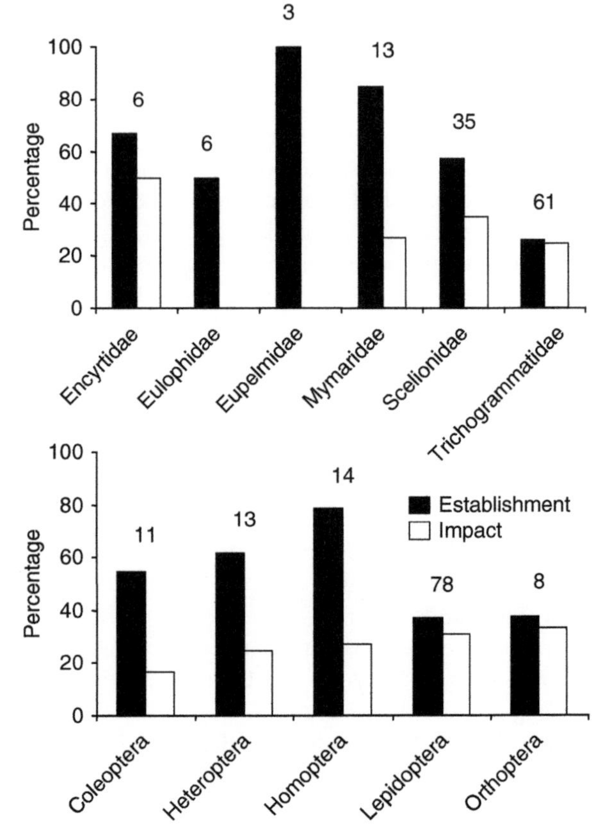

Fig. 15.1 The result of classical biological control introductions of egg parasitoids from the BIOCAT database summarized by (**a**) parasitoid family and (**b**) pest order. Results are presented for the single best outcome for each parasitoid species – pest species combination (after Mills 1994a), with percent establishment estimated from the number of unique combinations that at least established in relation to the total number of unique combinations with known outcome (inset numbers), and percent impact estimated as the number of unique combinations providing partial, substantial or complete control in relation to the number of unique combination that resulted in establishment

Examples of the success of classical introductions of egg parasitoids are considered for each parasitoid family below:

15.2.1 Eulophidae

There are four different agricultural pests for which eulophid egg parasitoids have been introduced as classical biological controls. *Ootetrastichus beatis* from Fiji and *Ootetrastichus formosanus* Timberlake from Taiwan, were introduced to Hawaii for control of the sugarcane leafhopper *Perkinsiella saccharicida* Kirkaldy, but failed to improve on the impact seen earlier from the mymarid egg parasitoid *Anagrus optabilis* (Perkins) (Pemberton 1964, Lai and Funasaki 1986, Funasaki et al. 1988). Another agricultural pest for which eulpohid egg parasitoids have been used is the taro leafhopper *Tarophagus proserpina* (Kirkaldy) in Hawaii (Pemberton 1964, Lai and Funasaki 1986, Funasaki et al. 1988). Two egg parasitoids *Ootetrastichus* sp. and *Ootetrastichus megameli* Fullaway were introduced from the Philippines, and although *O. megameli* became established it is not known to have contributed to the successful control of this pest by the mirid egg predator *Cyrtorhinus fulvus* Knight that was imported from the Philippines at the same time. A more recent example of the use of an eulophid egg parasitoid is that of *Edovum puttleri* Grissell from Columbia that has been introduced into the USA and Italy for control of the Colorado potato beetle *Leptinotarsa decemlineata* (Say) (Hare 1990, Colazza and Bin 1992). While this parasitoid can kill up to 80% of the eggs in an egg mass, it has no overwintering diapause, and so has failed to successfully establish. Nonetheless, it has shown good potential in the form of augmentative releases as a component of integrated pest management (Hamilton and Lashomb 1996). Other recent introductions include *Aprostocetus vaquitarum* (=*A. gala*) (Wolcott) and *Quadrastichus haitiensis* Gahan from the Caribbean for control of the citrus weevil *Diaprepes abbreviatus* (L.) in Florida (Hall et al. 2001). The first of these two eulophids is an egg predator rather than an egg parasitoid, and while both have become established in the southern part of the state it is too early to tell whether they will have a significant impact on the abundance of citrus weevil (Castillo et al. 2006, Ulmer et al. 2006, Jacas et al. 2008) (*see also* Chapter 13).

15.2.2 Elasmidae

The only example of classical biological control of an agricultural pest with an elasmid parasitoid is that of the coconut stick insect *Graeffea crouanii* (Le Guillou) in Tonga, Western Samoa and Tokelau using the egg parasitoids *Paranastatus verticalis* Eady and *Paranastatus nigriscutellatus* Eady from Fiji. While both parasitoids established in Western Samoa and the former species also established in Tonga, levels of parasitism have remained low despite observations of up to 80% parasitism in the region of origin (Waterhouse and Norris 1987).

15.2.3 Encyrtidae

The most successful example of classical biological control with an encyrtid egg parasitoid is that of the banana skipper *Erionota thrax* L. which was completely controlled in both Hawaii and Mauritius following introduction of *Ooencyrtus erinotae* from Malaysia and Thailand respectively (Lai and Funasaki 1986, Waterhouse and Norris 1989). Other examples of the use of encyrtid egg parasitoids in classical biological control include *Ooencyrtus malayensis* (Ferriere) that was introduced from Java for control of the coreid *Amblypelta cocophaga* China an indigenous pest of coconut with unknown results (Rao et al. 1971), *Ooencyrtus johnsoni* (Howard) from the USA that established in Hawaii, but not in Bermuda, for control of the pentatomid brassica pest *Murgantia histrionica* (Lai and Funasaki 1986, Funasaki et al. 1988), and *Ooencyrtus submetallicus* (Howard) and *Ooencyrtus trinidadensis* Crawford from the Caribbean that failed to establish in Hawaii for control of the green stinkbug *Nezara viridula* (L.) (Lai and Funasaki 1986).

15.2.4 Mymaridae

Mymarids have proved to be very successful egg parasitoids for use in the biological control of leafhoppers in agricultural crops, as well as the classic example of complete control of the eucalyptus weevil *Gonipterus scutellatus* (Gyllenhal) in plantations in southern Africa and many other parts of the world by *Anaphes nitens* (Girault) from Australia (Greathead 1971, Hanks et al. 2000). Three complete successes in agriculture have been the introduction of *Gonatocerus ashmeadi* Girault from southeastern USA for control of the glassy-winged sharpshooter *Homalodisca vitripennis* Germar a pest of perennial agricultural crops and ornamental plants in Tahiti (Grandgirard et al. 2008), *Anaphes optabilis* Perkins from Australia for control of the sugarcane leafhopper *Perkinsiella saccharicida* in Hawaii (Pemberton 1964, Lai and Funasaki 1986, Funasaki et al. 1988), and *Anagrus armatus* (Ashmead) from New Zealand for control of *Edwardsiana froggatti* (Baker) an apple pest in Australia (Wilson 1960). In the case of *G. ashmeadi*, parasitism of egg masses averaged 80–100% following its introduction to Tahiti in 2005 (Grandgirard et al. 2008). As a result, population densities of the glassy-winged sharpshooter declined by more than 90% between April, just before the first parasitoids were released, and December 2005 when parasitoids had become generally distributed around the island (Fig. 15.2).

Other mymarids that have successfully established, but with minimal impact, include four *Anaphes* species, *Anaphes calendrae* (Gahan) for control of the sugarcane weevil *Rhabdoscelus obscurus* (Boisduval) in Hawaii (Beardsley 2000), *Anaphes flavipes* Förster from Europe for control of the cereal leaf beetle *Oulema melanopus* (L.) in the USA (Lampert and Haynes 1985), *Anaphes frequens* Perkins

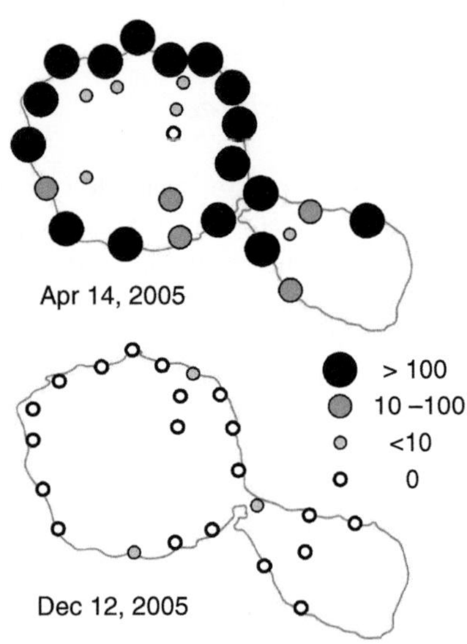

Fig. 15.2 Classical biological control of the glassy-winged sharpshooter *Homalodisca vitripennis* in Tahiti, showing the abundance of nymphs collected from 1 min sweep net samples from Hibiscus at 22 sites around the island, both in April, before the release of *Gonatocerus ashmeadi*, and in December, 7 months after the first parasitoid release (from Grandgirard et al. 2008)

from Australia for control of the sugarcane leafhopper *P. saccharicida* in Hawaii (Lai and Funasaki 1986), and *Anaphes yawi* Fullaway from Mexico for control of the mirid *Pycnoderes quadrimaculatus* Guérin-Méneville a pest of vegetable crops in Hawaii (Lai and Funasaki 1986). Similar results have been obtained with *Polynema striaticorne* Girault from the USA for control of the membracid *Ceresa bubalus* (Fabr.) a pest of fruit trees in Italy (Alma et al. 1987).

15.2.5 Platygastridae

Few platygastrid egg parasitoids are known to be associated with agricultural pests, as they are primarily parasitoids of beetles or Fulgoroidea. However, one current introduction is that of *Fidiobia dominica* Evans and Peña from the Caribbean into Florida for control of the citrus weevil *D. abbreviatus* (Jacas et al. 2008).

15.2.6 Scelionidae

One of the most well known examples of classical introductions of scelionid egg parasitoids is that of *Trissolcus basalis* (Wollaston) from Egypt for control of the green stink bug *Nezara viridula* in Australia, New Zealand and Hawaii (Clarke 1990, Jones 1995). While often heralded as a complete success (Clausen 1977, Caltagirone 1981, Waterhouse and Norris 1987), it appears that this vegetable pest remains a pest of annual concern in soybeans and other pulses in Australia (Loch and Walter 1999, Knight and Gurr 2007) and of macadamia nuts in Hawaii (Jones 1995). It is evident that this parasitoid has also had impacts on non-target pentatomid bugs both in Australia (Loch and Walter 1999) and Hawaii (Johnson et al. 2005). Although *Scelio* species have been introduced four times for control of orthopteran pests, only the introduction of *Scelio pembertoni* Timberlake from Malaysia to Hawaii for control of the grasshopper *Oxya chinensis* (Thunberg) in rice has been considered successful (Greathead 1992). Other scelionid egg parasitoids introduced against heteropteran pests include *Trissolcus painei* (Ferriere) from the Solomon islands that established on the coconut coreid *Axiagastus campbelli* Distant in Vanuatu (Cochereau 1972), *Trissolcus murgantiae* Ashmead from California that established on the brassica pentatomid *Murgantia histrionica* (Hahn) in Hawaii (Lai and Funasaki 1986, Funasaki et al. 1988), and the interesting example of a *Telenomus* sp. that was introduced to Hawaii from Mexico in an effort to minimize the non-target impact of the predatory reduviid *Zelus renardii* Kolenati that had earlier been introduced as a biological control agent (Lai and Funasaki 1986, Funasaki et al. 1988). One of the few introductions against flatid pests involved the use of *Aphanomerus pusillus* Perkins from Australia that has had substantial success in suppressing the abundance of *Siphanta acuta* (Walker) on tree fruit and ornamentals in Hawaii (Lai and Funasaki 1986).

There has also been considerable interest in the use of scelionid egg parasitoids as classical biological controls for lepidopteran pests. In an agricultural context, the most successful example is that of *Telenomus remus* Nixon, introduced from Papua for the control of *Spodoptera* armyworms in corn and vegetable crops in India and subsequently in the Caribbean, where substantial control was recorded (Cock 1985). Other introductions of *Telenomus* species that have led to establishment, but no apparent impact, are *Telenomus nawaii* Ashmead from Hawaii to Guam, the Philippines and the Caribbean for control of *Spodoptera* armyworms, *Telenomus alecto* Crawford from Colombia for control of sugarcane stemborer *Chilo infuscatellus* (Snellen) in India and Pakistan, *Telenomus beneficians* from Java for control of the sugarcane borer *Scirpophaga nivella* F. in Taiwan, and *Telenomus tirathabae* (Wilkinson) from Java for control of the coconut spike moth *Tirathaba complexa* (Butler) in Fiji (Rao et al. 1971).

15.2.7 Trichogrammatidae

Many of the genera of eggs parasitoids in the Trichogrammatidae remain poorly known, and while some may be more specialized parasitoids of agricultural pests the focus of attention for almost all classical biological control introductions has been on two genera *Trichogramma* and *Trichogrammatoidea*, which both include very generalist egg parasitoids of lepidopteran pests. Of the 113 introductions of *Trichogramma* and *Trichogrammatoidea* species listed in the BIOCAT database (Greathead and Greathead 1992), most have been targeted against *Chilo*, *Diatraea*, *Helicoverpa* and *Spodoptera*, and more frequently on islands rather than mainland locations. While many such introductions have failed, a number of establishments have occurred and these no doubt have resulted in numerous non-target impacts while contributing little to the control of the target pest. Exceptions where substantial control of the target pest has been recorded are the impact of *T. evanescens* from Europe on the imported cabbageworm *Artogeia rapae* (L.) in the USA (Clausen 1977), *Trichogramma minutum* Riley and *Trichogramma perkinsi* Girault from the USA on noctuid pests in Chile (Zuniga 1985), and *Trichogramma* spp. from the USA on the potato tuberworm *Phthorimaea operculella* (Zeller) in Spain (Greathead 1976). Since indigenous *Trichogramma* species occur in each of these mainland environments it seems hard to believe that introduced species would have been more effective than local species, and thus these records should be treated with caution, particularly as species within this genus can be difficult to identify with certainty.

In addition to the introductions of generalist trichogrammatids, some introductions of more specialized species have also been made. These include the establishment in Hawaii of *Uscana semifumipennis* Girault from mainland USA on the bruchid pest *Algarobius prosopis* LeConte (Lai and Funasaki 1986, Funasaki et al. 1988), and the establishment of *Aphelinoidea anatolica* Novicky and *Aphelinoidea turanica* Trjapitzin from Turkmenistan and Iran on the beet leafhopper *Circulifer tenellus* (Baker) in California (Walker et al. 1997, 2005). In addition, an ongoing project in Florida for control of the citrus weevil *D. abbreviatus* involves introductions of *Brachyufens osborni* (Dozier), *Ceratogramma etiennei* Delvare and *Haekeliania sperata* Pinto from the island of Dominica (see Chapter 13). While the first of these three species has been recovered from another citrus weevil *Pachnaeus opalus* Schoenherr, the second has so far failed to establish, and releases of the third are currently in progress (Evans and Peña 2005, Jacas et al. 2008). Then finally, while some initial interest in trichogrammatid egg parasitoids of proconiine sharpshooters was generated from the invasion of California by the glassy-winged sharpshooter *H. vitripennis*, none were selected for introduction (Triapitsyn 2003).

15.3 Egg Parasitoids in Augmentative Biological Control and IPM

As for all types of natural enemies, the success in using egg parasitoids for augmentative releases is dependent upon a number of criteria (Heimpel and Mills 2011):

- Technical effectiveness – the ability of the natural enemy to suppress the pest population below the desired level of abundance over the period of interest, and how consistent, or at least predictable, a release can be under all circumstances and environments in which the pest occurs
- Public good – the extent to which the use of natural enemies can improve the quality of human life in terms of benefits to the environment and human health and well being, such as through reduction in pesticide use
- Ease of use – an unpredictable shelf life, poor formulation, ineffective delivery, a need for frequent applications, or a requirement for complex instructions or application techniques can all be important barriers to adoption
- Commercial viability – many natural enemies are more specialized, which creates niche markets with specific and limited periods of annual production, but sales must remain competitive with chemical insecticides
- Safety – although natural enemies are often more specialized there is a need to consider potential impacts on non-target organisms in the environment, and a need to assess the implications of inundative releases on human health

While egg parasitoids readily satisfy the criteria of public good and safety, technical effectiveness, ease of use, and commercial viability, remain important obstacles to their more widespread adoption as a component of integrated pest management.

15.3.1 Trichogramma

Not all families of egg parasitoids have been mass reared for use in augmentative biological control, and clearly the greatest emphasis has been placed on the single genus *Trichogramma*. There are more than 200 known species of *Trichogramma* around the world (Universal Chalcidoidea Database, http://www.nhm.ac.uk/jdsml/research-curation/projects/chalcidoids/), but only 19 species have been mass reared for use in augmentative biological control (Li 1994). In fact, for many years, only one or two species dominated field trails and commercial applications in each of the major continents, *Trichogramma chilonis* Ishii in Asia, *Trichogramma dendrolimi* Matsumura in China, *Trichogramma evanescens* Westwood in Europe and the former USSR, and *Trichogramma pretiosum* Riley in both North and South America (Hassan 1993, Smith 1996, van Lenteren 2000, Parra and Zucchi 2004). These species appear to have dominated *Trichogramma* trails based primarily on ease of rearing on factitious stored product or silkworm hosts (*see* Chapter 10). While commercial viability is clearly enhanced by ease of rearing and a broad host range of target pests, technical effectiveness can be an important trade-off for this strategy.

Trichogramma releases have been used against pests on a wide variety of agricultural crops including industrial crops (beet, cotton, soybean and sugarcane), cereals (corn, rice and sorghum), vegetables (cabbage, pepper and tomato) and tree crops (apple, citrus and olive) (Hassan 1993, Li 1994, Smith 1996). It is notable, however, that the two main groups of target pests for *Trichogramma* have been stem borers in graminaceous crops, and *Heliothis*/*Helicoverpa* spp. in corn, cotton and tomato. In an earlier review, Smith (1996) noted several difficulties in assessing the success of *Trichogramma* augmentation programs. These included inconsistency in the reporting of application rates, variability in the extent to which the results were monitored and the measure of outcome used (percent egg parasitism or pest larval densities rather than crop yield), and the absence of suitable control plots for valid comparisons of the release treatments. As a result, solid documentation of success has often been lacking, and this has lead to much skepticism in the use of *Trichogramma* augmentation.

Some of the most consistent successes have been against graminaceous stem borers. It is interesting to note that, in some cases, these programs have been based on more selective *Trichogramma* species that are not the most widely used in the region, such as *Trichogramma brassicae* Bezdenko in Europe (Hassan and Zhang 2001), *Trichogramma galloi* Zucchi in Brazil (Parra and Zucchi 2004) and *Trichogramma nubilale* Ertle and Davis in the U.S.A. (Losey et al. 1995), while in other cases the regional generalist predominates, such as *T. chilonis* in India (Rao et al. 2006) and *T. dendrolimi* in China (Wang et al. 2005). However, even among these applications that are used annually on large acreages of corn, documentation of efficacy is limited. Factors that may have contributed to the apparent success are the simplified plant architecture of graminaceous crops for parasitoid search, the close spacing of plants in the field that facilitates parasitoid dispersal from release points, the limited number of generations per year of the target pest (typically 1–3), and the use of formulations (parasitized eggs of different ages) that extended the period of adult parasitoid emergence. One of the most detailed recent accounts of success is from China, where *T. dendrolimi* was used for the control of the Asian corn borer *Ostrinia furnicalis* (Guenee) on 4.1 million ha of corn in Jilin Province from 1990 to 2002 (Wang et al. 2005). Parasitoids were reared on *Antheraea pernyi* Guer. with an optimal number of 60 wasps per egg producing a sex ratio of 80% female. Inundative releases of 150,000–300,000 wasps/ha at 45 release points/ha generated egg parasitism rates of 60–85% with an associated reduction in crop damage of 65–92% (Piao and Yan 1996, Liu et al. 2000).

Other examples that provide good evidence of the success of inundative releases of *Trichogramma* are the use of *T. pretiosum* in tomato for control of *Heliocoverpa zea* (Boddie) in Mexico and for control of *Tuta absoluta* (Meyrick) in Brazil, and initial trials with indigenous *Trichogramma* species for control of olive pests in Egypt. Releases of 610,000 wasps/acre per week over a period of 5–9 weeks in Mexico, reduced tomato fruitworm populations by 80–90% and fruit damage was often comparable to plots treated with conventional insecticides (Trumble and Alvarado-Rodriguez 1993). Similarly, for glasshouse tomatoes in Brazil, a release rate of 800,000 wasps/ha has reduced fruit damage by tomato borer to levels that are comparable to the use of conventional insecticides (Parra and Zucchi 2004). In Egypt, 11 two-weekly releases of 3 million female wasps/ha through the season proved to be very effective in suppressing two key lepidopteran pests of olive (Hegazi et al. 2007). Using comparative releases of different *Trichogramma* species, egg parasitism for each of the two pests reached 91% for the most effective indigenous wasp species, with pest larval densities reduced by as much as 83%, and fruit damage maintained below 5%. *Trichogramma bourarachae* Pintureau and Babault proved most effective against the olive moth *Prays oleae* Bernanrd, while *Trichogramma cordubensis* Vargas and Cabello was more effective against the jasmine moth *Palpita unionalis* (Hübner).

In most cases, however, recent trials of *Trichogramma* augmentation in agricultural crops, whether used inundatively (Andow et al. 1995, Glenn and Hoffmann 1997, Scholz et al. 1998, Mills et al. 2000, Suh et al. 2000a, Hommay et al. 2002, Lundgren et al. 2002) or as a seasonal inoculation (Wang et al. 1988, Hoffmann et al. 2002, Gardner et al. 2007), have not provided sufficient control as a stand alone treatment. While targeting the egg stage is often an effective strategy for preventing crop damage (Hassan 1993), for some pests increased egg mortality can result in reduced larval mortality with the result that there is no reduction in crop damage (Van Hamburg and Hassell 1984). This also appears to be the reason for lack of success in using inundative releases of *Trichogramma exiguum* Pinto and Platner against heliothines in cotton in North Carolina, U.S.A. (Suh et al. 2000b). Plots in which *T. exiguum* were released experienced a significantly smaller number of hatching neonate larvae, but density dependent compensatory mortality subsequently occurred in the plots such that final instar larval densities were comparable (Fig. 15.3). There are, however, many other reasons for lack of technical effectiveness that include the following:

- Lack of compatibility with fungicides (Thomson et al. 2000, Manzoni et al. 2006) or insecticides (Cônsoli et al. 1998, 2001, Carvalho et al. 2003, Giolo et al. 2007) that are being used simultaneously in the crop
- Reduction in quality of the *Trichogramma* produced, including general vigor and representation of females (Losey and Calvin 1995, Hassan and Zhang 2001)
- Use of *Trichogramma* species that are not indigenous to the crop and region (Losey and Calvin 1995, Mills 2003, Herz and Hassan 2006)

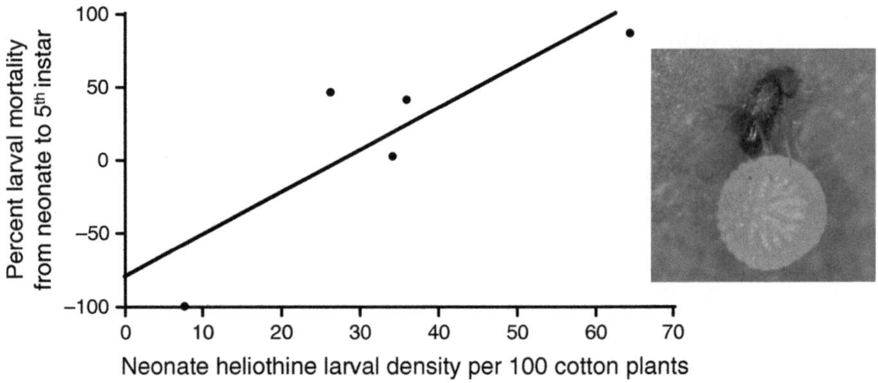

Fig. 15.3 Density dependent compensatory mortality among heliothine larvae on cotton following a 59% reduction in neonate larval densities in cotton in plots in which 9 releases of 100,000+ female of *Trichogramma exiguum*/ha in 1996 (after Suh et al. 2000b)

- Incompatible weather that can severely limit *Trichogramma* activity and performance in the field (Bourchier and Smith 1996, Wang et al. 1999, Gardner et al. 2007)
- Poor release timing that generates a mismatch between the period of female *Trichogramma* activity and the presence of eggs in the crop (Glenn and Hoffmann 1997)
- Too low a release rate may result in no apparent benefit as an optimum must be determined for each pest and crop (Mills et al. 2000)
- Insufficient number of release points such that female *Trichogramma* are unable to disperse throughout the crop (Mills et al. 2000, Parra and Zucchi 2004)
- Generalist predators that can greatly reduce the number of adult *Trichogramma* emerging from egg cards or capsules (Suh et al. 2000b, Pereira et al. 2004)
- Refuges from parasitism caused by the foraging behavior of *Trichogramma* wasps in relation to the architecture of the crop plant, which result in an asymptotic dose-response curve with an upper limit to effectiveness (Mills et al. 2000)

Of the limitations noted above, refuges from parasitism are the least well appreciated and understood. One of the generally accepted principles of augmentative releases is that the release of greater numbers of individuals can compensate for any limitation in the effectiveness of mass reared parasitoids. However, asymptotic dose-response curves are a common feature of augmentative biological control (Heimpel and Mills 2011) and represent an upper limit to the technical effectiveness of the application of a single natural enemy species. An example of this relationship comes from our experimental field trials using *Trichogramma platneri* Nagarkatti for inundative biological control of codling moth *Cydia pomonella* (L.) in California. We used a series of different weekly release rates during the oviposition period of each of the three generations of the pest, in two unsprayed walnut

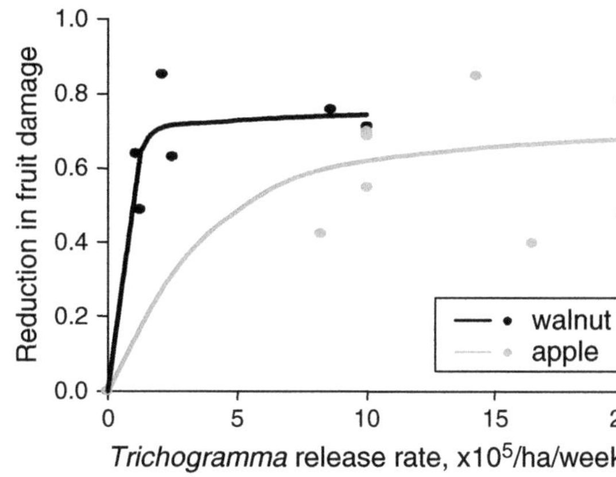

Fig. 15.4 Dose- response curves for inundative releases of *Trichogramma platneri* used for control of codling moth *Cydia pomonella* in apple and walnut orchards in California (after Mills et al. 2000)

orchards in 1994 and three unsprayed apple orchards in 1993 and 1994, and measured the reduction in fruit damage at harvest as the response variable. These trails show that in both apples and walnuts there is an upper limit to the reduction in fruit damage that can be achieved by the inundative releases of *T. platneri* (Fig. 15.4). The asymptote is not only higher in walnuts (0.74) than in apples (0.67), indicating that codling moth eggs have less of a refuge from parasitism in walnuts than in apples, but is also reached at a much lower release rate than in apples. Thus a release rate of 1 million wasps/ha/wk may be needed to achieve a 2/3 reduction in fruit damage in apples, while a release rate of 200,000 wasps/ha/wk is sufficient to gain a 3/4 reduction in nut damage in walnuts. The difference between apples and walnuts lies in the surface architecture of the foliage. Codling moth only lays eggs on smooth surfaces and while eggs can be found on both surfaces of walnut leaves they are absent from the lower surface of apple leaves which are covered in small hairs. Adult *T. platneri* forage on both upper and lower surfaces of the foliage in both crops which results in reduced effectiveness in finding codling moth eggs in apple orchards.

In earlier reviews of augmentative biological control, both Stinner (1977) and Smith (1996) suggested that greater emphasis should be placed on the integration of *Trichogramma* releases with other control options. Despite these recommendations there continue to be few attempts to test combination programs that include *Trichogramma* releases in agricultural crops. Perhaps the most widely explored combination involves the use of *Bacillus thuringiensis* (Bt) with *Trichogramma* releases. Trials have shown that this combination can provide greater suppression of stem borers in corn in the USA (Mertz et al. 1995), Taiwan (Cheng et al. 1995), India (Jalali and Singh 2006) and China (Wang et al. 2005) and deserves greater attention as a viable tactic in integrated pest management. Surprisingly little research has been done on combination programs using *Trichogramma* and mating disruption. In Pakistan, *T. chilonis* plus pheromones provided better control of cotton bollworms than either treatment alone (Ahmad et al. 2005), but whether this is true for other

Trichogramma-pest situations remains untested despite the increased interest in mating disruption in recent years. Similarly, a combination of *T. platneri* and sterile insect release reduced fruit damage by codling moth in apples in field cages to a greater extent than either tactic used alone (Bloem et al. 1998). An interesting field trial in which two different *Trichogramma* species, *T. nubilale* and *T. ostriniae*, were released either alone or in combination provided the surprising result that parasitism of European corn borer eggs was lower for the combined release than for either of the two species alone (Wang et al. 1999). In contrast, a combination of egg and larval parasitoids was more successful for control of sugarcane borer in Brazil (Parra and Zucchi 2004). In this case, three releases of 200,000 *T. galloi*/ha and a single release of 6,000 *Cotesia flavipes* (Cameron)/ha provided more effective control than either species alone.

The combination of *Trichogramma* releases with behavior-modifying infochemicals, including both synomones and kairomones, has also been explored. One of the first observations that plant volatiles can increase parasitism rates under field conditions came from studies where soybean plants were sprayed with a water extract of *Amaranthus* spp. (Altieri et al. 1981). Similarly, a hexane extract of tomato leaves on corn plants has been shown to increase parasitism by *T. pretiosum* (Nordlund et al. 1985), and both cotton and tomato leaf extracts increased parasitism by *T. chilonis* (Shanmugam et al. 2005). However, as noted by Romeis et al. (2005), not all plants produce extracts of green leaf volatiles that elicit a positive response in *Trichogramma* wasps, as some may have no influence, while others may act as deterrents. In addition, a water extract of eggs or adult scales of lepidopteran hosts can act as kairomones for *Trichogramma* and this has lead to increased parasitism rates by *T. pretiosum* in cotton fields (Lewis et al. 1979). However, field trials in cotton with a combination of a synomone (*Amaranthus* extract) plus kairomone (eggs and scales of *Helicoverpa zea*) spray and aerial application of wasps did not lead to increased parasitism by *T. pretiosum* (Lewis et al. 1985). Since *Trichogramma* species are poor fliers it seems likely that these responses are mediated through wasp arrestment rather than attraction and that increased rates of parasitism are achieved either through wasps spending longer in the crop or being stimulated to search individual plants more efficiently. As inundative releases are implemented on a small localized scale, behavior-modifying infochemicals appear to provide a particularly feasible approach to increasing the technical effectiveness of *Trichogramma* releases and it is surprising that they have not received greater attention in recent years.

In an integrated pest management context, an ability to combine synthetic insecticides with *Trichogramma* releases could also lead to more effective management. Cônsoli et al. (2001) tested the compatibility of insecticides used for sugarcane borer management in Brazil with *T. galloi*. While most were incompatible, the ecdysone-agonist tebufenozide was an exception, and this was suggested to result from the fact that *Trichogramma* complete their larval development in a single instar without the need for ecdysis before the pupal stage. In general, however, many organophosphate, carbamate, pyrethroid, neonicotinyl and IGR insecticides are incompatible with *Trichogramma* either through acute toxicity or through sublethal effects on fecundity (Brunner et al. 2001, Cônsoli et al. 2001, Carvalho et al. 2003, Bastos et al. 2006, Giolo et al. 2007). Thus combination programs of *Trichogramma* releases with synthetic insecticides needs to be assayed very carefully using topical, residual and oral routes of entry, and testing for sublethal as well as acute effects on the wasps. In the absence of compatible products it may be possible to select either naturally from the field or through laboratory rearing for tolerance to key insecticides. In this regard Jalali et al. (2006) report on the selection of a strain of *T. chilonis* that is tolerant of field concentrations of endosulfan for use against *Helicoverpa armigera* (Hübner) on cotton in India.

15.3.2 Other Egg Parasitoid Taxa

Few other taxa of egg parasitoids have been used in augmentative biological control due primarily to the technical difficulty of mass production. Unless host eggs can be produced with ease at low cost, commercial viability for inundative releases is highly unlikely and even production for inoculative releases may not be feasible. Nonetheless, several other taxa have been tested for augmentative releases including some mymarid and scelionid species.

The mymarid *Anaphes iole* Girault has been mass reared for inundative releases against the mirid *Lygus hesperus* (Knight) in both cotton and strawberries (Jones and Jackson 1990). A high release rate of 37,000 parasitoids/ha resulted in 50% parasitism of *Lygus* eggs, a 43% reduction in *Lygus* nymphs, and a 22% reduction in fruit damage (Norton and Welter 1996). Further study of this system (Udayagiri and Welter 2000) found that *Lygus* primarily lays its eggs directly into the fruit and that parasitism by *A. iole* in glasshouse cages was much lower in young strawberry fruit (25%) than in larger fruit (77%) or other parts of the plant (85% or more). Thus in strawberries *Lygus* eggs have a refuge from parasitism in young fruit that limits the impact of inundative releases of *A. iole*. It remains unknown whether a refuge would exist in cotton, but the development of mass production technology for *A. iole* continues and field trials are expected in cotton in the future (Smith and Nordlund 2000). Augmentative releases of the mymarid *Anagrus atomus* L. have been used in combination with the IGR buprofezin for management of both the greenhouse leafhopper *Hauptidia maroccana* (Melichar) a sporadic pest of protected tomato crops in Europe (Jacobson et al. 1996), and for control of the green leafhopper *Empoasca decipiens* Paoli a pest of protected vegetables in Europe (Schmidt and Rupp 1997).

Orr (1988) reviewed experimental trials using scelionid egg parasitoids in augmentative biological control, and while several lepidopteran and pentatomid pests had been targeted for egg parasitoid releases, none appeared to show much success. Subsequently, Hassan (1993) noted the use of augmentative releases of *T. remus* in corn and sorghum in Honduras, presumably against armyworms, and of *Platytelenomus hylas* Nixon against sugarcane stem borers in Iran. In addition, van Lenteren and Bueno (2003) note the use of augmentative releases of *T. alecto* against sugarcane borer in Guyana, although Cock (1985) notes that there is no evidence for the success of this long standing practice in the Caribbean. *Telenomus* species are often effective egg parasitoids of lepidopteran pests, but as they are frequently more specialized, the development of a cost effective mass production system is the limiting factor for augmentative biological control. In this regard the greatest focus has been on *T. remus* which can be reared on eggs of *Anagasta kuehniella* (Zeller), although parasitoid size and fecundity are compromised as is the case for *Trichogramma* (Martinez-Martinez and Bernal 2002). It appears that augmentative releases of *T. remus* may currently be used against armyworms in India, as both release rates (Jalali et al. 2005) and packaging for egg cards (Ballal et al. 2006) have recently been explored.

In Brazil, inoculative releases of *Trissolcus basalis* have contributed significantly to the integrated pest management of insect pests in soybean (Corrêa-Ferreira et al. 2000). In an initial four year study in the mid 1980s in which 15,000 parasitoids/ha were released into trap crops of early-maturing soybean, stink bug densities [*Nezara viridula*, *Piezodorus guildinii* (Westwood) and *Euschistus heros* (F.)] were reduced by 54% in the trap crop and by 58% in the main crop (Corrêa-Ferreira and Moscardi 1996). The success of this initial project led to a subsequent four year project that began in the 1994/95 field season, in which 4,600 ha of soybean in the Rio do Campo basin were treated with softer insecticides and biological controls where possible (Fig. 15.5). Inoculative releases of 300,000 *T. basalis* were made annually for control of green stink bug *N. viridula*, while velvetbean caterpillar *Anticarsia gemmatalis* Hübner was treated with baculovirus, allowing total insecticide applications to be reduced from 2.8 to 1.2 per season and applications for stink bug to be reduced from 0.81 to 0.19 per season over the four year period. The success

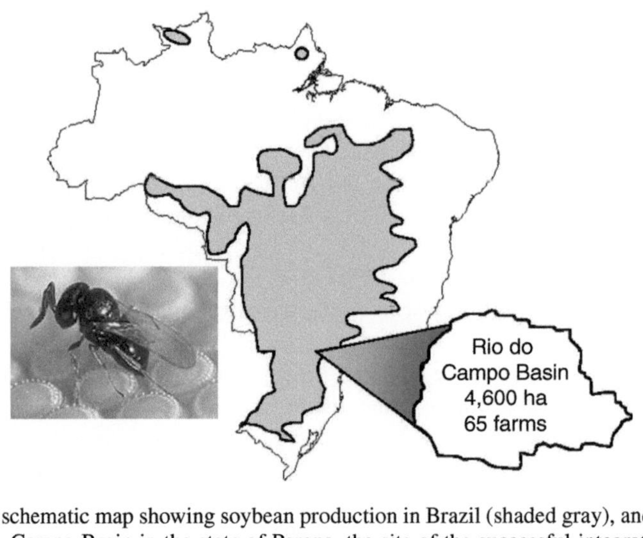

Fig. 15.5 A schematic map showing soybean production in Brazil (shaded gray), and the location of the Rio do Campo Basin in the state of Parana, the site of the successful integrated pest management project for stink bug involving inoculative releases of *Trissolcus basalis* and compatible insecticides (after Corrêa-Ferreira et al. 2000)

of this program has subsequently led to the adoption of these control tactics by 35% of the 2.7 million ha of soybean production in Brazil (Corrêa-Ferreira et al. 2000). While it remains unclear from this project the extent to which the inoculative releases of *T. basalis* contributed to stink bug control versus the change in selective insecticides, this is an excellent example of the potential to reduce insecticide use through integrating augmentative releases of egg parasitoids into a successful integrated pest management program.

While *Edovum puttleri*, an egg parasitoid of the Colorado potato beetle, failed to establish successfully in the USA and Italy due to an absence of overwintering diapause, it has shown some promise for early season inoculative releases (Hamilton and Lashomb 1996). For a number of years the New Jersey Department of Agriculture reared *E. puttleri* for augmentative application in eggplant until growers switched to the newly available insecticide imidacloprid (Tipping et al. 1999). *Edovum puttleri* kills eggs of *L. decemlineata* through probing as well as parasitism and field trials in small plots resulted in up to 70–90% parasitism of egg masses and 68–70% mortality of the eggs in an egg mass (Lashomb et al. 1987).

15.4 Concluding Remarks

In the context of classical biological control introductions, egg parasitoids exhibit a greater rate of establishment than other parasitoid guilds, with the exception of Trichogrammatidae, but have not proved to be as effective in suppressing the abundance of the target pest. Mymarid introductions have provided the most spectacular results and such opportunities should always be considered for new invasive pests, with particular emphasis on cicadellids and curculionids. Encyrtid egg parasitoids have also proved to be effective as introductions in some instances, whereas *Trichogramma* introductions have very seldom been effective and should not be undertaken under circumstances where indigenous species already exist.

The success of augmentative biological control with egg parasitoids has been mixed, which highlights the need to know enough about their ecology and behavior to be able to select effective indigenous species, to mass produce them as a quality product at a competitive price, and to formulate and apply them effectively. This leads to the difficulty that effective implementation of mass produced egg parasitoids for field release often requires sufficient training. It is no surprise that the best field results are frequently those obtained from trials carried out by experienced researchers rather than from implementation by end users, as there are many steps from formulation and shipment from the insectary, through application in the field, to emergence in the crop, where the viability and performance of egg parasitoids can be compromised. While combination programs involving egg parasitoid releases plus behavior-modifying infochemicals or compatible insecticides are likely to provide the greatest advances for augmentative biological control in the future, there remains plenty of scope for further research into formulation and application technology to facilitate successful adoption by end users.

References

Ahmad N, Wagan MS, Fatima B, Khan GC (2005) Significance and cost benefit of using pheromones in conjunction with parasitoids for the management of cotton bollworms. Pak J Zool 37:43–47

Alma A, Arno C, Vidano C (1987) Particularities on *Polynema striaticorne* as egg parasite of *Stictocephala bisonia* (Rhynchota Auchenorrhyncha). In: Proceedings of the 6th Auchenorrhyncha Meeting, Turin, Italy, 7–11 Sept. 1987, National Research Council, Rome, pp 597–603

Altieri MA, Lewis WJ, Nordlund DA, Gueldner RC, Todd JW (1981) Chemical interactions between plants and *Trichogramma* wasps in Georgia, USA soybean fields. Prot Ecol 3:259–264

Andow DA, Klacan GC, Bach D, Leahy TC (1995) Limitations of *Trichogramma nubilale* (Hymenoptera, Trichogrammatidae) as an inundative biological control of *Ostrinia nubilalis* (Lepidoptera, Crambidae). Environ Entomol 24:1352–1357

Ballal CR, Lylla KR, Joshi S, Lakshmi L (2006) Appropriate packaging for transportation of *Telenomus remus* Nixon (Hymenoptera: Scelionidae) egg cards. J Biol Control 20:219–223

Bastos CS, de Almeida RP, Suinaga FA (2006) Selectivity of pesticides used on cotton (*Gossypium hirsutum*) to *Trichogramma pretiosum* reared on two laboratory-reared hosts. Pest Manage Sci 62:91–98

Beardsley JW (2000) The introduction and establishment of *Anaphes* (*Patasson*) *calendrae* (Gahan) in Hawaii (Hymenoptera: Mymaridae). Proc the Hawaiian Entomol Soc 34:209–211

Bloem S, Bloem KA, Knight AL (1998) Oviposition by sterile codling moths, *Cydia pomonella* (Lepidoptera: Tortricidae) and control of wild populations with combined releases of sterile moths and egg parasitoids. J Entomol Soc B C 95:99–110

Bourchier RS, Smith SM (1996) Influence of environmental conditions and parasitoid quality on field performance of *Trichogramma minutum*. Entomologia Experimentalis et Applicata 80:461–468

Brunner JF, Dunley JE, Doerr MD, Beers E (2001) Effect of pesticides on *Colpoclypeus florus* (Hymenoptera: Eulophidae) and *Trichogramma platneri* (Hymenoptera: Trichogrammatidae), parasitoids of leafrollers in Washington. J Econ Entomol 94:1075–1084

Caltagirone LE (1981) Landmark examples in classical biological control. Ann Rev Entomol 26:213–232

Carvalho GA, Reis PR, Rocha LCD, Moraes JC, Fuini LC, Ecole CC (2003) Side-effects of insecticides used in tomato fields on *Trichogramma pretiosum* (Hymenoptera, Trichogrammatidae). Acta Scientiarum Agron 25:275–279

Castillo J, Jacas JA, Pena JE, Ulmer BJ, Hall DG (2006) Effect of temperature on life history of *Quadrastichus haitiensis* (Hymenoptera: Eulophidae), an endoparasitoid of *Diaprepes abbreviatus* (Coleoptera: Curculionidae). Biol Control 36:189–196

Cheng WY, Wang ZT, Chen SM (1995) Biological control of Asian corn borer in sweet corn fields. Report of the Taiwan Sugar Research Institute No. 148, 11–29

Clarke AR (1990) The control of *Nezara viridula* L. with introduced egg parasitoids in Australia. A review of a 'landmark' example of classical biological control. Aust J Agric Res 41:1127–1146

Clausen CP (1977) Introduced parasites and predators of arthropod pests and weeds: a world review. U.S.D.A. Agricultural Handbook No. 480

Cochereau UP (1972) Biological control in the Pacific. Cahiers O.R.S.T.O.M. (Office de la Recherche Scientifique et Technique Outre-Mer) Serie Biologie 16:89–104

Cock MJW (1985) A Review of Biological Control of Pests in the Commonwealth Carribean and Bermuda up to 1982. Commonwealth Agricultural Bureaux, Farnham Royal, UK

Colazza S, Bin F (1992) Introduction of the oophage *Edovum puttleri* Griss. (Hymenoptera: Eulophidae) in Italy for the biological control of Colorado potato beetle. Redia 75:203–225

Cônsoli FL, Botelho PSM, Parra JRP (2001) Selectivity of insecticides to the egg parasitoid *Trichogramma galloi* Zucchi, 1988, (Hym., Trichogrammatidae). J Appl Entomol 125: 37–43

Cônsoli FL, Parra JRP, Hassan SA (1998) Side-effects of insecticides used in tomato fields on the egg parasitoid *Trichogramma pretiosum* Riley (Hym., Trichogrammatidae), a natural enemy of *Tuta absoluta* (Meyrick) (Lep., Gelechiidae). J Appl Entomol 122:43–47

Corrêa-Ferreira BS, Moscardi F (1996) Biological control of soybean stink bugs by inoculative releases of *Trissolcus basalis*. Entomologia Experimentalis et Applicata 79:1–7

Corrêa-Ferreira BS, Domit LA, Morales L, Guimarães RC (2000) Integrated soybean pest management in micro river basins in Brazil. Integr Pest Manage Rev 5:75–80

Evans GA, Peña JE (2005) A new *Fidiobia* species (Hymenoptera: Platygastridae) reared from eggs of *Diaprepes doublierii* (Coleoptera: Curculionidae) from Dominica. Fla Entomol 88:61–66

Flanders SE (1930) Mass production of egg parasites of the genus *Trichogramma*. Hilgardia 4:465–501

Funasaki GY, Lai PY, Nakahara LM, Beardsley JW, Ota AK (1988) A review of biological control introductions in Hawaii, USA 1890 to 1985. Proc Hawaiian Entomol Soc 28:105–160

Gardner J, Hoffmann MP, Cheever SA, Seaman AJ, Westgate P, Hazzard RV (2007) Large-scale releases of *Trichogramma ostriniae* to suppress *Ostrinia nubilalis* in commercially grown processing and fresh market sweet corn. J Appl Entomol 131:432–440

Giolo FP, Gruetzmacher AD, Manzoni CG, De Lima CAB, Noernberg SD (2007) Toxicity of pesticides used in peach orchard on adults *Trichogramma pretiosum*. Bragantia 66: 423–431

Glenn DC, Hoffmann AA (1997) Developing a commercially viable system for biological control of light brown apple moth (Lepidoptera: Tortricidae) in grapes using endemic *Trichogramma* (Hymenoptera: Trichogrammatidae). J Econ Entomol 90:370–382

Grandgirard J, Hoddle MS, Petit JN, Roderick GK, Davies N (2008) Engineering an invasion: classical biological control of the glassy-winged sharpshooter, *Homalodisca vitripennis*, by the egg parasitoid *Gonatocerus ashmeadi* in Tahiti and Moorea, French Polynesia. Biol Invasions 10:135–148

Greathead DJ, Greathead AH (1992) Biological control of insect pests by parasitoids and predators: the BIOCAT database. Biocontrol News and Information 13:61 N-68 N

Greathead DJ (1971) A Review of Biological Control in the Ethiopian Region. Commonwealth Agricultural Bureaux, Farnham Royal, UK

Greathead DJ (1976) A Review of Biological Control in Western and Southern Europe. Commonwealth Agricultural Bureaux, Farnham Royal, UK

Greathead DJ (1992) Natural enemies of tropical locusts and grasshoppers: their impact and potential as biological control agents. In: Lomer CJ, Prior C (eds) Biological Control of Locusts and Grasshoppers. CAB International, Wallingford, pp 105–121

Hall D, Peña J, Franqui R, Nguyen R, Stansly P, McCoy C, Lapointe S, Adair R, Bullock B (2001) Status of biological control by egg parasitoids of *Diaprepes abbreviatus* (Coleoptera: Curculionidae) in citrus in Florida and Puerto Rico. BioControl 46:61–70

Hamilton GC, Lashomb J (1996) Comparison of conventional and biological control intensive pest management programs on eggplant in New Jersey. Fla Entomol 79:488-496

Hanks LM, Millar JG, Paine TD, Campbell CD (2000) Classical biological control of the Australian weevil *Gonipterus scutellatus* (Coleoptera: Curculionidae) in California. Environ Entomol 29:369–375

Hare JD (1990) Ecology and management of the Colorado potato beetle. Annu Rev Entomol 35:81–100

Hassan SA, Zhang WQ (2001) Variability in quality of *Trichogramma brassicae* (Hymenoptera: Trichogrammatidae) from commercial suppliers in Germany. Biol Control 22:115–121

Hassan SA (1993) The mass rearing and utilization of *Trichogramma* to control lepidopterous pests: Achievements and outlook. Pestic Sci 37:387–391

Hegazi E, Herz A, Hassan SA, Khafagi WE, Agamy E, Zaitun A, El-Aziz GA, Showeil S, El-Said S, Khamis N (2007) Field efficiency of indigenous egg parasitoids (Hymenoptera, Trichogrammatidae) to control the olive moth (*Prays oleae*, Lepidoptera, Yponomeutidae) and the jasmine moth (*Palpita unionalis*, Lepidoptera, Pyralidae) in an olive plantation in Egypt. Biol Control 43:171–187

Heimpel GE, Mills NJ (2011) Biological Control: Ecology and Application. Cambridge University Press, Cambridge

Herz A, Hassan SA (2006) Are indigenous strains of *Trichogramma* sp (Hym., Trichogrammatidae) better candidates for biological control of lepidopterous pests of the olive tree? Biocontrol Sci Technol 16:841–857

Hoffman MP, Wright MG, Pitcher SA, Gradner J (2002) Inoculative releases of *Trichogramma ostriniae* for suppression of *Ostrinia nubilalis* (European corn borer) in sweet corn: Field biology and population dynamics. Biol Control 25:249–258

Hommay G, Gertz C, Kienlen JC, Pizzol J, Chavigny P (2002) Comparison between the control efficacy of *Trichogramma evanescens* westwood (Hymenoptera: Trichogrammatidae) and two *Trichogramma cacoeciae* marchal strains against grapevine moth (*Lobesia botrana* Den. & Schiff.), depending on their release density. Biocontrol Sci Technol 12:569–581

Jacas JA, Peña JE, Duncan RE, Ulmer BJ (2008) Thermal requirements of *Fidiobia dominica* (Hymenoptera: Platygastridae) and *Haekeliana sperata* (Hymenoptera: Trichogrammatidae), two exotic egg parasitoids of *Diaprepes abbreviatus* (Coleoptera: Curculionidae). BioControl 53:451–460

Jacobson RJ, Chambers RJ, van Lenetern JC (1996) Control of glasshouse leafhopper (*Hauptidia maroccana*: Homoptyera, Cicadellidae) within an IPM programme in protected tomatoes. Bulletin OILB-SROP 19:67–70

Jalali SK, Singh SP (2006) Biological control of *Chilo partellus* using egg parasitoid *Trichogramma chilonis* and *Bacillus thuringiensis*. Indian J Agric Res 40:184–189

Jalali SK, Singh SP, Venkatesan T, Murthy KS, Lalitha T (2006) Development of endosulfan tolerant strain of an egg parasitoid *Trichogramma chilonis* Ishii (Hymenoptera: Trichogrammatidae). Indian J Exp Biol 44:584–590

Jalali SK, Venkatesan T, Murthy KS, Biswas SR, Lalitha Y (2005) Influence of temperature and host density on functional response of *Telenomus remus* Nixon, an egg parasitoid of Spodoptera litura Fabricius. Entomon 30:193–199

Johnson MT, Follett PA, Taylor AD, Jones VP (2005) Impacts of biological control and invasive species on a non-target native Hawaiian insect. Oecologia 142:529–540

Jones VP (1995) Reassessment of the role of predators and *Trissolcus basalis* in biological control of southern green stink bug (Hemiptera: Pentatomidae) in Hawaii. Biol Control 5: 566–572

Jones WA, Jackson CG (1990) Mass production of *Anaphes iole* for augmentation against *Lygus hesperus*: effects of food on fecundity and longevity. Southwest Entomol 15:463–468

Knight KMM, Gurr GM (2007) Review of *Nezara viridula* (L.) management strategies and potential for IPM in field crops with emphasis on Australia. Crop Prot 26:1–10

Lai PY, Funasaki GY (1986) List of beneficial organisms purposely introduced for biological control in Hawaii: 1890 to 1985. Plant Pest Control Branch, Plant Industry Div., Hawaii Department of Agriculture

Lampert EP, Haynes DL (1985) Population dynamics of the cereal leaf beetle Oulema melanopus (Coleoptera: Chrysomelidae) at low population densities. Environ Entomol 14:74–79

Lashomb J, Ng YS, Jansson RK, Bullock R (1987) *Edovum puttleri* (Hymenoptera: Eulophidae) an egg parasitoid of Colorado potato beetle (Coleoptera: Chrysomelidae): development and parasitism on eggplant. J Econ Entomol 80:65–68

Lewis WJ, Beevers M, Nordlund DA, Gross HR, Hagen KS (1979) Kairomones and their use for management of entomophagous insects. IX. Investigations of various kairomone treatment patterns for *Trichogramma* spp. J Chem Ecol 5:673–80

Lewis WJ, Gross HR, Nordlund DA (1985) Behavioral manipulation of *Trichogramma* (Hymenoptera: Trichogrammatidae). Southwest Entomol Suppl No. 8:49–55

Li LY (1994) Worldwide use of *Trichogramma* for biological control on different crops: a survey, In: wajnberg E, Hassan SA (eds) Biological control with egg parasitoids. CAB International, Oxon, UK, pp 37–51

Liu ZC, Liu JF, Zhang F, Li DS, Feng XX (2000) Production and field application techniques of *Trichogramma*. Golden Shield Press, Beijing

Loch AD, Walter GH (1999) Multiple host use by the egg parasitoid *Trissolcus basalis* (Wollaston) in a soybean agricultural system: biological control and environmental implications. Agric For Entomol 1:271–280

Losey JE, Calvin DD (1995) Quality assessment of four commercially available species of *Trichogramma* (Hymenoptera: Trichogrammatidae). J Econ Entomol 88:1243–1250

Losey JE, Fleischer SJ, Calvin DD, Harkness WL, Leahy T (1995) Evaluation of *Trichogramma nubilalis* and *Bacillus thuringiensis* in management of *Ostrinia nubilalis* (Lepidoptera: Pyralidae) in sweet corn. Environ Entomol 24:436–445

Lundgren JG, Heimpel GE, Bomgren SA (2002) Comparison of *Trichogramma brassicae* (Hymenoptera: Trichogrammatidae) augmentation with organic and synthetic pesticides for control of cruciferous Lepidoptera. Environ Entomol 31:1231–1239

Manzoni CG, Gruetzmacher AD, Giolo FP, de Lima CAB, Noernberg SD, Mueller C, da R Haerter W (2006) Susceptibility of *Trichogramma pretiosum* Riley (Hymenoptera: trichogrammatidae) adults to fungicides used to control apple diseases. Neotrop Entomol 35:223–230

Martinez-Martinez L, Bernal JS (2002) *Ephestia kuehniella* Zeller as a factitious host for *Telenomus remus* Nixon: Host acceptance and suitability. J Entomol Sci 37:10–26

Mertz BP, Fleischer SJ, Calvin DD, Ridgway RL (1995) Field assessment of *Trichogramma brassicae* (Hymenoptera: Trichogrammatidae) and *Bacillus thuringiensis* for control of *Ostrinia nubilalis* (Lepidoptera: Pyralidae) in sweet corn. J Econ Entomol 88:1616–1625

Mills NJ (1994a) Biological control: some emerging trends. In: Leather SR, Watt AD, Mills NJ and Walters KFA (eds), Individuals, Populations and Patterns in Ecology. Intercept, Andover, pp 213–222

Mills NJ (1994b) Parasitoid guilds: defining the structure of the parasitoid communities of endopterygote insect hosts. Environ Entomol 23:1066–1083

Mills NJ (2003) Augmentation in orchards: improving the efficacy of *Trichogramma* inundation. In: R. van Driesche (ed) 1st international symposium on biological control of arthropods. USDA Forest Service FHTET-03-05, pp 130–135

Mills NJ, Pickel C, Mansfield S, McDougall S, Buchner R, Caprile J, Edstrom J, Elkins R, Hasey J, Kelley K, Krueger W, Olson W, Stocker R (2000) *Trichogramma* inundation: integrating parasitism into management of codling moth. Calif Agric 54(6):22–25

Nordlund DA, Chalfant RB, Lewis WJ (1985) Response of *Trichogramma pretiosum* to volatile synomones from tomato plants. J Entomol Sci 20:372–376

Norton AP, Welter SC (1996) Augmentation of the egg parasitoid *Anaphes iole* (Hymenoptera: Mymaridae) for *Lygus hesperus* (Heteroptera: Miridae) management in strawberries. Environ Entomol 25:1406–1414

Orr BD (1988) Scelionid wasps as biological control agents–a review. Flo Entomol 71:506–528

Parra JRP, Zucchi RA (2004) *Trichogramma* in Brasil: feasibility of use after twenty years of research. Neotrop Entomol 33:271–281

Pemberton CE (1964) Highlights in the history of entomology in Hawaii 1778-1963. Pac Insects 6:689–729

Pereira JA, Bento A, Cabanas JE, Torres LM, Herz A, Hassan SA (2004) Ants as predators of the egg parasitoid *Trichogramma cacoeciae* (Hymenoptera: Trichogrammatidae) applied for biological control of the olive moth, *Prays oleae* (Lepidoptera: Plutellidae) in Portugal. Biocontrol Sci Technol 14:653–664

Piao YF, Yan S (1996) Progress of mass production and field application of *Trichogramma dendrolimi*. In: Zhang ZL, Piao YF, Wu JW (eds) Proceedings of the national symposium on IPM in China. China Agricultural Scientech Press, Beijing, pp 1135–1136

Quicke DLJ (1997) Parasitic Wasps. Chapman and Hall, London, UK

Rao CVN, Rao NV, Bhavani B (2006) Efficacy of *Trichogramma chilonis* ishii against sugarcane early shoot borer, *Chilo infuscatellus* snellen under sugar factory operational areas of coastal Andhara Pradesh. J Biol Control 20:225–228

Rao VP, Ghani MA, Sankaran T, Mather KC (1971) A Review of the Biological Control of Insects and Other Pests in South-East Asia and the Pacific Region. Commonwealth Agricultural Bureaux, Farnham Royal

Romeis J, Babendreier D, Wackers FL, Shanower TG (2005) Habitat and plant specificity of *Trichogramma* egg parasitoids – underlying mechanisms and implications. Basic Appl Ecol 6:215–236

Schmidt U, Rupp J (1997) Zikadenschäden an Gurke auf der Insel Reicenau. Gemüse 12/97: 691–692

Scholz BCG, Monsour CJ, Zalucki MP (1998) An evaluation of selective *Helicoverpa armigera* control options in sweet corn. Aust J Exp Agric 38:601–607

Shanmugam PS, Thenesh KK, Satpute US (2005) Synomonic effects of plant extracts on parasitisation of *Corcyra* eggs by *Trichogramma chilonis* Ishii. J Appl Zool Res 16:13–14

Smith RA, Nordlund DA (2000) Mass rearing technology for biological control agents of *Lygus* spp. Southwest Entomol Suppl 23:121–127

Smith SM (1996) Biological control with *Trichogramma*: advances, successes, and potential of their use. Annu Rev Entomol 41:375–406

Stinner RE (1977) Efficacy of innundative releases. Ann Rev Entomol 22:515–531

Suh CPC, Orr DB, Van Duyn JW (2000a) *Trichogramma exiguum* (Hymenoptera: Trichogrammatidae) releases in North Carolina cotton: Evaluation of heliothine pest suppression. J Econ Entomol 93:1127–1136

Suh CPC, Orr DB, Van Duyn JW (2000b) *Trichogramma* releases in North Carolina cotton: Why releases fail to suppress heliothine pests. J Econ Entomol 93:1137–1145

Thomson LJ, Glenn DC, Hoffmann AA (2000) Effects of sulfur on *Trichogramma* egg parasitoids in vineyards: Measuring toxic effects and establishing release windows. Aust J Exp Agric 40:1165–1171

Tipping PW, Holko CA, Abdul-Baki AA, Aldrich JR (1999) Evaluating *Edovum puttleri* Grissell and *Podisus maculiventris* (Say) for augmentative biological control of Colorado potato beetle in tomatoes. Biol Control 16:35–42

Triapitsyn SV (2003) Taxonomic notes on the genera and species of Trichogrammatidae (Hymenoptera) egg parasitoids of the proconiine sharpshooters (Hemiptera: Clypeorrhyncha: Cicadellidae: Proconiini) in southeastern USA. Trans Am Entomol Soc 129:245–265

Trumble JT, Alvarado-Rodriguez B (1993) Development and economic evaluation of an IPM program for fresh market tomato production in Mexico. Agric Ecosyst Environ 43: 267–284

Udayagiri S, Welter SC (2000) Escape of *Lygus hesperus* (Heteroptera: Miridae) eggs from parasitism by *Anaphes iole* (Hymenoptera: Mymaridae) in strawberries: Plant structure effects. Biol Control 17:234–242

Ulmer BJ, Jacas JA, Peña JE, Duncan RE (2006) Effect of temperature on life history of *Aprostocetus vaquitarum* (Hymenoptera: Eulophidae), an egg parasitoid of *Diaprepes abbreviatus* (Coleoptera : Curculionidae). Biol Control 39:19–25

van Hamburg H, Hassell MP (1984) Density dependence and the augmentative release of egg parasitoids against graminaceous stalk borer. Ecol Entomol 9:101–108

van Lenteren JC (2000) Success in biological control of arthropods by augmentation of natural enemies, In: Gurr G, Wratten S (eds) Biological Control: measures of success. Kluwer, Dordrecht, The Netherlands, pp 77–103

van Lenteren JC, Bueno VHP (2003) Augmentative biological control of arthropods in Latin America. BioControl 48:123–139

Walker GP, Bayoun IM, Triapitsyn SV, Honda JY (2005) Taxonomy of *Aphelinoidea* (Hymenoptera: Trichogrammatidae) species attacking eggs of the beet leafhopper, *Circulifer tenellus* (Hemiptera: Cicadellidae), in California. Zootaxa 1068:1–25

Walker GP, Zareh N, Bayoun IM, Triapitsyn SV (1997) Introduction of western Asian egg parasitoids into California for biological control of beet leafhopper, *Circulifer tenellus*. Pan-Pacific Entomol 73:236–242

Wang B, Ferro DN, Hosmer DW (1999) Effectiveness of *Trichogramma ostriniae* and *T. nubilale* for controlling the European corn borer *Ostrinia nubilalis* in sweet corn. Entomol Exp Appl 91:297–303

Wang F, Zhang S, Hou S (1988) Inoculative release of *Trichogramma dendrolimi* in vegetable gardens to regulate populations of cotton pests. Les Colloques de lINRA 43:613–620

Wang Z, He K, Yan S (2005) Large-scale augmentative biological control of Asian corn borer using *Trichogramma* in China: a success story. In: Hoddle MS (ed) Proceedings of the 2nd international symposium on biological control of arthropods. USDA Forest Service Publication FHTET-2005-08, pp 487–494

Waterhouse DF, Norris KR (1987) Biological control: Pacific prospects. Inkata Press, Melbourne, Australia

Waterhouse DF, Norris KR (1989) Biological control – Pacific prospects. Supplement 1. Aust Centre Int Agric Res, Canberra, Australia

Wilson F (1960) A Review of the Biological Control of Insects and Weeds in Australia and Australian New Guinea. Commonwealth Agricultural Bureaux, Farnham Royal

Zuniga SE (1985) Eighty years of biological control in Chile: historical review and evaluation of the projects undertaken 1903-1983. Agricultura Tecnica 45:175–184

Chapter 16
Risk Assessment and Non-target Effects of Egg Parasitoids in Biological Control

Franz Bigler, Dirk Babendreier, and Joop C. van Lenteren

Contents

16.1 Introduction

In the past 100 years many exotic natural enemies have been imported, mass reared and released as biological control agents for pest control (Albajes et al. 1999, van Lenteren 2000, 2003, Lynch et al. 2000, USDA 2001, Mason and Huber 2002, Copping 2004). Although the majority of these releases have not resulted in unwanted side effects, some serious cases of non-target effects by exotic biological control agents against insects and weeds have been recently reported (e.g. Boettner et al. 2000, Follett and Duan 2000, Wajnberg et al. 2000, Louda et al.

F. Bigler (✉)
Agroscope Reckenholz-Taenikon Research Station ART, Reckenholzstr. 191, 8046 Zurich, Switzerland
e-mail: franz.bigler@art.admin.ch

F.L. Consoli et al. (eds.), *Egg Parasitoids in Agroecosystems with Emphasis on Trichogramma*, Progress in Biological Control 9, DOI 10.1007/978-1-4020-9110-0_16,

2003, van Lenteren et al. 2006a). Due to the current popularity of biological control, many new invertebrate biological control agents will become available. To reduce the chance of releasing exotic natural enemies that might pose a risk for the environment, guidelines and methods are being developed to assist in environmental risk assessment.

Various organizations have developed standards, including guidelines for the export, import, shipment, evaluation and release of biological control agents (e.g. EPPO 2002, IPPC 2005). Environmental effects of biological control agents form a central element of these guidelines and a growing number of countries already apply risk assessment procedures prior to the import and release of a new natural enemy. Earlier, we collected, studied and summarized procedures to assess risks of natural enemies currently used by about 25 countries and codes of conduct or guidelines produced by various organizations (van Lenteren and Loomans 2006). Within an EU funded project (van Lenteren et al. 2003), an OECD working group (Anonymous 2004) and an IOBC Commission (Bigler et al. 2005), guidelines have been developed to harmonize information requirements for import and release of invertebrate biological control agents. Based on all this information, we designed a new comprehensive method for environmental risk assessment. Subsequently, we also developed a quick scan to be used for natural enemies that are already in use (van Lenteren and Loomans 2006). In this way, we hope to provide biological control experts and risk assessors the tools for a proper and uniform evaluation of the information provided in the application. In this chapter, we summarize the development of risk assessment procedures for natural enemies, we then describe a stepwise risk assessment procedure, and we will apply this procedure to evaluate the environmental risks in a case study of *Trichogramma brassicae* Bezdenko.

16.2 Environmental Risk Assessment of Natural Enemies

Risk assessment procedures for biological control agents are usually characterized by questions on four issues: (i) characterization and identification of the biological control agent; (ii) health risks; (iii) environmental risks; and (iv) efficacy.

The kind of information needed to evaluate these issues are addressed in Anonymous (2004), van Lenteren et al. (2003) and Bigler et al. (2005), and information on the methods to be used to assess non-target effects are addressed in Babendreier et al. (2005) and Bigler et al. (2006). In this chapter we will concentrate on the third issue, but also shortly address the other issues. Assessment of risks related to releases of natural enemies demands integration of many aspects of their biology, as well as information on ecological interactions. A comprehensive risk assessment comprises the following steps: (i) identification and evaluation of potential risk of releasing a natural enemy, (ii) a plan to minimize risk and mitigate unwanted effects of biological control agents (e.g. Moeed et al. 2006), and (iii) a risk/benefit analysis of the proposed release of the natural enemy, together with risk/benefit analyses of current and alternative pest management methods (e.g. Bigler and Kölliker-Ott 2006).

The last step is essential, because the risk/benefit posed by the release of an exotic natural enemy might particularly be considered acceptable in comparison with the risks posed by other control methods. For definitions of terms used in this chapter, we refer to Anonymous (2003) and Bigler et al. (2006).

16.2.1 Risk Identification and Calculation of Risk Index

Normally, for a risk assessment, one will identify and evaluate the potential negative effects and determine the probabilities that these will materialize (e.g. Moeed et al. 2006, Bigler et al. 2006). The negative impacts of a biological control agent can be defined as any negative effect, which can be named and measured, such as direct and indirect negative effects on non-target organisms and negative effects on the environment. The risk of negative effects of the release of a biological control agent is the product of the likelihood (L) of impact and the magnitude (M) of impact. The likelihood and magnitude of five groups (ecological determinants) of risks are usually considered: establishment, host range, dispersal, direct effects, and indirect non-target effects. Next, qualitative scales for likelihood and magnitude need to be described (for likelihood from very unlikely to very likely, and for magnitude from minimal to massive), after which one may quantify the scales for likelihood and magnitude (see e.g. Tables 15.2 and 15.3 in van Lenteren and Loomans 2006). In an early version of an environmental risk assessment, a numerical value was added to each descriptor of likelihood and magnitude to be able to quantify risk (see van Lenteren et al. 2003). The overall risk index for each natural enemy was obtained by first multiplying the values obtained for likelihood and magnitude, followed by summing-up the resulting values obtained for establishment, dispersal, host range, direct and indirect effects. Based on an evaluation of 31 cases of natural enemy introductions, the following risk categories were proposed (van Lenteren et al. 2003): *i*) Low risk category: for organisms falling in this category, a proposal of no objection against release of the agent can usually be issued; *ii*) Intermediate risk category: for organisms falling in this category, the advise will be issued to come up with specific additional information before a conclusion concerning release can be drawn; and *iii*) High risk category: for organisms falling in this category, generally a proposal to not release the agent will be issued.

Low risk indices were found for many parasitoids, several predatory mites and one predatory insect. Intermediate risk indices were found for all guilds of natural enemies: parasitoids, predatory insects, predatory mites, parasitic nematodes and entomopathogenic fungi. The highest risk indices were found for several predatory insects and parasitoids, among which *T. brassicae*. At the time of this first risk evaluation exercise, we already realized that *T. brassicae* would end up in the high risk category because it was said to have a broad physiological host range and that it could disperse to non-target habitats. However, very limited data were available for the actual host range and for direct or indirect effects caused by *T. brassicae* for Europe. We will show later in this chapter that newly collected information has lead to a modification of the risk categorization of *T. brassicae*. Because this was the first quantitative risk assessment developed for natural enemies, it was foreseen that the quantification system might have to be adapted based on growing experience. The main problems encountered with this risk assessment were (i) information for the likelihood and magnitude of all five areas of assessment needed to be available before an evaluation could be made. This makes the assessment in a number of cases unnecessarily costly; (ii) the assessment did not identify candidate natural enemies that appear to be clearly unacceptable for import and release based on data for one group of risks (e.g. establishment or host range) early in the process. This should be improved to prevent unnecessary data collection; (iii) the numerical values calculated with this assessment did not allow a very clear separation between risk categories. This may result in interpretation and decision making that can easily be manipulated; (iv) the overall risk index was obtained by adding five different categories which are in fact not completely independent from each other and should not be rated equally; and (v) the overall score of a certain species for a certain eco-region might lead to establishing an absolute value and unnecessary strict administrative need for measures.

Therefore, we designed a new environmental risk assessment (see Section 16.3), which is now a stepwise procedure and includes weight factors to solve the problems mentioned above (van Lenteren et al. 2006a).

16.2.2 Risk Management

The next step of a risk assessment process is to discuss risk management, including risk mitigation and risk reduction. If an exotic biological control agent is expected to cause significant adverse effects on non-target organisms, a permit for releases will not be issued. In some cases, risks may be minimized by imposing restrictions concerning for example the types of crops on which the use of the organism is or is not allowed (e.g. treatment of flowering plants with a myco- insecticide), by requesting specific application techniques (e.g. soil incorporation only for insect pathogenic nematodes), or by specifying the eco-regions were the organism is allowed for use (e.g. use of tropical natural enemies in greenhouses in temperate climates).

16.2.3 Risk/Benefit Analysis

The last step in making a justified environmental risk analysis for a new biological control agent is to conduct a risk/benefit analysis which should include a comparative performance of pest management methods. The environmental benefits of use of the proposed biological control agent should be compared to environmental effects of currently used and other alternative control methods. Then, the environmental risk analysis is used in the overall risk/benefit assessment where the data concerning characterization, health risks, environmental risks and efficacy of all the control methods for a specific pest will be compared (for details see van Lenteren et al. 2003, 2006a, Bigler and Kölliker-Ott 2006).

16.3 A New Stepwise Risk Assessment Procedure

Recently, as a follow up to the first quantitative risk assessment, an environmental risk assessment method was developed consisting of a stepwise procedure which can be used for all types of invertebrate biological control agents in augmentative and classical biological control, for relevant species or biotypes (e.g. in the case of biotypes that diapause or not, or biotypes with and without wings), whether they are native, established exotics or not yet established exotics (van Lenteren and Loomans 2006, van Lenteren et al. 2006a). Contrary to the first quantitative risk assessment summarized in the previous section, here the decision to advise release or not is taken after each relevant step in the process, thus preventing unnecessary research and resulting in early elimination of clearly risky natural enemies (Fig. 16.1).

At *step 1*, exotic and native natural enemies are distinguished. For native natural enemies only one more step (step 6) in the procedure needs to be followed. Dispersal (step 5) of native agents may be an important issue to be considered in order to address step 6 accordingly. For example, direct and indirect effects of a polyphagous biological control agent may be limited because of very limited dispersal (e.g. often relevant for *Trichogramma*). However, because experimental procedures to establish the dispersal potential of natural enemies might be quite lengthy, this is not included

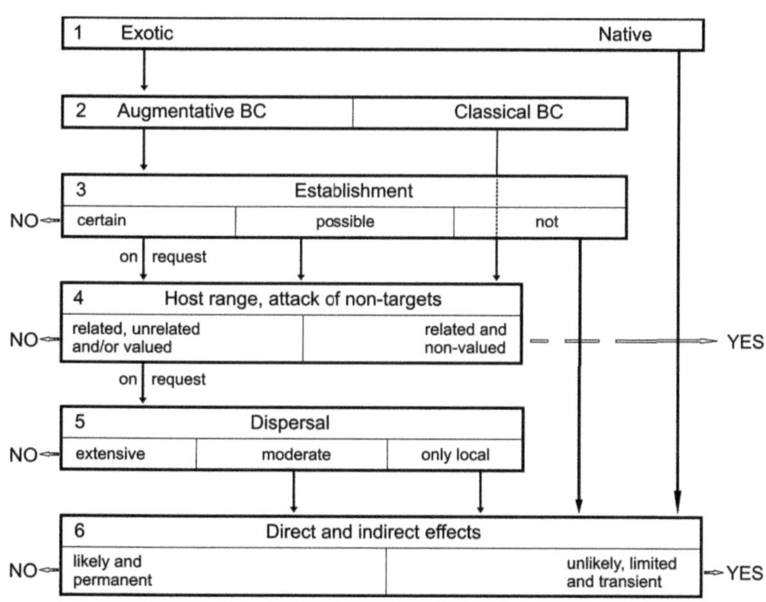

Fig. 16.1 Environmental risk assessment scheme for arthropod biological control agents. NO: release is not recommended; YES: release is recommended. On request: when the applicant desires, information about following issue(s) in the risk assessment scheme can be provided to allow reconsideration of the decision not to release the species (after van Lenteren and Loomans 2006)

here as a standard procedure for native natural enemies. For exotic natural enemies, whether already present or absent in the target area, more steps need to be followed.

At *step 2*, natural enemies that are aimed for augmentative biological control (ABC) programs where establishment of the organism in the area of release is not intended, are separated from natural enemies aimed for classical biological control (CBC) where establishment is the aim. For ABC natural enemies one then needs to demonstrate that they cannot establish in step 3.

If the natural enemy cannot establish (*step 3*), one more step of the procedure (step 6) needs to be followed. However, if it can establish, the Environmental Risk Index (ERI= Likelihood (L) × Magnitude (M)) should be calculated for establishment (van Lenteren and Loomans 2006). If the risk threshold is crossed, the natural enemy should not be released, and is thus eliminated early in the evaluation process. However, if the applicant desires, he can provide data from studies on host range (step 4), dispersal (step 5) and direct/indirect non-target effects (step 6) and ask for the decision to be reconsidered. If the risk threshold is not crossed, the same procedure needs to be followed as for CBC natural enemies in step 4.

At *step 4*, the host range issue (see van Lenteren et al. 2006b) is addressed. If the ABC or CBC agent is either monophagous or oligophagous/polyphagous and attacks only related AND no valued non-targets, i.e. species not of conservation concern, it should be considered for release. On the other hand, if the agent is oligophagous/polyphagous and does attack related and unrelated non-targets AND/OR valued non-targets, the agent should not be considered for release. However, if the applicant desires, he can provide data from studies on dispersal (step 5) and direct/indirect non-target effects and ask for the decision to be reconsidered. In that case, continue with step 5. On request, dispersal can be considered relevant for risk assessment of augmentative releases (see Mills et al. 2006).

At *step 5*, questions about dispersal of ABC and CBC (where appropriate and on request) agents are addressed. If dispersal is local and mainly in the area of release, the procedure can be continued at step 6. But if dispersal is outside the target area AND is extensive, the agent should not be released.

At *step 6*, issues related to direct and indirect non-target effects are addressed as releases of exotic agents may negatively affect the abundance of native non-target species or other natural enemies that exploit the same resource (see Messing et al. 2006). If direct and indirect effects inside the 'dispersal area' are unlikely AND at most transient and limited, the agent can be released. However, if direct and indirect effects inside the 'dispersal area' are likely OR permanent, the agent should not be released.

To calculate risk levels for establishment, dispersal and direct/indirect non-target effects, the criteria are applied as given in van Lenteren et al. (2003), but weight factors are added, and the resulting values can be obtained from information given in van Lenteren et al. (2006a). Threshold values as indicated in van Lenteren et al. (2006a) are currently still largely based on expert judgement, and therefore these values need justification and further fine-tuning. Here, accuracy and stringency are likely to increase as more data become available through experimental research. The final part of this new risk assessment, i.e. the risk management and the risk/benefit analysis, is the same as described in the previous section.

The stepwise risk assessment procedure has successfully been applied to the 150 species of natural enemies that are currently commercially available in Northwest Europe (producer's information on the web; producer's price lists in Loomans 2004). The following conclusions could be drawn after this exercise (van Lenteren and Loomans 2006): (i) all native species that were evaluated are considered safe for release, (ii) exotic species intended for use in augmentative biological control that are likely to establish and cross the risk threshold are detected very early in the evaluation process, and will be excluded from release without the need to study host range, dispersal and direct/indirect non-target effects, (iii) exotic species that are monophagous or oligophagous/polyphagous with attack of only related and no attack of valued non-targets are also detected early in the evaluation without the need to study dispersal and direct/indirect non-target effects; they can be released, and (iv) exotic species that are oligophagous/polyphagous and attack related and unrelated non-targets and/or valued non-targets will be excluded from release without the need to study dispersal and direct/indirect non-target effects.

The early elimination of obviously risky species, and the acceptance of other species that scored – erroneously – a high index in the previous assessment by van Lenteren et al. (2003), like *T. brassicae*, clearly show improvements of the stepwise assessment proposed in van Lenteren and Loomans (2006).

16.4 A Case-Study with *Trichogramma brassicae* in Western Europe

The egg parasitioid *T. brassicae* was collected in Moldavia (northern Black Sea area) in 1973 and introduced to France for rearing and studying its potential for augmentative biological control against the European corn borer, *Ostrinia nubilalis* Huebner. The original founder population (strain number 16 in the collection of INRA, Antibes) consisted of only one female and three male individuals (Pintureau et al. 1981). The species was first determined as *Trichogramma maidis* Pintureau and Voegele (Pintureau and Voegele 1980) and later recognized as being *T. brassicae*, formerly described in Russia by Bezdenko in 1968 (Pintureau 1990). The species was first reared and studied at INRA in Antibes, France, and material of the same strain was given to Switzerland in 1975 where the first experimental field releases were performed in the same year (Suter and Baebler 1976). First field releases were made with the same strain in Germany in 1976 (Neuffer 1981). After development of the first mass rearing methods in the mid-seventies (Bigler et al. 1989), the first commercial inundative field applications were carried out in 1978 (Bigler 1986).Continuous improvement of mass rearing, diapause storage and release strategies were the keys to make *T. brassicae* a successful biocontrol agent in maize (and to a minor extent on vegetables in glasshouses) with a constantly growing market since. Commercial releases were made in 2007 in France, Germany, Czech Republic and Switzerland on a total of 140,000 hectares of maize (Jacques Frandon, Biotop, pers. comm.).

Environmental effects of *T. brassicae* were studied in detail from 1998 to 2002 in the frame of a European Union funded research project "Evaluating Environmental Risks of Biological control Introductions into Europe, ERBIC. *Trichogramma brassicae* was chosen for this project as an example of a generalist egg parasitoid. Laboratory and field studies were carried out in Switzerland where the climate is characterized by cold winters and mild summers.

16.4.1 Potential for Establishment of Trichogramma brassicae

To permanently establish *T. brassicae* in Central and Western Europe, a few conditions must be met. First, abiotic factors such as temperature and humidity must allow overwintering. Second, suitable host eggs must be available during the vegetation period allowing the parasitioid to permanently propagate and finally to overwinter in diapause. Since annual crops are not suited for permanent establishment, other habitats such as perennial crops or non-crop habitats must be within dispersal distance. Finally, appropriate plants providing shelter and adult food increase longevity and dispersal capacity and thus the likelihood of finding suitable hosts.

Overwintering

It is known that *T. brassicae* overwinters in most regions of its distribution as diapausing pre-pupae (Pizzol and Voegele 1988, Voegele et al. 1988). All other stages die out during winter under Central and Western European conditions. Hence, it is obvious that *T. brassicae* needs suitable host eggs towards the end of the vegetation period when diapause induction occurs. Because egg-laying of the target host, *O. nubilalis,* ends too early in the season and maize plants are destroyed when harvested, eggs for overwintering must be found outside maize fields. Most suited are undisturbed habitats where non-target lepidopterans lay their eggs.

In our project, eggs of six lepidopteran species were selected for overwintering studies based on availability and the potential to serve as an overwintering hosts in Switzerland under natural conditions. *T. brassicae* was allowed to parasitize host eggs under laboratory conditions between September and November and thereafter eggs were exposed to natural climate under outdoor conditions and reared in the laboratory as a control. Emergence was checked in the following spring and emergence rates determined (for details of the methods see Babendreier et al. 2003a). In order to evaluate the fitness of overwintering females, fecundity of females was measured which emerged from parasitized eggs that passed the winter outdoors.

Results of the overwintering studies showed that host eggs parasitized before mid-September did not enter diapause and all adults emerged in fall of the same year. Host eggs that were parasitized in late September and exposed under outdoor conditions showed emergence rates next spring (mid-April to mid-May) between

75% and 100% for all tested host species (Babendreier et al. 2003a). Emergence from eggs parasitized at the beginning of October was variable but still high while on later exposure dates, overwintering abilities decreased as the parasitoid failed to reach the pre-pupal stage before winter. Consequently, spring emergence was low (0–42%) for eggs exposed in late October and no overwintering was observed from host eggs exposed in November. Our results show that overwintering rates of *T. brassicae* are highest in Switzerland if host eggs are parasitized between mid-September and early October We found that *T. brassicae* was able to overwinter north of the Alps on eggs of various host species despite the fact that temperatures dropped below −10°C during 14 days and reached −15 to −20°C on four days during the winter 1998/1999. Survival at such low temperatures indicates considerable cold hardiness of *T. brassicae*.

Finally, the fecundity of females which emerged after about five month in diapause did not differ from fecundity of females reared under laboratory conditions.

These results obtained during two winter periods demonstrate that *T. brassicae* can overwinter in different host eggs in Central and Western Europe. Surveys of host egg availability under natural conditions in Switzerland have shown that a number of lepidopterans overwinter in the egg stage and may serve *T. brassicae* as hosts (Babendreier et al. 2003a).

Persistence and Establishment in Non-target Habitats

Field studies were carried out in different regions (south and north of the Alps) of Switzerland to evaluate whether and to what extent *T. brassicae* disperses out of maize fields into adjacent non-target habitats and whether the parasitoids persist and establish in such habitats (Kuske et al. 2003, Babendreier et al. 2003d). The European corn borer accomplishes one generation per year north and two generations south of the Alps and the release strategy of *T. brassicae* differs accordingly. Egg cards baited with either *Anagasta* (= *Ephestia*) *kuehniella* (Zeller) or *Mamestra brassicae* eggs were used to monitor *T. brassicae* adults in non-target habitats in the surroundings of farmer's maize fields. Non-target habitats considered were field edges, extensively managed meadows with high biodiversity (no fertilizer, low amount of manure, late first cut and 2–3 cuts in total per year only), sown wildflower strips, hedgerows and reeds which are expected to provide food, hosts, shelter and overwintering sites for emigrating *T. brassicae*. Egg baited cards were attached to wild plants and to maize plants during two seasons at regular intervals and collected after a few days and incubated in the laboratory until emergence of adult *T. brassicae*.

Sticky trap surveys were made to quantify dispersal of adult *T. brassicae* from maize fields into adjacent non-target habitats. Traps consisted of transparent plastic sheets that were sprayed with insect glue and placed directly above the vegetation or inside the vegetation in maize, reeds and hedgerows. Species identification of individual *Trichogramma* emerged from egg baited cards and collected on sticky traps was conducted by PCR according to the method of Stouthamer et al. (1999).

In both years, *T. evanescens* was the most abundant species observed (Fig. 16.2). With respect to the environmental risk assessment it is important to note that *T. brassicae* was already present from previous years prior our inundative releases, and repeatedly found throughout the season in both years. The number of egg cards parasitized by *T. brassicae* in non-target habitats was slightly increased after commercial release in adjacent maize fields in 1999, but not in 2000. These results prove that *T. brassicae* can survive and overwinter under natural conditions in southern Switzerland and persist all year round in non-target habitats. Earlier surveys in the same region (Bigler et al. 1990), prior to the first releases of *T. brassicae* in the study area indicated absence of this species in maize fields with high egg mass densities of the European corn borer. It is thus evident that *T. brassicae* was established in southern Switzerland through consecutive commercial releases in this area. The surveys also show that *T. brassicae* enters non-target habitats that are exploited by native *Trichogramma* species.

Similar results were obtained from sticky trap surveys in the same area. Prior to the inundative commercial releases of *T. brassicae* a very low proportion of this species was found in wildflower strips, but not in maize and reed. After the first and second release, the proportion increased strongly for a period of one to two weeks and dropped afterwards to numbers found prior to release (for details see Kuske et al. 2003).

Host Plant and Habitat Exploitation

Trichogramma species are known to have habitat and host plant preferences and exploit habitats with different success (Rabb and Bradley 1968, Thorpe 1985, Romeis et al. 1998, 2005). Such preferences may also affect the impact of *T. brassicae* on non-target lepidopterans living on different plants in non-crop habitats. Laboratory and field experiments were carried out to investigate searching efficiency of *T. brassicae* on different plants and its impact on parasitism. Field experiments were carried out in meadows, in sown wildflower strips and in hedges. These habitats were chosen because many butterfly species, including endangered ones, may find their host plants there. These habitats are important for biodiversity conservation in agricultural landscapes and they are of growing importance in Switzerland and other European countries. Egg clusters of *A. kuehniella* were glued to the lower side of leaves of typical plants in the respective non-target habitats and on leaves of maize plants (for details see Babendreier et al. 2003d).

Searching efficiency of *T. brassicae* on common meadow plants was compared to maize by measuring in laboratory experiments walking speed of females, turning rates and time spent on plants. All plants chosen are known hosts of indigenous common and endangered butterfly species. The plants differed in characters such as trichome density and length, leaf surface and shape and wax layer (Babendreier et al. 2003d). Parasitism was assessed in glasshouse and field cages by exposing egg clusters on different plant mixtures to a known number of females and in cages installed in non-target habitats and compared to parasitism on maize plants. Eggs of *A. kuehniella* were glued in clusters on the plants and later incubated in the laboratory until adult emergence.

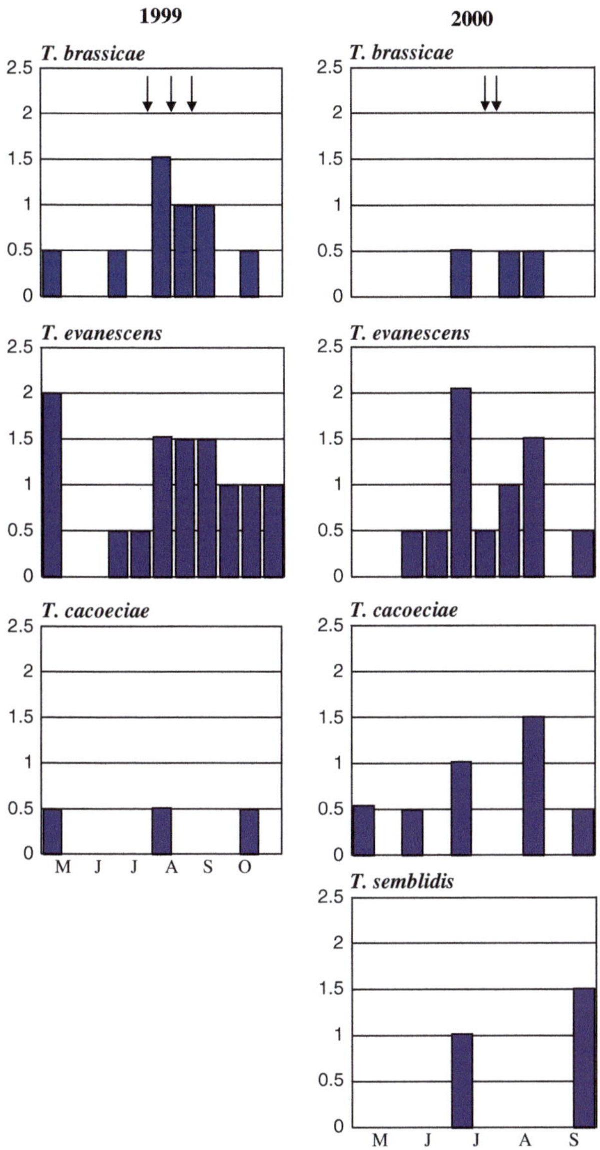

Fig. 16.2 Seasonal distribution of *Trichogramma* species in non-target habitats in southern Switzerland. *Trichogramma* were monitored by using egg baited cards. The survey was carried out from mid-May to the beginning of October in 1999 and to the end of September in 2000. Inundative commercial releases of *T. brassicae* were carried out in adjacent maize fields (indicated by arrows) (Kuske et al. 2003)

Our results confirm known effects of leaf morphological characters on walking speed of females and on turning angles (Babendreier et al. 2003d) and we conclude from our data that architecture, size and leaf characters of common plants from non-target habitats may have strong implications on host searching efficiency and parasitism of *T. brassicae*. Parasitism of glued host eggs on different plant species exposed in glasshouse cages to *T. brassicae* females confirmed our findings on plant preference and searching efficiency. Parasitism of eggs glued on maize plants was always higher than on any plant from non-target habitats and among them large differences were observed as well.

Parasitism in non-target habitats under caged conditions confirmed the results shown above. Again, parasitism was significantly lower in all three non-target habitats compared to maize (Fig. 16.3).

In field release experiments we were able to show that parasitism was significantly reduced when *T. brassicae* was searching for hosts in meadows or in flower strips compared with maize. Parasitism in a meadow and in wildflower strips was between 1% and 4%, but reached 20% to 67% in maize (Fig. 16.4a, b). The same experiment in the hedgerow failed because unknown predators preyed heavily upon the sentinel egg clusters of *A. kuehniella* leaving virtually no eggs for the analyses.

Our results are in line with findings of Orr et al. (2000), who released *T. brassicae* into non-target habitats such as wetlands, old fields and forests and found parasitism of sentinel egg masses between 2.1 and 5.5%, which was significantly lower than in maize and very close to our parasitism rates. Meadows, wildflower strips and hedgerows have a much higher structural complexity than maize due to plant characters and architecture of the plant community. These data suggest that mass releases of *T. brassicae* into maize with dispersal into adjacent non-target habitats poses a low risk to populations of butterflies and other potential hosts.

In summary, the results show that *T. brassicae* can overwinter and establish in different non-target habitats in Central and Western Europe. The data suggest that host searching on plants other than maize is much reduced due to unfavourable characters of wild plants and to high structural complexity of non-target habitats. This leads to very low parasitism of non-target hosts in such habitats and consequently to a low risk for non-target species.

Fig. 16.3 Parasitism of *Trichogramma brassicae* on eggs of *Anagasta kuehniella* in maize and non-target habitats in field cages. Parasitism was compared with the Mann-Whitney U-test (***= $P < 0.001$) (Babendreier et al. 2003b)

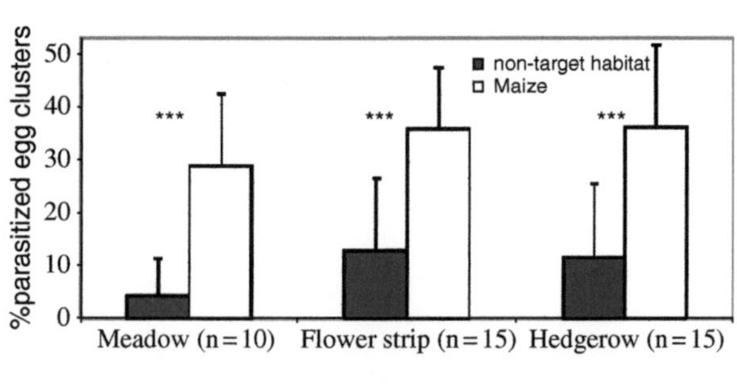

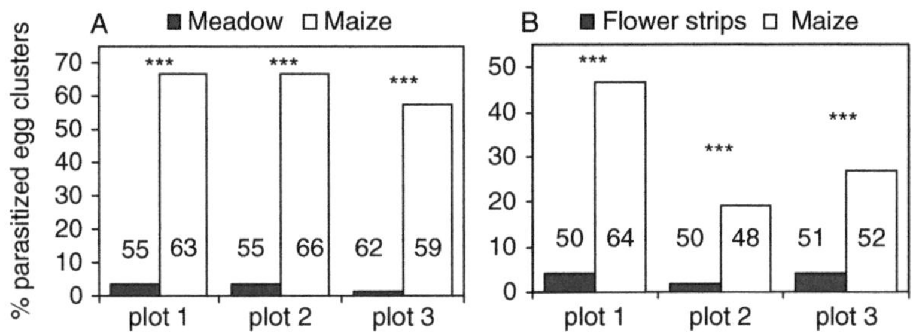

Fig. 16.4 Parasitism of *Trichogramma brassicae* on eggs of *Anagasta kuehniella* in two non-target habitats (A: meadows, B: flower strips) compared to maize. Indicated is the percentage of egg masses found parasitized. Number of egg masses included in the analyses is shown above or inside the bars. Bars marked with asterisks indicate a significant difference between parasitism rate on eggs of *A. kuehniella* between non-target habitats and maize (χ^2 test; $^{*}P < 0.05$; $^{**}P < 0.01$; $^{***}P < 0.001$) (Babendreier et al. 2003b)

16.4.2 Host Range Testing of Non-target Lepidoptera and Aphid Predators

Evaluation of the host range of a natural enemy is a central issue in any biological control programme and it is generally hypothesized that polyphagous agents have the potential for non-target effects (Howarth 1991, McEvoy 1996, Kuhlmann et al. 2006). The vast majority of *Trichogramma* species are known to be polyphagous attacking a wide range of lepidopterans and species of other insect orders (e.g. Clausen 1940, Thomson and Stinner 1989, Pinto and Stouthamer 1994).

Within the ERBIC project, a sequential test procedure was worked out (Fig. 16.5) which is described in more detail by van Lenteren et al. (2003, 2006a). The general idea is to start with simple Petri dish experiments in the laboratory with or without direct observations to see whether hosts are attacked at all. In subsequent steps one may proceed under semi-natural conditions (e.g. in cages) with those hosts that have been attacked in earlier tiers. Finally, field experiments (e.g. field cages) may be needed if results from the previous tier still indicate that potential risks could not be ruled out.

16.4.2.1 Non-target Lepidoptera

From the literature it is known that *T. brassicae* is a polyphagous parasitoid which has been found on a wide range of lepidopterans (Fulmek 1955, Orr et al. 2000). However, information on butterflies attacked by *T. brassicae* in Central and Western Europe is scarce and, therefore, we embarked upon a host-range study. The selection from the 180 butterfly species known from Switzerland (SBN 1987) was at first based on spatial and temporal overlap with *T. brassicae* mass releases (Babendreier et al. 2003b). After overwintering had been confirmed on some butterfly species in Switzerland, other lepidopteran species were included given that *T. brassicae*

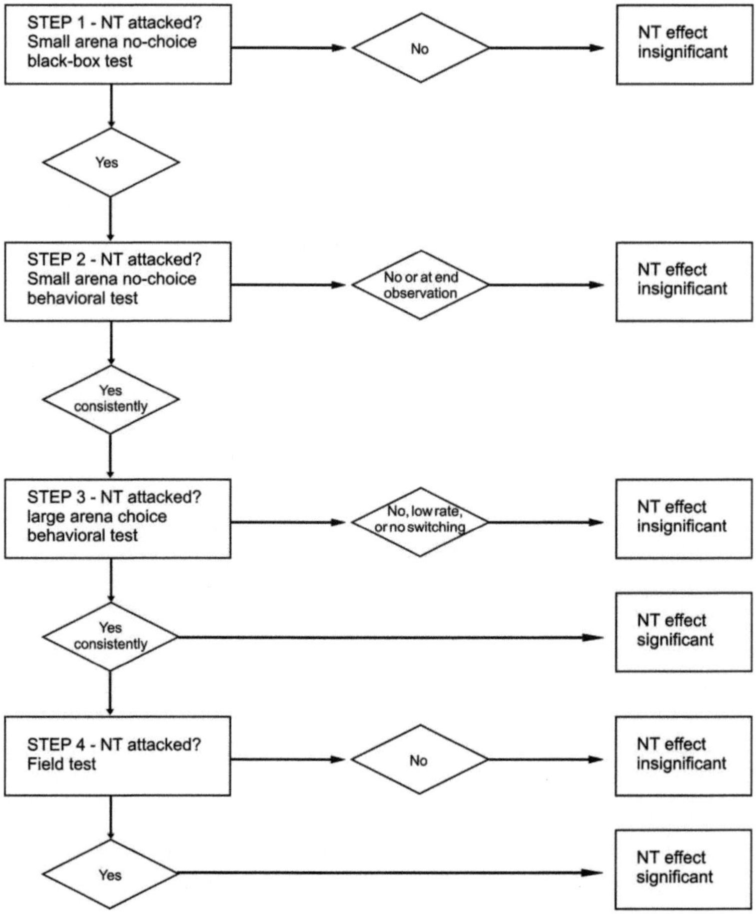

Fig. 16.5 Flow chart summarizing host range assessment of arthropod biological control agents. NT: non-target (van Lenteren et al. 2006a)

could potentially be present also far from release fields through consecutive dispersal over years. Major Lepidoptera families were covered and, for the larger families, several species were chosen. Since endangered species were of special concern, some butterflies from the red list of Switzerland (Duelli 1994) were also included.

For our laboratory experiments we selected a total of 23 non-target lepidopteran species from major families of butterflies and moths. *Anagasta kuehniella* (the rearing host) and *O. nubilalis* (the target) were also tested (for a full list of species tested see Babendreier et al. 2003b). For subsequent testing under field cage or field conditions, a subset of species was used based on acceptance and parasitism in the no-choice tests carried out in the laboratory.

Special attention was given to two lepidopteran hosts of the indigenous larval parasitoid *Lydella thompsoni* Herting, which is an abundant tachinid parasitoid of the European corn borer, *O. nubilalis* throughout southern Europe (Galichet et al. 1985, Manojlovic 1985, Eizaguirre et al. 1990, Grenier et al. 1990) and was found to be the most important natural enemy of *O. nubilalis* in southern Switzerland as well (Kuske et al. 2004). To complete its first (spring) generation, *L. thompsoni* parasitizes larvae of lepidopteran stem borers in pristine habitats, such as *Archanara geminipuncta* Haworth and *Chilo phragmitellus* Hüebner that live on common reed. Later in the season, when European corn borer larvae become available in maize fields, *L. thompsoni* adults migrate into maize fields where they develop the second and third generation on *O. nubilalis* larvae. We hypothesised that inundative releases of *T. brassicae* might affect populations of *L. thompsoni* either directly through competition within maize fields and/or indirectly through population reduction of the tachinid's spring hosts *A. geminipuncta* and *C. phragmitellus* in adjacent non-target habitats. We assumed that a strong population reduction of the two above mentioned spring hosts could lead to severe population density impacts for *L. thompsoni* (see also under indirect effects).

Laboratory Testing

Single eggs or egg masses of the 23 non-target butterfly and moth species were used for the laboratory experiments. Single, non-experienced *T. brassicae* females were continuously observed on host eggs which were placed in a 40 mm Petri dish. An egg was regarded as being accepted when the female penetrated or tried to penetrate the ovipositor continuously for longer than 60 s. Parasitism was calculated from parasitized (black) eggs after 5 days and emergence of parasitoid offspring was evaluated. Percent egg parasitism was compared between each non-target and the target *O. nubilalis* by a χ^2 test.

Eggs of the majority of non-target species were readily accepted with acceptance rates between 75 and 100%. Eggs of only one species were totally rejected by *T. brassicae* females very early in the host selection process after drumming the egg for a few seconds or even at 1–3 mm distance without coming into contact with the eggs. Eggs of another species were accepted at a low level (below 10%) and eggs of one species were accepted at an intermediate level (60%). Twelve out of the 23 species tested had parasitism rates that were not significantly different from that of the target *O. nubilalis.* Four species had parasitism rates higher and six lower than the target. Parasitism was recorded for all egg species that were accepted showing that accepted host eggs are in general suited for offspring development. Eggs with acceptance rates below 10% yielded normal offspring as all the other parasitized eggs did.

Acceptance and parasitism of *A. geminipuncta* and *C. phragmitellus* were tested under laboratory conditions in no-choice experiments (Kuske et al. 2004). As eggs of *A. geminipuncta* are hidden underneath the leaf sheaths of common reed plants, they are not accessible by *T. brassicae* and therefore parasitism was practically zero. Eggs of *C. phragmitellus* are deposited freely on leaves of common reed plants and are thus well accessible by *T. brassicae*. These eggs were though not attractive, hardly accepted and parasitism remained thus very low.

Semi-field Testing

In a next tier, an experiment was carried out in field cages (2 × 2 × 2 m, situated outdoors within a meadow) to investigate parasitism under more natural conditions (Babendreier et al. 2003c). Eggs of six non-target butterfly species that were well accepted and parasitized in the laboratory tests were glued on their respective host plant (four species) or naturally laid by adult females on their host plant (two species). In addition, eggs of *A. kuehniella* were glued on these plants as a control, and high numbers of *T. brassicae* adults (780 ± 273 cage) were released and exposed to the eggs for 24 h at ambient weather conditions. Adult density in the cages was thus 16 times the number that is normally released in maize in central Europe against the European corn borer. Eggs were then incubated under optimal laboratory conditions and parasitism assessed after 5 days. Percent parasitism was calculated from all eggs of each butterfly species exposed in the cage and analyzed by ANOVA.

Mean 24 h parasitism rates of the six butterfly species ranged from 2.5% to 18.7% depending on the species and the host plant, and were similar to the mean parasitism of eggs of the mass rearing host *A. kuehniella* glued on the same host plants. Based on the daily parasitism and the 16 fold adult density in the cages compared to the normal field release, we calculated a 24 h parasitism rate ranging for the non-target hosts from 0.15 to 1% and for *A. kuehniella* from 0.45 to 1%. These figures show that parasitism of non-target butterfly eggs are low at normal field release density of *T. brassicae* and depend mainly on the host plants on which the eggs are deposited.

Field Testing

As a final tier in the assessment of attack and parasitism of non-target butterflies, field trials were carried out in maize fields and in adjacent extensively managed meadows. We released 30,000 *T. brassicae* females as commercially recommended in a plot of 50 × 50 m. All release plots were situated inside maize fields but bordering the meadows. We then exposed eggs of two non-target hosts which have been attractive and highly parasitized in earlier tiers [i.e. *Hipparchia alcyone* (Denis and Schiffermuller) and *Melanargia galathea* (L)] using *A. kuehniella* eggs as a control. Eggs were exposed for three days at 2 m distance inside the maize field and at 2 and 20 m distance from the edge of the maize field in the meadow. In the meadow as well as in the maize, eggs were attached to the underside of leaves of the plants.

Average parasitism in maize fields was 40% for *A. kuehniella* eggs, and it was significantly lower for eggs of the non-target hosts (Fig. 16.6). At 2 m distance from the edge of the maize field parasitism dropped below 10% for *A. kuehniella* and was again lower for the two non-target host eggs. At 20 m distance from the edge of the maize field, no single egg of either host was parasitized.

Eggs of *A. geminipuncta* and *C. phragmitellus* were exposed in a common reed stand located in-between seed maize fields where *T. brassicae* was released at a high density (1.2 million per hectare and season). Egg clusters of *A. geminipuncta* with a total of 3775 eggs were exposed for eight days on reed plants or on pieces of reed stalks at 10–25 m distance from the maize field border. Egg masses of *C. phragmitellus* with a total of 2997 eggs were exposed for seven days on reed

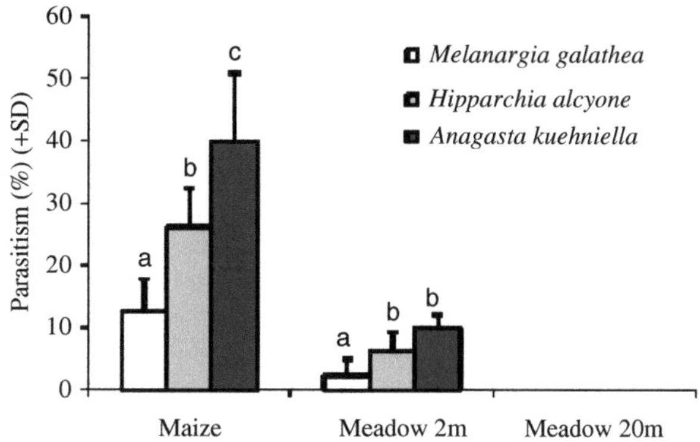

Fig. 16.6 Parasitism of *Anagasta (Ephestia) kuehniella* and non-target eggs inside of maize and inside of a meadow at 2 and 20 m distance from the edge of a maize field where *Trichogramma brassicae* had been released (Babendreier et al. 2003d)

plants or pieces of reed stalks at about 30 m distance from the borders of release fields. After collection, the eggs were incubated in the laboratory to observe parasitism (see details in Kuske et al. 2004).

Percent parasitism of eggs of *A. geminipuncta* on common read plants was generally high; however, all parasitoids belonged to the genus *Telenomus* and were assigned to the *T. busseola* species-complex. Not a single *T. brassicae* specimen or any other *Trichogramma* species was present. The exposure of *C. phragmitellus* egg masses yielded not a single egg parasitized by *Trichogramma*.

From our studies with non-target lepidopterans, we concluded that *T. brassicae* successfully parasitizes eggs of a broad range of species under no-choice laboratory conditions. Under cage and field conditions with different plants and more complex habitat structures however, parasitism drops drastically and within a short distance away from the release sites, there is hardly any non-targets parasitized. The two most important alternative hosts for the larval parasitoid *L. thompsoni, A. geminipuncta* and *C. phragmitellus* were unaffected by mass releases of *T. brassicae*.

16.4.2.2 Non-target Natural Enemies

We investigated whether *T. brassicae* might parasitize eggs of natural enemies present in maize. Predators of the European corn borer and other maize pests were included as non-target beneficial insects into the risk assessment (Babendreier et al. 2003e). As important natural enemies in maize fields we selected predators which are regularly present in maize and mainly keep aphids under natural control. Eggs of the green lacewing *Chrysoperla carnea* (Stephens), the syrphid *Episyrphus balteatus* (Degeer) and the two coccinellids *Coccinella septempunctata* L and *Adalia bipunctata* (L) are freely accessible for *T. brassicae* and were therefore chosen for testing (Babendreier et al. 2003e).

Laboratory Testing

In a first tier, a no-choice black box experiment was carried out. Single *T. brassicae* females were put in glass vials (75 × 13.5 mm) together with eggs of *C. carnea, A. bipunctata, C. septempunctata* or *E. balteatus* for 24 h. As a control,

T. brassicae females were allowed to parasitize 30–50 eggs of the mass rearing host, *A. kuehniella*, glued on a cardboard strip.

In order to get more detailed information on acceptance behavior, the same species were observed in direct observations. Eggs of *C. carnea* were offered in two variants: (a) on top of their stalks, as they are naturally deposited and (b) laid onto the substrate without stalks. The aims were to assess to what extent eggs are protected from parasitization when deposited on stalks and how well they are accepted by *T. brassicae* when found. Each observation started by introducing an individual female *T. brassicae* into a small plexiglass ring covered with a glass lid. The arena was then placed over the egg batch and the wasp behavior was recorded using the Observer Video-Pro software package.

The results indicate that *T. brassicae* parasitized eggs of *C. carnea* and *E. balteatus* with viable offspring produced whereas eggs of the two coccinellids did not yield any offspring. Direct observations showed that all eggs of the four predators are less accepted than eggs of lepidopterans. However, eggs of *C. carnea* and *E. balteatus* are still quite attractive. Half of the eggs of *A. bipunctata* were attacked by *T. brassicae* females, but without producing offspring, whereas eggs of *C. septempunctata* were not attacked.

Semi-field Testing

Only eggs of *C. carnea* and *E. balteatus* successfully parasitized by *T. brassicae* under laboratory conditions were further tested in cages under glasshouse conditions. Three maize plants with a total of 35 *C. carnea* (naturally laid) or 36 eggs of *E. balteatus* (glued on maize plants) were placed for 24 h in one cage (40 × 40 × 70 cm). In each cage ten *T. brassicae* females were released. As a control we exposed three maize plants of the same size with two egg clusters of *A. kuehniella* on each plant (30–50 eggs per egg cluster) in a cage together with ten *T. brassicae* females for 24 h. The eggs where checked for parasitism after 5 days.

Parasitism reached 7% for *C. carnea* eggs and 0.4% only for *E. balteatus* eggs compared to 21–27% for eggs of *A. kuehniella.* The data indicated and confirmed direct observation results that syrphid eggs are not attractive and are hardly attacked by *T. brassicae* on the plant. *C. carnea* eggs seem to have some attractiveness and could be used as a host by *T. brassicae*.

Field Testing

Since only *C. carnea* appeared to be potentially at risk, we tested this species in a final tier under field conditions. Experimental plants with naturally laid *C. carnea* eggs were exposed in a maize field where *T. brassicae* has been released at commercial field rate. The plants together with their pots were exposed in the maize fields to make up three treatments: (1) maize plants with *C. carnea* eggs and two *A. kuehniella* egg clusters; (2) maize plants with only *C. carnea* eggs; (3) maize plants with only two *A. kuehniella* egg clusters. One replicate consisted of 24 treated plants, eight of each treatment. All treated plants were removed from the field after 48 h and checked for parasitism 5 days later. Trichogrammatids emerging from *C. carnea* eggs were subjected to PCR for species identification.

The field experiment indicated that only 3.1% of the *C. carnea* eggs were parasitized by *T. brassicae* and 1.3% was parasitized by a naturally occurring *Telenomus* species. In comparison to the 30% loss of eggs for unknown reasons in the experiments, parasitism by *T. brassicae* is negligible and thus not a risk to *C. carnea.*

From our studies we concluded that none of the four aphid predators is at risk through commercial *T. brassicae* releases and does not seem to have any negative impact on natural aphid control (Babendreier et al. 2003e).

16.4.3 Dispersal

Dispersal in biological control with egg parasitoids has often been studied to optimize distance between release points that allow the parasitoids to "inundate" the crop completely within an expected time laps to give a good control of the pest (Keller et al. 1985, Suverkropp 1997). Information on the dispersal capacity of biological control agents is also important to answer questions such as: What is the proportion of the released population leaving the crop in what time? What is the distance travelled in time? Is it a directed or undirected movement? Is there a preference for specific habitats? Obviously, this information is also needed to analyse dispersal of mass released *Trichogramma* as a basis for assessing risks posed to non-target insects in non-crop habitats.

Earlier studies have shown that dispersal of *Trichogramma* species released in different habitats can vary strongly depending on a multitude of environmental factors (Heiningen et al. 1985, Keller et al. 1985, Andow and Prokrym 1991, McDougall and Mills 1997). Voegele et al. (1975) have shown that *T. brassicae* migrated out of release fields and could be found in other crop fields where it parasitized the same pest. Bigler et al. (1990) showed that a considerable number of the released population moved into other maize fields following inundative releases. These findings suggest that *T. brassicae* may also enter adjacent non-target habitats in which they might parasitize non-target host eggs.

16.4.3.1 Dispersal Studies in Switzerland

To quantify emigration of *T. brassicae* from maize fields, sticky traps were placed along transects leading from the inside of a release field into adjacent non-target habitats. Traps consisted of transparent plastic sheets (21 × 30 cm) that were sprayed with insect glue on both sides. They were attached to metal sticks and placed directly above the vegetation. Exposure periods of the traps were the week preceding the first release, as well as the week following the first and the second *T. brassicae* release, and three weeks after the second release. The mean number of parasitoids per date and distance was subjected to ANOVA. Species identification of individual *Trichogramma* was done by PCR (Kuske et al. 2003).

Prior to inundative releases, we found significant differences in mean numbers of wasps per trap between *T. brassicae* and native *Trichogramma* species (Fig. 16.7). *Trichogramma brassicae* was virtually absent, while native *Trichogramma* species were distributed throughout maize, wildflower strip and reeds. Following the first and second inundative releases of *T. brassicae* in maize, an immediate increase in numbers of trapped *T. brassicae* was measured in maize and non-target habitats. However, native *Trichogramma* species were still evenly distributed throughout the target and non-target habitats. After the second inundative release *T. brassicae* was the predominant species in maize, and to some extent in wildflower strips and in reeds. Three weeks after the second release, the *T. brassicae* population was strongly decreased both in maize, in wildflower strips and in reeds and native species were predominant again similarly to the pre-release period. The major native species found in the non-target habitat survey were in the order of abundance *T. evanescens* Westwood, *T. cacoeciae* Marchal and *T. semblidis* Aurivilius. The PCR identification failed for at least two other species, one being close to the North American species *T. aurosum* Sugonjaev and Sorokina. No identification was possible for the

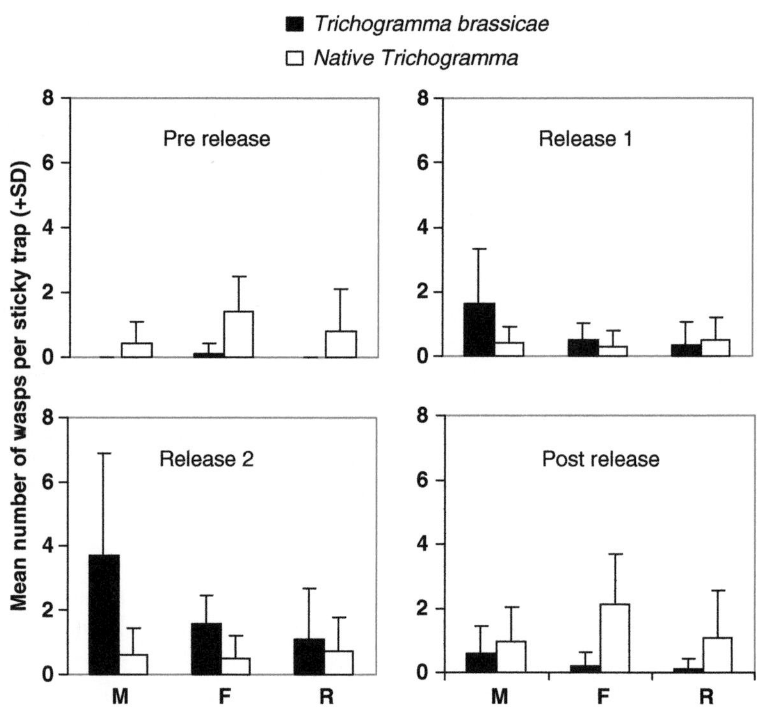

Fig. 16.7 Emigration of *Trichogramma brassicae* from maize into non-target habitats and presence of native *Trichogramma*. Parasitoids were caught on sticky traps along transects leading from maize (M) into wildflower strips (F), and reeds (R). Survey periods lasted one week and included the week prior to first inundative release, the weeks following first and second release (release 1 and 2) and three weeks after the second release (post release) (Kuske et al. 2003)

other species which was trapped mainly in wildflower strips, and to a minor extent also in maize and reed.

In dispersal experiments carried out in Southern Switzerland, an immediate and distinct increase of *T. brassicae* was found in the wildflower strip and the natural common reed stand up to 40 m distance from maize field borders both after the first and the second inundative release. No consistent decrease in numbers of trapped adult wasps was found within this scale. Thereafter, *T. brassicae* represented only a minor part of the *Trichogramma* community in non-target habitats and even in maize (Kuske et al. 2003).

Another experiment was set up to evaluate how far adult *T. brassicae* disperse after having left the maize field and when foraging within non-target habitats. Sticky traps (see above) were installed in an extensively managed meadow above the vegetation at distances of 2, 4, 8, 16, 32 and 64 m from a point source release. One trap was set up at each distance for each of the four main directions. Thereafter, 100,000 adults were released on the first day of the experiment. Traps were replaced daily for a period of nine days. In order to investigate whether *T. brassicae* adults also disperse to higher strata, we set up traps in two different heights, 40–70 cm and 100–130 cm. Mills et al. (2006) found in their analysis that females had a limited dispersal distance, and 67% of the dispersing individuals did not travel further than 3.3–3.7 m within four days and 95% were contained within radii of 7.6–11.0 m. Estimated disappearance and diffusion rate indicated that the dispersal distance of *T. brassicae* was limited and characterized by a very low rate of dispersal and a high rate of disappearance. Based on experience with other *Trichogramma* species, the authors estimated that disappearance was probably due to mortality, as the longevity of *Trichogramma* adults under field conditions is often short. The results also confirm that the sticky traps had a very low trapping efficiency as would be expected from an intercept trap without any attractant. Mills et al. (2006) concluded that it is important to note that estimates of dispersal distance and density can be strongly influenced by the type of landscape, the intervening landscape matrix and the climatic conditions prevailing at the time of the experiments. Based on their experience, the authors made specific recommendations on how to proceed for an assessment of dispersal by natural enemies in the context of non-target impacts from augmentative release of biological control agents.

In other dispersal experiments, it was found that a large fraction of the released population did not leave the maize fields (Babendreier and Bigler 2002). Similar results were obtained by Keller and Lewis (1985) in cotton, who did not find any significant dispersal of *T. pretiosum* out of the fields where they had been released. Wind may be an important parameter under certain conditions, and influence of strong wind has been shown to affect dispersal especially in open habitats. In only one of our experiments we found dispersal to be dependent on wind conditions while no such effect was observed in the other experiments.

We concluded from our dispersal studies that a low fraction of the released *T. brassicae* population flies into nearby non-target habitats after release where it was trapped in higher proportions to native *Trichogramma* for only a short period. The number of *T. brassicae* trapped outside maize dropped to pre-release levels within three weeks after the second release. The very low captures in non-target habitats prior to release indicates that *T. brassicae* has not become a predominant *Trichogramma* species even after many years of massive release in this area.

16.4.4 Direct and Indirect Effects

A general appreciation of direct and indirect effects of releasing exotic biological control agents in the context of non-target risk assessment was made by van Lenteren et al. (2006a). Massive releases of *Trichogramma* species in a crop can result in direct and indirect effects on other insects living in the crop and in off-crop habitats. Such effects depend primarily on the establishment, host range and dispersal characteristics, but other properties of the introduced agent and the receiving environment may influence the degree of direct and indirect effects, e.g. host plant and habitat preference, competition patterns in different habitats and type and structures of landscape.

We have shown in our overwintering studies that *T. brassicae* can survive in Switzerland under harsh winter conditions in a number of non-target host eggs. Our surveys in southern Switzerland demonstrated that it has established in non-target habitats in southern Switzerland in areas where repeated and massive releases were carried out. From this we conclude that *T. brassicae* must have the potential to establish in many other regions of Europe as well. The question then arises how much direct and indirect effects on non-target insects might be the result of massive releases and the establishment in non-target habitats.

Direct effects can occur in maize and in non-crop habitats in which *T. brassicae* adults disperse after release and parasitize eggs of non-target herbivore species living on maize and/or weeds, e.g. butterflies and moths, and predatory insects that naturally regulate other herbivores on maize, e.g. aphid predators like syrphids, coccinellids and chrysopids. Our results show that, despite the wide host range identified for *T. brassicae* under laboratory conditions, the ecological host range is relatively narrow under semi-field and field conditions. This is due mainly to the fact that the parasitoid shows obviously a clear host plant and habitat preference and, as a consequence, has low success in attacking host eggs on other plants than maize. From this we conclude that direct effects on non-target insects in maize and in off-crop habitats has a negligible to low population and community effect, and that quantitative effects will be temporary (see Lynch et al. 2000 for indirect effects). Our data on prevalence of native *Trichogramna* species as a function of mass releases of *T. brassicae* show that increased captures disappear shortly after releases have ceased and that *Trichogramma* species composition is back to the pre-release status indicating that there are little or no negative interactions with native *Trichogramma* species. Possible interspecific competition among *Trichogramma* species in maize and in non-target habitats may be transient and last only for the release period. We therefore conclude that inundative releases of *T. brassicae* are unlikely to have permanent and severe population level impacts on native *Trichogramm*a species.

Competition with other parasitoids of the target like *L. thompsoni* occurs in maize because *T. brassicae* reduces the larval population of the second generation of *O. nubilalis* by approximately 70–80% (Bigler et al. 1990), resulting in a reduced supply of host and prey on which parasitoids and predators can forage. Since the innate intention of pest control is to reduce pest organisms below an injury level, we do not consider the lower larval and pupal populations as being a negative effect for other parasitoids and predators, but an accepted effect of pest control whatever method applied.

The indigenous larval parasitoid *L. thompsoni* is an abundant parasitoid of *O. nubilalis* throughout southern Europe, including southern Switzerland (see previous discussion on this issue). If *T. brassicae* would disperse into reed stands and heavily parasitize eggs of the two moths, *A. geminipuncta* and *C. phragmitellus*, we would expect a reduced population density of *L. thompsoni* migrating back to maize resulting in a lower parasitism of the target. Our studies demonstrate that this hypothesized drawback was not observed after more than 10 years of massive release of *T. brassicae* close to the main reed stands in the area (Kuske et al. 2003). Laboratory and field studies of Kuske et al. (2004) demonstrated that eggs of the two hosts were not attacked by *T. brassicae*. We therefore conclude that massive releases of the egg parasitoid have no or only insignificant negative effects on populations of *L. thompsoni*.

16.5 New Risk Assessment and Rating of *Trichogramma brassicae*

Based on the four year's studies with *T. brassicae* and the experience in non-target risk assessment gained over a couple of years, we are now able to revisit the first risk assessment and to make it more accurate. Van Lenteren and Loomans (2006) proposed a new and more comprehensive risk assessment method consisting of a stepwise procedure that can be applied to all types of biological control agents used in augmentative and classical biological control programmes (see in Section 16.3). The new method gives more precise decision criteria and endpoints and the advice whether to release or not is taken at relevant points in the risk evaluation process. This method was applied to 92 species of natural enemies listed in the EPPO positive list of 2002 (EPPO 2002). For the new risk evaluation of *T. brassicae* we follow the procedure outlined in Fig. 16.1 (van Lenteren et al. 2006a).

Step 1: *Trichogramma brassicae* was collected in 1973 in Moldavia (Black sea region) and must be considered as being exotic to Central and Western Europe (for more details see first section in paragraph 4 of this chapter).

Step 2: *Trichogramma brassicae* is used for augmentative (inundative) releases.

Step 3: Our data on overwintering have shown that *T. brassicae* can successfully overwinter in diapause under harsh conditions in Switzerland where temperatures dropped below –10°C for 14 days and reached –20°C on four consecutive days (Babendreier et al. 2003a). These are rather exceptionally low temperatures for Western and Central Europe and we therefore assume that overwintering under extreme conditions is not a limiting factor for the parasitoid. Survey data in Southern Switzerland have shown that *T. brassicae* has established and is now a species belonging to the fauna there (Kuske et al. 2003). Based on these data we assume that *T. brassicae* would establish in most regions of Europe if non-target hosts are available. This information on establishment is now evaluated with the Risk Index Matrix (RIM) published in van Lenteren and Loomans (2006) where the likelihood (very unlikely to very likely) and magnitude (% area of establishment) is assessed. Since *T. brassicae* would not be able to establish permanently in annual crops, the total percentage area of permanent establishment in landscapes would mainly consist of non-crop habitats like hedgerows, forest edges, extensively managed meadows, nature conservation areas, and other pristine habitats. Perennial crops such as orchards and vineyards are most often intensively managed and hence, do not normally harbour significant amounts of potential hosts of *T. brassicae*. Considering these facts, we conclude that permanent establishment would be possible in maximum 10–25% of the total area and we therefore rank *T. brassicae* with 20 points in the RIM for establishment. Going back to the risk assessment scheme in Fig. 16.1, this would lead to the conclusion not to recommend release of the parasitoid. However, on request of an applicant (e.g. biocontrol company), further evaluation can be performed.

Step 4: We have assessed the physiological host range of *T. brassicae* under laboratory conditions by testing 25 butterfly and moth species and four aphid predators that are common in maize fields and contribute substantially to the natural regulation of aphid populations all over Europe. Eggs of only one species were totally rejected, two other species were hardly parasitized and all other species were parasitized, some of them even better than the target or mass rearing host (Babendreier et al. 2003d). Based on this information, the ecological host range was then assessed under semi-field and field conditions (Babendreier et al. 2003d). These studies clearly demonstrated that all non-target species tested were parasitized to a low degree when eggs were laid or glued on non-host plants. In comparison to egg parasitism of the mass rearing host on different plants, eggs of non-target hosts were less parasitized in the field. When non-target plants with non-target eggs were potted and exposed to *T. brassicae* in non-target habitats, parasitism became very low to insignificant. This information indicates that the ecological host range is quite narrow and by far overestimated by laboratory no-choice tests. This might motivate an applicant to perform dispersal studies on request as non-target effects will largely depend on the dispersal in time and space and the fraction of the released population dispersing into non-target habitats.

Step 5: Our dispersal studies have shown that only a small fraction of the released *T. brassicae* population disperses into non-target habitats and when there, the great majority of the individuals remain within an activity radius of less than 11 m (Mills et al. 2006). Catches with sticky traps and other surveys in release areas demonstrated that *T. brassicae* density increases during the mass releases, but drop to pre-release densities 2–3 weeks after the last release (Kuske et al. 2003). This indicates that only a small fraction (<10%) of the released population disperses out of maize fields to non-target areas. The number of *T. brassicae* adults captured in sticky traps in non-target habitats was relatively low and in general not at long distances (<10 m) away from the release sites. A very small fraction of dispersed individuals may fly longer distances and contribute to the build up of small populations in more distant non-target habitats. During a full vegetation season, *T. brassicae* can develop 4–5 generations in Switzerland and dispersal takes place with each generation from several breeding sites. This may finally contribute to a considerable distribution of the parasitoid in non-target habitats within a season. If values for distance travelled and percent dispersing individuals that we obtained from our investigations are entered in the RIM proposed by van Lenteren and Loomans (2006), we see that the threshold indicating a "no release scenario" is not reached and we evaluate the dispersal of *T. brassicae* as being minor to moderate.

Step 6: The final question remains whether mass releases of *T. brassicae* may have ecologically significant direct and/or indirect effects on non-target insects. We concluded from our studies that inundative releases of *T.* brassicae were unlikely to have permanent and severe population level impacts on native *Trichogramma* species living in maize and in non-crop habitats. Our results showed that the impact on Lepidoptera and aphid predators was very low under semi-field and field conditions (Babendreier et al. 2003c) because *T. brassicae* has a clear host plant and habitat preference and, as a consequence, is not successful in attacking host eggs on plants other than maize. We concluded that mortality effects caused by *T. brassicae* in off-crop habitats were likely, but the magnitude was below 40% and of transient nature. From this we concluded that direct effects on non-target insects in maize and in off-crop habitats has a minor population and community effect. Indirect effects of *T brassicae* on *L. thompsoni* were hypothesized if severe impact on populations of the two spring hosts *A. geminipuncta* and *C. phragmitellus* would be observed. However, laboratory and field studies carried out by Kuske et al. (2004) showed that eggs of the two hosts *A. geminipuncta* and *C. phragmitellus* were not attacked by *T. brassicae*. We therefore concluded that massive releases of the egg parasitoid have no or only insignificant negative effects on populations of *L. thompsoni*. If our estimates on direct and indirect effects are entered in the RIM of van Lenteren and Loomans (2006), we conclude that the overall effects are below the indicated threshold.

Here we have revised the first risk assessment based on many more data from our studies in the ERBIC project. It shows that the overall effects on non-target insects are low to insignificant and we conclude that the release of *T. brassicae* is an environmentally safe method to control the European corn borer in Central and Western Europe.

16.6 Concluding Remarks

Over the last ten years, a number of new and improved tools for environmental risk assessment for biological control agents have been developed and applied to many invertebrate biological control agents. Experience has shown that these methods can be used for agents intended for classical and augmentative control. Egg parasitoids of the genus *Trichogramma* are the most frequently used invertebrates in augmentative biological control and known to have a rather broad host range. Because of this broad host range, the question is often raised whether exotic species of this egg parasitoid should be used in biological control programs as they may show potential adverse effects on non-target invertebrates. *T. brassicae* has been commercially released in Central and Western Europe over the last 30 years and the area of application is still increasing as a result of the expanding geographical range of its major target host, the European corn borer, and due to the efficient control of this pest. It was only 20 years after the first commercial releases in Switzerland that we started to study in detail the potential non-target effects of *T. brassicae*. Our retrospective study and application of risk assessment methods has provided us an in-depth view of the complexity of direct and indirect effects on non-target invertebrates. During the past ten years we have learned to apply the new tools of risk assessment with increasing accuracy.

As a result of our risk assessment we conclude that *T. brassicae* is a nice example of a relatively safe biological control agent used in augmentative biological control despite the fact that it can establish and has a broad physiological host range. In our study we were able to demonstrate that *T. brassicae* has a limited ecological host range due to host plant and host- habitat preferences and that permanent establishment occurs only in limited areas of semi-natural and natural habitats. Our investigations yielded evidence that dispersal is very limited in time and space and that only a small fraction of the released populations do emigrate to non-target habitats. Thus, impacts on native *Trichogramma* populations are insignificant. Our studies on direct and indirect effects on non-target invertebrates have shown that such effects are small and transient and hence, inundative releases of *T. brassicae* can be considered safe and environmentally benign.

The case study of *T. brassicae* has substantially contributed to developing methods and tools for environmental risk assessment in biological control. We have experienced that large scale and time-consuming studies were needed to develop and test the first risk assessment methods. These in-depth studies also allowed us to design the stepwise risk assessment which guides risk assessors to more advanced studies only if natural enemies clearly show potential risks. In addition to applying the stepwise risk assessment as an efficient tool, we propose to use existing expert knowledge and to compile information of case studies in databases which indicate what relevant information is needed to perform accurate risk assessments. Such databases will assist biological control companies and regulators of national authorities dealing with import and release of exotic organisms in quicker and more reliable decision making.

Acknowledgements The development of risk assessment methods and tools was supported by the EU research grant FAIR5-CT97-3489 ERBIC). We are thankful to all partners of the ERBIC project who have contributed to develop the principles of environmental risk assessment of natural enemies, especially Antoon J. M. Loomans, Heikki M.T. Hokkanen, Paul C. J. van Rijn, Matt B. Thomas, Giovanni Burgio and Stefan Kuske.

References

Albajes R, Gullino ML, van Lenteren, Elad Y (1999) Integrated pest and disease management in greenhouse crops. Kluwer, Dordrecht

Andow DA, Prokrym DR (1991) Release density, efficiency and disappearance of *Trichogramma nubilale* for control of European Corn Borer. Entomophaga 36:105–113

Anonymous (2003) Glossary of terms. The second report on harmonisation of risk assessment procedures. Appendix 2. Scientific Steering Committee, European Commission, Brussels

Anonymous (2004) Guidance for information requirements for regulation of invertebrates as biological control agents. OECD Series on Pesticides 21. http://www.oecd.org/dataoecd/6/20/28725175.pdf. Cited 4 August 2007

Babendreier D, Bigler F (2002) Evaluating environmental risks of biological control introductions into Europe. In: Hokkanen HMT (ed) ERBIC final report. FAIR5-CT97-3489, Brussels

Babendreier D, Bigler F, Kuhlmann U (2005) Methods used to assess non-target effects of invertebrate biological control agents of arthropod pests. Biol Control 50:821–870

Babendreier D, Kuske S, Bigler F (2003a) Overwintering of the egg parasitoid *Trichogramma brassicae* in Northern Switzerland. Biol Control 48:261–273

Babendreier D, Kuske S, Bigler F (2003b) Non-target host acceptance and parasitism by *Trichogramma brassicae* Bezdenko (Hymenoptera: Trichogrammatidae) in the laboratory. Biol Control 26:128–138

Babendreier D, Kuske S, Bigler F (2003c) Parasitism of non-target butterflies by *Trichogramma brassicae* Bezdenko (Hymenoptera: Trichogrammatidae) under field cage and field conditions. Biol Control 26:139–145

Babendreier D, Schoch D, Kuske S, Dorn S, Bigler F (2003d) Non-target habitat exploitation by *Trichogramma brassicae* (Hym. Trichogrammatidae): what are the risks for endemic butterflies? Agric Forest Entomol 5:199–208

Babendreier D, Rostas M, Höfte MCJ, Kuske S, Bigler F (2003e) Effects of mass releases of *Trichogramma brassicae* on predatory insects in maize. Entomol Exp Appl 108: 115–124

Bigler F (1986) Mass production of *Trichogramma maidis* Pint. et Voeg. and its field application against *Ostrinia nubilalis* Huebner in Switzerland. J Appl Entomol 101: 23–29

Bigler F, Kölliker-Ott UM (2006) Balancing environmental risks and benefits: a basic approach. In: Bigler F, Babendreier D, Kuhlmann U (eds) Environmental impact of invertebrates for biological control of arthropods: methods and risk assessment. CAB International, Wallingford, CT, pp 273–286

Bigler F, Bosshart S, Waldburger M (1989) Bisherige und neue Entwicklungen bei der biologischen Bekämpfung des Maiszünslers, *Ostrinia nubilalis* Huebner mit *Trichogramma maidis* Pint. et Voeg. in der Schweiz. Landwirtschaft Schweiz 2:37–43

Bigler F, Bosshart S, Waldburger M, Ingold M (1990) Dispersal of *Trichogramma evanescens* Westw. and its impact on parasitism of eggs of *Ostrinia nubilalis* Hbn. Mitteilungen der Schweizerischen Entomologischen Gesellschaft 63:381–388

Bigler F, Bale JS, Cock MJW, Dreyer H, GreatRex R, Kuhlmann U, Loomans AJM, van Lenteren JC (2005) Guidelines on information requirements for import and release of invertebrate biological control agents (IBCAs) in European countries. Biocontrol News Inf 26(4): 115 N-123 N

Bigler F, Babendreier D, Kuhlmann U (2006) Environmental impact of invertebrates for biological control of arthropods: methods and risk assessment. CAB International, Wallingford, CT

Boettner GH, Elkinton JS, Boettner CJ (2000) Effects of a biological control introduction on three nontarget native species of saturniid moths. Conserv Biol 14:1798–1806

Clausen CP (1940) Entomophagous insects. McGraw-Hill, New York

Copping LG (2004) The manual of biocontrol agents. BCPC Publications Sales, 7 Omni Business Centre, Omega Park, Alton, Hampshire

Duelli P (1994) Rote Liste der gefährdeten Tierarten der Schweiz. BUWAL-Reihe Rote Listen, Bern

Eizaguirre M, Albajes R, Galichet PF (1990) A note on the presence in Catalonia Spain of a parasitic system bound to the tachinid fly *Lydella thompsoni* Herting, parasitoid of corn borers. Investigaciön Agraria – Producciön y Protección Vegetales 5:345–348

EPPO (2002) List of biological control agents widely used in the EPPO region. EPPO Standard PM6/3(2). Bull OEPP/EPPO Bull 32(3):447–461.

Follett PA, Duan JJ (2000) Nontarget effects of biological control. Kluwer, Dordrecht

Fulmek L (1955) Wirtsbereich von *Trichogramma evanescens* Westw. und *Tr. minutum* Ril. Anzeiger Schädlingskunde 8:113–116

Galichet PF, Riany M, Agounke D, Tavernier J, Cousin M, Magnin H, Radisson A (1985) Bioecology of *Lydella thompsoni* Herting (Dip., Tachinidae) within the Rhone Delta in southern France. Entomophaga 30:315–328

Grenier S, Anglade P, Naibo B, Galichet PF, Hawlitzky N (1990) Survey on distribution of *Tachinaria* (Dip., Tachinidae), parasitoids of corn moths *Ostrinia nubilalis* (Lep., Pyralidae) in France (1985–1987). Entomophaga 35:485–492

Heiningen TG, van Pak GA, Hassan SA, van Lenteren JC (1985) Four year's results of experimental releases of *Trichogramma* egg parasites against lepidopteran pests in cabbage. Mededelingen-van-de-Faculteit-Landbouwwetenschappen Rijksuniveriteit-Gent 50: 379–388

Howarth FG (1991) Environmental impacts of classical biological control. Annu Rev Entomol 36:485–509

IPPC (2005) Revision of ISPM No. 3. Guidelines for the export, shipment, import and release of biological control agents and beneficial organisms. International Plant Protection Convention. Available at.https://www.ippc.int/IPP/En/default.jsp. Cited 4 August 2007

Keller MA, Lewis WJ (1985) Movements by *Trichogramma pretiosum* (Hymenoptera: Trichogrammatidae) released into cotton. Southwest Entomol Suppl 8:99–109

Keller MA, Lewis WJ, Stinner RE (1985) Biological and practical significance of movement by *Trichogramma* species: A review. Southwest Entomol Suppl 8:138–155

Kuhlmann U, Schaffner U, Mason PG (2006) Selection of non-target species for host specificity testing. In: Bigler F, Babendreier D, Kuhlmann U (eds) Environmental impact of invertebrates for biological control of arthropods: methods and risk assessment. CAB International, Wallingford, CT, pp 15–37

Kuske S, Widmer F, Edwards PJ, Turlings TCJ, Babendreier D, Bigler F (2003) Dispersal and persistence of mass released *Trichogramma brassicae* (Hymenoptera: Trichogrammatidae) in non-target habitats. Biol Control 27:181–193

Kuske S, Babendreier D, Edwards PJ, Turlings TCJ, Bigler F (2004) Parasitism of non-target Lepidoptera by mass released *Trichogramma brassicae* and its implication for the larval parasitoid *Lydella thompsoni*. Biol Control 49:1–19

Loomans AJM (2004) Biologische bestrijders en de Flora- en Faunawet: criteria voor risicoinschatting en toelating biologische bestrijders in Nederland. Gewasbescherming 35:33–37

Louda SM, Pemberton RW, Johnson MT, Follett PA (2003) Nontarget effects: the Achilles heel of biological control? Retrospective analyses to reduce risk associated with biocontrol introductions. Annu Rev Entomol 48:365–396

Lynch LD, Hokkanen HMT, Babendreier D, Bigler F, Burgio G, Gao ZH, Kuske S, Loomans AJM, Menzler-Hokkanen I, Thomas MB, Tommasini G, Waage J, van Lenteren JC, Zeng QQ

(2000) Indirect effects in the biological control of arthropods with arthropods. In: Wajnberg E, Scott JC, Quimby PC (eds), Evaluating indirect ecological effects of biological control. CAB International, Wallingford, CT, pp 99–125

Manojlovic B (1985) Some biological properties of *Lydella thompsoni* Hert. (Dip.: Tachinidae) – An important parasite of the European corn borer. Zastita Bilja 36:93–100

Mason PG, Huber JT (2002) Biological control programmes in Canada, 1981–2000. CAB International, Wallingford, CT

McDougall SJ, Mills NJ (1997) Dispersal of *Trichogramma platneri* Nagarkatti (Hym., Trichogrammatidae) from point-source releases in an apple orchard in California. J Appl Entomol 121:205–209

McEvoy PB (1996) Host specificity and biological pest control. Biol Sci 46:401–415

Messing R, Roitberg B, Brodeur J (2006) Measuring and predicting indirect impacts of biological control: competition, displacement and secondary interactions. In: Bigler F, Babendreier D, Kuhlmann U (eds) Environmental impact of invertebrates for biological control of arthropods: methods and risk assessment. CAB International, Wallingford, CT, pp 64–77

Mills NJ, Babendreier D, Loomans AJM (2006) Methods for monitoring the dispersal of natural enemies from point source releases associated with augmentative biological control. In: Bigler F, Babendreier D, Kuhlmann U (eds) Environmental impact of invertebrates for biological control of arthropods: methods and risk assessment. CAB International, Wallingford, CT, pp 114–131

Moeed A, Hickson R, Barratt BIP (2006) Principles of environmental risk assessment with emphasis on the New Zealand perspective. In: Bigler F, Babendreier D, Kuhlmann U (eds) Environmental impact of invertebrates for biological control of arthropods: methods and risk assessment. CAB International, Wallingford, CT, pp 241–253

Neuffer G (1981) Zur biologischen Bekämpfung des Maiszünslers *Ostrinia nubilalis* Huebner mit *Trichogramma evanscens* Westwood in Speisemaiskulturen. Mitteilungen der Deutschen Gesellschaft für allgemeine Angewandte Entomologie 2:208–213

Orr DB, GarciaSalazar C, Landis DA (2000) *Trichogramma* nontarget impacts: a method for biological control risk assessment. In: Follett PA, Duan JJ (eds) Nontarget Effects of Biological Control. Kluwer, Norwell, USA, pp 111–125

Pinto JD, Stouthamer R (1994) Systematics of the Trichogrammatidae with emphasis on *Trichogramma*. In: Wajnberg E, Hassan SA (eds) Biological control with egg parasitoids. CAB International, Wallingford, CT, pp 1–36

Pintureau B (1990) Polymorphisme, biogéographie et spécificité parasitaire des Trichogrammes européens. Bull Soc Entomol France 95:17–38

Pintureau B, Voegele J (1980) Une nouvelle espèce proche de *Trichogramma evansescens: T. maidis* (Hym. Trichogrammatidae). Entomophaga 25:431–440

Pintureau B, Babault M, Voegele J (1981) Etudes de quelques facteurs de variation de la fécondité chez *Trichogramma maidis* Pintureau at Voegele (Hym. Trichogrammatidae). Agronomie 1:315–322

Pizzol J, Voegele J (1988) The diapause of *Trichogramma maidis* Pintureau and Voegele in relation to some characteristics of its alternative host *Ephestia kuehniella* Zell. Colloques l'INRA 48:93–94

Rabb RL, Bradley JR (1968) The influence of host plants on parasitism of eggs of the tobacco hornworm. J Econ Entomol 61:1249–1252

Romeis J, Shanower TG, Zebitz CPW (1998) Physical and chemical plant characters inhibiting the searching behaviour of *Trichogramma chilonis*. Entomol Exp Appl 87:275–284

Romeis J, Babendreier D, Wäckers F, Shanower TG (2005) Habitat and plant specificity of *Trichogramma* egg parasitoids – underlying mechanisms and implications. Basic Appl Ecol 6:215–236

SBN (1987) Schweizerischer Bund für Naturschutz. Tagfalter und ihre Lebensräume: Arten-Gefährdung-Schutz. Band 1. Rotofotar AG, Egg-Zürich, Switzerland

Stouthamer R, Hu J, van Kan FJPM, Platner GR, Pinto JD (1999) The utility of internally transcribed spacer 2 DNA sequences of the nuclear ribosomal gene for distinguishing sibling species of *Trichogramma*. Biol Control 43:421–440

Suter H, Baebler M (1976) Möglichkeiten einer biologischen Bekämpfung des Maiszünslers mit *Trichogramma*-Eiparasitoiden. Mitteilung Schweiz Landwirtschaft 2:41–51

Suverkropp BP (1997) Host-finding behaviour of *Trichogramma brassicae* in maize. PhD Thesis no.1823, University of Wageningen, NL, 249p

Thomson MS, Stinner RE (1989) *Trichogramma spp.* (Hymenoptera: Trichogrammatidae): field hosts and multiple parasitism in North Carolina. J Entomol Sci 24:232–240

Thorpe KW (1985) Effects of height and habitat type on egg parasitism by *Trichogramma minutum* and *T. pretiosum* (Hymenoptera: Trichogrammatidae). Agric, Ecosyst Environ 12:117–126

USDA (2001) The ROBO Database. Available at http://www.ars-grin.gov/nigrp/robo.html. Cited 4 August 2007

van Lenteren JC (2000) Measures of success in biological control of arthropods by augmentation of natural enemies. In: Gurr G, Wratten S (eds) Measures of success in biological control. Kluwer, Dordrecht, pp 77–103

van Lenteren JC (2003) Quality control and production of biological control agents: theory and testing procedures. CAB International, Wallingford, CT

van Lenteren JC, Loomans AJM (2006) Environmental risk assessment: methods for comprehensive evaluation and quick scan. In: Bigler F, Babendreier D, Kuhlmann U (eds) Environmental impact of invertebrates for biological control of arthropods: methods and risk assessment. CAB International, Wallingford, CT, pp 254–272

van Lenteren JC, Babendreier D, Bigler F, Burgio G, Hokkanen HMT, Kuske S, Loomans AJM, Menzler-Hokkanen I, van Rijn PCJ, Thomas MB, Tommasini MG, Zeng QQ (2003) Environmental risk assessment of exotic natural enemies used in inundative biological control. Biol Control 48:3–38

van Lenteren JC, Bale J, Bigler F, Hokkanen HMT, Loomans AJM (2006a) Assessing risks of releasing exotic biological control agents of arthropod pests. Annu Rev Entomol 51:609–634

van Lenteren JC, Cock MJW, Hoffmeister TS, Sands DPA (2006b) Host specificity in arthropod biological control, methods for testing and interpretation of the data. In: Bigler F, Babendreier D, Kuhlmann U (eds) Environmental impact of invertebrates for biological control of arthropods: methods and risk assessment. CAB International, Wallingford, pp 38–63

Voegele J, Stengel M, Schubert G, Daumal J, Pizzol J (1975) Les Trichogrammes. V(a). Premiers résultats sur l'introduction en Alsace sous forme de lâchers saisonniers de l'écotype moldave de *Trichogramma evanescens* Westwood contre la pyrale du maïs, *Ostrinia nubilalis*. Ann Zool Ecol Anim 7:535–551

Voegele J, Pizzol J, Babi A (1988) The overwintering of some *Trichogramma* species. Trichogramma and Other Egg Parasites. Colloques l'INRA 43:275–282

Wajnberg E, Scott JC, Quimby PC (2000) Evaluating indirect ecological effects of biological control. CAB International, Wallingford, CT

Chapter 17
Genetically Modified Crops and Biological Control with Egg Parasitoids

Julio S. Bernal

Contents

17.1 Introduction

After one decade and a half from their commercial release, genetically-modified (GM) crop varieties have become central components of pest management strategies for crops (and in countries) in which they are commercially available, particularly for soybean, maize, cotton, and canola. The most predominant types of commercial GM crops worldwide include insect-resistant varieties, which express transgenes encoding for insect-active toxins from the bacterium *Bacillus thuringiensis* (*Bt*),

J.S. Bernal (✉)
Department of Entomology, Texas A&M University, College Station, TX 77843-2475, USA
e-mail: juliobernal@tamu.edu

F.L. Consoli et al. (eds.), *Egg Parasitoids in Agroecosystems with Emphasis on Trichogramma*, Progress in Biological Control 9, DOI 10.1007/978-1-4020-9110-0_17,

and herbicide-tolerant (*HT*) varieties, which tolerate post-emergent applications of particular herbicides, such as glyphosate or glufosinate. Other common, insect-resistant GM crop varieties, albeit not yet commercialized, are those that express plant transgenes (*PTg*) from plant species other than the GM crop, and that encode mainly for insect-active lectins or enzyme inhibitors, e.g., GM sugarcane (*Saccharum officinarum*, Poaceae) that expresses a lectin-producing transgene from snowdrop lily (*Galanthus nivalis*, Amaryllidaceae) (e.g., Sétamou et al. 2002). Hereafter, "GM crop protein(s)" is used to collectively refer to the insect-active transgene products of GM crops (toxins, lectins, enzyme inhibitors, etc.).

Modern pest management strategies for crops, including integrated pest management (IPM), minimize crop damage and losses due to pests by combined application of control tactics that reduce crop colonization by pests, increase pest mortality rates, or decrease pest natality rates. Such control tactic combinations are applied against a backdrop of pest mortality caused by naturally-occurring biological control by indigenous and non-indigenous natural enemies, which include parasitoids, predators, and pathogens. Thus, pest management strategies complement pest mortality due to naturally-occurring biological control with mortality, or decreased natality and colonization, from various control tactics, including applied biological control, host plant resistance, chemical control, behavioral manipulations, and others. The first two of those control tactics, applied biological control and host plant resistance, are common and important components, and naturally-occurring biological control is a fundamental component of pest management strategies (see below, and Bernal et al. 2004). In a broad perspective, naturally-occurring and applied biological control, host plant resistance, and interactions among them, are the foci of this chapter: In particular, the chapter focuses on interactions between egg parasitoids, which are important natural enemies in the contexts of both naturally-occurring and applied biological control, and insect-resistant GM crop varieties, which are a form of host plant resistance. *HT* crops are discussed also in this chapter because of the potential impacts of growing herbicide use on egg parasitoids.

Insect-resistant GM crops, including *Bt* and *PTg* varieties, were developed to target the feeding stages of pests, i.e. insect larvae or nymphs, and adults, so any impacts on their eggs, and consequently on egg parasitoids have been expected to be minimal, particularly in comparison with impacts on parasitoids of the feeding stages of pests specifically targeted by GM crops. That any impacts of insect-resistant GM crops on egg parasitoids are expected to be practically inconsequential is clear from the paucity of scientific literature addressing interactions among egg parasitoids, their hosts, and GM crops, particularly in comparison to the literatures addressing the corresponding interactions for other parasitoids (see, e.g., Bernal et al. 2004, Lövei and Arpaia 2005, Hilbeck and Schmidt 2006, Romeis et al. 2006). Similarly, because *HT* varieties target weeds (through herbicides directed at weeds present in *HT* crop fields), rather than insects, any impacts on egg parasitoids – and insects and their natural enemies generally – have been expected to be negligible. While the expectation of fewer and weaker impacts of insect-resistant GM crops on egg parasitoids relative to other parasitoids may be warranted in part, the various known or plausible routes of egg parasitoid exposure (e.g., honeydew, pollen; see below) to GM crop proteins warrant closer examination of any potential impacts. Also, *HT* crop acreage predominates over that of *Bt* crops, occupying nearly two-thirds of the worldwide acreage planted with GM crops (James 2007). Moreover, data available from the USA suggest that, over time, herbicide use (average rate per-hectare, kg/ha) in *HT* crops may exceed use in conventional crops (Benbrook 2004). For example, herbicide use in 1996 and 1997 was lower (−0.36 and −0.29%, respectively) in *HT* soybeans compared to conventional soybeans, but was higher (+0.67%) in the former by 2004, a trend that was evident also in *HT* cotton (+0.94% greater use compared to conventional cotton) (Benbrook 2004). Egg parasitoids may be affected, directly or indirectly, by herbicides used in *HT* crops (see below). Therefore, an examination of the effects of herbicides on egg parasitoids in the context of *HT* crops is warranted.

17.2 Pest Management, Biological Control and GM Crop Varieties

Modern pest management strategies for many crops worldwide depend on biological control, whether naturally-occurring or applied, or both, to minimize crop injury and economic losses due to arthropod pests. In particular, naturally-occurring biological control is fundamental to IPM: The foundation of IPM strategies typically consists of three components: (i) close monitoring of the density dynamics of pest populations, (ii) decision rules based on those density dynamics (i.e. economic or other action thresholds), and (iii) naturally-occurring biological control (Fig. 17.1). Usually, the contribution of naturally-occurring biological control to IPM strategies is an important, though frequently unappreciated, amount of pest mortality from natural enemies present within or in the vicinity of crops, and it is upon this mortality that scientists construct effective IPM strategies. Other control tactics, such as applied biological control, host plant resistance (including GM crops), chemical control, behavioral manipulations, among others, are incorporated to IPM strategies upon the three-component foundation. Incorporating host plant resistance, i.e. insect-resistant varieties, to IPM strategies is especially attractive both to farmers and scientists. For farmers, host plant resistance represents a typically inexpensive, effective, and user-friendly pest management technology, while for scientists it represents a broadly sustainable technology that is for the most part compatible with biological control (Hare 2002). Thus, usually, insect-resistant crop varieties are quickly incorporated into IPM strategies, and more so if they are highly effective against key crop pests, as are GM crop varieties. Moreover, where they are available and their use is economically justified, insect-resistant varieties, particularly GM crops, rapidly become regionally dominant crop varieties, and central components of pest management strategies. For example, the percentage of crop acreage planted with all GM maize varieties (viz., *Bt*, *HT*, and stacked *Bt+HT* varieties) in the USA increased from 25% in 2000 to 45% in 2004, while the corresponding increase for all GM cotton varieties was from 61 to 76% (Benbrook 2004). Consequently, the frequency of interactions between biological control (both naturally-occurring

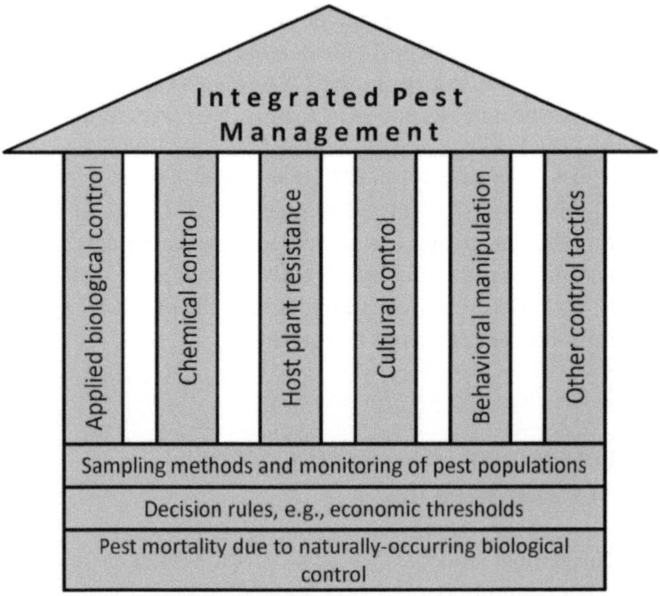

Fig. 17.1 Naturally-occurring biological control is a fundamental component of IPM strategies. Effective IPM strategies build upon the pest mortality caused by naturally-occurring biological control by adding mortality and reducing natality and crop colonization through application of a variety of control tactics. Modified from Gonzalez (1971)

and applied) and GM crop varieties should be high compared to the corresponding frequency of interactions with conventional, insect-resistant varieties. Moreover, because of their regional predominance GM crops should substantially modify the composition and dynamics of regional agroecosystems. Overall, the pervasiveness of naturally-occurring biological control – and to a lesser extent applied biological control – coupled with regional predominance of GM crops points to an abundance of interactions between these two important components of pest management strategies, and so to the necessity of closely examining the outcomes of such interactions.

Interactions between biological control and host plant resistance, including insect-resistant GM crops, can be additive (neutral), synergistic (positive), or antagonistic (negative) (Bernal et al. 2004). Antagonism between host plant resistance and biological control has been documented, though infrequently (Hare 2002, but see Groot and Dicke 2002). Such antagonism, coupled with growing regional predominance of GM crop varieties over conventional varieties and their increasingly central role in pest management strategies, are the bases for concerns over potentially negative impacts on biological control from the novel and particularly aggressive form of host plant resistance represented by GM crops. Naturally-occurring biological control is present by definition in all agroecosystems, and it maintains most herbivores within crops at non-pest levels (i.e. at secondary or potential pest statuses) and contributes to controlling key pests. In contrast, GM crops specifically target one or few key pests and do not contribute to controlling the balance of key, secondary, or potential pests present in cropping systems. A key question, then, is whether extensive planting of GM crop varieties modifies agroecosystems to the extent that natural enemy populations are negatively affected, and so compromises the broad contributions of biological control to pest management generally within crops (Bernal et al. 2004). At present, this question is unresolved, though a number of studies, mostly laboratory-based, point to plausible, though untested, predictions concerning potential negative impacts of GM crops on natural enemies and biological control (see Bernal et al. 2004, Lövei and Arpaia 2005, Hilbeck and Schmidt 2006, Romeis et al. 2006).

As noted above, GM crops are a particularly attractive pest control technology, and where available they quickly dominate pest management strategies and regional agroecosystems. Such domination greatly increases the frequency and intensity of interactions – whether synergistic, additive, or antagonistic – with biological control (Bernal et al. 2004). Clearly, synergistic or additive interactions between GM crops and biological control should be sought, and antagonistic interactions avoided, when such crops are developed and then incorporated into IPM strategies. So, it should be advantageous for pest management strategy development to examine interactions between GM crops and natural enemies during these crop's developmental processes rather than in hindsight. In particular, interactions between GM crops and parasitoids should be examined closely. Parasitoids are the natural enemies most frequently relied upon in applied biological control (Gordh et al. 1999) and, because of their life history, especially their close relationship with their hosts and relatively narrow host range, are more likely than predators or pathogens to suffer significant negative impacts of GM crops (Bernal et al. 2004). Consequently, a growing number of studies (mostly laboratory-based) have sought to uncover any interactions between GM crops and parasitoids. One review of such studies found antagonistic interactions in ~40% of cases (57% of cases specifically involving *Bt* crop varieties; 32% involving *PTg* crop varieties), and fewer cases of additive (~34%) or synergistic (~6%) interactions, though *PTg* crop varieties were involved in the majority of cases of additive or synergistic interactions (Lövei and Arpaia 2005). However, most research addressing interactions between GM crops and parasitoids to date has focused on parasitoids other than egg parasitoids, so prior analyses may not apply closely to egg parasitoids. Biological differences (including behavioral and ecological differences) between egg parasitoids, and their hosts, relative to other parasitoids and their hosts should translate into differences between the parasitoid types in routes and degrees of exposure to GM crops. These differences include, among others, that egg parasitoids, compared to other parasitoids, typically have broader host ranges, and their host stage (= egg) does not consume, so should not contain nor be directly affected by GM crop proteins. These differences, in turn, may signify, for example, that populations of egg parasitoids should suffer less than those of other parasitoids if populations of their principal hosts are decimated by GM crops, and that developing egg parasitoids, unlike developing larval parasitoids, should not be directly exposed to GM crop proteins present inside their hosts (more below). Altogether, the biological differences between egg parasitoids and parasitoids of subsequent host stages indicate that analyses of their interactions with GM crops should be conducted independently for egg and other parasitoids.

17.3 Routes of Egg Parasitoid Exposure to GM Crop Proteins

GM crops can affect parasitoids at two levels, viz. as individuals and populations (Bernal et al. 2004). However, independently of any effects, exposure of parasitoids, generally, to GM crop proteins will occur at the individual level, and at this level may occur in the larval or adult stages, or both, and less likely in the egg or pupal stages (all parasitoids are in Endopterygota). In the case of egg parasitoids, specifically, exposure should be in the adult stage because transfer of GM crop proteins to insect eggs, within which the eggs, larvae, and pupae of egg parasitoids typically occur, has not been documented. Thus, eggs, larvae, and pupae of egg parasitoids should not be exposed directly to GM crop proteins within their hosts. However, while direct exposure of egg parasitoids to GM crop proteins is unlikely, GM crops may indirectly affect egg parasitoids via other mechanisms, e.g. through qualitative changes or decimation of host populations (see below).

Plausible routes of exposure to GM crop proteins for adult egg parasitoids include both direct and indirect routes, both associated to their feeding activities. Indirect routes are likely limited to insect honeydew, while direct routes should include floral and extra-floral nectar, pollen, and other plant fluids, such as phloem and xylem saps and fluids emanating from damaged tissues. Though egg parasitoids may engage in host-feeding, exposure to GM crop proteins via host-feeding is unlikely because such proteins have not been detected in host eggs, as noted above.

17.3.1 Floral and Extra-floral Nectar

Parasitoids generally (Jervis et al. 1996, Koptur 2005, Wäckers 2005), including egg parasitoids (e.g., Begum et al. 2004, Rahat et al. 2005, Irvin et al. 2007), consume and benefit from consuming floral and extra-floral nectar. Moreover, nectar availability may improve biological control generally, and by egg parasitoids in particular (e.g., Treacy et al. 1987, Heimpel and Jervis 2005), though accessing nectar may be risky due to predation (Koptur 2005, Wäckers 2005). Thus, egg parasitoids may consume GM crop proteins if they occur in floral or extra-floral nectar produced by GM crops. However, nectar should only rarely contain GM crop proteins. Plant nectar is a sugar-rich excretion derived from phloem sap, or phloem and xylem sap, rather than a tissue, so lacks cellular content, including DNA and RNA (Malone 2002, Wäckers 2005). Moreover, its composition includes mostly sugars, though nectar may also contain free amino acids, lipids, and secondary metabolites, among other components, at minor concentrations; also, the composition of floral and extra-floral nectar may differ within plant species, and the latter may contain more sugar than the former (Baker and Baker 1975, 1983, Koptur 2005, Wäcker 2005). Proteins, in general, occur in nectar in a few cases, albeit at very low levels

(<1%) (e.g., Peumans et al. 1997, Carter et al. 1999), so it is plausible that GM crop proteins may occur in nectar, though at low concentrations. Consequently, very few studies addressed the presence of GM crop proteins in plant nectar, and they did not detect such proteins in nectar (e.g., Jouanin 1998, Anonymous 2001). It follows, then, that the results of studies showing effects on parasitoids of purified GM crop proteins administered via (artificial) sugar- or honey-based solutions (e.g., Romeis et al. 2003, Bell et al. 2004) should not be interpreted to mean that parasitoids in the field invariably will be directly affected by, or exposed to, GM crop proteins via consumption of plant nectar.

Because the occurrence of GM crop proteins in plant nectar is likely infrequent, and at low concentrations, generalizations concerning egg parasitoid exposure to GM crop proteins through this route should be avoided. However, case-by-case analyses of any impacts of GM crop proteins via floral or extra-floral nectar on egg parasitoids are still warranted because of the known importance of carbohydrate sources, including floral and extra-floral nectars, to egg parasitoids in the field (and laboratory), and the potential predominance of such nectars, relative to other carbohydrate sources, within GM crops.

Unplanned phenotypic changes resulting from genetic modification (e.g., due to pleiotropy, epistasis, or insertional mutagenesis) may affect the amount and quality of floral or extrafloral nectar produced by GM crops (see below). These changes may affect biological control by egg parasitoids if nectar production is modified in an ecologically significant manner in GM crops (cf. Treacy et al. 1987).

17.3.2 Pollen

Pollen, unlike nectar, is a plant tissue, so is capable of protein synthesis, and its protein content may exceed 60% (Roulston and Cane 2000, Malone 2002, Wäckers 2005). Moreover, pollen is rich in free amino acids, sterols, lipids, and carbohydrates (Solberg and Remedios 1980, Dobson 1988, Nepi and Franchi 2000). Thus, pollen may contribute to filling the protein needs, and other dietary needs, of parasitoids. However, pollen may contain secondary plant compounds, such as alkaloids, phenolics, and terpenes, or mannose sugars, which are known to be toxic to honeybees (Roulston and Cane 2000, Wäckers 2005).

Egg parasitoids, such as *Trichogramma*, may consume pollen, though it is unclear whether any benefits are gained from pollen consumption, or whether any such benefits are mediated by concurrent access to a carbohydrate source (e.g., floral nectar) or water. For example, one study failed to find evidence of pollen feeding by *Trichogramma* (Wellinga and Wysoki 1989), while another study demonstrated pollen feeding, and showed that consumption of pollen alone (with access to water) increased fecundity and longevity of females, though in the presence of honey, consumption of pollen did not lead to measurable increases in these parameters (Zhang et al. 2004). Other studies, however, showed that pollen consumption produced an increase in fecundity and longevity of female *Trichogramma* only in the presence of honey (e.g., Geng et al. 2006), or did not produce an increase in the presence of honey or water (Wang et al. 2007). It is presently unclear whether egg parasitoids other than *Trichogramma* consume pollen, or whether they benefit from pollen consumption, because studies are unavailable. Nonetheless, pollen consumption may be precluded in most egg parasitoids because their adult size is not conducive to pollen feeding, given the size of pollen grains (Jervis et al. 1996). Altogether, the available data show that under laboratory conditions *Trichogramma* consume pollen, though it is unclear whether they benefit from its consumption, and whether any benefits are dependent on concurrent access to a carbohydrate source.

GM crop proteins may be present in pollen, but depends on the tissue specificity of the promoter sequence used for transgene expression, and their content level is dependent upon a variety of factors, including plant species, transformation event, and environmental conditions, among others (Schuler et al. 1998). However, independently of the tissue specificity of promoter sequences, expression levels of GM crop proteins in pollen appear to be invariably lower than in vegetative tissues. For example, *Bt* toxin levels in pollen did not exceed 0.04% of total soluble protein when a pollen-specific promoter was used for transgene expression in maize (Kozeil et al. 1993, Fearing et al. 1997), and 0.01% when a constitutive promoter was used in potato (Ferry et al. 2007), while levels in the range ~0.1–0.2%, or higher, typically are present in vegetative tissues (Sachs et al. 1998, Schuler et al. 1998, Ferry et al. 2007, Alvarez-Alfageme 2008). However, case-by-case analyses of any impacts of GM crop proteins via pollen on egg parasitoids are warranted because of the potential importance of this food resource for adult parasitoids, and known presence of GM crop proteins in pollen, albeit at lower levels than in vegetative tissues, and because complete exclusion of GM crop proteins from pollen seems beyond current capabilities.

17.3.3 Insect Honeydew

Honeydew is a sugar-rich waste product that is typically excreted by phloem-feeding Hemiptera, particularly Sternorrhyncha, so its composition usually reflects that of the phloem sap ingested by the producing insect (Malone 2002, Wäckers 2005). While xylem-feeding Auchenorrhyncha also excrete sugary waste products, the sugar contents of their excretions are substantially lower than those of sternorrhynchan honeydews (cf. Anderson et al. 1989, Irvin et al. 2007). Phloem sap is usually rich in sugars, but contains also amino acids, though mostly non-essential but at levels typically higher than present in plant nectars, secondary plant compounds, including lectins and enzyme (proteinase) inhibitors, and proteins in some cases (Malone 2002, Wäckers 2005). However, to fill their dietary needs phloem-feeding insects selectively assimilate and excrete nutritional and other components of the sap they ingest (Wäckers 2005). Thus, honeydews commonly are a nutritionally-poor food resource for parasitoids, though they may be important due to their availability, particularly within crop fields (Wäckers et al. 2008). In contrast to floral nectars, which may be available only transitorily within crop fields, and extra-floral nectars, which are produced only by plant species or varieties with extra-floral nectaries, honeydew may be produced season-long by sternorrhynchous hemipterans feeding in a crop or associated vegetation. However, in the case of egg parasitoids in particular, access to honeydew requires deliberate food-foraging because their hosts frequently (e.g., lepidopteran and auchenorrhynchan hosts of Trichogrammatidae, Mymaridae and Scelionidae), and their host stage (= egg) invariably, do not produce honeydew. So, honeydew foraging in egg parasitoids usually entails directed energetic and time costs, and predation risks from ants that sustain mutualistic relationships with honeydew-producing insects.

Honeydew consumption enhances the longevity, and in some cases also the fecundity of parasitoids in the field, as well as in the laboratory, including of egg parasitoids (McDougall and Mills 1997, Romeis et al. 2005, Faria et al. 2007, Fuchsberg et al. 2007, Irvin et al. 2007, Wäckers et al. 2008). Thus, honeydews should be important for enhancing biological control by egg parasitoids, particularly where availability or access to other carbohydrate sources is limited. However, because phloem sap may contain secondary plant compounds, particularly lectins and proteinase inhibitors, which are the transgene products of numerous GM crops (Schuler et al. 1998, Gatehouse 1999), GM crop proteins may be present in hemipteran honeydews. Thus, GM crop proteins, including lectins (Shi et al. 1994, Kanrar et al. 2002) and *Bt* toxins (Bernal et al. 2002) have been found in phloem sap and honeydew, though not invariably (Head et al. 2001, Raps et al. 2001). Importantly, recent studies showed that the longevity and fecundity of parasitoids, including an egg parasitoid, *Trichogramma*, were negatively affected by consumption of purified lectins in sugar solutions (Romeis et al. 2003, Bell et al. 2004). Thus, the potential impacts of insect honeydew consumption on egg parasitoids merit closer examination in the context of GM crops and biological control because of the importance of egg parasitoids for biological control, the potential importance of honeydews as food resources for adult egg parasitoids, and the known presence of GM crop proteins in honeydew and impacts on egg parasitoids.

17.3.4 Plant Sap and Other Fluids

Plant fluids other than nectars, such as phloem and xylem saps, and fluids emanating from damaged tissues, may be accessible to egg parasitoids, and phloem sap may contain GM crop proteins (Shi et al. 1994, Bernal et al. 2002, Kanrar et al. 2002). While such fluids may be important for predators that actively damage plant tissues for access to plant fluids (cf. Eubanks and Styrsky 2005), their importance for parasitoids, generally, is unclear. Very few observations are available of parasitoids in general, including egg parasitoids (e.g., Keller et al. 1985, Wellinga and Wysoki 1989), feeding on fluids from damaged tissues in the field, and these did not assess the importance of feeding on such fluids. Moreover, in experimental studies, *Trichogramma* females given access to maize phloem sap parasitized fewer hosts on average than females given access to honeydew (from maize-fed aphids) or maize pollen (Zimmermann et al. 2004). Generally, plant saps should be a nutritionally-poor food resource for parasitoids because of their frequently low sugar content, compared to nectars and honeydew, and the frequent presence of secondary plant compounds (Malone 2002, Wäckers 2005), including lectins of GM crops (Shi et al. 1994, Kanrar et al. 2002). It seems unlikely, on the basis of the few data that are available, that plant sap and fluids from damaged tissues are important food resources for egg parasitoids in the field. Therefore, examining plant sap and fluids from damaged tissues of GM crops for impacts on egg parasitoids presently seems unwarranted. Moreover, while GM crop proteins are known to occur in phloem sap (Shi et al. 1994, Bernal et al. 2002, Kanrar et al. 2002), their presence there is either necessary for protection against phloem-feeding pests, so it is deliberate, or is unavoidable and dependent upon the promoter sequence used for transgene expression (cf. Bernal et al. 2002).

17.4 Interactions Between GM Crops and Egg Parasitoids

To date, most research addressing interactions between GM crops and parasitoids, generally, has focused on parasitoids other than egg parasitoids, particularly parasitoids whose hosts are in actively-feeding stages, as noted above. The lesser focus on egg parasitoids reflects the predicted absence of GM crop proteins in insect eggs, the host stage of egg parasitoids, and, seemingly, as well, the low expectation that the quality of insect eggs as hosts for egg parasitoids would be affected by consumption of GM crop proteins by the ovipositing insect. The occurrence of GM crop proteins in eggs hosting developing egg parasitoids does not appear to have been investigated, but seems improbable. Similarly, whether ingestion of GM crop proteins by adults, or the preceding immature stages, affects the quality of their eggs as hosts for egg parasitoids has not been investigated. However, herbivore egg size, which is an important egg (i.e. host)-quality parameter, is affected by numerous factors, such as host plant quality, including quality changes due to induced herbivore resistance (van Huis and Rooy 1998, Agrawal and Klein 2000, Moreau et al. 2006, 2007), competition (Spitzen and van Huis 2005), and female adult size (Fox 1993), among other factors (Fox and Czesak 2000). Moreover, egg parasitoid fitness parameters are affected by host (egg) quality, including size (Spitzen and van Huis 2005, Da Rocha et al. 2007), and host plant quality (van Huis and Rooy 1998, Senthil Nathan et al. 2006). Thus, research addressing potential host quality-mediated impacts of GM crops on egg parasitoids is warranted. A substantial amount of research identifying negative, host-quality mediated interactions of GM crops with larval, and other, parasitoids, is available, and such interactions are argued to be the most likely bases of any field-level impacts of GM crops on biological control by parasitoids generally (Bernal et al. 2004, Lövei and Arpaia 2005, Romeis et al. 2006). Similarly, assessing host-quality mediated impacts of GM crops on egg parasitoids should be important for anticipating any field-level impacts of GM crops on biological control by these parasitoids, though, presently, the question of whether GM crops affect aspects of herbivore egg quality that are relevant to egg parasitoids remains to be addressed. In reference to this, the results of one study are suggestive of host-quality mediated effects of GM crops on an egg parasitoid (Manachini and Lozzia 2004), though because the study was not designed to address such effects its results cannot be considered conclusive. In that study, the adult emergence rate of *Trichogramma* was lower when it developed on eggs laid by *Ostrinia nubilalis* (Hübner) females recovered from *Bt* maize fields compared to eggs laid by females recovered from conventional maize fields (see below).

The available studies addressing interactions between egg parasitoids and GM crops can be grouped in several categories for the purpose of discussion. These include interactions between egg parasitoids and insect-resistant GM crops addressed in laboratory studies, and in field (or semi-field) studies, and between egg parasitoids and *HT* crops via herbicides used in those crops. With minor exceptions, all available studies have focused on *Trichogramma*, presumptively because of their importance in biological control of lepidopteran pests, which are frequent targets of GM crops, especially of the commercially-available *Bt* crops. Still, other egg parasitoid genera from the families Scelionidae, Mymaridae, and Trichogrammatidae, are important in biological control (see Chapter 13), so should be included in studies addressing impacts of GM crops on egg parasitoids, as warranted. Some studies that focused on interactions between GM crops and Hymenoptera other than egg parasitoids, such as larval or nymphal parasitoids or bees (e.g., exposure of adults via nectar and pollen feeding), or GM crops and target or non-target pests (e.g., aphid abundance in *Bt* crops) are relevant to interactions between GM crops and egg parasitoids, and are discussed below, as warranted.

17.4.1 GM Crop–Egg Parasitoid Interactions Under a Microscope: Laboratory Studies

Most studies addressing impacts of GM crops on egg parasitoids have been conducted in the laboratory. With few exceptions, such studies focused on exposure routes for adults, i.e. pollen, nectars, honeydew, and phloem sap. One study compared *Trichogramma* parasitism and adult emergence ($\approx$successful development) rates between eggs laid by hosts collected from *Bt* maize or conventional maize fields (Manachini and Lozzia 2004).

At least three studies addressed impacts on *Trichogramma* of *Bt* toxins present in GM crop pollen. Invariably, these studies did not detect significant effects of *Bt* toxins present in pollen on *Trichogramma* reproductive and adult lifespan parameters, even while one study evaluated a *Bt* crop variety (*Bt* maize, transformation event 176) with a high transgene expression level in pollen, and another evaluated a combined (i.e., stacked) *Bt+PTg* (CpTI) crop variety (Zimmermann et al. 2004, Geng et al. 2006, Wang et al. 2007).

Several studies addressed potential impacts of GM crop nectars, or honeydew, on egg parasitoids, though indirectly. Two studies showed that various purified *PTg* products (GNA, ConA, CpTI) administered to adult parasitoids in sugar solutions (water and honey or sucrose solutions), to emulate exposure via nectar or honeydew, negatively affected parasitoid adult longevity and reproduction, though only one study included an egg parasitoid, *Trichogramma* (Romeis et al. 2003,

Bell et al. 2004). However, neither study addressed whether the *PTg* products were present in GM crop nectar, or honeydew, or whether the levels evaluated (0.01–1%) were biologically realistic. Other studies addressing interactions between GM crop nectars and bees showed that nectar production differed quantitatively and qualitatively between GM and conventional (isogenic) plants, which, presumably, reflected unplanned plant phenotypic changes associated with genetic transformation (Picard-Nizou et al. 1993, 1995, Tesoriero et al. 2004). The quantitative and qualitative changes in nectar production evident in those studies could alter any nectar-mediated interactions between GM crops and egg parasitoids, thus merit further examination (see below). Only one study addressed potential impacts of phloem sap or honeydew (or pollen, see above) from *Bt* maize plants on *Trichogramma*, but did not detect an effect of either on reproduction (Zimmermann et al. 2004).

One study addressed potential plant host, *Bt* or conventional maize, effects mediated by the insect host, *O. nubilalis*, on *Trichogramma*, as noted above (Manachini and Lozzia 2004). In that study, *O. nubilalis* obtained from *Bt* or conventional maize fields were allowed to oviposit on maize leaves, their eggs were then exposed to *Trichogramma* females, and parasitism and adult emergence rates were subsequently assessed. While a significant host plant effect on parasitism rate was not detected, adult emergence rate was lower from eggs oviposited by *O. nubilalis* obtained from *Bt* maize. Because parasitism rates were assessed on the basis of earliest external evidence of parasitism, viz. blackening of host eggs, the results indicated that there was no host plant effect on early parasitoid development (nor on host selection by the *Trichogramma* females). In contrast, because adult emergence rates were assessed on the basis of adult emergence frequencies from evidently parasitized host eggs, they reflected host plant effects on latter parasitoid development and actual emergence of adults (i.e. emergence of pharate adults from pupae, and subsequently from host eggs). The effect of *Bt* maize on adult emergence rates thus can be interpreted, preliminarily, as evidence that egg (host)-quality was impaired so by *Bt* maize that complete parasitoid development and/or adult emergence were correspondingly impaired. However, further and more detailed studies are necessary to confirm, or refute, this interpretation, and, if warranted, to identify causal mechanisms.

17.4.2 GM Crop–Egg Parasitoid Interactions Under a Loupe: Field Studies

Few studies have addressed parasitoid-GM crop interactions in the field (cf. Bernal et al. 2004, Obrycki et al. 2004, Romeis et al. 2006), and a smaller number focused on egg parasitoids, though in some cases only tangentially or as part of larger studies. Most studies focused on *Bt* maize or *Bt* cotton, a reflection of their widespread planting, particularly in Australia, China, Europe and USA. With minor exceptions, all studies involving egg parasitoids were focused on *Trichogramma*. A small number of studies focusing on crops other than GM cotton or maize, e.g. *Bt* canola, or on GM crops not yet used commercially in the country of study, but not on egg parasitoids are discussed herein because they are relevant to analyses of GM crop–egg parasitoid interactions. Broadly, the available studies compared egg parasitoid abundance between conventional and GM crop fields (or small experimental plots) using proxies, such as rates of parasitism of pest eggs, or recovery frequencies of egg parasitoids within surveys of arthropod diversity. Overall, a consistent effect of GM crops, whether positive or negative, on egg parasitoid abundance in the field is not evident from the available studies.

In cotton, two studies comparing the rates of parasitism of lepidopteran eggs by *Trichogramma* yielded conflicting results (Yang et al. 2005, Whitehouse et al. 2007). The results of one such study suggested an antagonistic interaction between biological control by parasitoids and GM cotton (*Bt* or *Bt+PTg* cotton varieties) (Yang et al. 2005). In that study, the rates of parasitism of *Helicoverpa armigera* (Hübner) eggs by *Trichogramma* were consistently lower in GM cotton compared to conventional cotton over three consecutive seasons, viz. 0–2.8 vs. 14.4–18.6%, 0–3.5% vs. 8.1–15.2%, and 0–5.3% vs. 22.0–30.1% in the first, second, and third seasons, respectively; a similar trend was evident in parasitism rates of *H. armigera* larvae by a larval parasitoid. In contrast, the second study yielded significantly higher rates of parasitism of *Helicoverpa* spp. eggs in GM cotton versus conventional cotton in one locality (95 vs. 61%, respectively), though not at another locality (6 vs. 3%, respectively) (Whitehouse et al. 2007). The results of both studies, however, were inconclusive because experimental plot sizes were small, replication in time (crop seasons) or space (localities) was inadequate, or pest egg sample sizes were small because *Helicoverpa* populations were small. Other studies compared arthropod communities between GM cotton and conventional cotton, and between insecticide use regimens (Whitehouse et al. 2005, Mansfield et al. 2006). Those studies found only subtle differences between arthropod community compositions of GM and conventional cotton crops, though not in egg parasitoids, which could be explained by differences in abundance of Lepidoptera (Whitehouse et al. 2005), or by differences in insecticide use regimen, rather than differences in crop type (GM or conventional) (Mansfield et al. 2006). One common observation (explicit or evident) among these studies is that *Trichogramma* and Lepidoptera abundances were low in GM plots compared to conventional plots; a relevant question arising from this observation is whether at a larger scale the lower abundance of *Trichogramma* in GM fields would affect egg parasitism rates in neighboring conventional fields in subsequent seasons (Bernal et al. 2004).

In maize, two studies compared rates of parasitism of lepidopteran eggs by *Trichogramma* between GM and conventional crops (Orr and Landis 1997, Fernandes et al. 2007). In one study, the rates of parasitism of *O. nubilalis* eggs and egg masses were low (~2–6%) in both *Bt* and conventional maize fields, and did not differ significantly between the fields (Orr and Landis 1997). In the second study, the rates of parasitism of *Helicoverpa zea* (Boddie) eggs were consistently ~30% in *Bt* and conventional maize fields, and did not differ significantly between the fields (Fernandes et al. 2007). However, as in the case of cotton, both studies employed small experimental plots, were not replicated in time, and plots were minimally replicated ($n = 3$), so their results should be considered inconclusive.

Additional studies, not focusing on egg parasitoids, yielded data that are relevant for anticipating possible outcomes of interactions between egg parasitoids and GM crops in the field. For example, a broad, long-term study focusing on *Bt* maize showed that aphid densities were ~23% higher in commercial *Bt* maize fields compared to conventional maize fields (Eizaguirre et al. 2006); in contrast, however, another study, which employed small, experimental plots, did not find differences in aphid densities between *Bt* and conventional maize (Bhatti et al. 2005). It would be interesting to examine whether higher aphid densities in *Bt* maize fields translate into greater availability of honeydew and higher rates of egg parasitism, particularly of pests not targeted by *Bt* maize. Greater rates of egg parasitism of such pests could dampen any increases in their populations, so lessen the likelihood that they become economically important pests (Bernal et al. 2004). A study addressing the increased abundance of aphids in *Bt* maize fields found that such greater abundance on *Bt* maize was due to changes in host plant quality due to pleiotropic effects of genetic transformation, rather than to a direct effect of the *Bt* toxin (Lumbierres et al. 2004). Similarly, other studies on GM crops have uncovered phenotypic trait changes that were presumably associated with genetic transformation, and that may be relevant to biological control and pest management generally; e.g., quantitative and qualitative changes in nectar production were observed in GM canola (see previous discussion) (Picard-Nizou et al. 1993, 1995, Tesoriero et al. 2004), and enhanced suitability for pests (i.e. greater pest reproduction) and antixenosis toward ovipositing adults were evident in GM sugarcane (Sétamou et al. 2002, Bernal and Sétamou 2003). Such trait changes are not rare and cannot be assumed inconsequential, so should be examined for their influence on GM crop–egg parasitoid interactions. It is tempting to predict, for example, that increased nectar and honeydew production should contribute to enhancing parasitism rates by egg parasitoids, particularly in light of evidence that both nectar and honeydew availability enhance biological control by and reproductive parameters of *Trichogramma* and other egg parasitoids (Treacy et al. 1987, Heimpel and Jervis 2005, Romeis et al. 2005, Faria et al. 2007, Fuchsberg et al. 2007, Wäckers et al. 2008). In other cases, however, negative impacts of phenotypic trait changes on egg parasitoids could be predicted, such as greater exposure to GM crop proteins if production of honeydew and nectar are enhanced; thus, the potential impacts of such trait changes should also be examined closely.

17.4.3 *HT* Crop–Egg Parasitoid Interactions: Effects of Herbicides on Egg Parasitoids

An array of *HT* crop varieties is planted extensively in countries where they are available. These varieties express herbicide tolerance traits for various herbicides, including glyphosate, glufosinate, bromoxynil, and ioxynil, and sulfonylurea or imidazolinone class herbicides, though tolerance to glyphosate or glufosinate are the most common traits (Knesevic and Cassman 2003, AgBios 2008). The main *HT* crops are maize, cotton, canola, soybeans, and alfalfa, though *HT* varieties are available for other crops, including creeping bentgrass, sugar beet, chicory, carnation, flax, linseed, tobacco, rice and wheat (James 2007, AgBios 2008). Despite the predominance of *HT* over insect-resistant crops worldwide, viz. nearly two-thirds of all GM crop acreage worldwide corresponds to *HT* crops (James 2007), very few studies have examined the toxicity of glyphosate or glufosinate, or other herbicides, towards egg parasitoids, though none of the available studies were conducted in the context of *HT* crops, and all focus solely on *Trichogramma* and were conducted in the laboratory. The available studies showed that at field-application rates, glufosinate was moderately harmful (category 3), and glyphosate was slightly (category 2) to moderately harmful to *Trichogramma*, according to IOBC/WPRS standards (category 1 = harmless, <30% reduction of egg parasitism; 2 = slightly harmful, 30–79%; 3 = moderately harmful, 80–99%; 4 = harmful, >99%) (Giolo et al. 2005a, b, 2007a, b, Manzoni et al. 2006). One study showed that glyphosate toxicity to *Trichogramma* was influenced by herbicide formulation, thus formulations containing potassium and ammonium salts were slightly harmful, while those containing isopropylamine salts were moderately harmful (Giolo et al. 2005b).

Clearly, additional research is needed addressing glufosinate and glyphosate toxicity to egg parasitoids in the context of *HT* crops, and other egg parasitoids important to biological control, and under field conditions, especially given the widespread use of *HT* crops worldwide. Also, because both glyphosate and glufosinate are non- selective herbicides, their use eliminates most non-*HT* plants, and dramatically shifts weed composition within *HT* crops fields (Hawes et al. 2003, Knezevic and Cassman 2003). The positive influence of plant diversity, including weed diversity that provides alternate hosts and food resources, for biological control is well known (e.g., Altieri 1999, Gurr et al. 1998, 2003, Landis et al. 2000, Nicholls and Altieri 2004), so will not be discussed here. Also, prior research showed that changes in weed composition and availability associated with use of *HT* crops negatively affected populations of natural enemies because of a reduced abundance of host insects (Hawes et al. 2003), as well as populations of nectar-feeding insects (e.g., bees, butterflies), likely because of a reduced availability of nectars (Roy et al. 2003). Overall, the paucity of research on *HT* crop–egg parasitoid interactions is conspicuous, especially given the predominance of *HT* over insect-resistant GM crops worldwide (James 2007), and the documented toxicity of both glyphosate and glufosinate to *Trichogramma* (Giolo et al. 2005a, b, 2007a, b, Manzoni et al. 2006).

17.5 Ignorance Is Not Bliss: Building Upon What We Know to Better Integrate GM Crop Use with Biological Control by Egg Parasitoids

Very few data are available concerning the interactions between egg parasitoids and GM crops, compared to the data on interactions involving other parasitoids. This paucity is probably a reflection, in part, of the improbability of GM crop proteins in the herbivore eggs used as hosts by egg parasitoids within GM crop fields. However, this inattention to egg parasitoids among the research addressing natural enemy-GM crop interactions ignores, for the most part, some important and established routes of exposure (e.g., honeydew, nectars), as well as plausible, but unexplored, indirect mechanisms for interactions of GM crops with egg parasitoids (e.g., host quality impairment, herbicide use in *HT* crops). A fuller understanding of any interactions between egg parasitoids and GM crops is particularly relevant in the context of analyses pointing to the importance of movement of natural enemy populations between crops and seasons within a landscape for pest management and biological control at a regional scale (Kareiva 1990, Tscharntke 2000, Bernal et al. 2004, Schmidt et al. 2004). A pertinent question in this context, for example, is whether near-elimination of several lepidopteran species that are susceptible to *Bt* maize in landscapes where *Bt* maize crops occupy substantial portions of the crop area (e.g., 49% in South Dakota and 46% in Minnesota, USA, in 2004, Benbrook 2004) affects biological control of lepidopteran pests in neighboring conventional maize or other crops, or in subsequent growing seasons. This question is particularly significant in light of data that suggest a decline of *O. nubilalis* populations in Minnesota, USA, since the introduction of *Bt* maize (Abrahamson 2007), and other data indicating that non-maize hosts are not important refugia for this species because it strongly prefers to oviposit on maize plants (Losey et al. 2001). A related question, but with much broader implications, asks whether biological control by egg parasitoids is affected at a regional scale, due to a scarceness of hosts or of hosts of suitable quality, in landscapes in which GM crops occupy the majority of crop area, e.g., together, *Bt*, *HT*, and stacked *Bt+HT* cotton crops occupied 91–97% of the cotton area in most cotton producing states of the USA in 2004, while the corresponding percentage for maize crops in South Dakota, USA, was 79% (Benbrook 2004). These and similar questions are probably the most important concerning the interactions between GM crops and egg parasitoids, and GM crops and parasitoids and biological control generally, though they are undoubtedly the most difficult to address. Tackling such questions, regrettably, does not appear to be a task that is accomplishable in the immediate future because so few data are presently available for egg parasitoids in relation to GM crops, and the difficulties of conducting research on parasitoids generally at a landscape level.

Though GM crops, specifically *Bt* and *HT* crops, are now widely available worldwide for commercial use and widely planted, opportunities remain for developing novel GM crops or novel ways of using existing GM crops so that their use complements or synergizes the impacts of egg parasitoids on pest populations. Prior analyses discussed some opportunities concerning GM crops and parasitoids generally (e.g., Bernal et al. 2004, Poppy and Sutherland 2004), or *Trichogramma* and biological control specifically (Romeis et al. 2005), including opportunities that should be conducive to enhancing biological control by egg parasitoids in GM crops. Such opportunities could be explored for deployment in combination with existing GM crops, or less desirably, as stand-alone approaches for achieving pest management goals via GM crops and biological control. Additionally, research focusing on honey bees and other hymenopteran pollinators should be useful for analyses of GM crop–egg parasitoid interactions because of the data available concerning impacts of GM crop proteins on those insects, as mediated by consumption of plant nectars, pollen, and honeydew (Malone 2002, Malone and Pham-Delègue 2002).

Two opportunities for minimizing any negative effects of GM crops on egg parasitoids involve strategic (spatial) deployment of GM crops in agroecosystems to favor egg parasitoid populations (Bernal et al. 2004), and other habitat management practices that provide refugia, including hosts and food resources, for egg parasitoids, as discussed previously for natural enemies in general and parasitoids in particular (Landis et al. 2000, Gurr et al. 1998, 2003, Altieri et al. 2004, Bernal et al. 2004, Romeis et al. 2005). Other opportunities involve developing GM crops with traits that enhance the activity of egg parasitoids. For example, developing glabrous GM crop varieties if trichomes are known to interfere with egg parasitoid foraging (Poppy and Sutherland 2004, Romeis et al. 2005), or nectariless or pollen- or flower-free varieties if GM crop proteins are known to occur in nectar or pollen at ecologically significant levels (Malone 2002); similarly, other physical traits (e.g., leaf waxes) known to negatively affect egg parasitoids could be addressed via genetic transformation of crops. A related opportunity involves deliberately avoiding promoter sequences, for genetic transformation, that are associated with high levels of transgene expression in pollen or nectars to minimize exposure of adult egg parasitoids to GM crop proteins (Malone 2002). Other examples involve developing GM crops with enhanced production of herbivore-induced leaf volatiles for more effective communication between plants and egg parasitoids (cf. Degenhardt et al. 2003, Colazza et al. 2004, Fatouros et al. 2008), or enhanced nectar production, including varieties with extra-flora nectaries, to enhance pest parasitism rates by egg parasitoids (cf. Treacy et al. 1987, Picard-Nizou et al. 1993, 1995, Tesoriero et al. 2004, Heimpel and Jervis 2005). Finally, other opportunities involve deploying host plant resistance based on GM crop proteins more rationally, so that their use is more sustainable and compatible with IPM strategies than has hitherto been the case. One approach to this is to develop GM crops that are partially resistant to pests, or promote the commercial release of *PTg* crops, to minimize effects on trophic levels above the targeted pests, including egg parasitoids (Bernal et al. 2004, Poppy and Sutherland 2004). Similarly, a more rational approach to herbicide use in *HT* crops so that weed composition and availability are affected less dramatically by herbicide use should help to conserve egg parasitoid populations by conserving nectars and alternate hosts (cf. Hawes et al. 2003, Roy et al. 2003), and would make *HT* crop use more sustainable and compatible with IPM.

A caveat, which should not go unsaid, is that any change introduced into existing pest management strategies should be preceded by carefully examined considerations of its effects on targeted pests, other pests (including key and secondary pests), other herbivores, and natural enemies to avoid aggravating current pest problems or creating new pest problems (Bernal et al. 2004). In the context of GM crops, this means that concerns should not overridingly emphasize maintenance of pest susceptibility (i.e. resistance management), but should extend to conserving natural enemy populations and finding compatibility with biological control, and avoiding target pest resurgences and secondary pest upsets (Bernal et al. 2004). GM crops are a powerful and effective technology in the current pest management arsenal so they must be preserved for continued use. Thus, novel GM crops should be developed or the use of existing GM crops should be modified with consideration to natural enemy populations, and within the confines of proven IPM concepts, particularly the established notion that multi-tactic pest management strategies are more sustainable and remain effective over a longer period of time than single tactic pest management strategies.

References

Abrahamson M (2007) Minnesota pest reports (16 Feb. 2007). Available at http://www.mda.state.mn.us/news/publications/pestsplants/insectsandpests/pestreport/2006summary.pdf. Accessed 23 April 2008

AgBios (2008) GM database. Available at http://www.agbios.com/dbase.php. Accessed 28 January 2008

Agrawal A, Klein CN (2000) What omnivores eat: direct effects of induced plant resistance on herbivores and indirect consequences for diet selection by omnivores. J Animal Ecol 69: 529–535

Altieri MA (1999) The ecological role of biodiversity in agroecosystems. Agric, Ecosyst Environ 74:19–31

Altieri MA, Gurr GM, Wratten SD (2004) Genetic engineering and ecological engineering: a clash of paradigms or scope for synergy. In: Gurr GM, Wratten SD, Altieri MA (eds) Ecological engineering for pest management: Advances in habitat manipulation for arthropods. Comstock Publishing Associates, Ithaca, NY, pp 13–31

Álvarez-Alfageme F, Ferry N, Castañera P, Ortego F, Gatehouse AMR (2008) Prey mediated effects of Bt maize on fitness and digestive physiology of the red spider mite predator *Stethorus punctillum* Weise (Coleoptera: Coccinellidae). Transgenic Res 17:943–954

Anderson PC, Broadbeck BV, Mizell RF (1989) Metabolism of aminoacids, organic acids and sugars extracted from the xylem fluid of four host plants by adult *Homalodisca vitripennis*. Entomologia Experimentalis et Applicatta 50:149–159

Anonymous (2001) Biopesticides registration action document for *Bacillus thuringiensis (Bt)* plant-incorporated protectants. U.S. Environmental Protection Agency, Office of Pesticide Programs, Biopesticides and Pollution Prevention Division (October 15, 2001), 27 p

Baker HG, Baker I (1975) Studies of nectar constitution and pollinator-plant coevolution. In: Gilbert LE, Raven PH (eds) Co-evolution of animals and plants. University of Texas Press, Austin, TX, pp 100–140

Baker HG, Baker I (1983) Floral nectar sugar constituents in relation to pollinator type. In: Jones CE, Little RJ (eds) Handbook of experimental pollination biology. Van Nostrand Reinhold, New York, pp 117–141

Begum M, Gurr GM, Wratten SD (2004) Flower color affects tri-trophic biocontrol interactions. Biological Control 30:584–590

Bell HA, Kirkbride-Smith AE, Marris GC, Edwards JP, Gatehouse AMR (2004) Oral toxicity and impact on fecundity of three insecticidal proteins on the gregarious ectoparasitoid *Eulophus pennicornis* (Hymenoptera:Eulophidae). Agric For Entomol 6:215–222

Benbrook CM (2004) Genetically engineered crops and pesticide use in the United States: The first nine years. BioTech InfoNet, Technical Paper Number 7, 38 p

Bernal CC, Aguda RM, Cohen MB (2002) Effect of rice lines transformed with *Bacillus thuringiensis* toxin genes on the brown planthopper and its predator *Cyrtorhinus lividipennis*. Entomol Exp Appl 102:21–28

Bernal JS, Sétamou M (2003) Fortuitous antixenosis in transgenic sugarcane: antibiosis-expressing cultivar deters oviposition by herbivore pests. Environ Entomol 32:886–894

Bernal JS, Prasifka J, Sétamou M, Heinz KM (2004) Transgenic insecticidal cultivars in integrated pest management: challenges and opportunities. In: Koul O, Dhaliwal GS, Cuperus GW (eds) Integrated pest management: Potential, constraints and challenges. CABI, Oxfordshire, pp 123–145

Bhatti MA, Duan J, Head GP, Jiang C, McKee M, Nickson TE, Pilcher CL, Pilcher CD (2005) Field evaluation of the impact of corn rootworm (Coleoptera: Chrysomelidae)-protected *Bt* corn on foliage-dwelling arthropods. Environ Entomol 34:1336–1345

Carter C, Graham RA, Thornburg RW (1999) Nectarin I is a novel, soluble germin-like protein expressed in the nectar of *Nicotiana* sp. Plant Mol Biol 41:207–216

Colazza S, Fucarino A, Peril E, Salerno G, Conti E, Bin F (2004) Insect oviposition induces volatile emission in herbaceous plants that attracts egg parasitoids. J Exp Biol 207:47–53

Da Rocha L, Kolberg R, Mendonça M De S, Jr Redaelli LR (2007) Body size variation in *Gryon gallardoi* related to age and size of the host. BioControl 52:161–173

Degenhardt J, Gershenzonz J, Baldwin IT, Kessler A (2003) Attracting friends to feast on foes: engineering terpene emission to make crop plants more attractive to herbivore enemies. Curr Opin Biotechnol 14:169–176

Dobson HEM (1988) Survey of pollen and pollenkit lipids: chemical cues to flower visitors? Am J Botany 75:170–182

Eizaguirre M, Albajes R, López C, Eras J, Lumbierres B, Pons X (2006) Six years after the commercial introduction of *Bt* maize in Spain: field evaluation, impact and future prospects. Transgenic Res 15:1–12

Eubanks MD, Styrsky JD (2005) The effects of plant feeding on the performance of omnivorous 'predators'. In: Wäckers FL, van Rijn PCJ, Bruin J (eds) Plant-provided food for carnivorous insects: a protective mutualism and its applications. Cambridge University Press, Cambridge, pp 148–177

Faria CA, Wäckers FL, Turlings TCJ (2007) Increased susceptibility of Bt maize to aphids enhances the performance of parasitoids of lepidopteran pests. PlosOne 2:1–11

Fatouros NE, Dicke M, Mumm R, Meiners T, Hilker M (2008) Foraging behavior of egg parasitoids exploiting chemical information. Behav Ecol 19:677–689

Fearing PL, Brown D, Vlachos D, Meghji M, Privalle L (1997) Quantitative analysis of CryIA(b) expression in *Bt* maize plants, tissues, and silage and stability of expression over successive generations. Mol Breed 3:169–176

Fernandes OA, Faria M, Martinelli S, Schmidt F, Carvalho VF, Moro G (2007) Short-term assessment of *Bt* maize on non-target arthropods in Brazil. Sci Agric 64:249–255

Ferry N, Mulligan EA, Majerus MEN, Gatehouse AMR (2007) Bitrophic and tritrophic effects of Bt Cry3A transgenic potato on beneficial, non-target beetles. Transgenic Res 16:795–812

Fox CW (1993) The influence of maternal age and mating frequency on egg size and offspring performance in *Callosobruchus maculatus* (Coleoptera: Bruchidae). Oecologia 96: 139–146

Fox CW, Czesak ME (2000) Evolutionary ecology of progeny size in arthropods. Annu Rev Entomol 45:341–369

Fuchsberg JR, Yong TH, Losey JE, Carter ME, Hoffmann AA (2007) Evaluation of corn leaf aphid (*Rhopalosiphum maidis*; Homoptera: Aphididae) honeydew as a food source for the egg parasitoid *Trichogramma ostriniae* (Hymenoptera: Trichogrammatidae). Biol Control 40: 230–236

Gatehouse AMR (1999) Biotechnological applications of plant genes in the production of insect-resistant crops. In: Clement SL, Quisenberry SS (eds) Global plant genetic resources for insect-resistant crops. CRC Press, BocaRaton, pp 263–280

Geng J, Shen Z, Song K, Zheng L (2006) Effect of pollen of regular cotton and transgenic Bt+CpTI cotton on the survival and reproduction of the parasitoid wasp *Trichogramma chilonis* (Hymenoptera: Trichogrammatidae) in the laboratory. Environ Entomol 35:1661–1668

Giolo FP, Grutzmacher AD, Procopio SO, Manzoni CG, Lima CAB, Nornberg SD (2005a) Side-effects of glyphosate formulations on *Trichogramma pretiosum* (Hymenoptera: Trichogrammatidae). Planta Daninha 23:457–462

Giolo FP, Grutzmacher AD, Manzoni CG, Fachinello JC, Nornberg SD, Stefanello GJ Jr (2005b) Side-effects of pesticides used in integrated production of peach on *Trichogramma pretiosum* Riley, 1879 (Hymenoptera: Trichogrammatidae). Revista Brasileira de Fruticultura 27:222–225

Giolo FP, Grutzmacher AD, Manzoni CG, De Lima CAB, Noernberg SD (2007a) Toxicity of pesticides used in peach orchard on adults *Trichogramma pretiosum*. Bragantia 66:423–431

Giolo FP, Grutzmacher AD, Manzoni CG, Harter WR, Castilhos RV, Muller C (2007b) Toxicity of pesticides used in peach production on the egg parasitoid *Trichogramma atopovirilia* Oatman and Platner, 1983 (Hymenoptera: Trichogrammatidae). Ciencia Rural 37:308–314

Gonzalez D (1971) Sampling as a basis for pest management strategies. In: Proceedings of the Tall Timbers Conference on Ecological Animal Control by Habitat Management. Tall Timbers Research Station, Tallahassee, USA, pp 83–101

Gordh G, Legner EF, Caltagirone, LE (1999) Biology of parasitic Hymenoptera. In: Bellows TS, Fisher TW (eds) Handbook of biological control. Academic, San Diego, CA, pp 355–381

Groot AT, Dicke M (2002) Insect-resistant transgenic plants in a multi-trophic context. Plant J 31:387–406

Gurr GM, van Emden HF, Wratten SD (1998) Habitat manipulation and natural enemy efficiency: implications for the control of pests. In: Barbosa P (ed) Conservation biological control. Academic, UK, pp 155–183

Gurr GM, Wratten SD, Luna JM (2003) Multi-function agricultural biodiversity: pest management and other benefits. Basic Appl Ecol 4:107–116

Hare JD (2002) Plant genetic variation in tritrophic interactions. In: Tscharntke T, Hawkins BA (eds) Multitrophic level interactions. Cambridge University Press, Cambridge, pp 8–43

Hawes C, Haughton AJ , Osborne JL, Roy DB, Clark SJ, Perry JN, Rothery P, Bohan DA, Brooks DR, Champion GT, Dewar MS, Heard MS, Woiwod IP, Daniels RE, Young MW, Parish AM, Scott RJ, Firbank LG, Squire GR (2003) Responses of plants and invertebrate trophic groups to contrasting herbicide regimes in the farm scale evaluations of genetically modified herbicide-tolerant crops. Philos Trans R Soc Lan B Biol Sci 358:1899–1913

Head G, Brown CR, Groth M, Duan JJ (2001) Cry1Ab protein levels in phytophagous insects feeding on transgenic corn: implications for secondary exposure risk assessment. Entomol Exp Appl 99:37–45

Heimpel GE, Jervis MA (2005) Does floral nectar improve biological control by parasitoids? In: Wäckers FL, van Rijn PCJ, Bruin J (eds) Plant-provided food for carnivorous insects: a protective mutualism and its applications. Cambridge University Press, Cambridge, pp 267–304

Hilbeck A, Schmidt JEU (2006) Another view on *Bt* proteins – how specific are they and what else might they do? Biopestic Int 2:1–50

Irvin NA, Hoddle MS, Castle SJ (2007) The effect of resource provisioning and sugar composition of foods on longevity of three *Gonatocerus* spp., egg parasitoids of *Homalodisca vitripennis*. Biol Control 40:69–79

James C (2007) Global status of Commercialized biotech/GM Crops: 2007. ISAAA Brief, No. 37. ISAAA: Ithaca, NY

Jervis MA, Kidd NAC, Heimpel GE (1996) Parasitoid adult feeding behavior and biocontrol – a review. Biocontrol News Inf 17:11 N–26 N

Jouanin L, Girard C, Bonadé-Bottino M, Le Metayer M, Picard Nizou A, Lerin J, Pham-Delègue M (1998) Impact of oilseed rape expressing proteinase inhibitors on coleopteran pests and honeybees. Cahiers Agric 7:531–536

Kareiva P (1990) The spatial dimension in pest-enemy interactions. In: Mackauer M, Ehler L and Roland J (eds), Critical issues in biological control. Intercept Press, Andover, UK, pp 213–226

Kanrar S, Venkateswari J, Kirti PB, Chopra VL (2002) Transgenic Indian mustard (*Brassica juncea*) with resistance to the mustard aphid (*Lipaphis erysimi* Kalt.). Plant Cell Rep 20: 976–981

Keller MA, Lewis WJ, Stinner RE (1985) Biological and practical significance of movement by *Trichogramma* species: a review. Southwest Entomol 8:138–155

Knesevic SZ, Cassman KG (2003) Use of herbicide-tolerant crops as a component of an integrated weed management program. Crop Manage J (online http://www.plantmanagementnetwork.org/pub/cm/management/2003/htc/)

Koptur S (2005) Nectar as fuel for plant protectors. In: Wäckers FL, van Rijn PCJ, Bruin J (eds) Plant-provided food for carnivorous insects: a protective mutualism and its applications. Cambridge University Press, Cambridge, pp 75–108

Kozeil MG, Beland GL, Bowman C, Carozzi NB, Crenshaw R, Crossland L, Dawson J, Desai N, Hill M, Kadwell S, Launis K, Lewis K, Maddox D, McPherson K, Meghji MR, Merlin E, Rhodes R, Warren GW, Wright M, Evola SV (1993) Field performance of elite transgenic maize plants expressing an insecticidal protein derived from *Bacillus thuringiensis*. Bio/Technology 11:194–200

Landis DA, Wratten SD, Gurr GM (2000) Habitat management to conserve natural enemies of arthropod pests in agriculture. Annu Rev Entomol 45:175–201

Losey JE, Calvin DD, Carter ME, Mason CE (2001) Evaluation of noncorn host plants as a refuge in a resistance management program for European corn borer (Lepidoptera: Crambidae) on *Bt*-corn. Environ Entomol 30:728–735

Lövei GL, Arpaia S (2005) The impact of transgenic plants on natural enemies: a critical review of laboratory studies. Entomol Exp Appl 114:1–14

Lumbierres B, Albajes R, Pons X (2004) Transgenic *Bt* maize and *Rhopalosiphum padi* (Hom., Aphididae) performance. Ecol Entomol 29:309–317

Malone LA (2002) Literature review on genetically modified plants and bee products. MAF Technical Paper No: 2002/05

Malone LA, Pham-Delègue M (2002) Effects of transgene products on honey bees (*Apis mellifera*) and bumblebees (*Bombus* sp.). Apidologie 32:287–304

Manachini B, Lozzia GC (2004) Studies on the effects of *Bt* corn expressing Cry1Ab on two parasitoids of *Ostrinia nubilalis* Hb. (Lepidoptera: Crambidae). Bulletin OILB/SROP - WPRS/SROP 27:109–116

Mansfield S, Dillon L, Whitehouse MEA (2006) Are arthropod communities in cotton really disrupted? An assessment of insecticide regimes and evaluation of the beneficial disruption index. Agric Ecosyst Environ 113:326–335

Manzoni CG, Grutzmacher AD, Giolo FP, Harter WR, Muller C (2006) Side effects of pesticides used in integrated production of apple in adults of *Trichogramma pretiosum*. Pesquisa Agropecuaria Brasileira 41:1461–1467

McDougall SJ, Mills NJ (1997) The influence of hosts, temperature and food sources on the longevity of *Trichogramma platneri*. Entomol Exp Appl 83:195–203

Moreau J, Benrey B, Thiery D (2006) Assessing larval food quality for phytophagous insects: are the facts as simple as they appear? Funct Ecol 20:592–600

Moreau J, Thiery D, Troussard JP, Benrey B (2007) Grape variety affects female but also male reproductive success in wild European grapevine moths. Ecol Entomol 32:747–753

Nepi M, Franchi GG (2000) Cytochemistry of mature angiosperm pollen. In: Dafni A, Hesse M, Pacini E (eds) Pollen and pollination. Springer, Vienna, pp 45–62

Nicholls CI, Altieri MA (2004) Agroecological bases of ecological engineering for pest management. In: Gurr GM, Wratten SD, Altieri MA (eds) Ecological engineering for pest management: Advances in habitat manipulation for arthropods. Comstock Publishing Associates, Ithaca, pp 33–54

Obrycki JJ, Ruberson JR, Losey JE (2004) Interactions between natural enemies and transgenic insecticidal crops. In: Ehler LE, Sforza R, Mateille T (eds) Genetics, evolution, and biological control. CAB International, Oxon, UK, pp 183–206

Orr DB, Landis DA (1997) Oviposition of European corn borer (Lepidoptera: Pyralidae) and impact of natural enemy populations in transgenic versus isogenic corn. J Econ Entomol 90:905–909

Peumans WJ, Smeets K, van Nerum K, van Leuven F, van Damme EJM (1997) Lectin and alliinase are the predominant proteins in nectar from leek (*Allium porrum* L.) flowers. Planta 201: 298–302

Picard-Nizou AL, Kerguelen V, Douault P, Marilleau R, Blight M, Jouanin L, Renard M, Pham-Delègue M (1993) Contribution to the study of honey bee-transgenic oilseed rape interactions. Apidologie 24:457–459

Picard-Nizou AL, Pham-Delègue M, Kerguelen V, Douault P, Marilleau R, Olsen L, Grison R, Toppan A, Masson C (1995) Foraging behavior of honey bees (*Apis mellifera* L.) on transgenic oilseed rape (*Brassica napus* L. var. oleifera). Transgenic Res 4:270–276

Poppy GM, Sutherland JP (2004) Can biological control benefit from genetically-modified crops? Tritrophic interactions on insect-resistant transgenic plants. Physiol Entomol 29:257–268

Rahat R, Gurr GM, Wratten SD, Mo J, Neeson R (2005) Effect of plant nectars on adult longevity of the stinkbug parasitoid, *Trissolcus basalis*. Int J Pest Manage 51:321–324

Raps A, Kehr J, Gugerli P, Moar WJ, Bigler F, Hilbeck A (2001) Immunological analysis of phloem sap of *Bacillus thuringiensis* corn and of the non target herbivore *Rhopalosiphum padi* (Homoptera: Aphididae) for presence of Cry1Ab. Mol Ecol 10:525–534

Romeis J, Babendreier D, Wäckers FL (2003) Consumption of snowdrop lectin (*Galanthus nivalis* agglutinin) causes direct effects on adult parasitic wasps. Oecologia 134:528–536

Romeis J, Babendreier D, Wäckers FL, Shanower TG (2005) Habitat and plant specificity of *Trichogramma* egg parasitoids - underlying mechanisms and implications. Basic Appl Ecol 6:215–236

Romeis J, Meissle M, Bigler F (2006) Transgenic crops expressing *Bacillus thuringiensis* toxins and biological control. Nat Biotechnol 24:63–71

Roulston TH, Cane JH (2000) Pollen nutritional content and digestibility for animals. In: Dafni A, Hesse M, Pacini E (eds) Pollen and pollination. Springer, Vienna, pp 187–211

Roy DB, Bohan DA, Haughton AJ, Hill MO, Osborne JL, Clark SJ, Perry JN, Rothery P, Scott RJ, Brooks DR, Champion GT, Hawes C, Heard MS, Firbank LG (2003) Invertebrates and vegetation of field margins adjacent to crops subject to contrasting herbicide regimes in the Farm Scale Evaluations of genetically modified herbicide-tolerant crops. Philos Trans R Soc Lan B Biol Sci 358:1879–1898

Sachs ES, Benedict JH, Stelly DM, Taylor JF, Altman DW, Berberich SA, Davis SK (1998) Expression and segregation of genes encoding CryIA insecticidal proteins in cotton. Crop Sci 38:1–11

Schüler TH, Poppy GM, Kerry BR, Denholm I (1998) Insect-resistant transgenic plants. Trends Biotechnol 16:168–175

Schmidt MH, Thies C, Tscharntke T (2004) Landscape context of biological control. In: Gurr GM, Wratten SD, Altieri MA (eds) Ecological engineering for pest management: Advances in habitat manipulation for arthropods. Comstock, Ithaca, NY, pp 55–63

Senthil Nathan S, Kalaivani K, Mankin RW, Murugan K (2006) Effects of millet, rice, and sorghum diets on development of *Corcyra cephalonica* (Stainton) (Lepidoptera: Galleriidae) and its suitability as a host for *Trichogramma chilonis* Ishii (Hymenoptera: Trichogrammatidae). Environ Entomol 35:784–788

Sétamou M, Bernal JS, Legaspi JC, Mirkov TE, Legaspi B (2002) Evaluation of lectin-expressing transgenic sugarcane against stalkborers (Lepidoptera: Pyralidae): effects on life history parameters and damage. J Econ Entomol 95:469–477

Shi Y, Wang MB, Powell KS, van Damme E, Hilder VA, Gatehouse AMR, Boulter D, Gatehouse JA (1994) Use of the rice sucrose synthase-1 promotor to direct phloem-specific expression of β-glucuronidase and snow drop lectin genes in transgenic tobacco plants. J Exp Botany 45:623–631

Solberg Y, Remedios G (1980) Chemical composition of pure and bee-collected pollen. Medlinger fra Norges Landbruksshoegskole 59:2–12

Spitzen J, van Huis A (2005) Effect of host quality of *Callosobruchus maculates* (Coleoptera: Bruchidae) on performance of the egg parasitoid *Uscana lariophaga* (Hymenoptera: Trichogrammatidae). Bull Entomol Res 95:341–347

Tesoriero D, Sgolastra F, Dall'Asta S, Venier F, Sabatini A, Burgio G, Porrini C(2004) Effects of Bt-oilseed rape on the foraging activities of honey bees in confined environment. Redia 87:195–198

Treacy MF, Benedict JH, Walmsley MH, Lopez JD, Morrison RK (1987) Parasitism of bollworm (Lepidoptera: Noctuidae) eggs on nectaried and nectariless cotton. Environ Entomol 16: 420–423

Tscharntke T (2000) Parasitoid populations in the agricultural landscape. In: Hochberg ME, Ives AR (eds) Parasitoid population biology. Princeton University Press, Princeton, NY, pp 235–253

van Huis A, Rooy Mde (1998) The effect of leguminous plant species on *Callosobruchus maculates* (Coleoptera: Bruchidae) and its egg parasitoid *Uscana lariophaga* (Hymenoptera: Trichogrammatidae). Bull Entomol Res 88:93–99

Wäckers FL (2005) Suitability of (extra-) floral nectar, pollen, and honeydew as insect food sources. In: Wäckers FL, van Rijn PCJ, Bruin J (eds) Plant-provided food for carnivorous insects: a protective mutualism and its applications. Cambridge University Press, Cambridge, pp 17–74

Wäckers FL, van Rijn PCJ, Heimpel GE (2008) Honeydew as a food source for natural enemies: Making the best of a bad meal? Biol Control 45:176–184

Wang Z, Wu Y, He K, Bai S (2007) Effects of transgenic *Bt* maize pollen on longevity and fecundity of *Trichogramma ostriniae* in laboratory conditions. Bull Insectol 60:49–55

Wellinga S, Wysoki M (1989) Preliminary investigation of food source preferences of the parasitoid *Trichogramma platneri* Nagarkatti (Hymenoptera, Trichogrammatidae). Anzeiger für Schädlingskunde Pflanzenschutz Umweltschutz 62:133–135

Whitehouse MEA, Wilson LJ, Fitt GP (2005) A comparison of arthropod communities in transgenic *Bt* and conventional cotton in Australia. Environ Entomol 34:1224–1231

Whitehouse MEA, Wilson LJ, Constable GA (2007) Target and non-target effects on the invertebrate community of Vip cotton, a new insecticidal transgenic. Aust J Agric Res 58:273–285

Yang YZ, Yu YS, Ren L, Shao YD, QianK, Zalucki MP (2005) Possible incompatibility between transgenic cottons and parasitoids. Aust J Entomol 44:442–445

Zhang GR, Zimmermann O, Hassan SA (2004) Pollen as a source of food for egg parasitoids of the genus *Trichogramma* (Hymenoptera: Trichogrammatidae). Biocontrol Sci Technol 14:201–209

Zimmermann O, Ren Z, Hassan SA (2004) Risk assessment of culturing transgenic crops: testing side effects of *Bt* corn on Microhymenoptera of the genus *Trichogramma* (Hym., Trichogrammatidae). Mitteilungen der Deutschen Gesellschaft für Allgemeine und Angewandte Entomologie 14:431–434

Subject Index

F.L. Consoli et al. (eds.), *Egg Parasitoids in Agroecosystems with Emphasis on Trichogramma*, Progress in Biological Control 9, DOI 10.1007/978-1-4020-9110-0,

Natural Enemy Species Index

MIX
Papier aus verantwortungsvollen Quellen
Paper from responsible sources
FSC® C105338

If you have any concerns about our products,
you can contact us on
ProductSafety@springernature.com

In case Publisher is established outside the EU,
the EU authorized representative is:
Springer Nature Customer Service Center GmbH
Europaplatz 3, 69115 Heidelberg, Germany

Printed by Libri Plureos GmbH
in Hamburg, Germany